11 BIOLOGY

PNG UPPER SECONDARY

Takis Solulu
Beatrix Marjen-Waiin
Diaiti Zure
Martin Hanson
Maria Sinclair

OXFORD

OXFORD
UNIVERSITY PRESS

Oxford University Press is a department of the University of Oxford. It furthers the University's objective of excellence in research, scholarship, and education by publishing worldwide. Oxford is a registered trademark of Oxford University Press in the UK and in certain other countries.

Published in Australia by
Oxford University Press
Level 8, 737 Bourke Street, Docklands, Victoria 3008, Australia

First published 2012
Reprinted 2012, 2013, 2014 (twice), 2017 (twice), 2019, 2020, 2021, 2022, 2024, 2025

This book was originally published by ESA Publications, Auckland, New Zealand. This edition, specially adapted for the Grade 11 syllabus in Papua New Guinea, is published by arrangement with ESA Publications. Authors of the original work were Martin Hanson and Maria Sinclair and this adaptation has been produced with the assistance of Takis Solulu, Beatrix Marjen-Waiin and Diaiti Zure.

ISBN 978 0 19 557877 5

Typeset by diacriTech
Printed in China by Golden Cup Printing Co. Ltd

Oxford University Press Australia & New Zealand is committed to sourcing paper responsibly.

Contents

Acknowledgments

There are many people to be thanked for their help and assistance in enabling the publication of this series to happen. First, the cooperation and generosity of Mark Sayes at ESA Publications in New Zealand who responded with interest and support when the proposal to adapt his Study Guide series was put to him.

The authors of the original edition were Martin Hanson and Maria Sinclair. This adaptation has been produced to provide Grade 11 Biology students in PNG with a book that they can use as a compact summary of content and skills related to the PNG Grade 11 syllabus. It was produced with the assistance of:

Takis Solulu, MSc Oxford, BScAg PNGUoT; Grade 12 Biology Examiner since 2007; Lecturer in Biology, UPNG.

Beatrix Marjen-Waiin, MSc Imperial UK, PGDSc UPNG, BSc-UPNG; Grade 12 Biology Examiner since 2008; Lecturer in Biology, UPNG.

Diaiti Zure, MSc PNGUoT, BScF PNGUoT; Forest Soils and Plant Biologist, Department of Forestry, PNG University of Technology.

We would also like to acknowledge many other individuals who have been happy to advise and assist in different ways: Joy Sahumlal, Michael Uglo, Greg Kapanombo and Anne Dandava Sangi.

Introduction

This book has been published to provide the information required by students in order to successfully complete the Upper Secondary course in Biology at Grade 11 in Papua New Guinea.

The book is written in a manner that best develops an overall understanding of the biological concepts and processes set out in the PNG Grade 11 Syllabus. The order of Units in this book follows the order of Units presented in the Biology syllabus for Grade 11:

Unit 11.1 Living Things

Unit 11.2 Nutrition

Unit 11.3 Transport systems

Unit 11.4 Respiration and Gas Exchange

Unit 11.5 Response to Stimuli

Unit 11.6 Reproduction

Supplementary Unit: Micro-organisms

Within each Unit there are a number of Topics which follow the main sub-headings and bullet points set out within each Unit in the syllabus. The aim is to provide structure and content in a concise and compact format for students to use as an effective resource to support the classroom experience. It is acknowledged that Biology is more interesting and meaningful when students are exposed to a variety of resources and materials and we encourage students and teachers not to rely on this book as a sole source of information.

For each of the above Units we look at plants, animals and humans but for micro-organisms, instead of treating them within those Units, we have instead created a Supplementary (7th) Unit.

Within this Supplementary Unit, we look at micro-organisms in relation to nutrition, transport systems, respiration and gas exchange, responses to stimuli, and reproduction. In this respect, the study of micro-organisms can show how interrelated these aspects of an organism's existence are. Much of the content in the Supplementary Unit is not in the Grade 11 Syllabus, but it is important – and an understanding of micro-organisms is extremely important in relating the study of Biology to real life, real issues and the local environment (as stated in the Secretary's Message on p. iv of the Biology Teacher Guide Upper Secondary (2009)).

If you have any suggestions about how this book might be improved in future editions, please make contact with Oxford University Press:

Fax: 00 61 3 9934 9100

Customer Service Email: exportsales.au@oup.com

We wish you every success in your studies.

The authors

Preliminary Unit

The importance of biology

Authors: David Blaker with Takis Solulu

The content of this Preliminary Unit is intended as an introduction to the importance of biology knowledge by looking at:

- How biology knowledge can be used.
- Carrying out an assignment – choosing a topic; collecting, sorting and interpreting the information; making a report or essay.

Biology knowledge

Biology knowledge is becoming increasingly important as the world faces some major issues and people demand practical solutions. Biology knowledge is about ideas. Its main aim is to find out how living things 'work' and to develop ideas about what has been discovered.

The biology learning outcomes set out in the PNG syllabus require you to process information you discover about biology knowledge and discuss how it is or can be used. Your teacher should provide you with direction with regards to the type of biology knowledge you are to find out about, how it is used and how to process it.

Biology knowledge can be used in several different ways. You may choose to focus on one of the following areas.

- *A technological application.*
  While biology knowledge is driven by human curiosity and a desire to learn, **technology** is driven by human needs and demands for such things as better safety, food quality or production, health or convenience. Technology is about finding practical uses for the biology knowledge that has been accumulated. Biology knowledge can have many different technological applications. For example, many types of food production (cheese, chocolate, wine, beer); improving health or fighting disease (kidney dialysis, managing diabetes, repairing damaged tissue organs, fertility treatments); biotechnology (including genetic engineering, cloning, tissue culture).
- *A management practice.*
  Biology knowledge may be used to better manage a useful or commercial plant crop or animal. For example, improving the fishing quota system so fishing is sustainable, or applying good horticultural practices to maximise the yield of a food crop or a forest.
- *Resolving an issue.*
  An issue is often something that is a problem in society and affects a large number of people. Issues can range from the best way to control an introduced pest, to addressing the growing problems of obesity, heart disease or diabetes, to better ways of dealing with waste.
- *Development of a theory.*
  A theory is the use of biology knowledge to explain an event or a phenomenon, eg the evolution of PNG's flightless birds such as the cassowary, climate change.
- *Development of a model.*
  Biology knowledge can be used to model or describe what is happening now. This model can then be used to make predictions about what could happen in the future. For example, the possible effects of climate change on agriculture or the spread of disease in developing countries.

Carrying out a research assignment

Finding out how biology knowledge is used is not original research. It will use **secondary data** – information already produced by others. However, the way information is selected and organised, and how the links are explained, should be original.

This chapter gives step-by-step directions on how to carry out a research assignment.

Most of the planning is set out below, but students will need to do some planning of their own.

Before starting a topic, it is essential to find out:

- The due date(s).
- That the topic is approved by the teacher.
- Whether it will be reasonably easy to get the information needed.
- What kind of report format is acceptable.
- How the final report will be evaluated.

Choosing a topic

It can be helpful to state the research title in the form of a question, as that can give more focus to the research.

Example questions for research titles

1. In what ways does a kidney dialysis machine imitate the normal working of a kidney?
2. Are current logging practices sufficient to ensure sustainable forestry?
3. What is the link between smoking and respiratory illnesses?
4. How has biology knowledge about nitrogen-fixing bacteria helped farming become more productive?
5. In what ways does the quality of paper depend on knowing about the cell structure of wood?
6. How are antibiotics developed and made, and why don't they work on viruses?
7. What is plant tissue culture, and what biological principles does it depend on?
8. The whole island of New Guinea is about one-half of one percent (0.5%) of the Earth's surface but it contains between 5% and 10% of the total species on the planet – why might this be so?
9. How might the production of food in the Highlands change if climate change is a reality?
10. What goes wrong when a person develops diabetes, and exactly how do insulin injections help them?

Collecting information

Library books and the internet will provide more information than a person can manage, so it helps to do some selection before even starting. It may be best to avoid heavy technical sources (too hard) and popular magazines (probably too simple).

Having found the right 'level', a researcher should aim to use *at least* three or more different sources, in order to get a *range* of information.

If a research starts with known websites, this can save a lot of time. Search engines such as *Google.com* find much more information, but can waste time if not used wisely. When using a search engine, it helps to know how to narrow down the number of websites accessed. The trick to saving time lies in choosing the best combination of search words.

Entering	Gives this many websites
"deaf"	1 800 000
"deafness"	269 000
"deafness" + "technology"	87 400
deafness + technology	87 400
deafness + technology + cochlear	7 560
deafness + technology + cochlear + implant	1 060
hearing + technology + cochlear + implant	2 140

Not all search engines need the "quote" marks, and in the case of *Google.com* they can be left out.

The example given is for the issue of hearing loss and cochlear implants, searched by *Google*.

Even when using the + sign correctly, there may be dozens of suitable websites. Instead of printing out whole sections, it helps to copy one single page from each site, then paste it to a ready-made document.

Guidelines for a report or essay

Performance criteria	Satisfactory achievement	High achievement	Very high achievement
Information used is appropriate to the topic	Yes	Yes	Yes
A range of information source types	Yes	Yes	Yes
Sources clearly recorded	Yes	Yes	Yes
Information sorted and arranged	Clearly processed	Clearly processed	Clearly processed
Kind and format and length of the report are all suitable	Yes	Yes	Yes
How are the links between biology knowledge and its use dealt with in the report?	Describes biology and use separately	Explains how/why biology knowledge applies to the use	Discusses how/why biology knowledge applies to the use

Sorting the information

A large collection of information on a topic may look impressive, but is not much use in its original form.

A first step is to make summary notes, and to highlight key bits from internet printouts.

The next step can be to sort the information into categories, such as:

- Too hard or too simple or not relevant – *so not wanted*.
- Biology facts and ideas and principles.
- Details of the use.
- Anything that shows how biology knowledge and use are related.

Processing the information

Assignments often ask for a situation to be 'discussed'. This needs a different kind of input to 'describing' something. 'Describe' usually means giving the reader a picture of how something works and what it looks like. The facts are left to speak for themselves.

A 'discussion' of the links between biology and a use could include the following:

- Saying which seemed to come first, the biology or the use.
- Clearly saying how the biology and the use interacted.
- Suggesting why the discoveries and links happened at that particular time.

While descriptions of biology knowledge or its use can be lists or bullet points, a discussion must be more than this. A discussion would be expected to involve paragraphs in which several ideas (descriptions and explanations) are linked. Furthermore, the discussion must clearly link the biology knowledge and its use, eg a discussion of how current knowledge of the orange roughy reproductive rate and methods have resulted in a change in the fishing quota.

The assessment

The assessment often takes the form of a report or essay, as these formats are better for writing a discussion than a *PowerPoint*™ presentation or a poster.

Before beginning the assignment, students need to find out how long the report needs to be, and which format – or combination of formats – is acceptable. One format that is *never* acceptable is a cut-and-paste of internet material.

Whichever format is used, the balance between visual material and text needs to be thought about. In the case of the topics in this chapter, the visual should not dominate. Words matter.

As a general guide to writing text, 'keep it simple'. It should be aimed at an average-to-better member of the class. Some guidelines for good quality writing are:

- Use short sentences.
- Use short paragraphs, with line breaks.
- Avoid listing or bullet points as these are difficult to integrate into a discussion.
- Cut out 'padding'.

Unit 11.1 Living Things

Topic 1: Cells, tissues and organs

Authors: Martin Hanson with Diaiti Zure

Millions of plant and animal species live in the world and PNG has over 5% of the total species on the planet. All these living things are made up of cells. Topic 1 in Unit 11.1 relates to living cells (ref. p. 9 in the Syllabus) which is important for understanding biological concepts throughout this book and includes:

- The discovery of cells.
- Cells, tissues and organs.
- How substances enter and leave cells.

Note that micro-organisms are covered in the Supplementary Unit (p. 329).

The discovery of cells

Robert Hooke in 1665, following the invention of the microscope, examined a thin slice of cork, noting thousands of 'little boxes' he called *cells*. Cork is dead – Hooke did not realise that the 'little boxes' had been produced by a transparent, jelly-like material that had died by the time the cork had fully developed. The importance of this jelly-like material was only appreciated many years later when biologists could see cells more clearly by *staining* them.

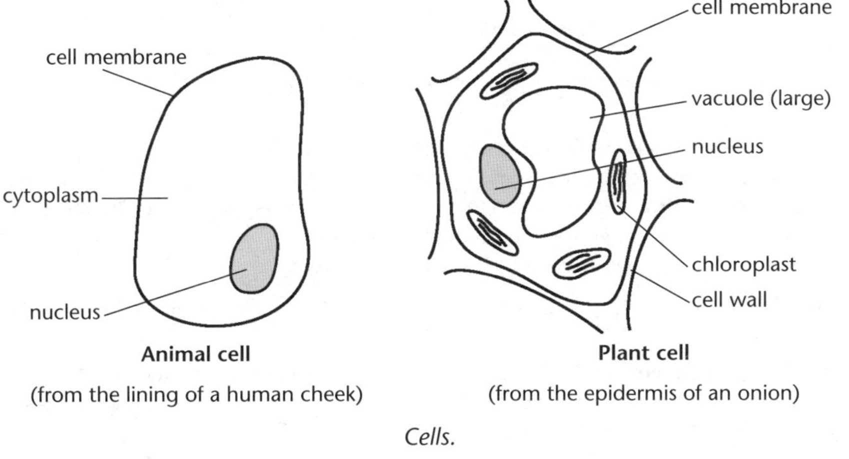

Cells.

The cell theory

During the 18th and 19th centuries, scientists gradually realised cells are the units of life, an idea that became known as the **cell theory**:

- All living things consist of cells. Some organisms are said to be **unicellular** because they consist of only one cell. Even **gametes** (eggs and sperms) are cells.
- Activities of living things are the outward signs of processes occurring in their cells (eg saliva poured into the mouth is made in cells of the salivary glands; the pumping of the heart is due to the contraction and relaxation of its muscle cells).
- New cells arise when a parent cell divides into two.

Cells

Most organisms seen with the naked eye are **multicellular** (consist of many cells). Many microscopic organisms are said to be **unicellular** (consist of a single cell). Animal and plant cells differ in several important ways.

Animal cells

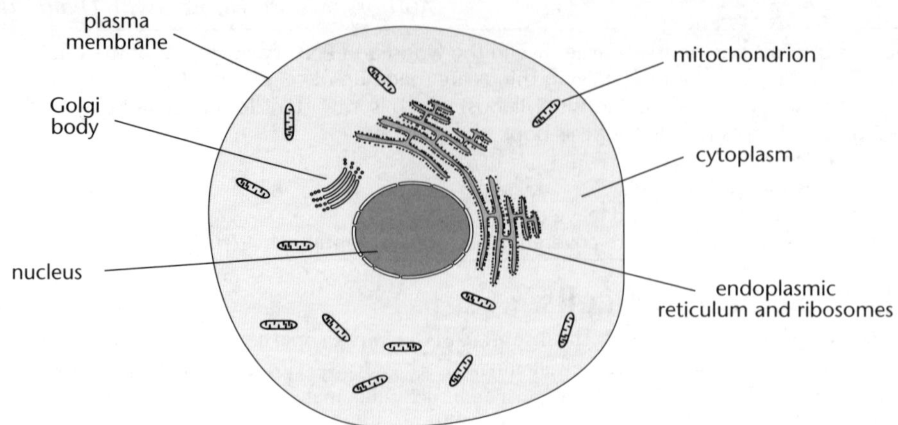

In reality, no cell looks just like this, because cells are *specialised* for carrying out particular functions.

General features of an animal cell.

A cell consists of a number of distinct parts or **organelles**, each specialised for carrying out a particular function.

- The **nucleus** – the *control centre*; contains 'instructions' the cytoplasm needs to perform its tasks. These instructions are *genetic information* inherited from the parent, stored in a chemical called **DNA**, organised into threads called **chromosomes**. Each chromosome consists of thousands of **genes**, arranged like beads on a necklace. The nucleus is separated from the surrounding **cytoplasm** by a **nuclear envelope**.
- The **plasma membrane** – the outer boundary of the cytoplasm. Controls the movement of substances into and out of the cell.
- **Mitochondria** (singular, mitochondrion) – the site of *respiration*, in which food is oxidised to generate useful energy for the cell. Particularly common in active cells such as those found in the liver, kidney and muscle.
- **Ribosomes** – these tiny particle-like organelles make proteins using information copied from the genes in the nucleus. Particularly abundant in cells making proteins, such as those cells that secrete digestive enzymes into the gut. Ribosomes are closely associated with a network of membrane-bound sacs called the **endoplasmic reticulum**.
- **Golgi body** – after being produced by the ribosomes, some proteins are further modified in the Golgi body. For example, the iron-containing part of haemoglobin is added in the Golgi body. Proteins destined for secretion (eg digestive enzymes), are assembled into 'packets' in the Golgi body.

Plant cells

Plant cells have a number of features that animal cells lack.

- The **cell wall** surrounds the plasma membrane and is made mainly of a carbohydrate called **cellulose**, which is strong enough to give the cell a firm shape. In some plant cells the living part (nucleus and cytoplasm) dies as the cell matures, leaving only the cell wall – eg cork cells and the water-conducting cells of the xylem.

- In leaves and other green parts of plants, cells contain small organelles called **chloroplasts**. These contain the green pigment **chlorophyll** and carry out photosynthesis.
- Most plant cells have a large cavity called a **vacuole**, containing *cell sap*, a solution of mineral salts and other substances. Because of the vacuole, the chloroplasts are close to the cell wall, so they are near the source of CO_2.

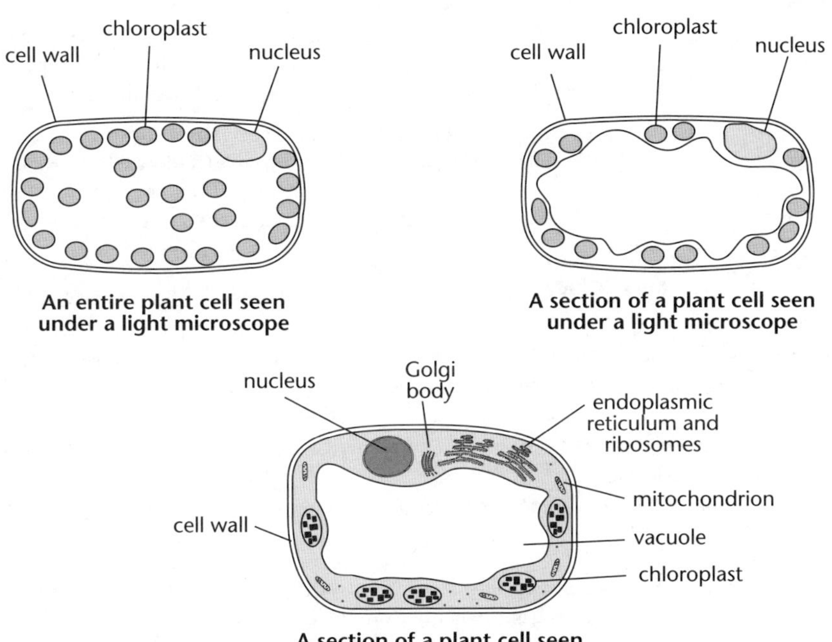

An entire plant cell.

Differences between animal and plant cells under the light microscope

Structure	Plant cells	Animal cells
Nucleus	Present	Present
Cell wall	Present	Absent
Large vacuole	Usually present	Absent
Chloroplasts	Present in photosynthetic cells	Absent

Comparison between animal and plant cells.

Prokaryotes and eukaryotes

Plants, animals and fungi are **eukaryotes** – ie the cells have chromosomes surrounded by a nuclear envelope to form a distinct compartment, the nucleus. In **prokaryotes**, which are bacteria, the DNA is not clearly separated and forms a closed loop (in eukaryotes, chromosomes are open-ended). Prokaryotes do not have any nuclear membrane or envelope surrounding their chromosomes.

Tissues

In multicellular organisms such as humans, all cells carry out certain basic processes, such as respiration and making proteins. Besides these fundamental processes, most cells are specially adapted for concentrating on a particular task, ie cells are *specialised* for certain functions.

In most cases cells act in groups called **tissues** (groups of cells specialised for carrying out a particular function). Usually, the cells are organised so the group works more effectively than the individual cells can do (eg, nerve cells are not randomly arranged but are organised into complex networks; adjacent muscle cells in the heart are aligned in the same direction and so pull together).

Examples of animal tissues include:

- Epithelia (singular, epithelium) – form coverings or linings and perform functions such as protection (eg the skin, and lining of the breathing passages). Glandular tissue consists of 'in-tuckings' of epithelium specialised for secretion of substances the cells produce (eg, glands in the stomach secrete gastric juice that helps digest food).
- Muscular tissue – consists of cells modified for changing chemical energy into mechanical energy for force and movement.
- Nervous tissue – consists of cells specialised for carrying electrical signals.

Examples of plant tissues include:

- The epidermis of a leaf – consists of cells that fit together like jigsaw pieces. They are covered by a continuous waxy *cuticle* that helps reduce water loss.
- The water-conducting cells of plants – consist of dead cells with no end walls and which fit together like drainpipe sections. Only by fitting together in this way can they carry water effectively.

Organs

An organ is a group of tissues that cooperate to perform a more complex function than the component tissues.

The heart consists of muscular tissue, epithelial tissue and nervous tissue, held together by connective tissue. The tissues of the heart work together to pump blood. The small intestine also contains the same types of tissue, but they are organised in a quite different way to carry out the completely different function of digesting and absorbing food.

Other examples of organs are the stomach and kidney, and in plants, leaves, stems, and roots.

How substances enter and leave cells

Cells are like miniature factories, constantly taking in raw materials and generating waste products.

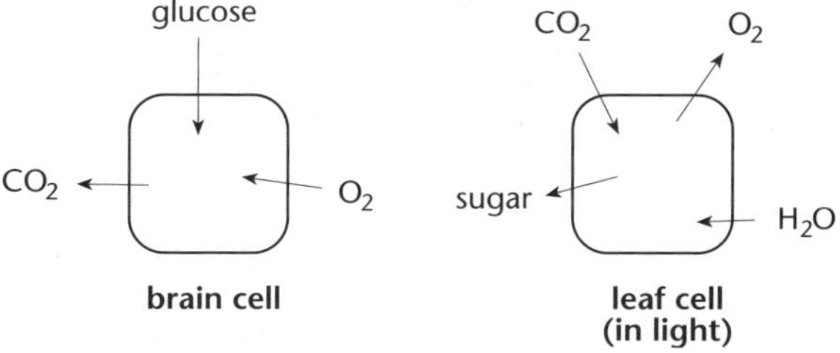

Movement of substances into and out of an animal cell and a plant cell.

Substances enter and leave cells by various kinds of process – the most important are diffusion, osmosis, and active transport.

Diffusion

Diffusion is the movement of a substance from where it is more concentrated to where it is less concentrated.

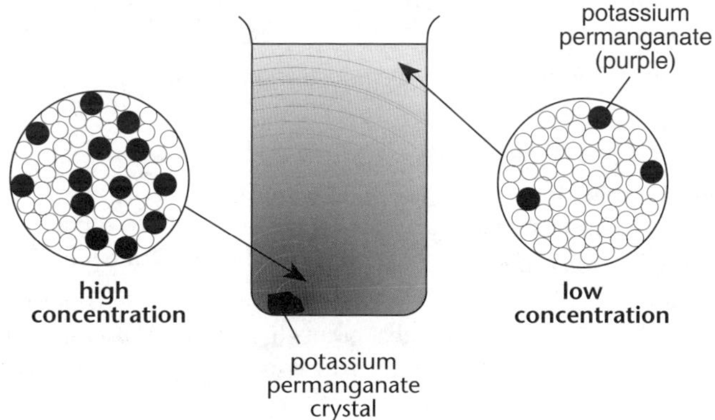

The effect of dropping a potassium permanganate crystal into a beaker of still water illustrates diffusion. As the crystal dissolves, the purple colour spreads very slowly outwards to where it is less concentrated.

Diffusion of potassium permanganate in water.

Diffusion depends on the random movement of particles (either molecules or ions).

Animal cells *use* oxygen, so oxygen is less concentrated inside than outside, causing it to diffuse *in*.

CO_2 is *produced* in an animal cell, so it is more concentrated inside than outside and diffuses *out*.

Osmosis

Osmosis is really a special kind of diffusion – it involves the movement of water from a dilute solution to a more concentrated solution through a **partially permeable membrane** (a partially permeable membrane is an extremely fine sieve, allowing water molecules to pass through, but not larger molecules such as sugar).

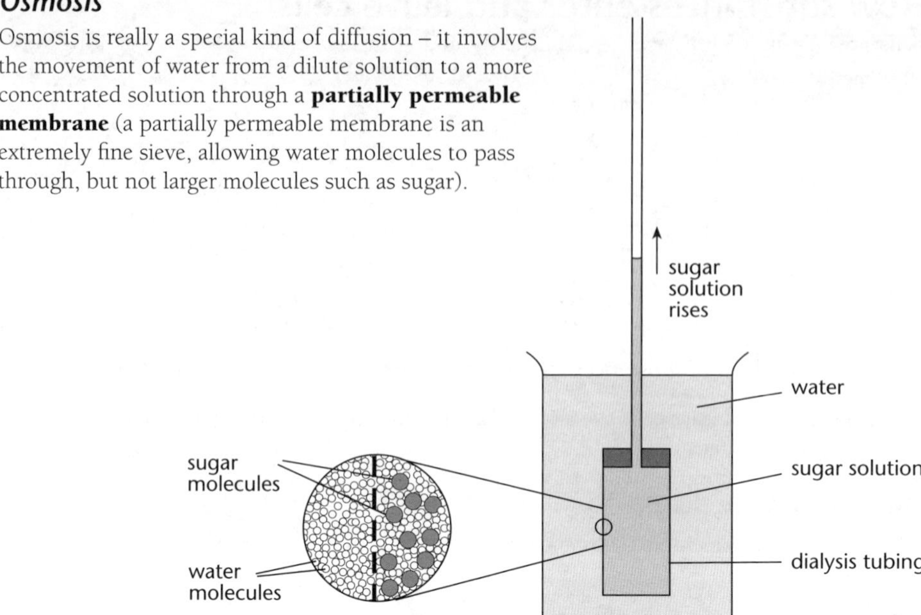

Demonstration of osmosis.

Active transport

The plasma membrane that forms the surface of the cytoplasm is permeable only to small substances like water, oxygen, and CO_2, which enter and leave cells by diffusion. Cells must absorb larger molecules such as glucose by a process called **active transport**.

Active transport moves substances from a low concentration to a higher concentration (ie in the opposite direction to the 'natural' direction that occurs from diffusion), so active transport uses *energy*. Energy is supplied by respiration (this is why it takes a lot of energy to absorb digested food after a meal).

Unit 11.1 Living Things

Topic 2: Cell structure and function

The content in Topic 2 includes explanation and description of the structure and function of cell organelles (ref. p. 9. in the Syllabus and p. 25 in the Teacher Guide). The Topic deals with:

- The structure and function of cellular components and organelles – cell wall, cell (plasma) membrane and nuclear membrane, nucleus, chromosomes, centrioles, cytoplasm, endoplasmic reticulum, ribosomes, mitochondria, chloroplasts, Golgi body, lysosomes, vacuoles, contractile vacuoles, cilia, flagella, eye spots.
- Factors that affect cell structures.
- Reasons for similarities and differences between cells.

Cells were first observed in 1665 by Hooke through one of the first microscopes. Nearly 200 years later (1839), the *cell theory* was proposed. This states that 'cells occur universally and are the basic units of living organisms'; the theory is still current.

Prokaryotes and eukaryotes

All cells belong to one of two categories – they are either **prokaryote cells** or **eukaryote cells**.

- Prokaryote cells are found only in bacteria and cyanobacteria (the Monera), and are distinguished by not having a true nucleus, only a central nuclear area containing a loop chromosome, and small circular DNA plasmids. The cytoplasm does not have membrane-bound organelles such as mitochondria.
- All other organisms are made of eukaryote cells. These have a true nucleus (one enclosed in a membrane) containing chromosomes and a variety of membrane-bound organelles in the cytoplasm.

The rest of this chapter refers to eukaryote cells only.

The *structure* of a cell is linked to its *function*. The *size* and *shape* of a cell and the **organelles** within it are linked to the way it carries out its function within the organism.

Examples

Animal and plant cells show differences in size, shape, structure:

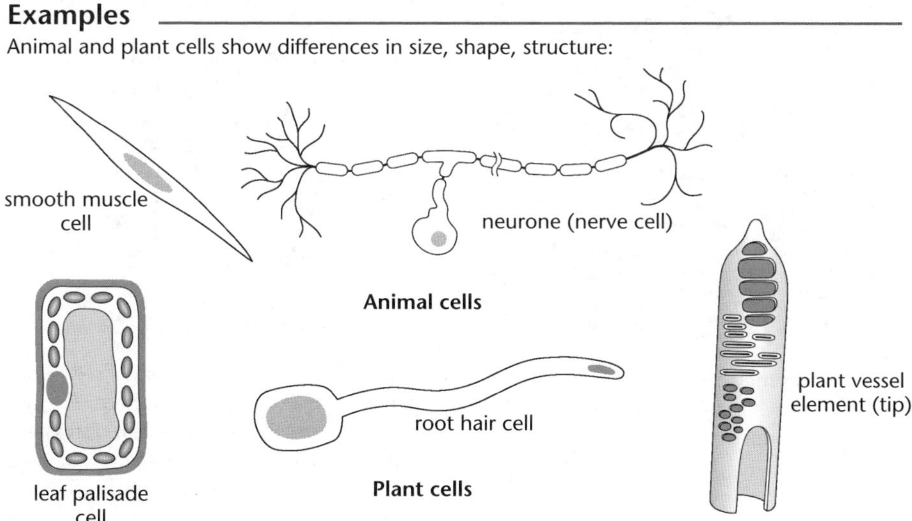

Basic cell structures of animal and plant cells

Basic cell structures of animal and plant cells can be seen using a light microscope.

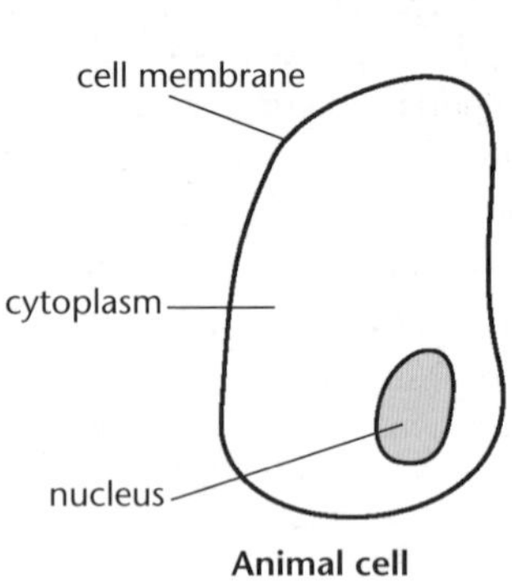

Animal cell

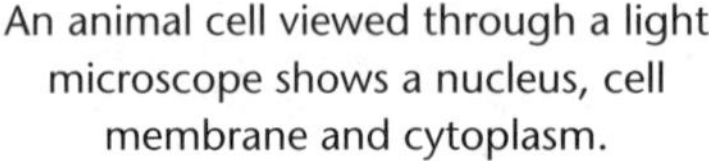

An animal cell viewed through a light microscope shows a nucleus, cell membrane and cytoplasm.

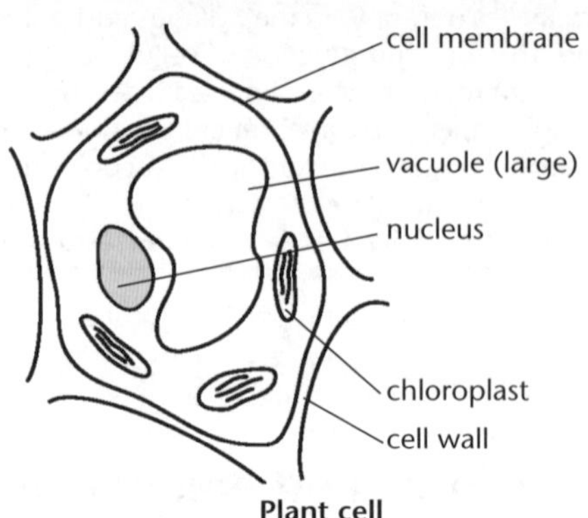

Plant cell

The light microscope shows that plant cells also contain chloroplasts, vacuoles and have a cell wall outside the cell membrane.

Cell structures visible under a modern compound light microscope

It is the presence of the *cell wall* that distinguishes plant from animal cells.

Plant cells taken from a leaf (or the outside of greens stems) will have *chloroplasts* and *large vacuoles* (which store the products of photosynthesis). Vacuoles are common in animal cells but are never large.

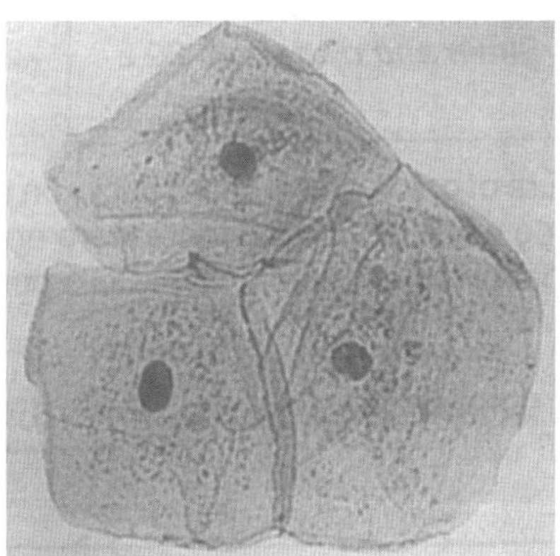

Cells lining the cheek of a human – visible in each is cell membrane, cytoplasm, nucleus.

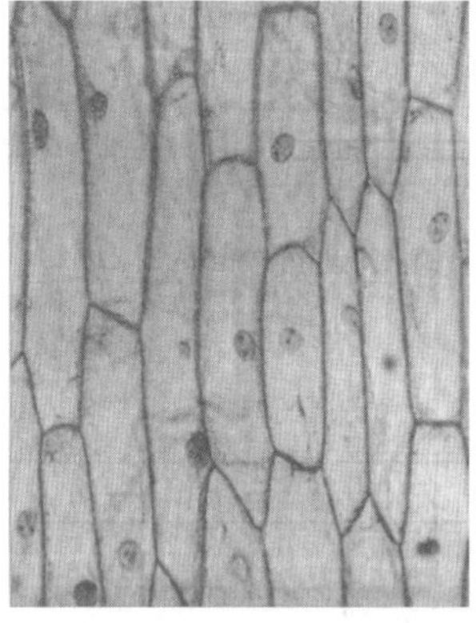

Cells from the epidermis of an onion cell – visible in each is cell wall, cytoplasm, nucleus. The cell membrane under the cell wall is not distinguishable from the cell wall. There are no (large) vacuoles or chloroplasts, as the cells are not photosynthetic.

Photomicrographs (photos taken from a light microscope) taken at 400 x magnification using a compound light microscope.

Cell detail

Cell detail is revealed under the (transmission) electron microscope or TEM, as this can magnify up to 400 000 times.

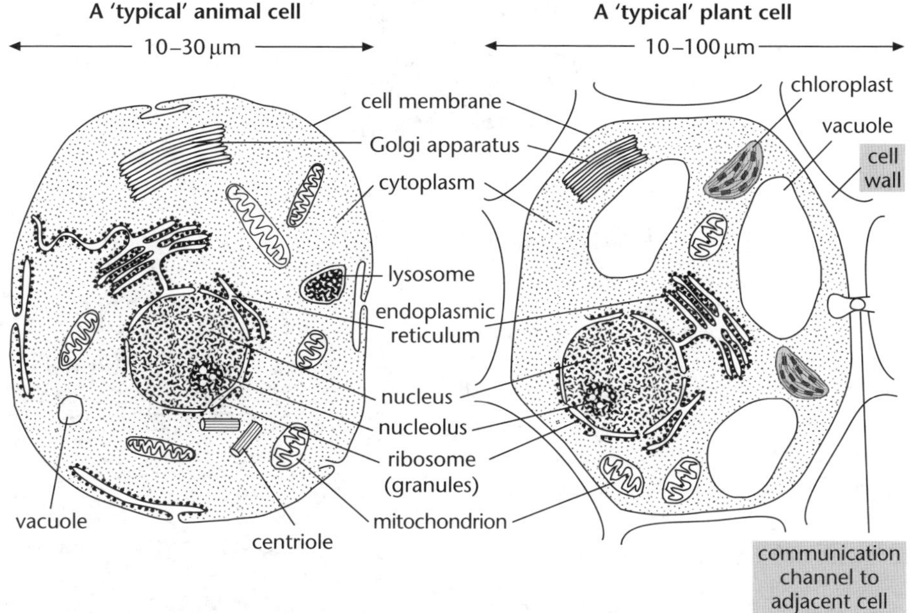

Cell detail of a 'typical' animal and a 'typical' plant cell as revealed by a TEM.

Organelles found in animal and plant cells

Cell (or plasma) membrane

All cells are bound by a **cell membrane**. In plant cells, this is enclosed by a rigid **cell wall** made of cellulose. Membranes are very thin (about 8 nm or 0.000008 mm) and act as a boundary between the cell and its environment (so maintaining the concentrations of substances inside and outside the cell). The membrane is made of a phospholipid bilayer embedded in which are many different proteins. The bilayer is fluid, allowing the proteins in it to move. The bilayer has 'heads' of glycerol-phosphate, which are hydrophilic ('water loving') and 'tails' of fatty acids, which are hydrophobic ('water hating'). This hydrophilic/hydrophobic arrangement (known as amphipathic) allows the membrane to assemble/reassemble itself and also to seal itself if damaged. The structure of the membranes of cell organelles is similar.

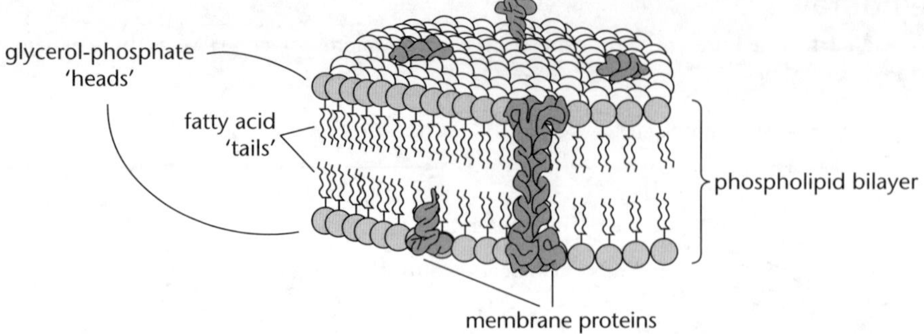

Membrane structure (of an animal cell).

Proteins on the outside of the membrane may be *receptors* for chemicals such as hormones and neurotransmitters. Other proteins on the surface allow the cell to recognise 'self' from 'foreign' cells.

The proteins that penetrate through the membrane are likely sites of transport of substances into/out of the cell (via facilitated diffusion and/or active transport).

Membranes are **semi-permeable**, meaning that they allow passage of only certain substances. Small molecules (eg O_2, CO_2, glucose) typically freely diffuse through the membrane, while large molecules (eg starch) are excluded or have to be actively transported across the cell membrane.

The membrane may be thrown up in many folds, known as *microvilli*, which greatly increase its surface area. Microvilli are found in cells that are very active in secretion (eg pancreatic cells) and/or absorption (eg cells lining the kidney tubules and the small intestine).

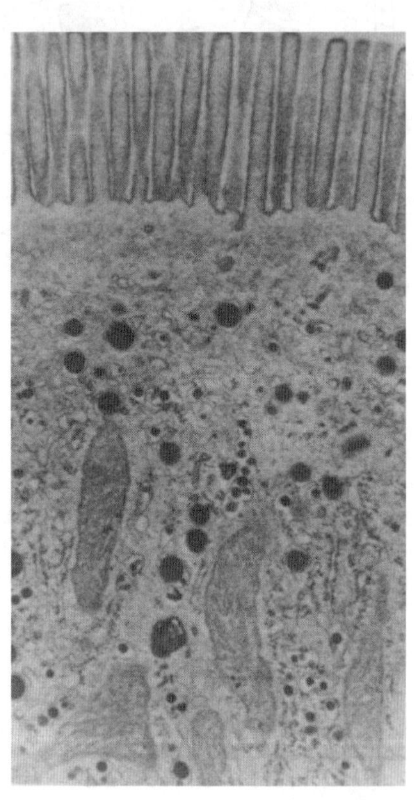

TEM micrograph of membrane of a cell lining the small intestine. The membrane is folded into microvilli to increase surface area for absorbing nutrients.

Cell nucleus

The **nucleus** is often called the *control centre* of the cell because it contains **DNA**, the genetic material that organises all cell processes. DNA is scattered throughout the nucleus as **chromatin**, which only forms into visible structures called **chromosomes** just before a cell is going to divide (ie mitosis, meiosis).

The **nucleolus** inside the nucleus produces the RNA component of **ribosomes**, which are involved in the production of protein.

The nucleus is contained in the nuclear envelope, a double membrane that has pores at intervals – these allow materials to move between the nucleus and cytoplasm.

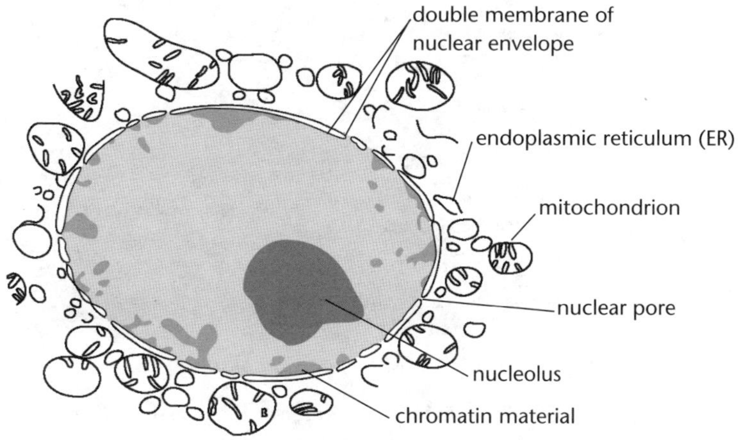

The nucleus with nuclear envelope showing pores.

Chromosomes

During the early stages of cell division, the *chromatin material* condenses to form chromosomes. The chromatin coils up tightly into looped structures made of two chromatids joined by a **centromere**.

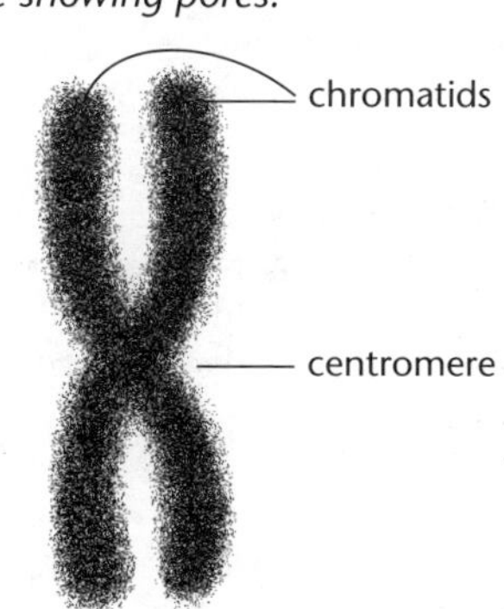

A chromosome.

Cytoplasm

Cell organelles are embedded in the **cytoplasm**. The cytoplasm is made of fluid called **cytosol**, which is mainly water but with many substances dissolved in it (eg sugars, amino acids, mineral ions), and it is where many of the chemical reactions of the cell occur (eg glycolysis).

Endoplasmic reticulum (ER)

The **endoplasmic reticulum** (**ER**) is a network of membranes running through the cytoplasm and takes up most of its space. The membranes enclose tubes, which, in certain places, may be enlarged to form flattened areas called cisternae (which may temporarily store substances). ER may have **ribosomes** attached (**rough ER**), concerned with *protein production*. Rough ER is therefore common in cells that make and secrete proteins (eg digestive cells that produce enzymes, white blood cells that produce antibodies).

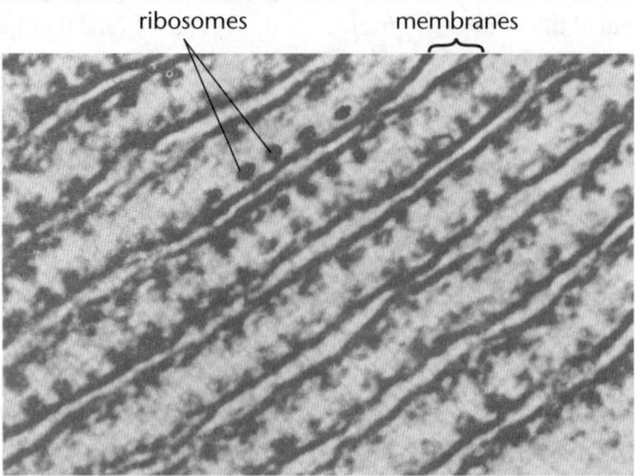

TEM micrograph of rough ER.

ER that does *not* have ribosomes attached, **smooth ER**, is associated with *production of lipids* – common in cells that produce steroid hormones.

ER also functions as a *transport system*, carrying materials from one part of the cell to another, as well as the nucleus and to the outside of the cell.

Ribosomes

Ribosomes are the site of *protein synthesis*. They are made in the nucleus (in the nucleolus) and pass out via the nuclear pores to the cytoplasm. They may attach to the ER (where they typically make protein for use inside the cell), or be free in the cytoplasm (where they typically make proteins for use outside the cell, ie are secreted), or are used in the cell's membranes.

Ribosomes are made of two subunits (one small and one large), which combine to form the protein-synthesising unit.

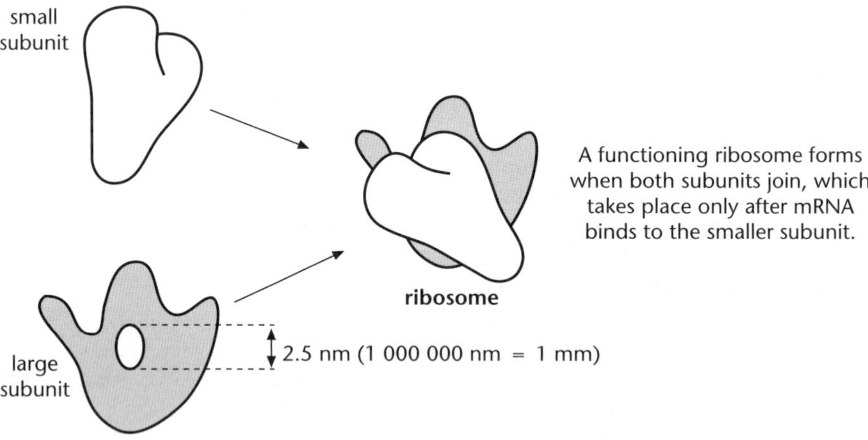

Golgi body

Most animals cells only have one **Golgi body**. It looks similar to the ER, but consists of a *stack* of membrane sacs called *cisternae*. (singular = cisterna).

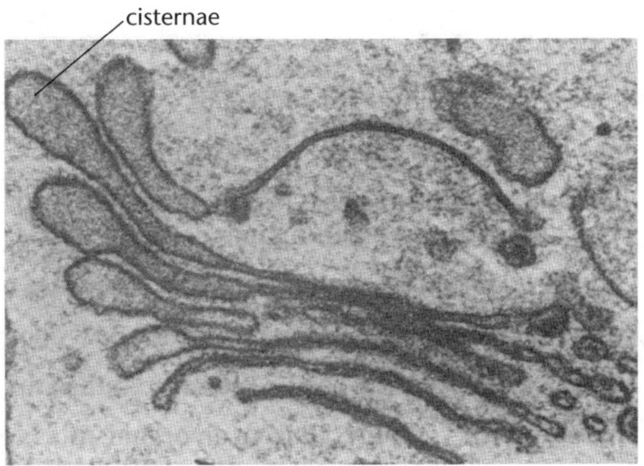

TEM micrograph of Golgi body.

After being synthesised in the ER, most molecules pass through the Golgi body. Transport vesicles carry the synthesised proteins from the rough ER to the nearest cisterna. A protein then moves from cisterna to cisterna and is modified as it goes by (eg having a carbohydrate added to form a glycoprotein). At the cisterna nearest the cell membrane, small vesicles with the modified protein pinch off, fuse with the cell membrane and discharge their contents to the outside.

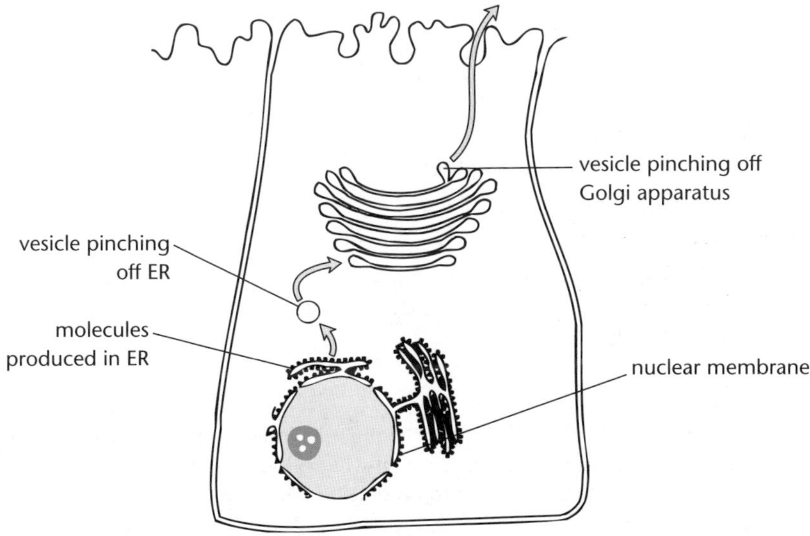

Secretion from a liver cell.

Golgi bodies are common in secretory cells (such as gland cells that produce hormones, pancreatic cells that produce enzymes).

Vacuoles

Vacuoles are membrane-bound sacs filled with fluid. In animal cells and unicellular organisms, vacuoles are typically small and may perform a variety of functions.

Example

In unicellular organisms, *contractile vacuoles* collect water entering and pump it to the outside. *Food vacuoles* form from food particles that enter the cell (these fuse with lysosomes for digestion).

Vacuoles are typically large in plant cells, occupying much of the centre of the cell.

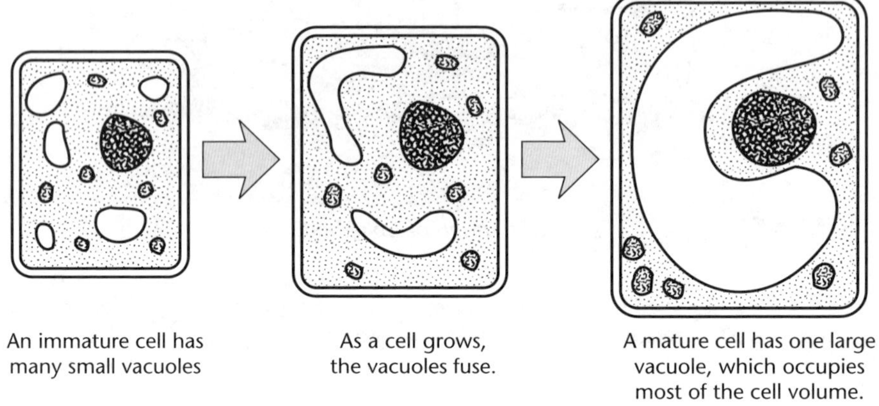

Vacuole development in a plant cell.

A large vacuole may be important as a *fluid skeleton*. Water entering the cell by osmosis collects in the vacuole, which swells and exerts pressure outwards on the cell membrane and cell wall. The cell becomes rigid or *turgid*. Turgid cells act as support to keep plants with non-woody stems upright.

Vacuoles are also important in *storage*, eg:

- Organic compounds such as sugars, amino acids.
- Inorganic ions, such as K^+ and Cl^-.
- Toxic wastes from metabolism.
- Toxic substances that stop herbivores eating the leaves.
- Pigments that colour flowers.

Mitochondria

Mitochondria (singular = mitochondrion) are commonly known as the 'powerhouses of the cell', as they are the site of **aerobic respiration**, in which glucose, $C_6H_{12}O_6$, is broken down in a *series of enzyme-controlled chemical reactions* to become CO_2 and H_2O, along with the production of the 'energy molecule' ATP.

Mitochondria are elongated ovals in shape (like sausages) and their inner membrane is thrown up into folds called **cristae** (singular = crista). Cristae provide a large surface area for the respiratory chemical reaction known as the *hydrogen transfer chain* to take place. The space between the cristae is known as the *matrix* and is the site of the respiratory chemical reaction known as the *Krebs cycle*.

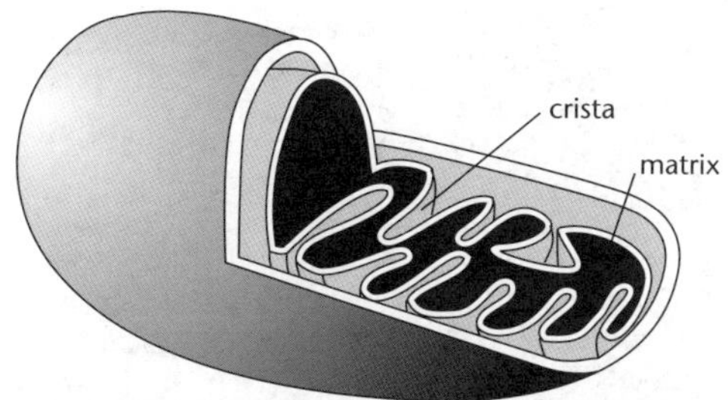

Cut-away view of a mitchondrion.

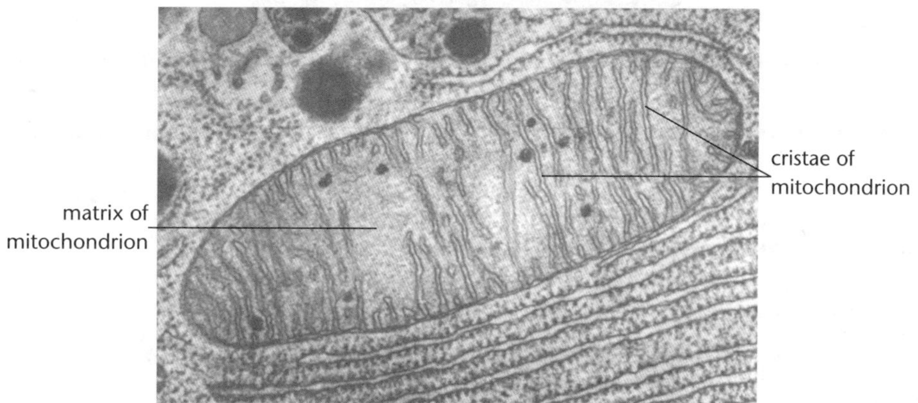

TEM photomicrograph of a mitochondrion in cytoplasm.

Mitochondria are more common in animal than in plant cells (as the energy demands of animal cells are typically higher), and are especially common in cells with high energy demands (eg sperm, muscle) – the most active of these cells may have close to 100 mitochondria. Cells lining the kidney tubules have large numbers of mitochondria, as they are very active in re-absorbing substances by *active transport*.

Mitochondria have their own DNA and protein-making machinery (ribosomes, RNA), and it is believed that they may once have been free-living organisms (possibly bacteria) that were 'captured' by other cells and then retained to become an organelle. Mitochondria can reproduce themselves.

Organelles found in plant cells (and some unicellular organisms)

Cell wall

The main distinction between plant and animal cells is that plant cells are surrounded by a **cell wall**. The cell wall is outside the cell membrane.

- **Primary cell walls** are formed in young plant cells and consist of **cellulose** molecules bundled together into *microfibrils*, which are oriented at right angles to one another so that they can provide strength to the cell.

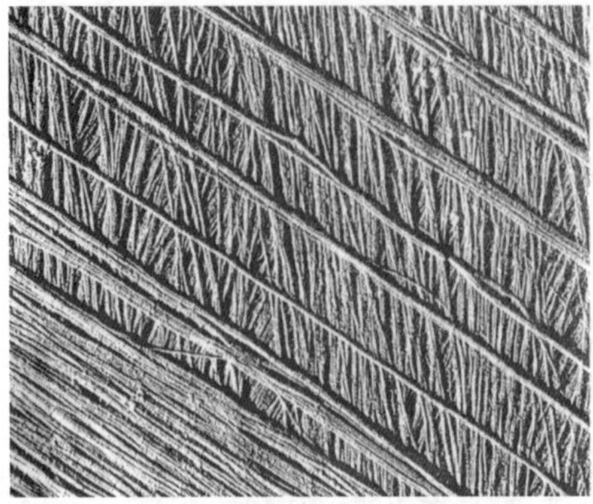

Each microfibril in the EM is about 20 nm wide; water and ions move freely through the mesh of microfibrils.

- **Secondary cell walls** are formed as the cell matures. They often contain a substance called *lignin*, which has stiffening properties. In such cells, the living material inside the cell often dies. These 'dead' cells form most of the wood in a plant.

Chloroplasts

Chloroplasts are large oval organelles found in leaf cells and cells in the outer layers of green stems (ie cells exposed to light). They are the site of **photosynthesis**, in which CO_2 and H_2O are joined together is a *series of enzyme-controlled chemical reactions* to become glucose, $C_6H_{12}O_6$. Solar energy is converted to chemical energy (ATP) to power the chemical reactions.

Within the chloroplast are flattened membranes called **thylakoids** arranged in stacks called **grana** (singular = granum). Embedded in these membranes are **chlorophyll** molecules, which 'catch' solar energy. These membranes provide a large surface area for the *light-dependent* chemical reactions of photosynthesis. Between the grana is the fluid matrix known as the **stroma** – this is where CO_2 and H_2 (H_2 is from water – the oxygen from water is the waste product) are joined to form glucose in the *light-independent* (also known as the *Calvin cycle*) chemical reactions of photosynthesis.

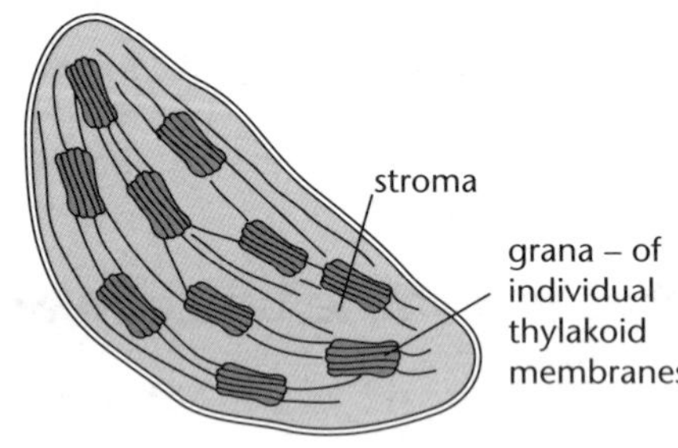

Diagram of a chloroplast.

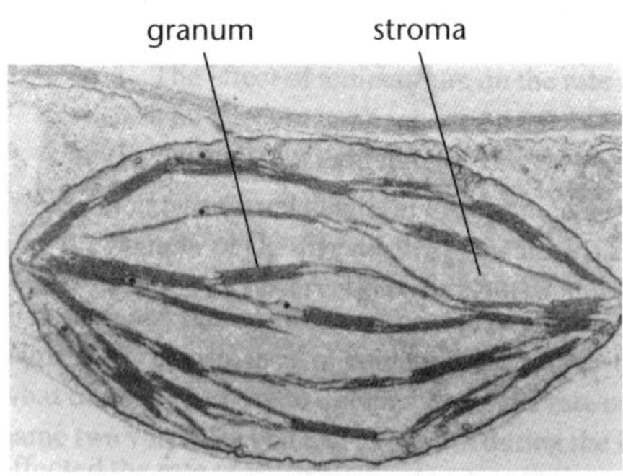

TEM micrograph of a chloroplast.

Chloroplasts (like mitochondria) have their own DNA and protein-making machinery (ribosomes, RNA), and it is believed that they may once have been free-living organisms (possibly bacteria) that were 'captured' by other cells and then retained to become an organelle. Chloroplasts can reproduce themselves.

Organelles found in animal cells (and some unicellular organisms)

Centrioles

Centrioles are found in all animal cells and most unicellular organisms, but in plants they are absent.

Centrioles appear as a 'tiny dot' next to the nucleus. They are made up of two identical cylindrical structures lying at right angles to each other. Each cylindrical structure is made up of **microtubules**. During cell division, these microtubules extend to form **spindles**, moving the chromosomes apart.

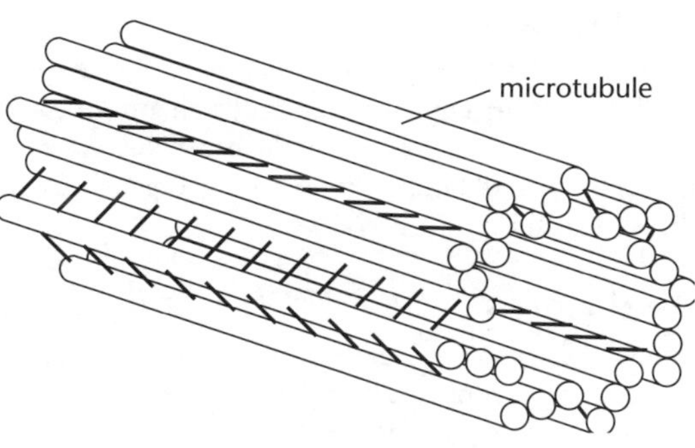

A centriole showing microtubules.

Cilia and flagella

These exist in animal cells and many unicellular organisms, but not in plants. Both are cellular projections and are almost identical in structure, but cilia are shorter and usually more numerous.

Their internal structure is similar to that of the centriole.

Cilia (singular = cilium) and **flagella** (singular = flagellum) are often associated with movement in unicellular organisms and some small animals.

Sperm cells of mammals move by a single flagellum.

Many of the cells that line the surfaces within our bodies are also ciliated. These cilia sweep substances across the cell surface.

Example

Cilia action

The cilia that line the cells of our respiratory tract beat upward, propelling a current of mucus that sweeps particles of dust, soot, pollen, etc, to our throats where they can be removed by swallowing.

Lysosomes

Lysosomes are vacuoles that contain **enzymes**. They are formed from vesicles produced by the Golgi body. Lysosome enzymes are used to:

- Break down worn-out organelles (eg mitochondria); the chemicals released are used to make new organelles or other needed products.

- Break down the cells of tissues during metamorphosis in insects and amphibians (eg tails of tadpoles).
- Digest bacteria that phagocytes/white blood cells engulf.
- Fuse with food vacuoles of unicells to digest the contents (eg *Amoeba*).
- Break down the membrane of the ovum to allow sperm entry (head of sperm releases lysosomes).

Organelles found in unicellular organisms

Oral grooves

Some unicellular organisms (eg *Paramecium*) have a permanent organelle that functions in feeding. This **oral groove** is a ciliated channel located on one side of the cell. Food particles are swept into the outer portion of the organelle. Water currents produced by beating the cilia move the food down the inner portion of the organelle, where a food vacuole forms around it so that digestion can begin.

Anal pores

The final stage in the digestive process in most unicellular organisms is the expelling of wastes through the anal pore. This is a specialised region of the cell surface where food vacuoles attach and rupture to the outside.

Eyespots

Some photosynthetic unicellular organisms (eg *Euglena*) have a small orange granule, usually at the anterior ('front') end, called an **eyespot**, which functions in light detection and phototaxic responses (ie movements that orientate the organism toward or away from the light).

Contractile vacoules

These are specialised vacuoles which are used to regulate the amount of water inside a unicellular organism. Each vacuole is surrounded by radiating canals that collect water from the cytosol and transport it to the vacuole. When the vacuole is full it expels the water to the outside of the cell.

Unit 11.1 Activity 2A: Structure and function of cells

1. Distinguish between plant and animal cells.
2. Distinguish between the following pairs of organelles:
   a. Flagella and cilia.
   b. Mitochondria and chloroplasts.
   c. Rough and smooth endoplasmic reticulum.
   d. Lysosomes and ribosomes.
   e. Vacuoles and contractile vacuoles.
3. Describe the cytoplasm.
4. a. The following are components and organelles that may be found in cells. Identify and name each component or organelle *and* describe its function.

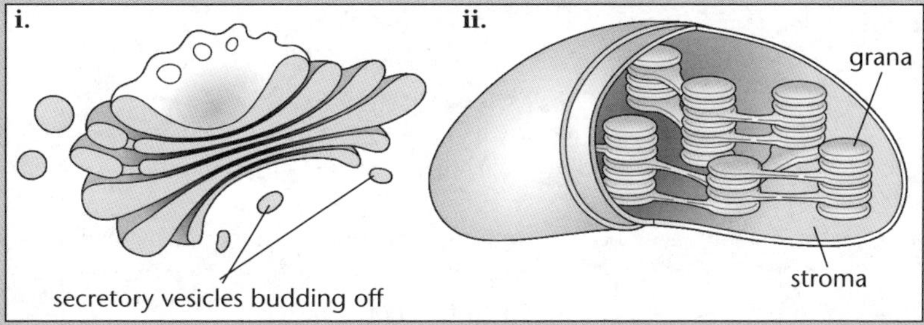

   b. Name and give the function of the oganelle represented by the small dots.

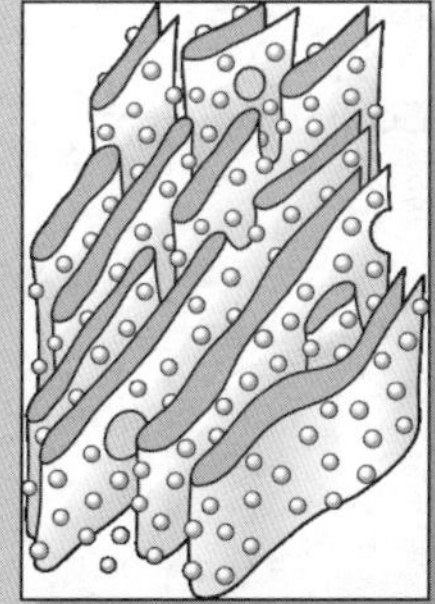

5. Cells can be specialised to carry out a specific role. Following are diagrams of two cells specialised for absorbing materials. The diagrams are not to scale.

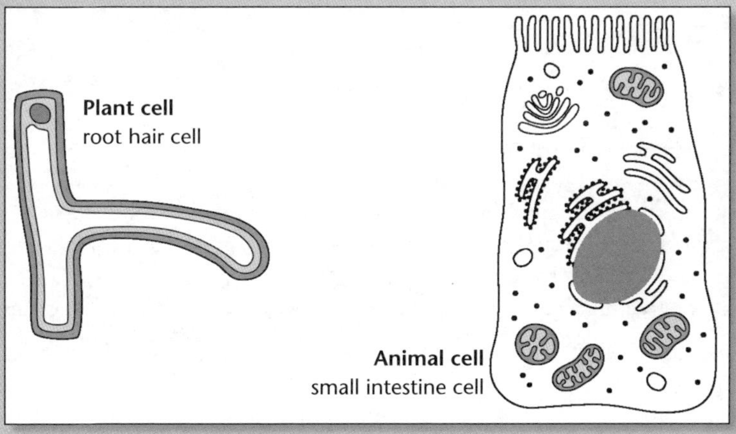

a. Describe *one* similarity (other than size) in the structure of these two cells for absorbing materials.

b. Explain how this similarity affects how these cells absorb materials.

6. a. Some unicellular organisms are able to move. Name two organelles that unicellular organisms use for movement.

b. Explain why the number of mitochondria found in a unicellular organism capable of self-propelled movement will differ from the number found in a non-moving unicellular organism.

7. Gamete-producing cells in the ovaries and testes contain large amounts of smooth endoplasmic reticulum, yet the cells in the human pancreas contain extensive systems of rough endoplasmic reticulum.

a. How does rough endoplasmic reticulum differ from smooth endoplasmic reticulum? You may use labelled diagrams in your answer.

b. Explain why the pancreas cells have extensive systems of rough endoplasmic reticulum, while the cells producing gametes contain large amounts of smooth endoplasmic reticulum.

8. The following diagram is of a typical cell from the lining of the human small intestine.

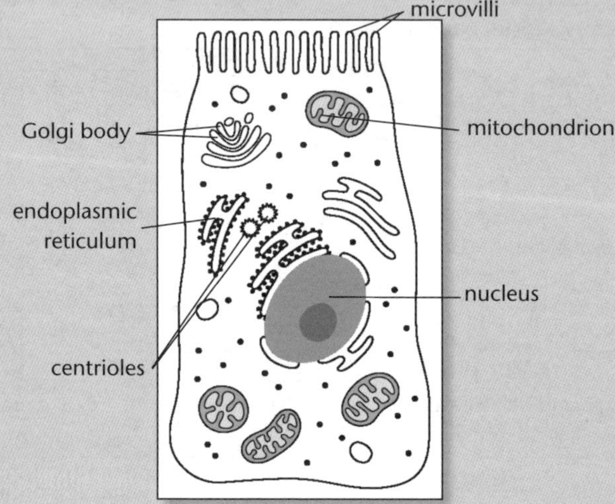

a. Describe where ribosomes are located in the cell shown.

b. Some cells in the human small intestine contain a relatively large number of ribosomes and Golgi bodies. Explain what 'a relatively large number of ribosomes and Golgi bodies' says about the function of these cells.

c. Explain why the cell membrane is thrown up into microvilli along one side.

d. Explain whether this cell is likely to have relatively large numbers of mitochondria.

9. Explain why plant cells have rigid cell walls and large, central, fluid-filled vacuoles.

10. Discuss the structure and function of the cell membrane. Clear, labelled diagrams may be used to assist with the answer.

Unit 11.1 Living Things

Topic 3: Cell processes – movement of materials

The material covered in Topic 3 continues the description and explanation of cell structure and function. The Topic deals with:

- Movement of materials into and out of the cell and organelles by passive and active transport.
- Secretion.
- Osmoregulation.
- Significance of SA:Volume ratios.
- Reasons for similarities and differences between cells.

Movement of materials

Movement of materials (such as oxygen, carbon dioxide, glucose, mineral ions, nutrients) across membranes may be by:

- **Passive transport** – process does not need energy (diffusion, facilitated diffusion, osmosis).
- **Active transport** – process needs energy (cytosis – endocytosis, pinocytosis and phagocytosis, exocytosis).

Passive diffusion

Diffusion refers to the random movement of particles in liquids and gases resulting in net movement *from an area of high concentration to an area of low concentration*. It does not need energy.

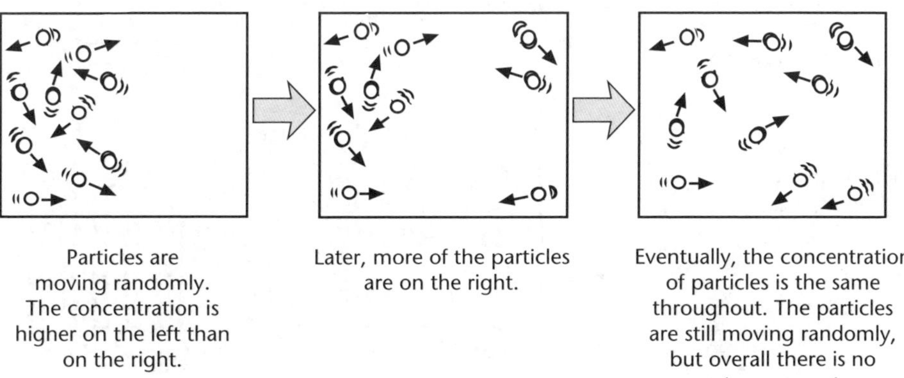

Particles are moving randomly. The concentration is higher on the left than on the right.

Later, more of the particles are on the right.

Eventually, the concentration of particles is the same throughout. The particles are still moving randomly, but overall there is no net movement.

The *difference* in concentration between two areas is called the **concentration gradient**. The higher the concentration gradient, the faster the rate of diffusion. Other factors that affect diffusion are:

- *Size* – smaller particles diffuse faster than larger ones.
- *Temperature* – particles diffuse faster in warmer than in colder temperatures.
- *State* – gas particles diffuse faster than particles in a liquid.

Small molecules (eg O_2, CO_2, glucose) diffuse freely across membranes, with the direction of movement being dependent on their concentration. *Large* molecules (eg starch) are prevented from diffusing through the membrane.

small molecule
high concentration (exterior)
phospholipid bilayer
low concentration (interior)

Facilitated diffusion

Molecules that cross the membrane faster than is possible from their concentration gradient do so by **facilitated diffusion**. Special transport or carrier proteins in the membrane provide channels for the process. Carrier proteins are specific (ie carry only one type of molecule).

Example

Both glucose and oxygen can be facilitated in their diffusion into cells.

- The protein cytochrome P450 can transport O_2 up to 1.8 times faster across the membrane than by simple diffusion.
- The diffusion of glucose into cells may be facilitated by the hormone insulin (a protein) which may activate transport channels.

Facilitated diffusion is a *passive* process, as the molecules can only diffuse from high concentration to low concentration across the membrane.

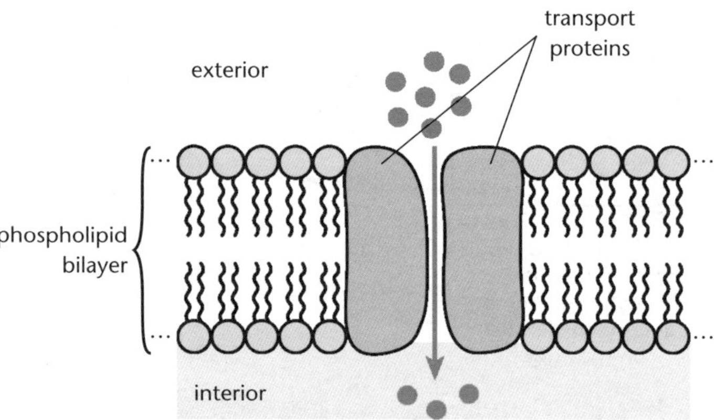

Osmosis

Osmosis is a *passive* process. Osmosis is a special case of diffusion. It is the movement of *water* across a *semipermeable membrane* (SPM) from where it is in *high concentration* to where it is in *low concentration*.

Water is in high concentration when it has few particles (solute) dissolved in it.

- Freshwater/tap-water will have few particles dissolved in it, so the concentration of water will be high.
- Sea water/marine water will have many particles (salts, ions) dissolved in it, so will have a comparatively low concentration of water.

Example

A **saline** solution is a salt solution (eg sea water).

A 5% saline solution would have a higher concentration of water than a 10% saline solution, therefore water would move across a membrane from a 5% saline solution into a 10% saline solution.

A weak or *dilute* solution (little dissolved solute) is a **hypotonic** solution.

A strong or *concentrated* solution (much dissolved solute) is a **hypertonic** solution.

Two solutions with the same concentrations (ie same concentrations of water and solute) are **isotonic** solutions.

Solute particles large (eg sucrose)

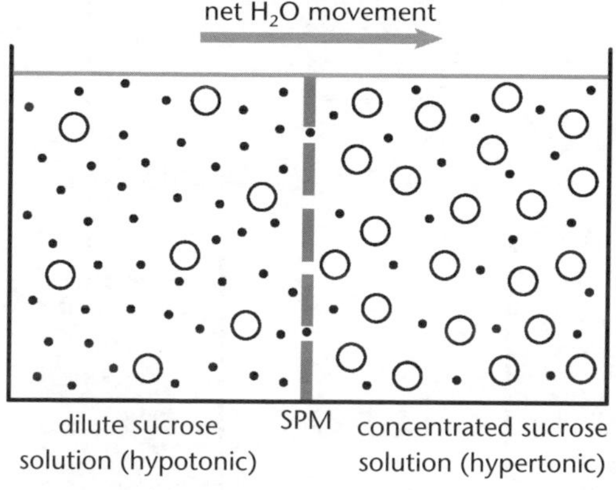

○ sucrose molecules

• water molecules

Sucrose molecules too large to pass through the SPM.

Net water movement is from LHS to RHS until concentrations are equal (ie isotonic).

As diffusion and osmosis result from *random* movement of particles, the particles will move in *both* directions across the membrane. However, the *net* movement will be towards the area of *lower* concentration.

Solute particles small (eg salt/ions)

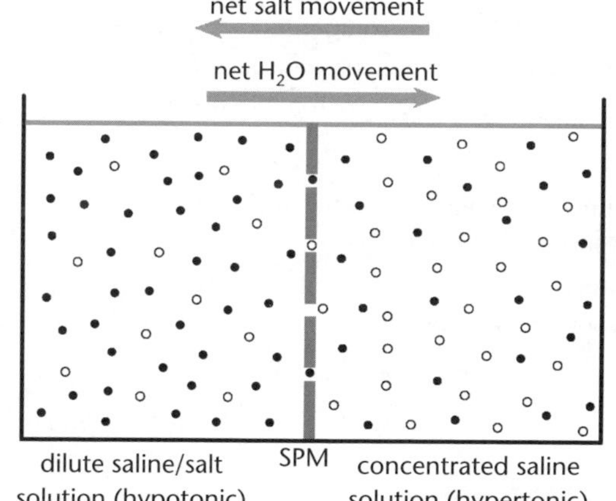

○ salt ions (Na^+, Cl^-)

• water molecules

Salt ions small enough to pass through the SPM.

Net salt movement is from RHS to LHS. Net water movement is from LHS to RHS. Net movement ceases when solutions are isotonic.

When both solute (eg salt/ions) and water are moving across a membrane, water moves *much more rapidly*, so its effect is more apparent than that of the solute diffusion (eg plant cells plasmolyse rapidly when placed in a concentrated salt solution because water tends to flood out of the plant cells).

Dialysis demonstration

A common way to demonstrate osmosis is to put a concentrated syrup/sugar solution in a dialysis bag (the dialysis bag represents a semipermeable membrane) tied to a long glass tube and suspend the bag in a beaker of tap water. The water moves into the dialysis bag (as it is more concentrated in the beaker than in the bag) and up the glass tubing, typically spilling over the top.

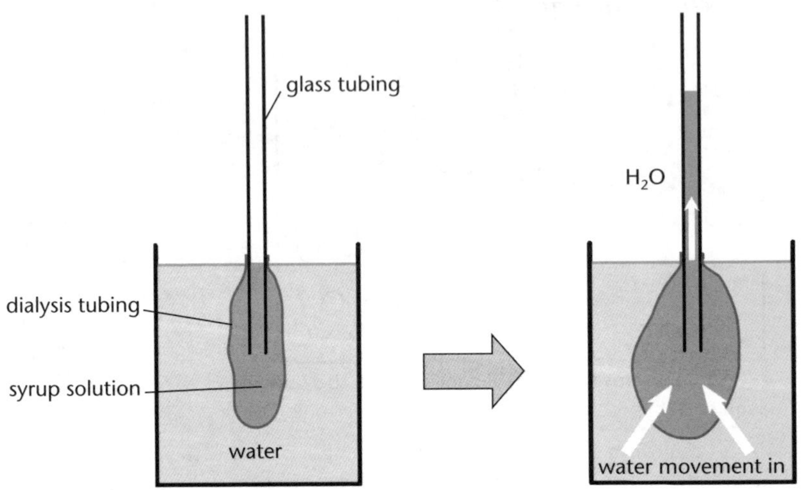

Osmoregulation

Osmoregulation is the control of water inside a cell/organism. Both plant and animal cells:

- Will have no *net* loss or gain of water in isotonic solutions.
- Will have a net loss of water through osmosis in hypertonic solutions.
- Will have a net gain of water through osmosis in hypotonic solutions.

Animal cells

Animal cells have no cell wall around their membrane. Therefore they will:

- Show no change in isotonic solutions.
- Shrivel up in hypertonic solutions.
- Expand and burst (lyse) in hypotonic solutions.

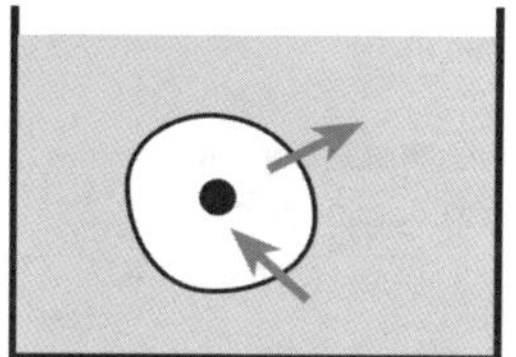

Normal animal cell in isotonic solution; no net loss or gain of water.

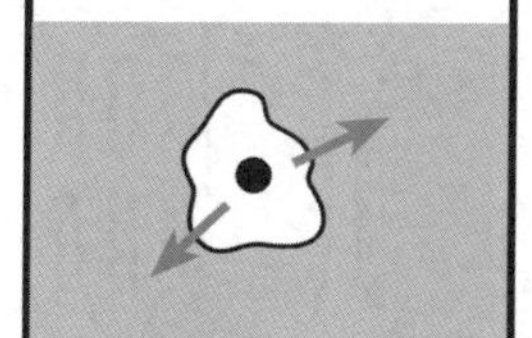

Dehydrated animal cell shrivels in hypertonic solution; net loss of water.

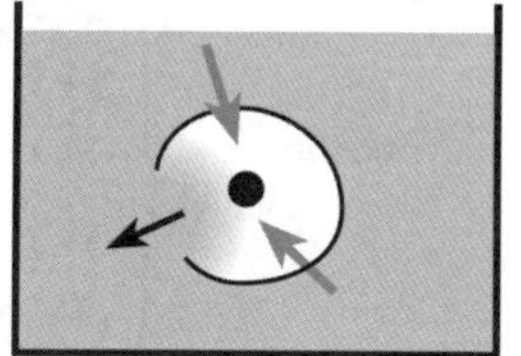

Burst animal cell in hypotonic solution; follows excessive net gain of water.

Plant cells

The presence of a cell wall around plant cells influences osmoregulation. Plant cells have a rigid cell wall surrounding the membrane. Therefore they will:

- Show no change in isotonic solutions.
- Become **plasmolysed** in hypertonic solutions – as water drains from the vacuole and the membrane pulls away from the cell wall. The cells become **flaccid** (or floppy) and the plant *wilts* if it is non-woody.
- Become firm in hypotonic solutions – as water fills the vacuole and the cell membrane presses against the wall; they are said to be **turgid**. The cell wall stops plant cells from bursting; no more water can enter. The resulting pressure from all cells being turgid (**turgor pressure**) acts to keep a plant upright if it doesn't have a woody stem.

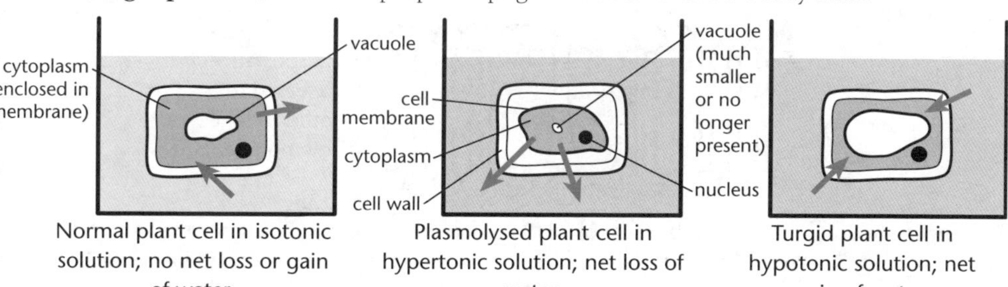

Normal plant cell in isotonic solution; no net loss or gain of water.

Plasmolysed plant cell in hypertonic solution; net loss of water.

Turgid plant cell in hypotonic solution; net gain of water.

Unicellular organisms

Unicellular organisms living in fresh water use their **contractile vacuoles** to osmoregulate. Contractile vacuoles collect the water that enters from osmosis and expel it to the outside. This requires energy, so is a form of active transport.

Active transport

Active transport moves substances (individual molecules/ions) across membranes *against a concentration gradient*, ie from low to high concentration, eg:

- Reabsorption of all glucose by cells of kidney tubules.
- Uptake of nitrates, NO_3^-, by root hairs.
- Removal of Na^+ ions from cells of gills in marine fish.

Research suggests that active transport of substances across membranes is via the large proteins embedded in the phospholipid bilayer. The substance temporarily combines with the carrier protein, which changes shape as it discharges the substance to the other side of the membrane. Such carrier proteins are likely to be specific (ie carry only one particular substance).

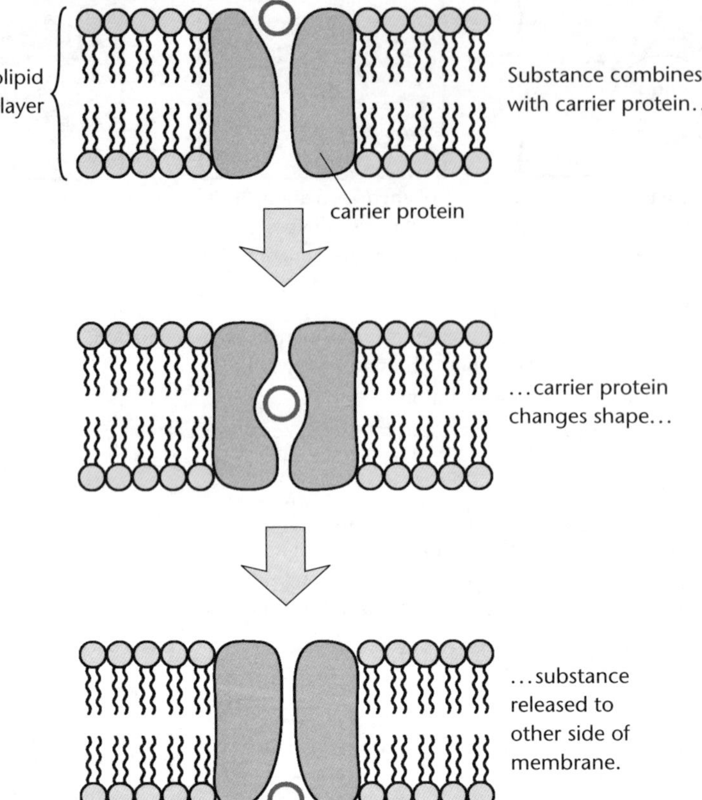

This process requires *energy* (in the form of ATP) – about a third of a cell's energy requirement may be needed for active transport. Cells undergoing a lot of active transport will have large numbers of mitochondria and use up large amounts of glucose and oxygen while producing large amounts of carbon dioxide and heat (in respiration).

- **Cytosis** is the movement of large amounts of substances into/out of cells by the *folding of membranes*.
- **Endocytosis** is the taking of substances into the cell by the infolding of the cell membrane. Fluids are taken in by **pinocytosis** ('cell drinking'), in which the membrane makes small infoldings which pinch off the liquid forming a vesicle – common in all cells. Large particles ('food') are taken in by **phagocytosis** ('cell eating'), in which the membrane appears to flow around the particles and close off to form a (food) vacuole (eg white blood cells/phagocytes consuming bacteria; *Amoeba* consuming food). Typically, lysosomes join with the food vacuole and the food is digested.
- **Exocytosis** is the removal of substances from the cell, and essentially is the reverse of endocytosis. It occurs when a cell needs to secret a substance (eg a hormone).

Endocytosis *removes* part of the cell membrane, while exocytosis *adds to* the cell membrane.

Energy is involved in the movement/removal/addition of membranes, so cytosis is active transport.

Pinocytosis

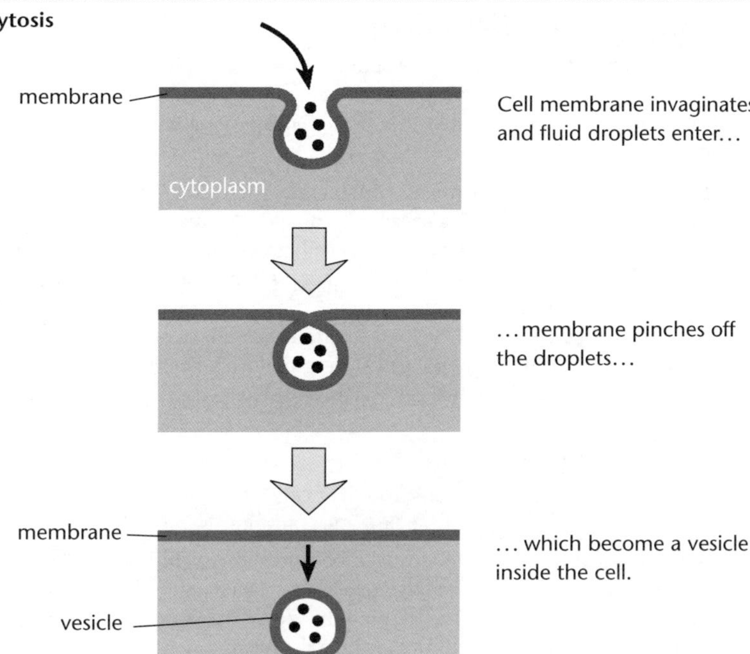

Process also works in reverse, *removing* droplets from a cell.

Phagocytosis

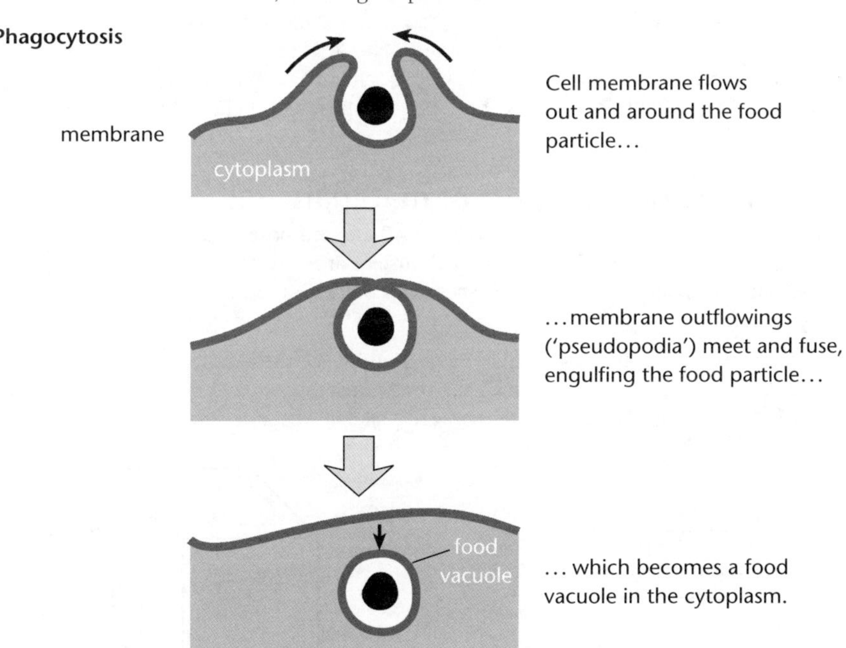

Cell size, shape and diffusion

Cells are typically microscopic (the exceptions are some eggs – technically, the yolk of birds' eggs is a cell; frogs' eggs are visible to the naked eye; the human egg is the size of a full stop). The need for cells to be so small relates to their *dependency on diffusion* for getting substances into and out of the cell.

When cells grow, their *volume* (the cytoplasm and contents) increases at a much faster rate than their *surface area* (the cell membrane). This is because volume (V) increases by a cube factor, while surface area (SA) increases as a square factor. As a cell grows, the *ratio* between surface area and volume (SA:V) *decreases*. Thus as a cell grows, there is comparatively less membrane for substances to diffuse through and comparatively more cytoplasm/organelles that need these substances. Diffusion gets less efficient, and, beyond a certain size, the centre of the cell does not receive the needed substances. At this stage, the cell stops growing. It may then divide to form two new, *smaller* cells, which will have a larger SA:V ratio so substances can diffuse efficiently throughout the cell.

Cells may also increase their SA:V ratio by having:

- An elongated shape – eg nerve cells/neurons, root hair cells.
- Having a biconcave shape – eg red blood cells for efficient diffusion of O_2.
- Folding of the cell membrane – microvilli occur in both secretory cells (eg pancreatic cells producing enzymes) and cells that absorb large quantities of substances (eg cells lining the small intestine absorbing nutrients, and cells lining the kidney tubules re-absorbing essential substances such as glucose) – allows for rapid diffusion.

Plant cells are usually larger than animal cells, because their centre is typically occupied by the large storage vacuole. The vacuole is centrally placed as it is *not* dependent on receiving substances diffusing in across the membrane. The organelles that are dependent on diffusion of substances – eg chloroplasts (CO_2 in, O_2 out) and mitochondria (O_2 in, CO_2 out) – must be close to the membrane. (Chloroplasts also have better exposure to light when closer to the membrane.)

Effect of cell size on transport of materials

Consider a cube (*Block 1*) of dimensions 2 cm × 2 cm × 2 cm. Suppose further that it takes too long for materials to be of any effective use if they diffuse or are transported over a distance > 1 cm. This means that materials *just reach* the centre of this block in time, since the centre of the cube is 1 cm in from each face.

Block 2 is 3 cm × 3 cm × 3 cm. It is bigger than Block 1 and its centre is now over 1 cm from the faces of the cube, so materials would not reach the centre area (shown as a grey dashed cube) in time to be effective. *This central area would thus need to be full of substances such as water or storage materials.*

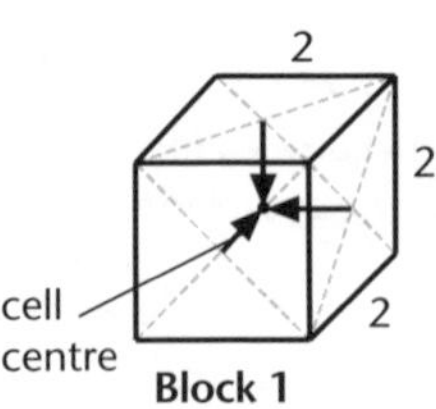

Block 1

2 × 2 × 2 cm cube.

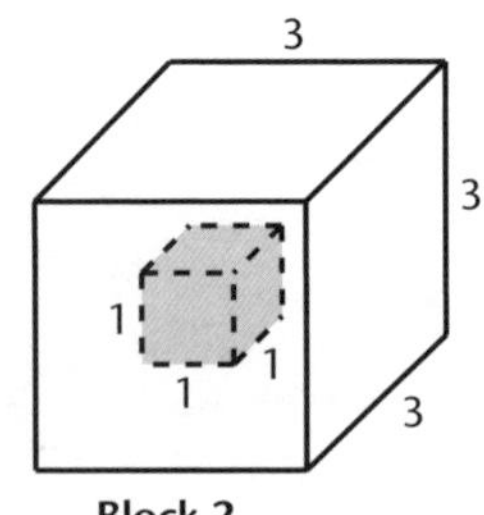

Block 2

3 × 3 × 3 cm cube.

Blocks 3 and 4 are the cubes formed when Block 2 is divided:

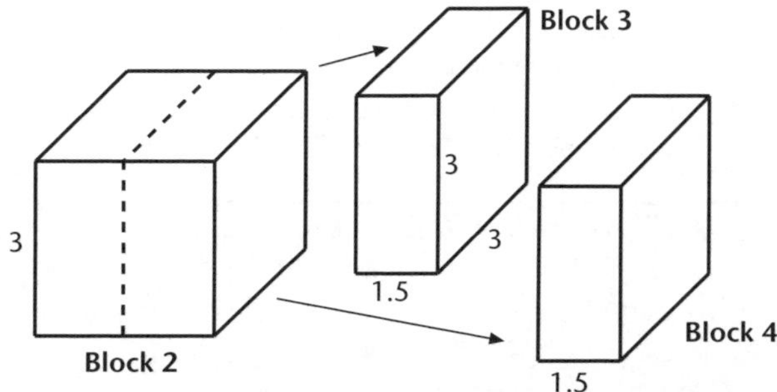

The centre of each new block is within 1 cm of the faces and would be easily supplied with materials. By dividing into two smaller 'cells', the distance materials must travel has been reduced.

Block 5 has the *same* volume as Block 2, but it is elongated.

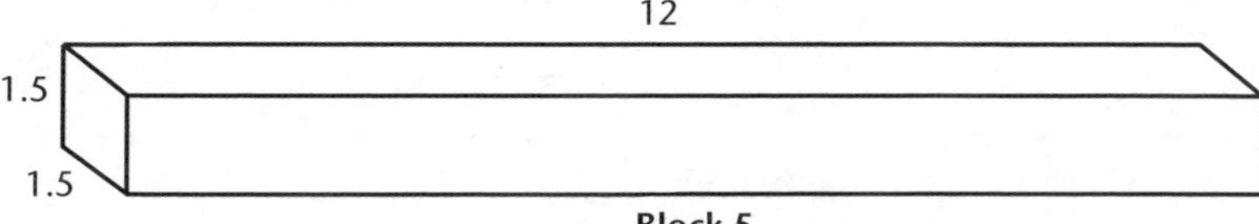

Despite having over three times the volume of Block 1, the centre of Block 5 is within 1 cm of the faces, and so would be readily supplied with materials.

Example

Effect of surface area to volume ratios

The surface area, volume, and SA:V ratios for Blocks 1–5 are as follows:

Blocks	SA (cm^2)	V (cm^3)	SA:V
1	$6 \times (2 \times 2) = 24$	$2 \times 2 \times 2 = 8$	$\frac{24}{8} = 3:1$
2	$6 \times (3 \times 3) = 54$	$3 \times 3 \times 3 = 27$	$\frac{54}{27} = 2:1$
3 and 4	$2 \times (3 \times 3) + 4 \times (3 \times 1.5)$ $= 18 + 18$ $= 36$	$1.5 \times 3 \times 3 = 13.5$	$\frac{36}{13.5} = 2.7:1$
5	$2 \times (1.5 \times 1.5) + 4 \times (12 \times 1.5)$ $= 4.5 + 72$ $= 76.5$	$1.5 \times 1.5 \times 12 = 27$	$\frac{76.5}{27} = 2.8:1$

A SA:V ratio that 'had' to be greater than 2.5 would explain why Block 2 'had' to 'divide'. (Small) Block 1 is more efficient at diffusing materials than Block 2. Block 3 and 4 are more efficient at diffusing materials than (larger) Block 2. 'Elongation' of Block 2 into Block 5 increases diffusion efficiency.

The principle of greater rate of activity with an increase in SA:V ratio occurs at all levels of biological organisation:

- Organ level – villi, small projections from the wall of the small intestine, project into the intestine for increased absorption of food.
- Organism level – native Africans are generally tall and thin, allowing rapid heat loss to stop overheating in the hot environment; elephants have large, thin ears to allow rapid heat loss.

Unit 11.1 Activity 3A: Movement of materials

1. Distinguish between the following pairs of terms:
   a. Passive and active transport.
   b. Diffusion and osmosis.
   c. Hypertonic and hypotonic.
   d. Dilute and concentrated solutions.
   e. Exocytosis and endocytosis.
   f. Phagocytosis and pinocytosis.

2. Describe the process of facilitated diffusion.

3. The diagram alongside represents a human red blood cell and the movement of sodium ions into the cell and potassium ions out of the cell. The relative concentrations of sodium ions and potassium ions in red blood cells and in the surrounding plasma are shown.

Ion	Plasma	Red blood cell
K^+	9 ←	152
Na^+	146 →	18

   a. Describe the method(s) of transport for the movement of Na^+ and K^+ ions between the red blood cell and the plasma. Give reasons for your choice of method(s).
   b. When red blood cells are separated from blood plasma and placed in distilled water, they burst open. Explain why this happens.

4. Explain the relationship between the SA:V ratio and diffusion in cells.

5. The diagram below represents two solutions separated by a semipermeable membrane (SPM).

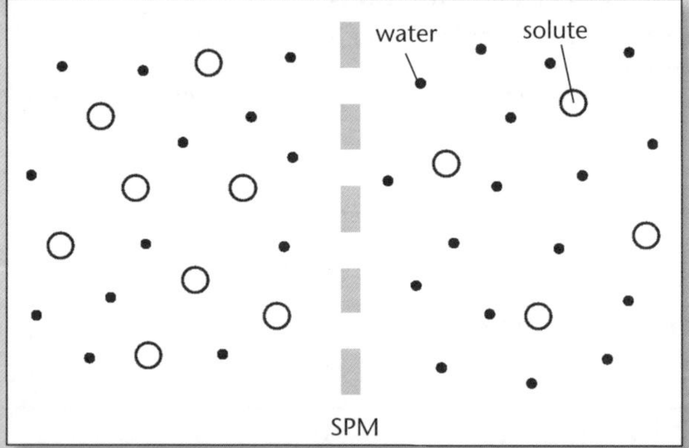

Explain the movement of the particles across the semipermeable membrane for both water and solute.

6. The three identical dialysis tubing bags (X, Y, and Z) suspended as shown contain equal quantities of a solution of 5% sucrose in water:

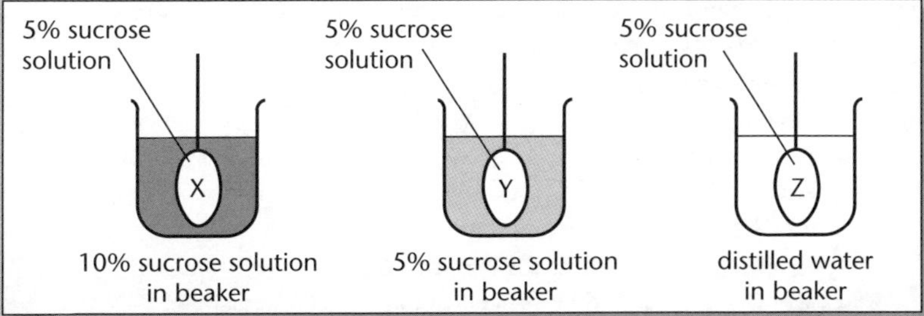

Dialysis tubing is permeable to water but not to sucrose.

Explain any changes that you would expect to see to each of the three bags (X, Y, Z) after 24 hours.

7. The diagram following shows an experimental set-up in which two solutions were placed in the arms of a U-tube separated by a semipermeable membrane (SPM). The semipermeable membrane is permeable to potassium chloride (K^+ and Cl^- ions), but not to sucrose.

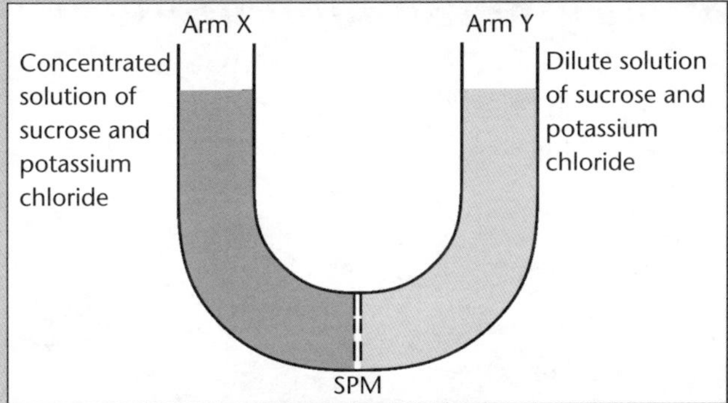

Discuss what would happen to the solutions.

8. The following graph shows the data obtained when 6 potato cubes of equal weight were placed in 6 different concentrations of salt water:

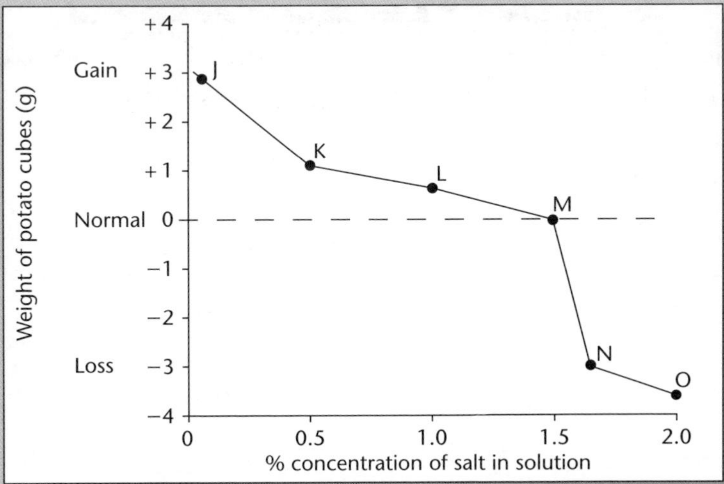

a. Draw a clear, labelled diagram of what a cell from one of the potato cubes from the 2.0% salt solution would look like under 400 × magnification of a school microscope.

b. Discuss the reasons for the gain and loss of weight by the potato cubes.

9. A unicellular organism that lives in a marine habitat was found to have much higher concentrations of iodine (a small molecule) in its cytoplasm than there was in the surrounding seawater.

a. Explain how the organism may maintain this high level of iodine.

b. Cyanide is a poison that inhibits respiration. Explain how the presence of cyanide in seawater would affect the iodine concentration in the organism.

10. The diagram following represents a cell from the lining of a human kidney tubule. A major role of the lining of the tubule is to absorb *all* the glucose from the fluid flowing along the tubule and pass it into the blood, as shown by the arrows on the diagram.

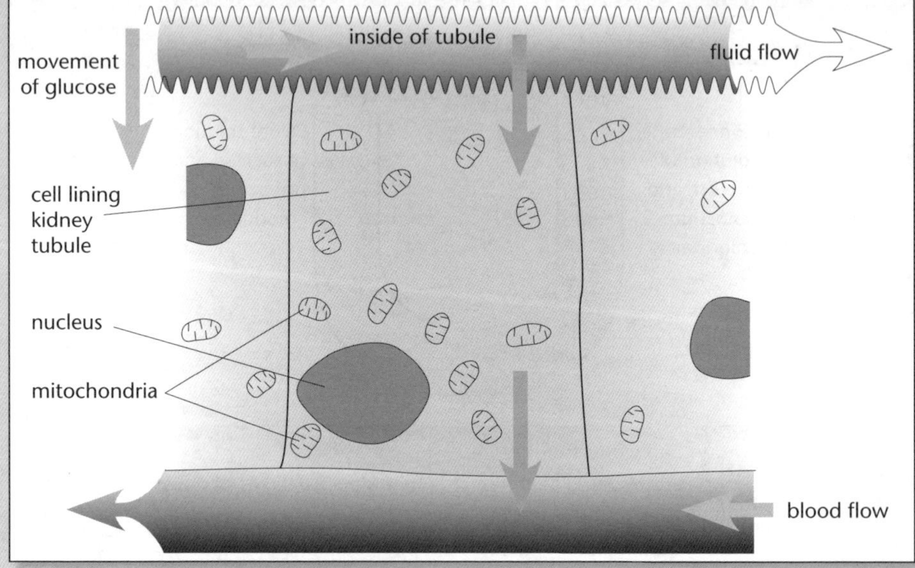

The tubule cell contains a large number of mitochondria. Explain the importance of mitochondria in the movement of *all* the glucose from the tubule to the blood.

Unit 11.1 Living Things

Topic 4: Cell processes – enzyme activity

The content presented in this Topic continues the description of the structure of cells and how they function. This Topic deals with:

- Structure of enzymes.
- 'Lock and key' and 'induced fit' models of enzyme function.
- Role of enzymes as biological catalysts.
- Enzyme specificity.
- Factors affecting enzyme activity.

Enzyme activity

Enzymes are part of the group of chemicals called **proteins**. Proteins, along with carbohydrates and fats, are the chemicals that make up most of the diet of animals. Proteins are made of amino acids, of which 20 different kinds exist in our cells. Essentially, there are two groups of protein:

- Fibrous – these are long and stringy and form much of the *structural* elements of the body (eg collagen in muscles and tendons; elastin in the connective tissue of the skin; keratin in hair, horns, nails).
- Globular – these are much-folded into a ball shape, and perform much of the *regulatory* processes of the body (eg hormones such as insulin, transporters such as haemoglobin, enzymes such as sucrase, antibodies, antitoxins).

It is these essential functions that make proteins so important. Proteins are coded for by DNA, with one gene of DNA coding for one specific protein.

Enzymes act as *biological* **catalysts**, controlling the speed of chemical reactions in all organisms. Enzymes increase the rate of chemical reactions – without them, **metabolism** would take place too slowly for life to exist.

Metabolism (metabolic reactions) refers to all the chemical reactions occurring in a cell. All the metabolic reactions of cells are controlled by enzymes.

- The reactions that synthesise large molecules from smaller ones (eg proteins from amino acids in protein synthesis) are **anabolism** (anabolic reactions).
- The reactions that break down large molecules into smaller ones (eg carbohydrates into glucose in digestion) are **catabolism** (catabolic reactions).

Enzymes are *specific*, ie one enzyme catalyses only one type of reaction. An enzyme is usually named after the substance it catalyses, by having the suffix '-ase' added, eg:

- *Lipase* catalyses the catabolism of lipids into fatty acids and glycerol.
- *Maltase* catalyses the catabolism of the sugar maltose into two glucose molecules.
- *Peroxidase* catalyses the breakdown of hydrogen peroxide, H_2O_2, into H_2O and O_2.

The reason that enzymes are specific relates to their shape. Each enzyme has a specific shape, determined by the sequence of amino acids that it is made of. The amino acids in a protein may be cross-linked by hydrogen bonds (in some cases sulfur bonds), and this is what gives the protein its distinctive shape. The shape of the enzyme in an area known as its **active site** corresponds to that of the substance(s) it catalyses – the *lock-and-key model* of enzyme action. In the way that a key fits into a particular lock, so the **substrate** fits into the active site of the enzyme. As the enzyme and substrate fit together (forming the *enzyme-substrate* complex), chemical bonds form or are broken. The enzyme's active site changes its shape slightly (through the weak hydrogen bonds that hold the shape) when combined with the substrate (called an *induced fit*). This distorts the substrate molecule(s), making reaction more likely.

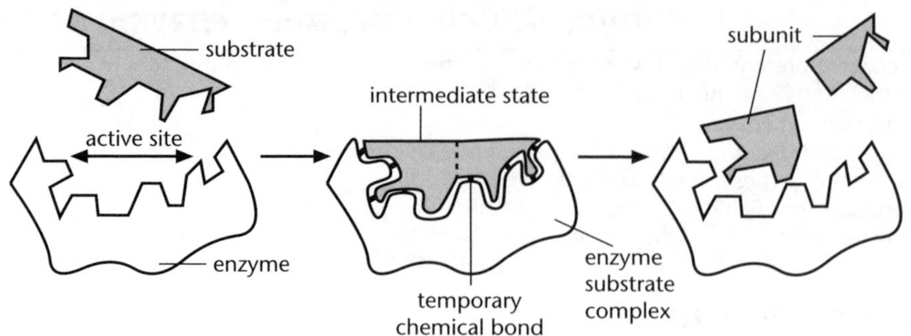

This is a catabolic reaction, ie the breaking down of a substance. A different enzyme would be used to reverse the reaction (ie bond the two smaller compounds together to synthesise the large one in an anabolic reaction).

Lock-and-key model of enzyme action.

Enzymes are *not* consumed or broken down in reactions, so one enzyme molecule can catalyse the same reaction many times and do so at a very fast rate (eg peroxidase in liver cells can catalyse the breakdown of several million hydrogen peroxide molecules into water and oxygen every minute).

Because their shape is maintained by weak hydrogen bonds, enzymes are susceptible to *denaturing* at high temperatures, as the hydrogen bonds are broken and the enzyme *loses its shape* – it can no longer catalyse the reaction.

Factors that affect enzyme activity

Temperature

The warmer the temperature (usually up to about 40–45°C), the faster enzymes will catalyse a reaction. This is because increasing the temperature increases the speed at which the reacting particles move, so they collide more often. It is necessary for particles to collide for them to react. However, above 40–45°C, enzymes are usually **denatured** and can no longer catalyse the reaction.

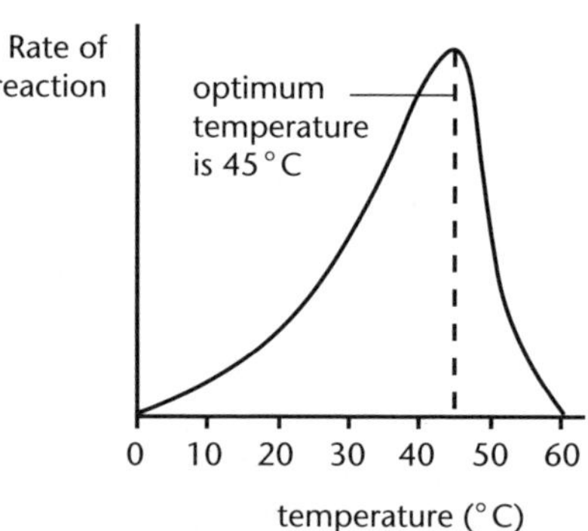

Enzyme reaction rate and temperature.

The temperature at which the reaction rate is fastest is the optimum temperature.

In homeotherms ('warm-blooded' animals – birds and mammals), the optimum temperature will be their core body temperature (37°C in humans).

In poikilotherms ('cold-blooded' animals/organisms), the optimum temperature will depend on their habitat; it may be low if the organism lives in cool habitats (eg some fish in the Antarctic seas have an optimum temperature approaching 0°C), and high in hot habitats (eg bacteria that inhabit geothermal hot pools have an optimum temperature approaching 100°C).

pH

Most enzymes work within cells, so their optimum pH will be approximately 7 (ie close to neutral); exceptions are the digestive enzymes (eg pepsin).

When the pH is outside the range for an enzyme, the enzyme denatures, and can no longer act as a catalyst.

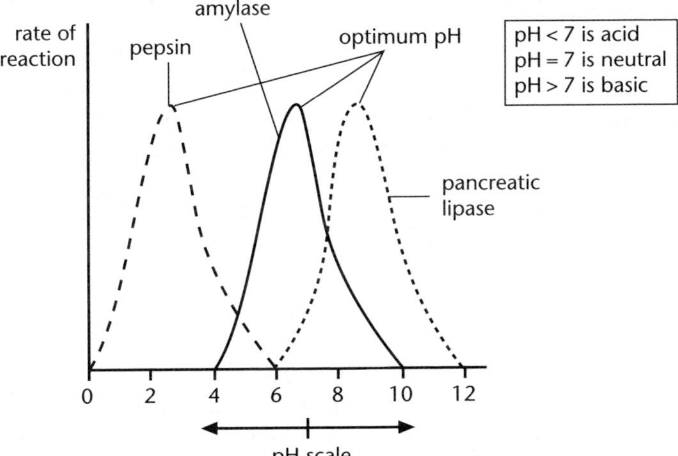

Pepsin breaks down protein in the stomach and works well in acidic conditions. The stomach contents have a low pH because of the presence of dilute hydrocholoric acid.

The optimum pH for *amylase* (an enzyme in saliva which breaks down starch) is about 7. Saliva has a neutral pH.

Pancreatic lipase breaks down lipids in the small intestine, where conditions are slightly basic (pH is greater than 7), because of the presence of bicarbonate ions.

Enzyme reaction rate and pH.

Substrate concentration

The rate of enzyme activity will increase as the concentration of the substrate increases up until a saturation point occurs (ie no free enzymes/active sites).

Co-factors

Many enzymes need another molecule to assist in catalysis. These co-factors may be (small) inorganic ions such as cobalt, Co, selenium, Se, or (large) organic molecules such as some vitamins. These organic co-factors are called **co-enzymes**.

- The carrier molecule NAD (required in respiration) is a co-enzyme derived from nicotine acid, a member of the vitamin B complex.
- Hydrolases, which control a transcription factor that turns on specific genes, need vitamin C to be activated.

Co-enzymes are necessary when only weak bonds form between the enzyme and substrate; the co-enzyme acting as a bridge, locking the enzyme and substrate more tightly together.

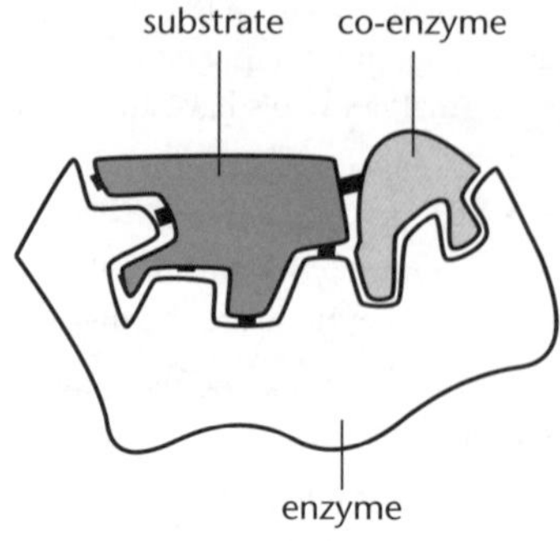

Inhibitors

Inhibitors are substances that prevent enzymes catalysing reactions, and so are *poisons*.

- Heavy metal ions (eg lead, Pb, and mercury, Hg) prevent enzymes in cells of the nervous system functioning.
- *Cyanide* prevents the action of an enzyme in the electron transfer chain of respiration (causing death, as respiration ceases).

An inhibitor can act by:

- Taking over the active site of the enzyme, so stopping the substrate from binding to the active site – eg the antibiotic penicillin inhibits an enzyme that bacteria use to make their cell walls.
- Bonding to another part of the enzyme and altering the shape of the active site so that it can no longer bind to the substrate – this is usually temporary, and can be a way in which the cell can control a metabolic pathway.

Unit 11.1 Activity 4A: Enzyme activity

1. Explain why enzymes are known as 'biological catalysts'.
2. Use the diagram following to help explain the 'lock-and-key' model of enzyme action.

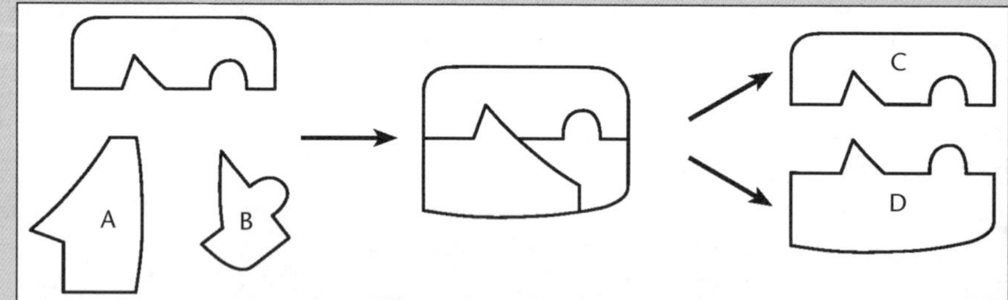

3. Explain why enzymes are essential to the survival of organisms.
4. Explain the role of co-factors in enzyme activity.

5. Explain the relationship between enzymes and pH.
6. Discuss the relationship between enzymes and temperature.
7. Bread can be made by mixing flour, water, salt, sugar and a small amount of yeast. Carbon dioxide is produced from the fermentation of simple sugars. Fermentation is controlled by enzymes from the yeast.

   a. Explain why the shape of an enzyme is important for the way it functions. (You may use a diagram to help with your explanation.)
   b. In relation to *enzyme structure*, explain why the enzymes involved in bread production will not function at or above 45°C.
   c. Researchers have investigated the effect of temperature on yeast activity. The following tables are a summary of their results.

Temperature	Fermentation
–20°C	No fermentation
27°C to 38°C	Optimum fermentation range
35°C	Optimum fermentation temperature

Temperature	Cell division
Less than 20°C, greater than 40°C	Cell division significantly reduced
20°C to 27°C	Most favourable range for yeast to multiply
26°C	Optimum temperature for multiplication of yeast
Greater than 60°C	Nil

   Between two and three hours are needed for yeast to ferment dough before it is baked in an oven.

   Discuss, with respect to the number of yeast cells and the fermentation rate, why it is important to have the temperature at:
   - 26°C for the first hour and;
   - 35°C for the next two hours, before the dough is baked.

   d. Heavy metals, such as mercury and lead, are enzyme inhibitors.
   Explain how an enzyme inhibitor affects enzyme activity. (You may use a diagram to help with your explanation.)

Unit 11.1 Living Things

Topic 5: Classification systems

The content presented in this Topic relates to biological classification systems including the binomial nomenclature and the Linnaean system of classification. The Topic deals with:

- Characteristics of the three evolutionary lineages or domains and the groups within them.
- The main taxonomic groups and simple classification of organisms.
- Specific names.
- Biological keys.
- Characteristics of life (MRS GREN).
- Characteristics of viruses.

Classification

There are millions of plant and animal species on Earth. In PNG we have over 20,000 species of plants, more than 200 species of mammals and almost 1000 species of birds. Scientists need to classify species in an orderly manner. How do they do this?

> In the Linnaean classification system, every living thing has a double name in Latin. This is called the 'binomial nomenclature'. The Linnaean classification system is a scientific method of classifying living things into increasingly smaller and specific groups with a hierarchy. It uses seven main categories: Kingdom, Phylum, Class, Order, Family, Genus, Species. Kingdom is the largest and Species is the smallest.

You are expected to know the main classification (taxonomic) groups and be able to place common organisms in them. Specific names should be identified and correctly written. You should be able to use keys to identify organisms or their groups.

Example

If an exam question included information about the 'endangered indigenous reptile the giant monitor lizard *(Veranus salvattorii)*', a student, besides knowing the meaning of the terms 'endangered' and 'indigenous', would be expected to:

- Know why the giant monitor lizard was placed in the group known as the 'reptiles'.
- Know that 'giant monitor lizard' was the common name, '*Veranus salvattorii*' was the scientific name, and '*Veranus*' referred to the genus while '*salvattorii*' referred to the species.

Classification systems

The first living creatures (**organisms**) appeared on Earth about 3.5 billion years ago. Since then, evolutionary processes have produced a large diversity of different organisms occupying a range of ecological **niches**. All organisms living today are related and share **common ancestry**; this is seen in the universality of **DNA**, the genetic code, found in the chromosomes of all life forms. The structure of all organisms is based on the **cell**, which carries out all the essential life processes (eg respiration, photosynthesis, protein synthesis).

The degree of relationship between organisms can be seen in their taxonomic (**classification**) groups. Classification is a changing area of biology, with DNA sequencing providing new evidence of the relationships between groups of organisms; therefore, differences in classification occur, depending on the source and date of the information.

Classification systems found in most textbooks place all organisms into one of two large **domains**, based on their cell type – **prokaryotes** (cells without a true nucleus) and **eukaryotes** (cells with a true nucleus), then into one of five large **kingdoms** – **monera**, **protista**, **fungi**,

plants and **animals**. The monera (bacteria and cyanobacteria) is the only kingdom within the prokaryote domain; the other four kingdoms are all within the eukaryote domain.

Within the last decade, DNA analysis has indicated that life represents not two but three evolutionary lineages (domains) – the Archaea, Eubacteria and Eukarya (these names may vary, depending on the reference used).

Organisms of the monera kingdom are now placed into either the archaea or the eubacteria. These organisms are all one-celled, with the cell lacking a true nucleus, as their genetic material/DNA/chromosome is not contained in a nuclear envelope (such cells are prokaryotic cells).

- **Archaea** includes the **archebacteria** – the oldest type of bacteria; inhabit environmental extremes (eg geothermal pools, methane vents in the deep oceans, salt marshes), so commonly called *extremophiles*.
- **Eubacteria** includes the **bacteria** and **cyanobacteria** – more recently-evolved forms; inhabit moderate environments, and include all common forms, such as those pathogenic to humans.

Bacteria may be *autotrophic* (make their own food in **chemosynthesis**), or *heterotrophic* (either parasites or saprophytes). The cyanobacteria are autotrophic, making their food in **photosynthesis**. These bacteria are believed to have been the first life forms that appeared on Earth.

Bacterial shape may be rod (bacilli), sphere (cocci), spiral (spirilli):

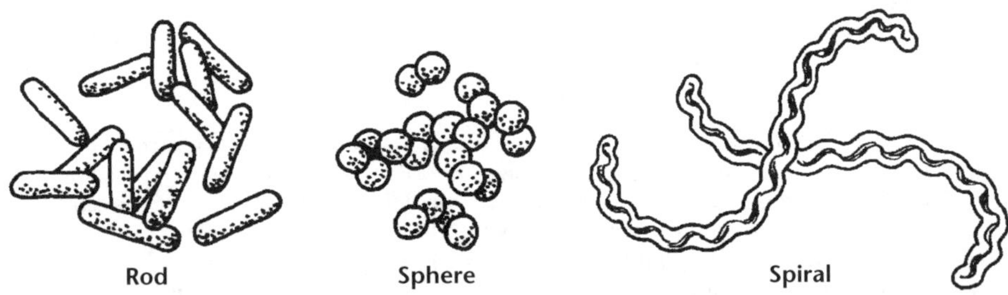

Protista, Fungi, Plants and Animals belong in the domain **Eukarya** because all are made of eukaryotic cells – cells with a true nucleus (ie one that has chromosomes contained in a nuclear envelope, and, in the cytoplasm, are membrane-bound organelles).

Protista (also known as Protoctista)

Most of these organisms are one-celled (eg *Euglena, Amoeba, Paramecium*). Some are simple threads of repeating cells (eg *Spirogyra*). Other are multi-celled but simplistic in structure, lacking systems (eg seaweeds such as *Hormosira, Corallina*).

Protists may be autotrophs (make their own food in photosynthesis) or heterotrophs (obtain their food from other organisms – herbivores, carnivores, omnivores, parasites).

Fungi

Fungi may be single-celled (eg yeasts), or made up of threads called *hyphae* (eg bread mould). These threads may be combined into a definite body (the *mycelium*), with reproductive and feeding parts (eg mushrooms – the 'mushroom' is the spore-producing reproductive structure; the feeding hyphae are in the ground). Feeding is *extra-cellular*, and fungi are all heterotrophs (parasites and saprophytes). Cells are enclosed in a cell wall, but it is made of chitin (not cellulose, as in plants).

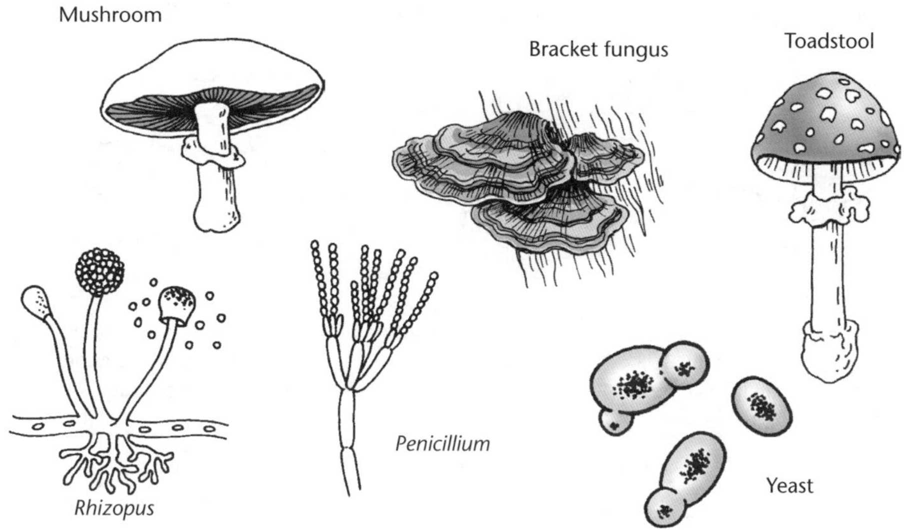

Plants

Plants are autotrophic, making their food in **photosynthesis** (exceptions are some parasitic species). All are multicellular, with the cells contained in cellulose cell walls; cytoplasm contains chloroplasts (with chlorophyll) for photosynthesis. Plants are classified on the basis of their reproduction (eg conifers have cones, angiosperms have flowers).

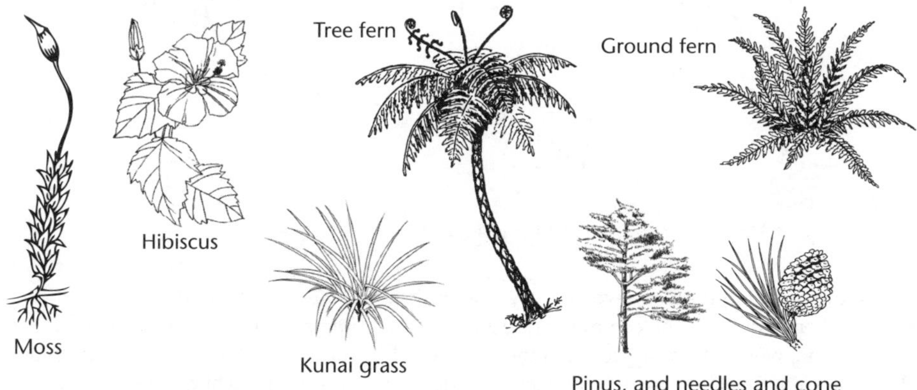

Animals

Animals are all heterotrophic (herbivores, carnivores, omnivores, parasites, scavengers, filter feeders) and multicellular. Cells are varied in shape and function; the cell membrane is *not* enclosed in a cell wall. Many have complex internal systems and may be highly mobile. A wide diversity of forms exist.

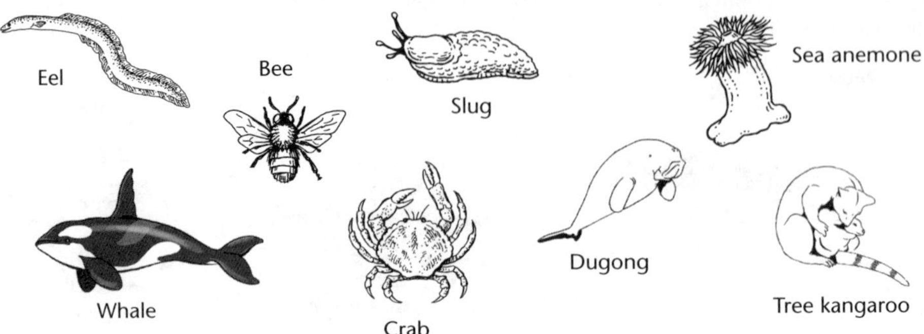

The evolutionary relationships of the living representatives of these groups is under constant review; classification will always be a work in progress, as we continue to learn more about organisms and their relationships. At the time of writing, the following diagram is the most accepted 'tree of life':

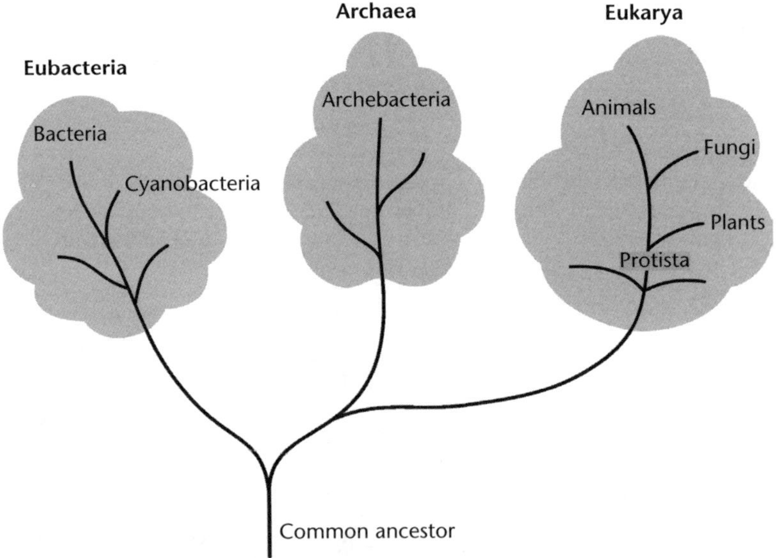

'Tree of Life' showing the likely evolutionary relationships between the three domains and their large taxonomic groups. The common ancestor refers to a simple cellular form of life that existed millennia ago, which, from mutations in the DNA and the process of natural selection, has evolved into the present-day life forms. The unnamed branches in the tree represent groups that have arisen, lived, then become extinct.

Taxonomic groups

Organisms are placed into the various taxonomic groups depending upon their relation to other organisms.

- 'Domain' is the largest grouping, so has the largest number of different organisms.
- 'Species' is the smallest grouping, and has only one kind of organism.

For animals, the most common taxonomic groups – from largest to smallest – are:

- Phylum.
- Class.
- Order.
- Family.
- Genus.
- Species.

Many of these groups may be subdivided again (eg 'super class', 'suborder'). The more groups that different organisms have in common, the more closely related they are.

The smallest taxonomic group, the **species**, is usually defined as 'a group of organisms that interbreed to produce fertile offspring'. They share a common gene pool and are reproductively isolated from other species. Sometimes, a species may have **subspecies**; these *are* capable of interbreeding. All species have a scientific name which is from Latin or Greek (many species also have a common name in the local language). The scientific name includes the **genus** followed by the species, a system devised by Linneaus nearly 250 years ago and known as the *binomial system of naming organisms*.

Both the genus and species names are written in *italics* or <u>underlined</u>. The genus name starts with an upper case letter, the species with a lower case letter, eg:

Common name	Scientific name
Human	*Homo sapiens* or <u>Homo</u> <u>sapiens</u>
Ship rat	*Rattus rattus* or <u>Rattus</u> <u>rattus</u>
Dog	*Canis familiaris* or <u>Canis</u> <u>familiaris</u>
Betel nut	*Areca catechu* or <u>Areca</u> <u>catechu</u>

Once the scientific name has been used, subsequent use may be abbreviated (eg *H. sapiens*, <u>R.</u> <u>rattus</u>). If the species being referred to is unknown or the reference is to all the species in the genus, then the abbreviations *Rattus sp* (singular) or *Rattus spp* (plural) may be used.

Although a species is typically reproductively isolated from other species, closely related species (eg members of cat families, dog families, horse families) may interbreed, forming **hybrids**. Hybrids typically happen in unnatural conditions (eg zoos, aquaria), where mating opportunities with the *same* species may be limited. When hybrids *do* breed, the offspring are **infertile** (ie are unable to breed), eg:

- Horse × Donkey → Mule; Horse × Zebra → Zebroid.
- Lion × Tiger → Liger or Tigroid (depends on sex of parents); Lion × Leopard → Leopon.

Keys

A key is a way of *identifying* organisms into their taxonomic groups (eg class, family, species).

Keys are typically *dichotomous* – each step in the key involves a choice of two (either/or). The choice leads to the next step and choice, until the organism is identified.

Classifying arthropods

Below are five arthropods found in a beech forest:

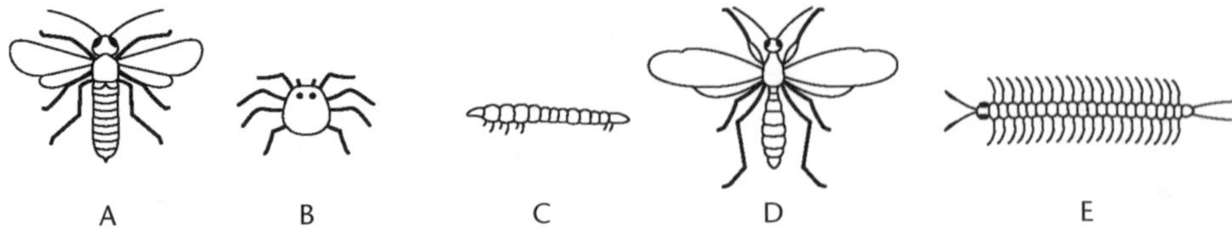

The key below is used to identify the five arthropods.

Key:

1.	Wings present	go to 2
	Wings absent	go to 3
2.	Abdomen thin when joins thorax	*Apantales* (D)
	Abdomen same thickness as thorax	*Tenthredo* (A)
3.	More than 10 legs	*Lithobius* (E)
	Fewer than 10 legs	go to 4
4.	Short fat body	*Pergamasus* (B)
	Long thin body	*Agriotes* (C)

Unit 11.1 Activity 5A: Classification or Dichotomous keys

1. Shown below are diagrams of the five native beeches (*Nothofagus*) which are found in New Zealand.

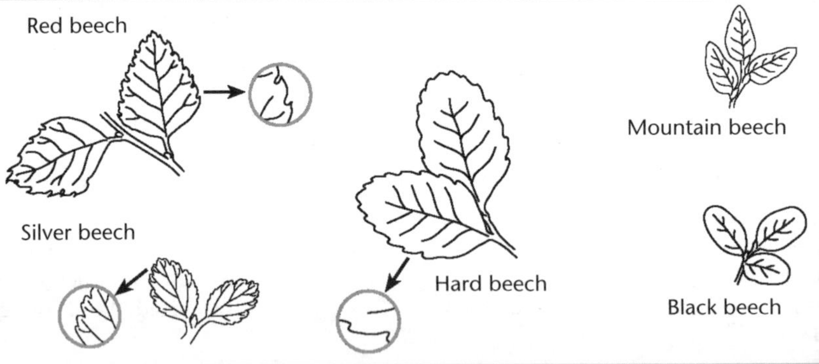

Use the key below to give the scientific names of the five beech species.

Key:

1.	Leaves toothed (on edges)	go to 2
	Leaves not toothed	go to 4

2. Teeth on leaves sharply pointed — *N. fusca*
   Teeth on leaves not sharply pointed — go to 3
3. Leaves 8–12 mm long and doubly toothed — *N. menziesii*
   Leaves 25–40 mm long and singly toothed — *N. truncata*
4. Leaves oval shaped — *N. solandri var solandri*
   Leaves triangular — *N. solandri var cliffortioides*

2. Use the following diagrams and key to find the scientific names of the six common grasses.
Note: Diagrams not to scale

Key to some common grasses:

1.	a.	Auricles present	go to 2
	b.	Auricles absent	go to 3
2.	a.	Ligule very short (0.5 mm) Sheath and leaf blade hairy	*Agropyron repens*
	b.	Ligule up to 2 mm long Sheath and leaf blade hairless	*Lolium perenne*
3.	a.	Ligule hairless	go to 4
	b.	Ligule hairy	go to 5
4.	a.	Sheath and leaf blade hairless	*Agrostis tenuis*
	b.	Sheath hairless; leaf blade also hairless except for tufts of long hairs on the collar	*Paspalum dilatatum*
5.	a.	Ligule 1–4 mm long and hairy	*Holcus lanatus*
	b.	Ligule 2–3 mm long with distinct fringe of airs	*Pennisetum clandestinum*

Viruses

All organisms (ie all members of the three domains) are able to carry out the seven life processes, given in the acronym *MRS GREN*:

- *M* – movement.
- *R* – respiration.
- *S* – sensitivity.
- *G* – growth.
- *R* – reproduction.
- *E* – excretion.
- *N* – nutrition.

Viruses are only able to carry out the process of reproduction, and can only do that using living cells, using the energy supplies and raw materials from the host cell (eg viruses called bacteriophages use *bacteria* as host cells).

Viruses do not fit into the three domains system (since they only show Reproduction in terms of *MRS GREN*); it is thus debatable whether they are living things. Their structure comprises a protein coat enclosing the genetic material of the virus. The genetic material of the virus is a nucleic acid, either DNA (eg Herpes virus) or RNA (eg retroviruses such as HIV).

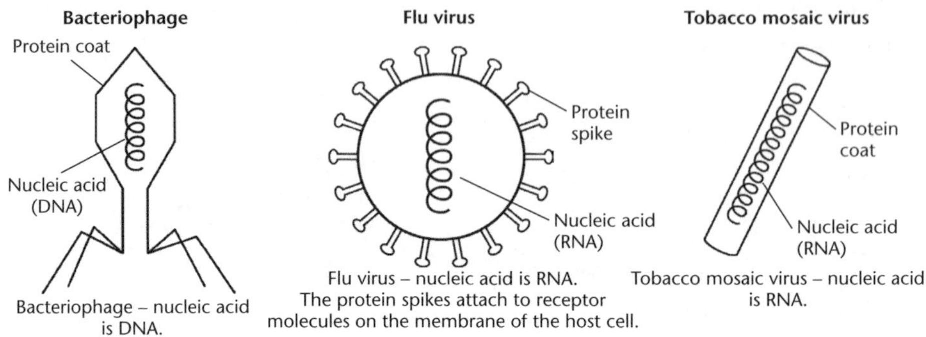

Common shapes of viruses.

All viruses can be considered as a type of obligate intracellular parasite; all are instrumental in causing disease in plants and animals, including humans.

Unit 11.1 Activity 5B: Classification

1. Describe the binomial system of naming organisms (also known as 'binomial nomenclature').
2. Describe how the scientific name of an organism is written.
3. Explain what is meant by a *taxonomic* group.
4. Explain the difference between prokaryotes and eukaryotes.
5. Name the main groups of:
   a. Prokayotes. **b.** Eukaryotes.
6. Older classification systems classified fungi as plants. Give the main reasons why fungi are no longer considered plants.
7. Distinguish between autotrophs and heterotrophs.
8. Define the following terms:
   a. Species. **b.** Hybrid. **c.** Infertile.
9. Explain what is represented by the acronym MRS GREN.
10. Explain why viruses are not classified as living organisms.

Unit 11.1 Living Things

Topic 6: Diversity of animals

The content presented in this Topic describes the diversity in the structure and function of animals and is a continuation of Topic 5 on the classification of living things.

Students of Biology at Upper Secondary level are expected to describe and give reasons for the *diversity* of aspects of the structure and function of *multicellular* animals in relation to a biological process:

- Internal transport.
- Sensitivity and co-ordination.
- Excretion.
- Support and movement.
- Nutrition.
- Gas exchange.
- Reproduction.

To fully appreciate the diversity of animals, there are many Topics that could be studied in conjunction with this Topic:

Unit 11.2 Nutrition, Topic 4: How humans feed – heterotrophic nutrition (p. 97)

Unit 11.2 Nutrition, Topic 5: How animals feed – heterotrophic nutrition (p. 111)

Unit 11.3 Transport Systems, Topic 3: Transport systems in animals (p. 143)

Unit 11.3 Transport Systems, Topic 5: Excretory systems in animals and humans (p. 179)

Unit 11.4 Respiration and Gas Exchange, Topic 2: Respiration, breathing and gas exchange (p. 197)

Unit 11.4 Respiration and Gas Exchange, Topic 3: Gas exchange in animals (p. 203)

Unit 11.5 Responses to Stimuli, Topic 2: Sensitivity and coordination in animals (p. 233)

Unit 11.6 Reproduction, Topic 1: How animals reproduce (p. 259)

Diversity

Diversity is about *differences*. All animals carry out the same seven essential life processes (*MRS GREN*), but they may do them in different ways and using different structures. These differences reflect the way that the animal/group has evolved to be successful in its *ecological niche*.

Classification of animals

Taxonomic groups refers to the groups into which an animal is classified (eg mammal, insect, annelid).

Example

Classification of humans

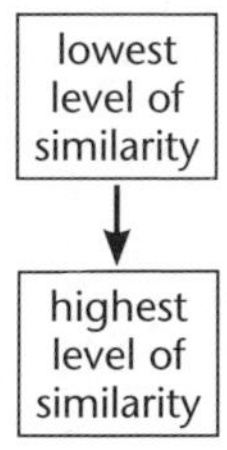

Classification group	Human
Kingdom	**Animalia** (animals)
Phylum	**Chordates** (have backbones)
Class	**Mammals** (warm-blooded, hairy, suckle young)
Order	**Primates** (monkeys, apes)
Family	**Hominins** (large brain, 'intelligent', walk upright)
Genus	*Homo* (human)
Species	*sapiens* ('intelligent')

The scientific name for humans is *Homo sapiens*, and this identifies us from all other species. It must be written in italics or underlined (eg Homo sapiens).

The taxonomic groups show the *degree of relatedness* to other animals.

Example

Humans belong to the *chordates* phylum – along with fish, amphibians, reptiles, birds – as we all have a spinal cord (housed in a spine/vertebral column). Within this phylum we belong to the class called mammals – along with many other groups – because we are 'warm blooded' (homeotherms), our bodies are hairy, our young are born alive and are suckled on milk from the mother.

The animal kingdom has 21 *phyla* (singular = *phylum*, the largest grouping in the kingdom). The following phyla are the most significant.

Phylum Cnidaria

Includes the jellyfish, sea anemones, corals, hydra (only freshwater member). All have a simple, sac-like body with one opening ('mouth') surrounded by tentacles. The body is of two cell layers, with a middle, jelly-like layer. The old term for cnidaria is *coelenterates*.

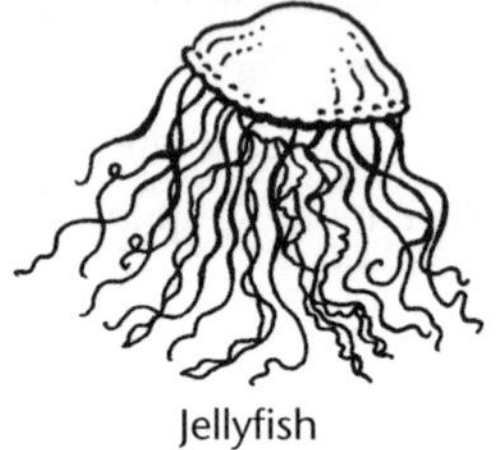

Jellyfish

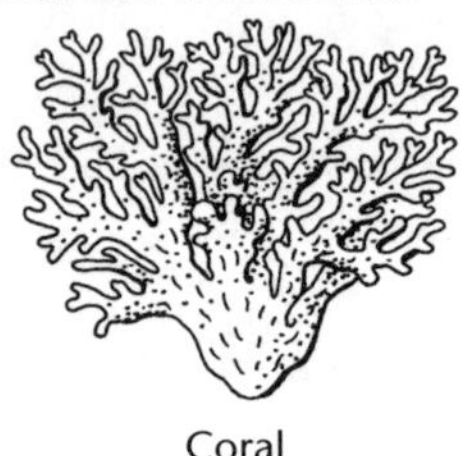

Coral

Phylum Platyhelminthes (flatworms)

Includes the free-living planarian worms and the parasitic flukes and tapeworms. All are flattened dorsal-ventrally, not segmented, and have one body opening ('mouth').

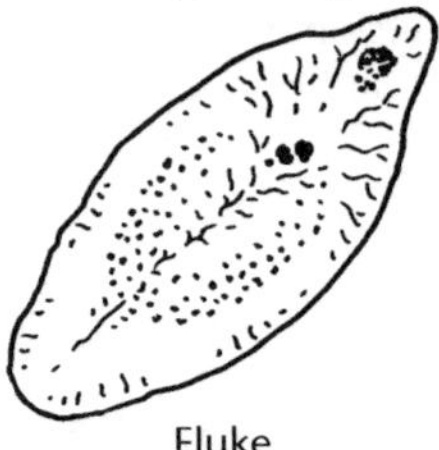

Fluke

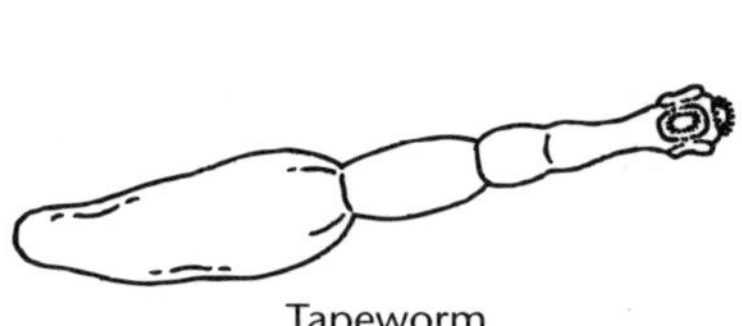

Tapeworm

Phylum Annelida (segmented worms)

Includes the earthworms, marine worms, leeches. Body is elongated and divided into many repeating segments. There are two openings (mouth and anus), and the body cavity is a fluid-filled coelom which acts as a hydrostatic skeleton.

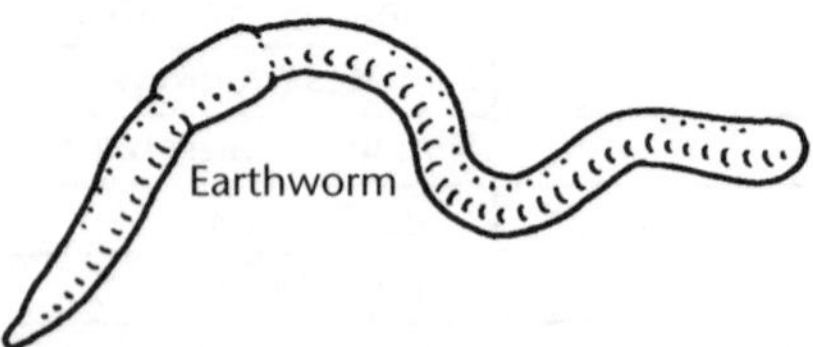

Earthworm

Phylum Mollusca

Includes all the snails, shellfish, octopuses. Body is soft and not segmented, often enclosed in a shell. Movement is by a muscular foot (except for octopus), feeding by a rasping radula (except for bivalves such as mussels, which are filter feeders). Mucus is an important lubricant. There are three major classes:

- **Gastropoda** – includes the slugs and snails. Soft body is enclosed in a (typically coiled) shell (slugs have lost their shell in their evolution). Includes the only terrestrial members of the molluscs.

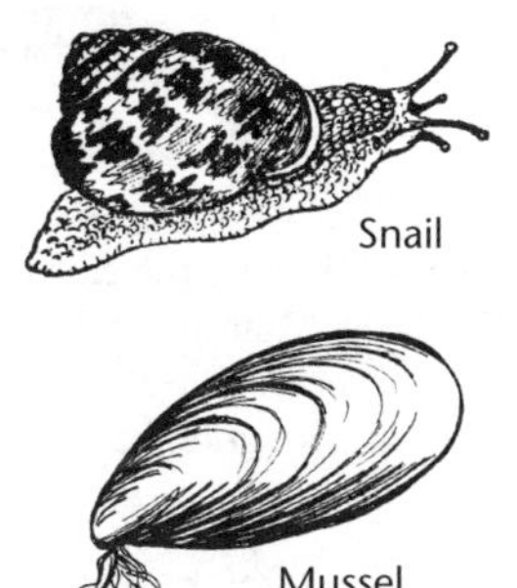

Snail

Mussel

- **Bivalvia** – includes the mussels, scallops, oysters, pipi. Soft body enclosed in two shells (valves); typically filter feeders.

- **Cephalopoda** – includes the octopus and squid. Soft body, shell no longer present (or greatly reduced). Tentacles present for moving and feeding; movement by 'jet propulsion'; eyesight well developed, as is a brain.

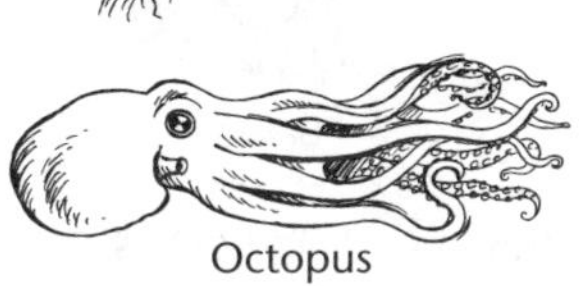

Octopus

Phylum Arthropoda

Includes the insects, spiders, centipedes and millipedes, crabs and shrimps. The most numerous group of animals, comprising over 75% of all species. Body has an exoskeleton – a cuticle of (mainly) chitin, with jointed limbs for movement. Exoskeleton is shed for growth. Body is segmented (not usually obvious), and divided into a head and one or two other regions (thorax, abdomen). Blood system is open. Four classes:

- **Insects** – includes grasshoppers, flies, moths, butterflies, bees, beetles. The most numerous of the arthropods, comprising over 75% of all arthropod species; the beetles are the most common insects. Body has three parts (head, thorax, abdomen) and three pairs of legs attach to the thorax. Many have wings (one or two pairs). Head has a pair of compound eyes and a pair of antennae. Tracheal system for gas exchange.

Housefly

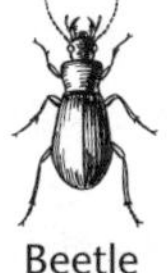

Beetle

Butterfly

- **Arachnida** – includes the spiders, ticks, mites, scorpions. Body of two parts (joined head and thorax, separate abdomen), with four pairs of legs; no antennae. Many sting.

Mite

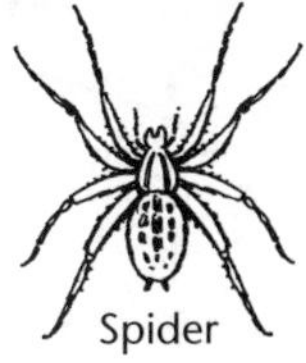

Spider

- **Myriapoda** – includes the centipedes and millipedes. Body of many segments, with a head (has antennae). Centipedes have one pair of legs per segment, and are carnivorous; millipedes have two pairs of legs per segment, and are herbivorous. Tracheal system of gas exchange. Centipedes have far fewer than 100 legs, millipedes have between 80 and 750 legs.

Millipede

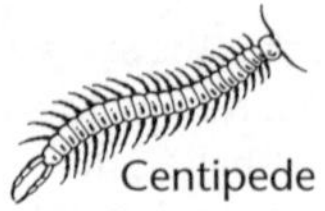
Centipede

- **Crustacea** – includes the crabs, crayfish, shrimps, slaters ('woodlice'), barnacles. Aquatic habitats (or need damp conditions, eg slaters). Often have a very hard exoskeleton (a carapace, eg crabs). Have many legs, which are often modified (eg for swimming, egg carrying, feeding). Distinct head with antennae. Gas exchange using gills.

Barnacle

Shrimp

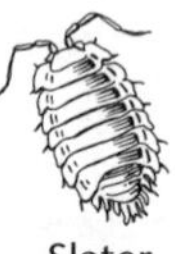
Slater

All the above phyla are **invertebrates**, ie the animals do *not* have a spinal column and spinal nerve cord. The nerve cord of invertebrates is either ventral (eg annelids, arthropods) or a network (eg cnidaria). The animals are also 'cold blooded' (poikilotherms), and cannot regulate their body temperature (which thus follows that of the environment).

Vertebrates are animals that have a spinal column (of **bones** called vertebrae). Technically, the vertebrates are a sub-phylum of the chordata. Apart from a very few primitive forms (eg sea squirts, lancelets), all chordates have a vertebral column enclosing their spinal cord at some stage during their life. Species of vertebrate animals are far less numerous than species of invertebrates, but typically much larger.

Phylum Chordata

The major classes of chordata are Pisces (fish), Amphibia, Reptilia, Aves (birds), Mammalia.

Fish/Pisces

The class 'fish' is more accurately given as several distinct classes, of which the two main ones:

- Osteichthyes (bony fish) – have a skeleton made of bone – the majority of fish, eg snapper, tuna, salmon, trout.
- Chondrichthyes (cartilaginous fish) – have a skeleton made of cartilage, eg sharks and rays.

All fish are aquatic. Outer covering of scales, gills are used for gas exchange, blood system is closed with single-loop circulation.

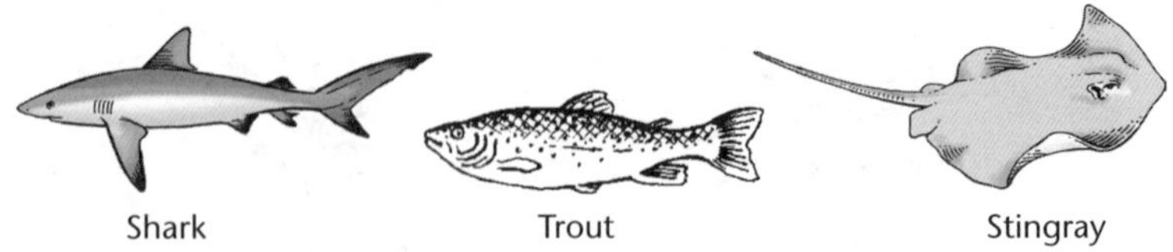
Shark Trout Stingray

Amphibia

Includes frogs, toads, and the lizard-like newts and salamanders. PNG is home to an abundance of frog species. Many are unusual (such as one which is just 1 cm in length and another that has vampire-like fangs) and are found nowhere else in the world. Amphibians are restricted to an aquatic or damp/humid environment by:

- Their skin – lacks scales and needs to be moist for gas exchange, though lungs also present.
- Reproduction – fertilisation is typically external, and the eggs do not have a hard shell.

Young may have a different form from adult and metamorphose into the adult (eg tadpole to frog).

Reptilia

A diverse class, including lizards (eg geckos and skinks), snakes, crocodiles and turtles. Skin is dry and scaly; reproduction is by shelled eggs, and fertilisation is internal, therefore reptiles are adapted fully to terrestrial living, unlike amphibians.

Birds/Aves

Birds are feathered homeotherms with beaks, scaly legs, wings for flight. However, some forms have lost power of flight – eg kiwi, ostrich, takahe (wings are vestigial), penguins have wings modified as flippers for swimming. Fertilisation is internal and hard-shelled eggs are laid. Energy needs are high, and blood system is closed, with a 4-chambered heart and complete, double circulation.

Bird of paradise

Cassowary

Mammalia

Mammals are hairy (hair lost in whales) homeotherms. Fertilisation is internal and young are born alive and suckled on mother's milk from mammary glands. Brain is large. A diverse group, divided into three large sub-classes, dependent on method of reproduction:

- **Monotremes –** the egg-laying platypus and echidna of Australia. Young hatch and suckle on mother's milk.

- **Marsupials** – young are born in embryonic state and complete development attached to teat in mother's pouch, eg kangaroo, possum.

- **Placentals** – the majority of mammals. Young retained in uterus and nourished by a placenta until birth, when they suckle from mother, eg human, cat, monkey, whale, bat.

Unit 11.1 Living Things

Topic 7: Diversity of plants

The content presented in this Topic describes the diversity in the structure and function of plants.

Students are expected to describe and give reasons for the *diversity* of aspects of the structure and function of *multicellular* plants in three different taxonomic or functional groups in relation to a biological process. The biological process is selected from:

- Nutrition (see Unit 11.2).
- Transport of materials (see Unit 11.3).
- Transpiration (see Unit 11.3).
- Reproduction (see Unit 11.6).

Classification of plants

Taxonomic groups refers to the groups into which a plant is classified (eg fern, moss, conifer).

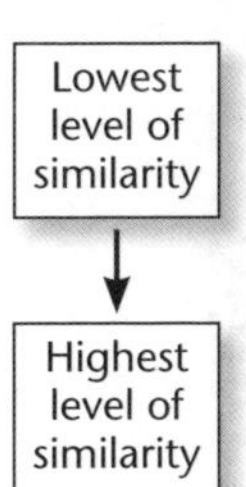

Classification group	Example – tomato
Kingdom	**Plantae** (plants)
Division	**Tracheophytes** (have conducting vessels)
Class	**Angiosperms** (flowering plants)
Order	**Scrophulariales** (flower structures similar)
Family	**Solanaceae** (flower structures very similar)
Genus	*Lycopersicon*
Species	*esculentum* (tomato)

Taxonomic scientific names must be written in *italics* or underlined. The scientific name identifies a species from all other species.

The taxonomic groups show the *degree of relatedness* to other plants.

Example

Tomato plant

The scientific name *Lycopersicon esculentum* (or Lycopersicon esculentum) identifies the tomato from all other species.

Tomatoes belong to the angiosperms, along with all other plants that reproduce using flowers.

Photosynthesising organisms are found in two groups:

- The **Protista** (also called **Protoctista**) – all the seaweeds – eg *Hormosira* (Neptune's necklace), *Corallina*, kelp, as well as the unicellular forms (eg *Euglena*).
- The plant kingdom – 'the plants'.

There are two major divisions (also known as phyla) of plants:

- **Bryophyta –** the *non-vascular* plants.
- **Tracheophyta –** the *vascular* plants. **Vascular tissue** is *transport* tissue – *xylem* and *phloem*.

Bryophyta

Bryophytes, the *non-vascular* plants, live in damp areas and are sufficiently small plants that they have no vascular tissue. Two groups exist – mosses and liverworts.

Mosses

All mosses are small, with no true roots (have hair-like rootlets called *rhizoids*). Mosses have thin, leaf-like structures surrounding a central, stem-like structure. Mosses have two distinct generations:

- Conspicuous leafy plant is the *gametophyte* (n) which is dependent on water for fertilisation.
- Rising out of the gametophyte is the spore containing capsule of the *sporophyte* (2n) generation.

Liverworts

Much simpler plants than the mosses, liverworts look like leaves lying flat on the ground. Liverworts are found only in wet areas.

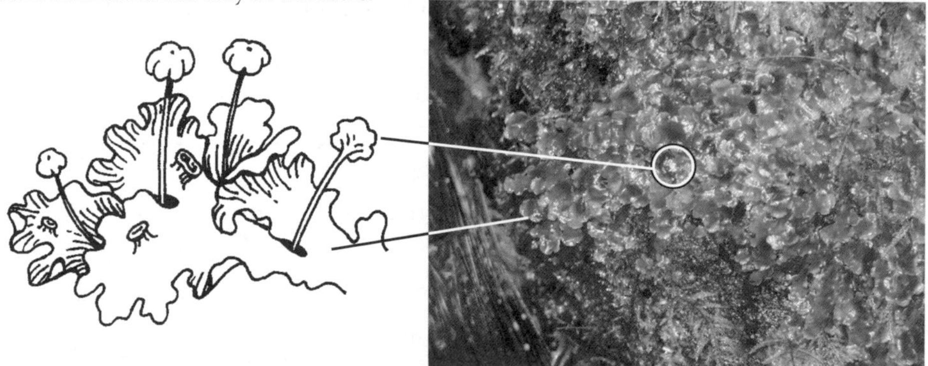

Tracheophyta

These are the *vascular* plants, ie they have transport tissue in the form of xylem and phloem. Vascular plants are typically much taller than non-vascular plants. Three major groups exist – ferns, conifers and angiosperms.

Ferns

Ferns have two distinct generations:

- The *sporophyte* (2n) – forms *spores* (not seeds) in capsules under the fronds. This is the fern plant we interpret as a 'fern'.

- The *gametophyte* (n) generation – a small, leafy structure, still dependent on water for fertilisation (which is why ferns are restricted to living in damp habitats). This is usually so inconspicuous it is not seen.

Conifers

Conifers (the *cone* bearers) are members of the larger group called the *gymnosperms* (= 'naked seeds', as their seeds do not develop within closed chambers).

The obvious tree is the *sporophyte* (2n); the gametophyte is microscopic. Seeds are produced in cones. Conifers are typically large trees – eg kauri, totara, rimu, and the introduced radiata pine. Conifers have true roots, stems ('trunks') and leaves (may be needle-like).

Large pine tree with needle-like leaves

Typical cones ('pine cones')

Angiosperms

In the flowering plants, the gametophyte generation is even smaller than in conifers and the seed develops within a closed chamber, the ovary, found in the *flower*. Seeds are contained in a *fruit*. Flowering plants have true roots, stems, leaves. Two groups exist – **monocotyledons** ('monocots') and **dicotyledons** ('dicots').

Angiosperm	Leaves	Stems	Flowers
Monocotyledons Includes all the grasses, lilies, orchids.	Leaves are typically long and thin, with parallel veins (the transporting xylem and phloem) along their length.	Vascular bundles (of xylem and phloem) are scattered throughout the width of the stem, and there is rarely secondary growth of the stem.	Flowers typically have parts arranged in threes (or multiples of three). Seed has only one seed leaf or *cotyledon*.
Dicotyledons Most of the angiosperms.	Leaves are typically broad and have a network of veins.	Vascular tissue is arranged in a circle around the stem, and secondary growth of the stem occurs.	Flowers typically have their parts in fours or fives (or multiples thereof). Seed has two seed leaves or *cotyledons*.

Diversity

Diversity is about *differences*. All plants carry out the same seven essential life processes (*MRS GREN*), but they may do them in different ways and using different structures. These differences reflect the way that the plant/group has evolved to be successful in its *ecological niche*.

Biologists study the diversity between different groups of plants (minimum of three groups). These groups may be:

- *Taxonomic* – ie a classification group such as those described above (eg mosses, ferns, conifers).
- *Functional* – ie a group that performs the same task or function (not bound by the taxonomic group), eg hydrophytes, mesophytes, xerophytes are all functional groups for the process of transpiration.

Unit 11.1 Living Things

Topic 8: Unicellular organisms – the protista

This Topic describes the cell structure and function of unicellular organisms. The Topic deals with:

- Structure and function of common unicellular organisms, eg *Amoeba, Paramecium, Euglena.*
- Cell processes – transport of materials, nutrition (ingestion, digestion, egestion), function of eyespot, photosynthesis, osmoregulation and excretion, locomotion (cilia and flagella), reproduction.
- Reasons for similarities and differences between unicellular organisms.

It would be helpful to refer to other Topics such as Topic 2 (p. 11) in this Unit and also Supplementary Unit, Topic 3: Respiration and gas exchange in micro-organisms (p. 353).

The **protista** kingdom (also known as the **protoctista** kingdom) is one of the five kingdoms of living things. It includes all the single-celled eukaryotic organisms.

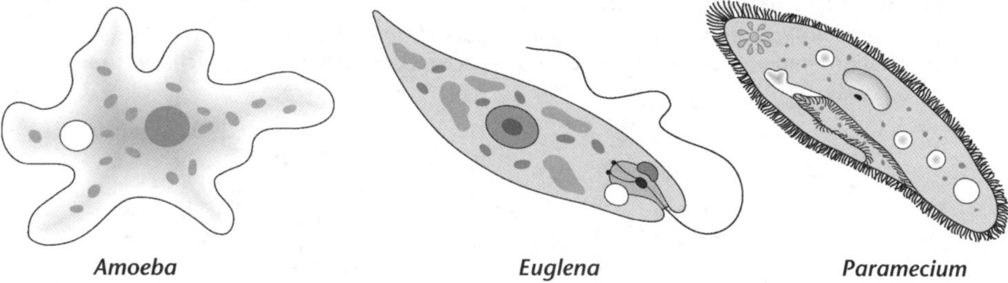

Other organisms placed in this kingdom include the slime moulds and all the algae such as the seaweeds (the latter are multicellular, but the overall structure/cell arrangement is simple, with little specialisation).

This chapter is only concerned with the *single-celled* (*unicellular*) members of the protista. Their single cell carries out all the life processes (*MRS GREN*).

Movement	Respiration	Sensitivity	Growth
Beating cilia, flagella or cytoplasmic streaming.	Aerobic, using oxygen, in mitochondria. O_2 diffuses in. CO_2 diffuses out.	Eyespot detects light.	Size restricted by diffusion of materials.

Reproduction	Excretion	Nutrition
Asexual reproduction by binary fission, sexual reproduction also occurs.	Contractile vacuole uses energy to expel excess water. CO_2 diffuses out of cell. Metabolic wastes may diffuse out across membrane (or active transport).	Oral groove ingests foods, or phagocytosis results in food being engulfed. Anal pore expels waste. Photosynthesis in unicells with chloroplasts.

Unicellular organisms carry out all their life functions inside a single cell. As a result, some of the cellular processes that occur in unicellular organisms are significantly different from similar processes in the cells of multicellular organisms.

Gas exchange, ingestion and digestion of food, excretion, water regulation, locomotion, responses to external stimuli and reproduction all need to be carried out by the single cell. (Most examples given refer to either *Paramecium* or *Amoeba*.)

Cell processes

Gas exchange

Gas exchange in almost all unicellular organisms is usually by diffusion across the cell membrane.

Example

Gas exchange in *Paramecium*

Gas exchange in *Paramecium* occurs by *diffusion* across the cell membrane in response to concentration gradient. Some carbon dioxide is actively removed by contractile vacuoles.

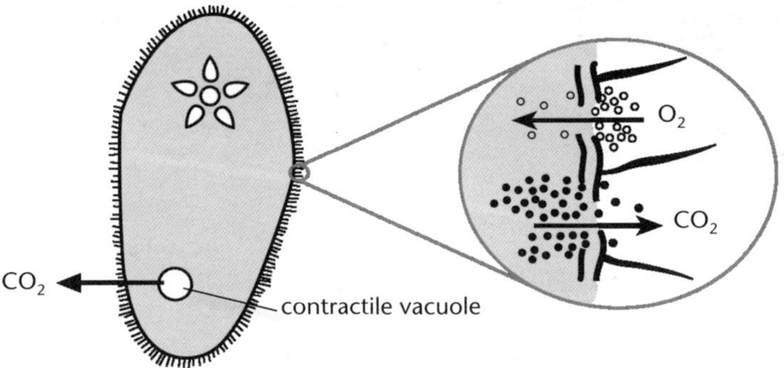

Gas exchange across the cell membrane in Paramecium.

To increase the efficiency of gas exchange, unicellular organisms are usually long and/or flat in shape, *increasing the surface area to volume ratio.*

Ingestion and digestion

All unicellular organisms that cannot photosynthesise must ingest small food particles as their food supply. *Euglena* is an example of a photosynthetic unicell. It has **chloroplasts** for photosynthesis, **eyespots** to locate favourable light conditions, *pyrenoids* to store starch.

In heterotrophic unicells, food particles cross the cell membrane by **phagocytosis** to form a **food vacuole**, which is then digested. Any indigestible material left in the food vacuole is discharged to the exterior, through the **anal pore**.

Example

Feeding in *Paramecium*

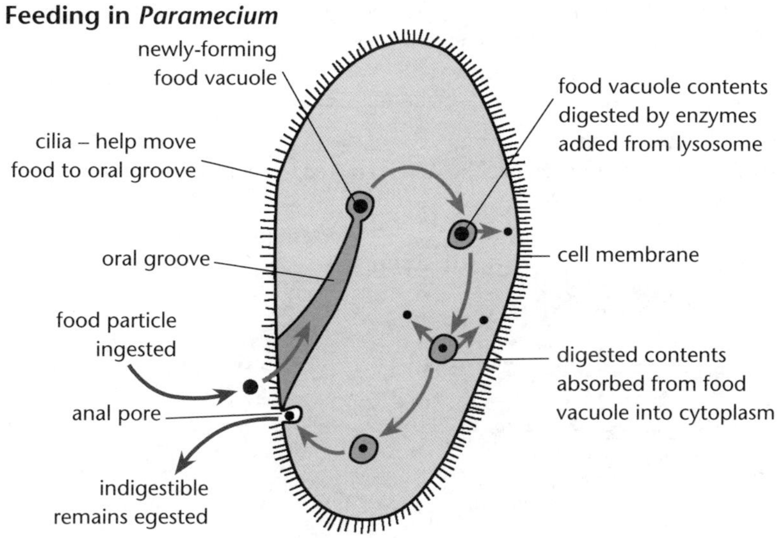

Feeding in unicellular Paramecium.

The part of the cell membrane specialised for ingestion in *Paramecium* is called the **oral groove**. Cilia lining the groove sweep food particles down the groove.

Example

Feeding in *Amoeba*

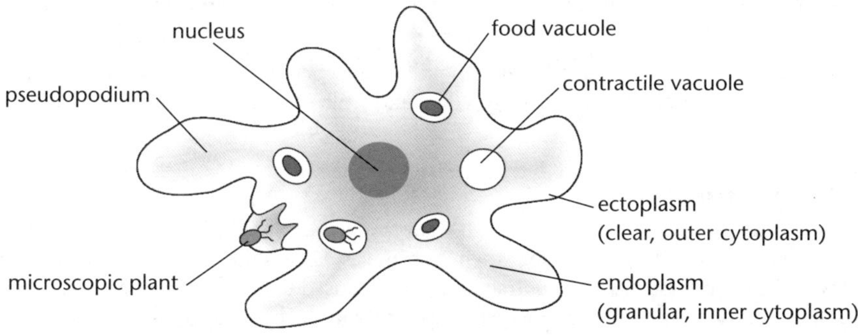

Feeding in unicellular Amoeba.

Amoeba do not have an oral groove to trap food, so they use pseudopodia ('false legs') to engulf the food particles before forming a food vacuole for digestion.

Excretion

In unicellular organisms, the main metabolic waste product formed is ammonia. Ammonia is very toxic, so it must be diluted by large volumes of water before it can be excreted.

Example

Excretion in *Paramecium*

Paramecium live in a freshwater habitat. The water required to dilute ammonia is provided from the environment by osmosis.

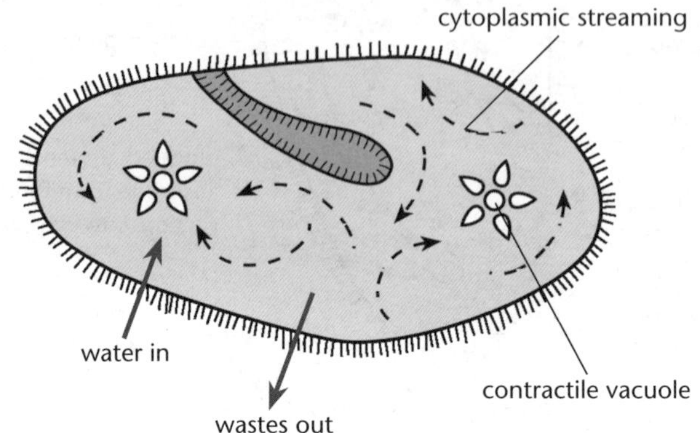

Excretion across the cell membrane of a unicellular organism.

The long, flat, body shape of *Paramecium* provides a large surface area, which allows rapid diffusion of wastes across the cell membrane. Contractile vacuoles collect and remove wastes as well; eg some CO_2, all excess H_2O.

Wastes are brought to the cell membrane and contractile vacuole by **cytoplasmic streaming** (the movement of cell contents).

Contractile vacuoles are surrounded by mitochondria (the energy organelles of cells), because *active transport* is used to move materials into the contractile vacuole then expel them to the outside.

Osmoregulation

Since many unicellular organisms live in fresh water and are enclosed by a semipermeable membrane, water continually moves into them in osmosis. Unless the excess water is removed, the cell would probably burst – the contractile vacuole is used to collect and remove water. Body fluids drain into the vacuoles through radiating channels in the cytoplasm; useful chemicals are reabsorbed back into the cytoplsm.

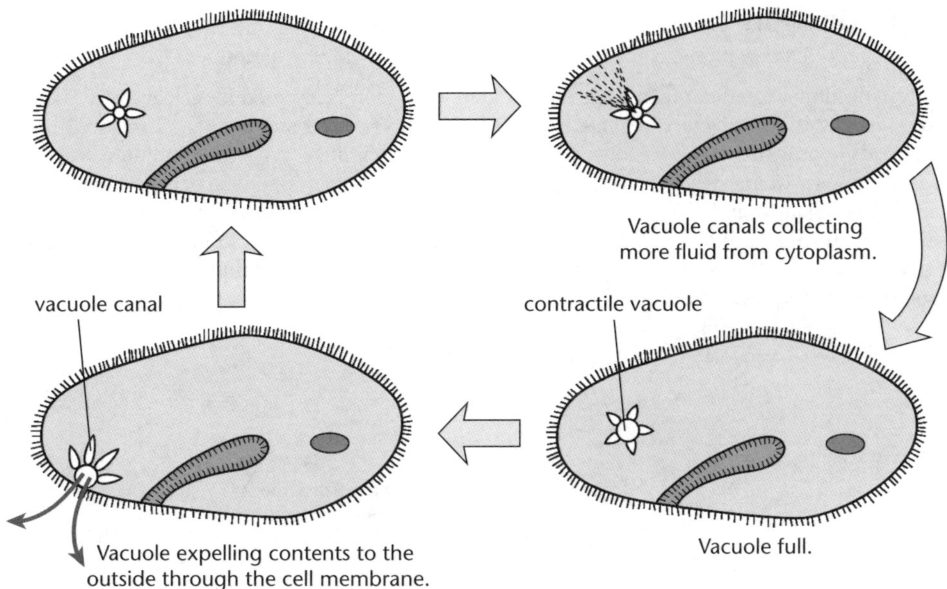

Osmoregulation in Paramecium.

Locomotion

In ciliate unicellular organisms like *Paramecium*, movement is achieved by the co-ordinated beating of the (approximately) 2 500 cilia which line the cell membrane. Co-ordination is achieved by a network of neurofibrils which connect the bases of the cilia. In flagellate unicellular organisms like *Euglena*, the flagellae spin like a ship's propeller to move the organism along. *Amoeba* use the pseudopodia used for feeding to move over a substrate, and rely on cytoplasmic streaming to move when floating in water.

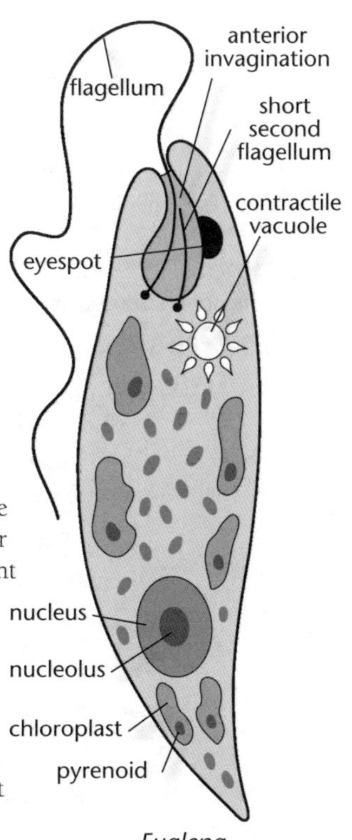

Euglena

Responses to external stimuli

Most unicellular organisms seem to respond to external stimuli as a result of changes to their cytoplasmic contents. Movement towards food or away from harmful substances often seems to be a simple process of 'trial and error'. In photosynthetic unicellular organisms such as *Euglena*, movement towards or away from light is co-ordinated by a special organelle known as an **eyespot** or **stigma**. The eyespot is able to detect the amount of available light and trigger a phototactic response. Movement up a water column (toward the light – *positive phototaxis*) during the day allows *Euglena* to harvest more sunlight for photosynthesis and movement down the water column (away from the light – *negative phototaxis*) during the night takes *Euglena* to regions that usually have high nutrient concentrations.

Reproduction

Asexual reproduction in unicellular organisms is achieved by binary fission.

Actively feeding unicellular organisms take in more food than is required for cell activity. The excess is used to manufacture new material, which is incorporated into the organism's body. This leads to growth – however, a body cannot enlarge indefinitely. When the optimum size is reached, growth in size is replaced by growth in number, ie reproduction occurs. When a unicellular organism has reached its full size, it stops growing; the nucleus divides mitotically, a second contractile vacuole develops, and eventually the cytoplasm divides into two more or less equal parts. When the division is complete, two cells are formed. This simplest form of asexual reproduction is known as **binary fission**.

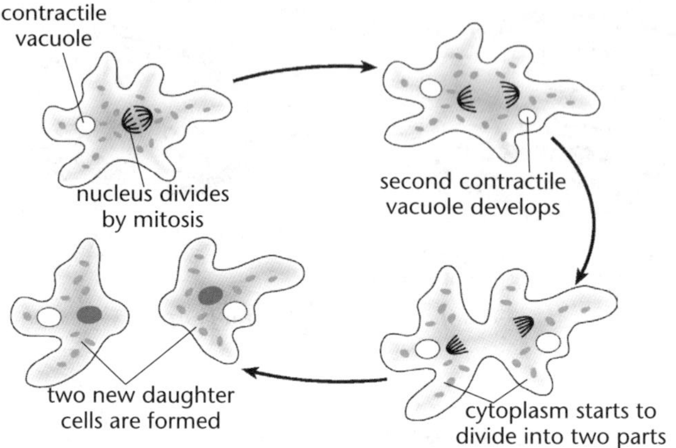

Binary fission in Amoeba.

Sexual reproduction, which involves the exchange of genetic material between two individuals, is rare amongst unicellular organisms, and usually occurs as a result of *unfavourable* living conditions.

Example

Paramecium is unusual in having two nuclei, a large (mega) and a small (micro).

Sexual reproduction in *Paramecium*

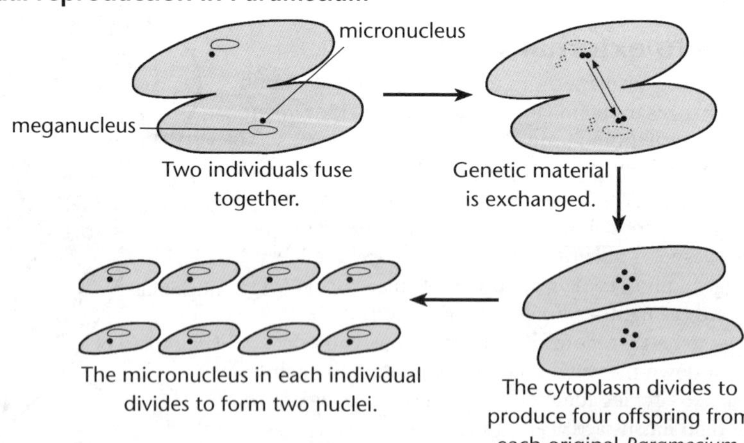

Unit 11.1 Activity 8A: Unicellular organisms

1. The diagram alongside shows a single-celled organism called *Euglena*.

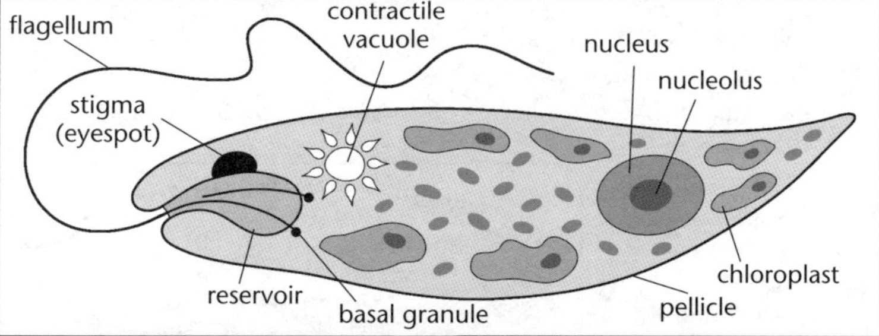

   a. Describe how a flagellum is used to move an organism.

   b. Ben looked up a website that had information on *Euglena*. He found the following information about the eyespot.

> The cell of *Euglena* is transparent, so a black, pigmented area covers the outside of the eyespot. The *Euglena* cell moves in reponse to the amount of shadow on the eyespot.

   Explain how the eyespot helps *Euglena*.

   c. *Euglena* has features of both plant and animal cells. This caused difficulties when early biologists tried to classify it as a plant (kingdom Plantae) or an animal (kingdom Animalia). *Euglena* is now classified in the kingdom Protista.

   Discuss why the features of *Euglena* (labelled on the diagram) made it difficult to classify *Euglena* as a plant or an animal.

2. A biology class collected a number of single-celled organisms called *Paramecium*. They placed them in beakers containing different salt solutions and recorded the number of times the contractile vacuole emptied its contents. The results are shown below.

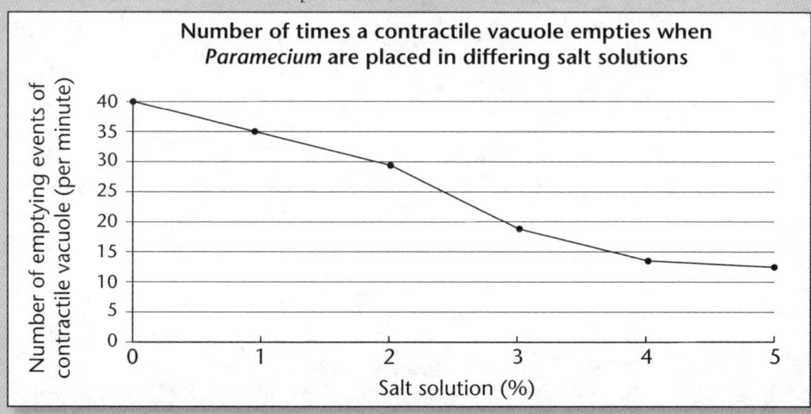

   a. Describe the role of the contractile vacuole.

   b. Explain why the contractile vacuole emptied more often when the organisms were placed in the 1% salt solution than when placed in the 4% salt solution.

   c. A student investigated the rate of contractile vacuole emptying in *Paramecium* placed in water samples containing different levels of dissolved oxygen – high levels and low levels. The contractile vacuole in the *Paramecium* in water with low levels of dissolved oxygen emptied fewer times per minute.

Explain why the rate of contractile vacuole emptying was lower when dissolved oxygen levels were low.

d. The processes of osmosis and diffusion are involved in moving water and other small materials in and out of cells.

Explain the difference between *osmosis* and *diffusion*. You may use diagrams in your explanation.

3. *Paramecium* is a common freshwater Protista, found in ponds or slow-moving streams.

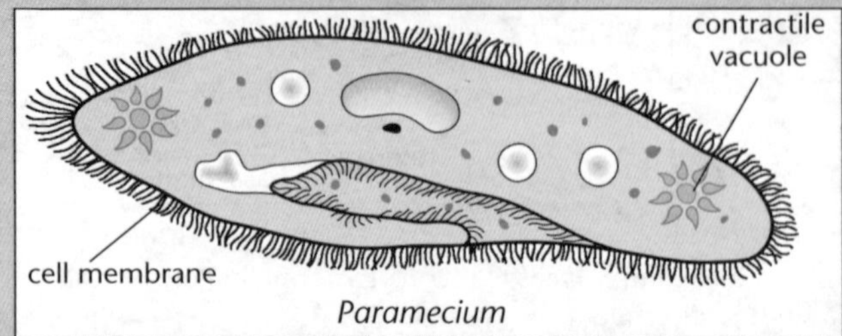

Paramecium

Discuss the relationship between osmosis and contractile vacuole activity in a *Paramecium*, in its freshwater environment.

4. Discuss the significance of the SA:V ratio to the life style of unicellular organisms.
5. Below is a diagram of *Chlamydomonas*, a small unicellular organism (usually less than 25 micrometres long) which inhabits freshwater ponds.

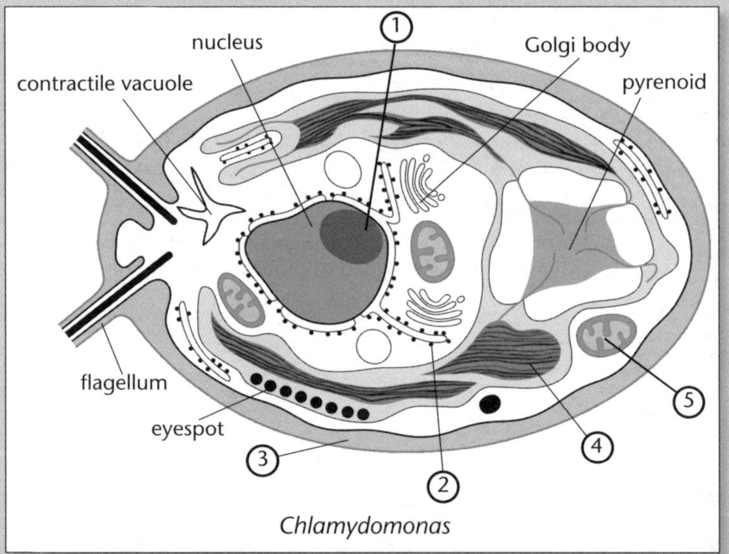

Chlamydomonas

a. Name each of the structures numbered 1 to 5, and describe their function.
b. Explain why *Chlamydomonas* is an autotroph and not a heterotroph in its nutrition.
c. Explain how *Chlamydomonas* osmoregulates.
d. Describe how *Chlamydomonas* moves.
e. Discuss the significance of the eyespot.

6. Explain why you would expect organisms such as *Paramecium* to have large numbers of mitochondria.

Unit 11.2 Nutrition

Topic 1: How plants feed – autotrophic nutrition

Author: Martin Hanson with Diaiti Zure

Have you wondered about how plants make their own food? This Topic provides an understanding of autotrophic nutrition and plant processes by looking at:

- Demonstrating photosynthesis.
- Leaf structure – adaptations to photosynthesis and water conservation.
- Photosynthesis and respiration.
- Effect of environmental factors on photosynthesis.

Note that nutrition in micro-organisms is covered in the Supplementary Unit (p. 329) (ref. Syllabus p.11).

The plant lifestyle

Plants and animals make their living in very different ways.

- Animals feed on complex organic compounds in the bodies of other organisms.
- Plants feed on simple inorganic substances such as CO_2, water and mineral salts.

Whereas animals usually consume their food as concentrated lumps, the CO_2 and minerals needed by plants are very dilute. To absorb them, a plant needs to have a *finely divided body* to expose a very large surface to its surroundings. Animals also have large surfaces to absorb raw materials, but these are usually in the form of intuckings, such as the lungs and gut, formed into a compact body shape.

Parts of a plant

A typical flowering plant is divided into two main parts:

- An above-ground *shoot system* – absorbs light and CO_2.
- A below-ground *root system* – absorbs water and minerals and provides anchorage.

The shoot is concerned with photosynthesis and sexual reproduction. It consists of a *stem* bearing *leaves* and sometimes *flowers* or fruits. Leaves carry out photosynthesis; the stem supports the leaves and transports substances to and from them.

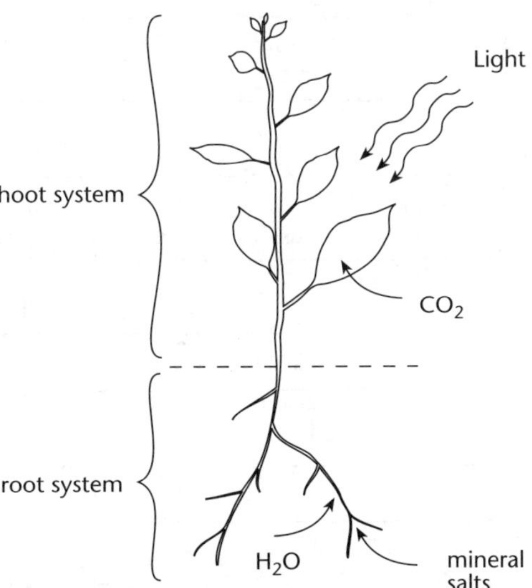

A typical flowering plant.

Photosynthesis

Plants are **autotrophic**, meaning that they make their own organic compounds from carbon dioxide and water. The energy for the process is light:

$$\text{carbon dioxide} + \text{water} + \text{light} \xrightarrow{\text{chlorophyll}} \text{sugar} + \text{oxygen}$$

$$6CO_2 + 6H_2O + \text{light} \xrightarrow{\text{chlorophyll}} C_6H_{12}O_6 + 6O_2$$

The entire process takes place within **chloroplasts**. These organelles are most abundant in the photosynthetic organs, the leaves.

- On a sunny day, sugar accumulates much faster than the leaf can export it, and it is temporarily converted into starch.
- At night, the starch is reconverted to sugar and transported to other parts of the plant.

$$\text{sugar} \underset{\text{night}}{\overset{\text{daytime}}{\rightleftharpoons}} \text{starch}$$

The starch production from sugar is not part of photosynthesis – potato tubers make starch from sugar in complete darkness!

Demonstrating photosynthesis

To show that photosynthesis has occurred, the presence of the photosynthetic product must be shown. Although this product is actually a sugar, it is much simpler to test a leaf for the starch that is made from the sugar.

Testing a leaf for starch

Starch turns deep blue with iodine, but to show the deep blue of the stained starch a leaf must first be made colourless by removing the chlorophyll.

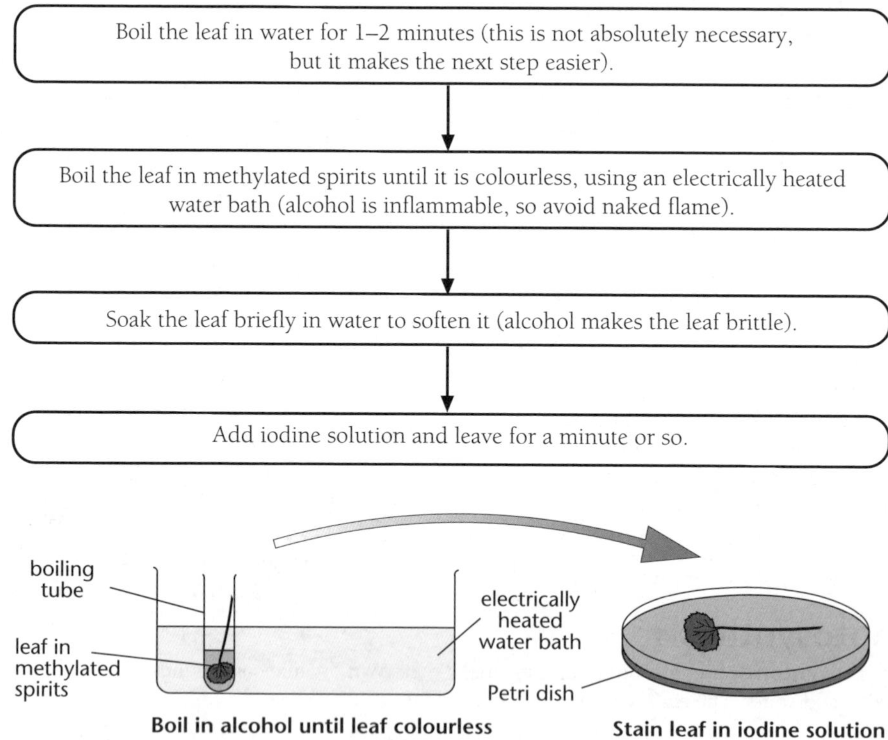

Testing a leaf for starch.

To show that starch has been made during the course of an experiment, there must be no starch at the beginning. The experimental plant must therefore be *de-starched* by leaving it in darkness for 24–36 hours beforehand.

De-starching a leaf before *demonstrating* the need for light is okay but, if you are doing an *experiment to find* out if light is needed, the plant must *not* be de-starched first. This is because it is unscientific to assume the results of an experiment – in other words, it would not be an experiment!

Demonstrating the need for light in photosynthesis

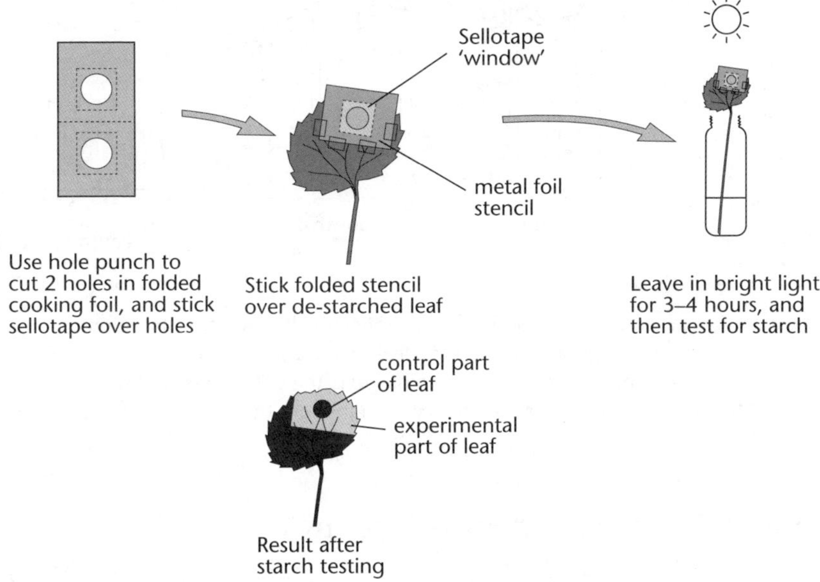

Demonstrating the need for light in photosynthesis.

Putting tinfoil over the leaf changes the conditions in *two* ways – shutting out light and reducing the air supply. The *Sellotape* 'window' ensures that the air supply to the part receiving light is equally restricted, so that the presence or absence of light is the *only* difference in the conditions. The part of the leaf that is the control is that under the 'window'.

Demonstrating the need for carbon dioxide

As in the case of light, it is essential to ensure that the environments of the two experimental leaves used differ in only *one* way – in this case, the presence or absence of CO_2.

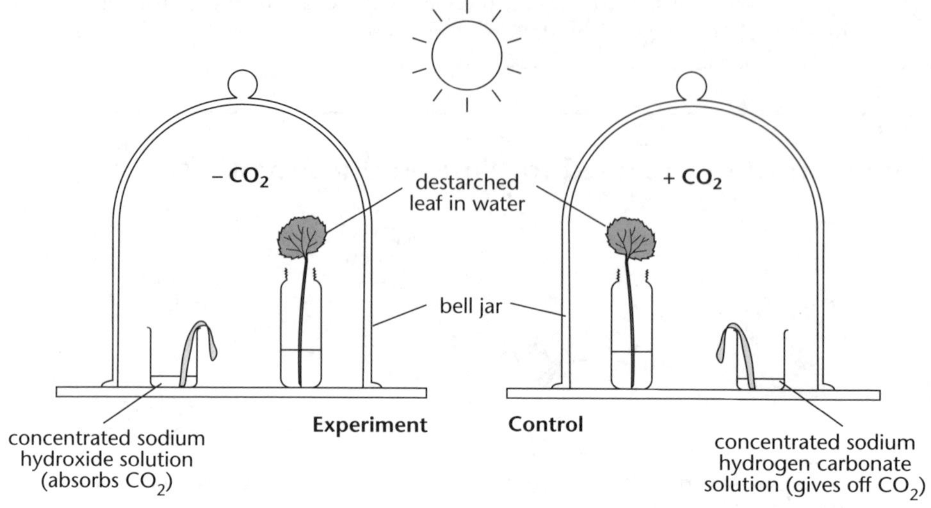

Placing a bell jar over the leaf raises the humidity and also probably the temperature. Hence the need for the control leaf to also be under a bell jar. The sodium hydrogen carbonate prevents the CO_2 being used up (it is only 0.035% normal air). After 3–4 hours in bright light the leaves are tested for starch.

Demonstrating the need for CO_2 in photosynthesis.

Demonstrating the need for chlorophyll

It is not possible to remove chlorophyll from a leaf without killing it. Fortunately, some plants have *variegated* leaves, in which some parts are naturally without chlorophyll. To show that chlorophyll is necessary for starch formation, a variegated leaf that has been exposed to light for some hours is tested for starch. If chlorophyll is essential, then the only areas that make starch should be the areas that previously had chlorophyll.

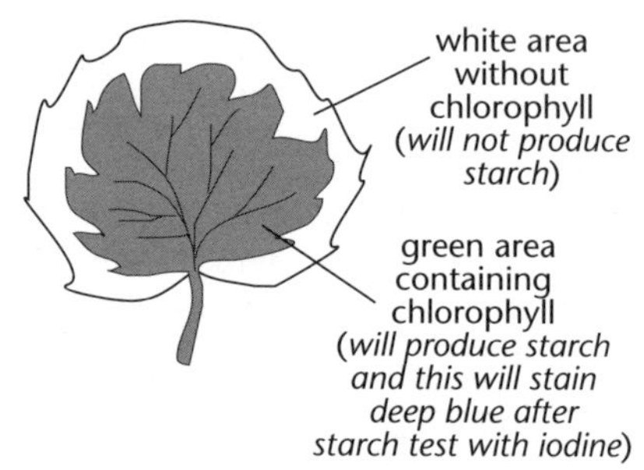

A variegated geranium leaf.

The production of oxygen

Bubbles rising from an illuminated pondweed can be collected. Provided that the gas is given off reasonably quickly, it should be sufficiently rich in oxygen to re-light a glowing splint.

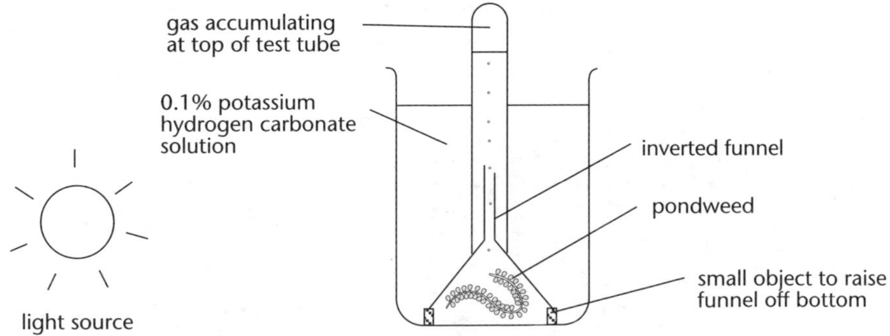

The funnel is kept off the bottom of the beaker to allow the water to circulate.

Oxygen production in photosynthesis.

The need for water

Most living plant cells are about 80% water, so if water is removed from a leaf by drying it, it dies. Long before leaf death occurs, however, the stomata close, shutting off the supply of CO_2. It is therefore not possible to restrict the supply of water to a leaf without shutting off the supply of CO_2. Using a special kind of oxygen as a 'label', it has been shown that the oxygen produced from photosynthesis comes from water, so water must therefore be used in photosynthesis.

The leaf as a photosynthetic organ

In a plant there are many different kinds of cells, each carrying out a different function. Groups of similar cells working together are called **tissues**. Different tissues work together to form an **organ**.

A leaf is a plant organ, whose tissues co-operate to carry out photosynthesis. Roots and stems are also organs.

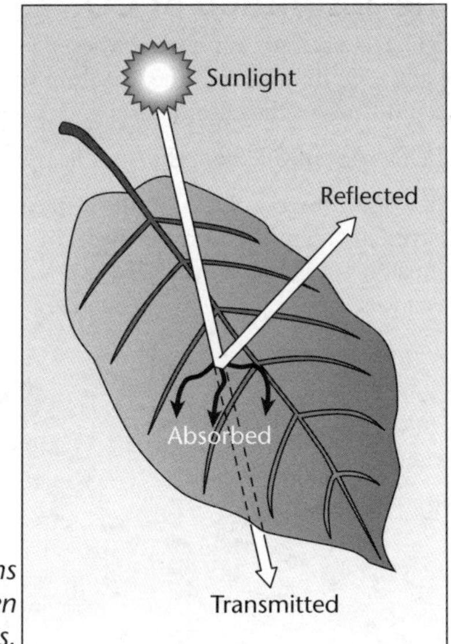

Leaves of the native tree rangiora. Veins (of xylem and phloem) are clearly seen forming a network over the leaves.

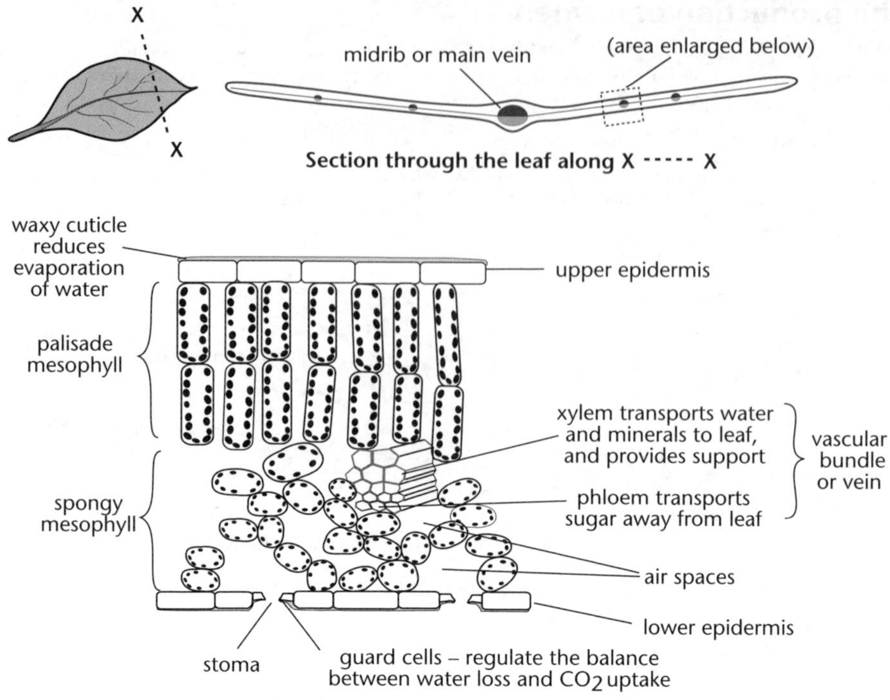

Structure of a leaf of a typical dicotyledonous plant.

To photosynthesise efficiently, leaf cells must have an adequate supply of light, CO_2, and water.

The absorption of CO_2

As CO_2 is used up in photosynthesis, its concentration in the chloroplasts is reduced. As a result, CO_2 diffuses into the leaf from the air outside (where it is more concentrated). Oxygen accumulates in the chloroplasts and so diffuses in the reverse direction.

The passage of CO_2 in and O_2 out is a process called **gas exchange**.

The concentration of CO_2 in the atmosphere is very low – only 0.035% (350 parts per million). Leaves are adapted to exploit even this meagre supply. The leaf makes use of two facts about diffusion – it is much faster (10 000 times) in gas than in solution, and it is faster over short distances.

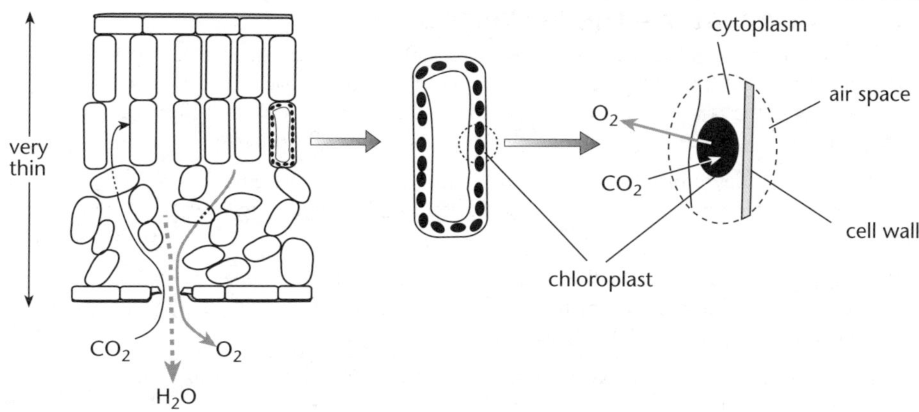

How leaf structure helps CO_2 uptake.

Features of the leaf help CO_2 to diffuse into it:

- Most of the diffusion pathway is through the system of air spaces between the cells. Although the last stage (through the cell walls and cytoplasm) is in liquid, the distance to travel is very short because the chloroplasts are very close to the cell walls.
- The thin shape of a leaf means that the distance CO_2 has to diffuse is very short – less than half a millimetre.
- A leaf has a large surface area compared with its volume.

The absorption of light

The diagram below shows that mesophyll is organised into two distinct layers – a **palisade layer** facing the light, and a **spongy layer** on the lower side.

- The palisade cells have so many chloroplasts that most of the light entering them is absorbed.
- The light passing *between* the palisade cells is scattered by the irregularly arranged spongy mesophyll cells. Some of this scattered light passes up through the palisade mesophyll, giving it 'a second chance' to be absorbed.

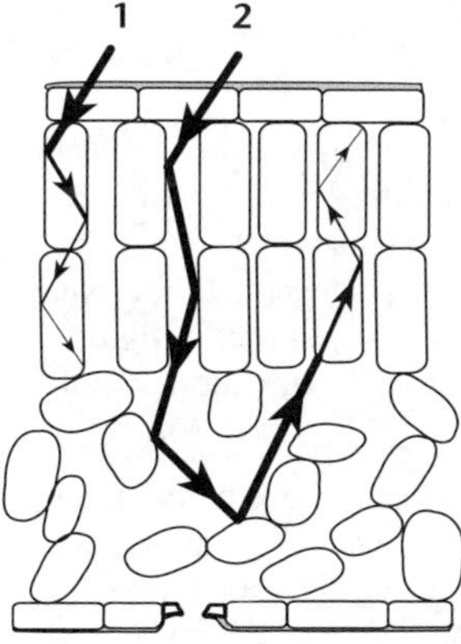

1 Ray passing *through* palisade cells – most is absorbed.

2 Ray reflected downwards *between* palisade cells – then reflected up by spongy mesophyll, giving it 'a second chance' to be absorbed by the palisade cells.

How mesophyll helps the absorption of light.

Conserving water – the epidermis

The features that make a leaf so good at absorbing CO_2 also make it good at losing water. Water diffuses *out* from the high concentration in the humid air spaces to the much lower concentration in the relatively dry air outside. This loss of water by evaporation is called **transpiration**, and is the inevitable price the plant has to pay for photosynthesis.

On a warm sunny day, a plant loses water over a hundred times faster than it is photosynthesising (ie, for every water molecule used in photosynthesis, over a hundred are lost in transpiration).

When soil is moist, roots are usually able to supply water as fast as the leaves lose it. But in dry conditions, water loss is a problem. The tissue responsible for dealing with water loss is the **epidermis**.

- The epidermis secretes a waxy **cuticle**, which resists water loss (the waxy cuticle makes many leaves look shiny).
- The lower epidermis is perforated by huge numbers of tiny holes called **stomata** (singular, stoma). Each stoma is surrounded by two guard cells. By changing their shape, guard cells are able to open or close the stoma. When the guard cells absorb water they bend, opening the stoma; when they lose water they straighten, closing it. Stomata generally close at night, and also when the soil is dry.

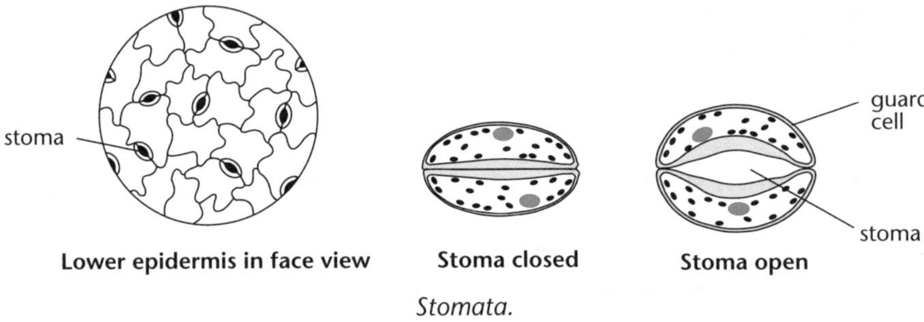

Stomata.

By closing stomata, a plant may survive drought conditions. Closing stomata shuts off the supply of CO_2, slowing photosynthesis – this is why plants adapted to dry climates usually grow slowly.

Water and sugar transport

Xylem tubes through the leaf stalk (petiole) deliver water. These tubes are formed by end-to-end joining together of highly modified cells which are dead when mature. As water evaporates from the mesophyll cells into the air spaces, it sets up a tension which is transmitted through the xylem tubes all the way down to the roots. Despite this tension in the xylem sap, xylem cells do not 'cave in' because their walls are reinforced by rings or spirals of **lignin**. This makes them stiff, so xylem strengthens the plant as well as transporting water and minerals. (Wood is simply xylem in bulk.)

Exporting sugar from the leaf is the job of the **phloem** cells. Together with the xylem cells, the phloem cells form the veins or **vascular bundles**. In a leaf they form such a dense network that no cell is more than a fraction of millimetre from the nearest vascular bundle.

The sugar made from photosynthesis

Eventually, all the sugar made in photosynthesis is used in one of two ways:

- About two thirds is used as *building materials* for growth; eg glucose molecules can be linked together to make the cellulose of new cell walls, some is combined with nitrogen from soil nitrates to form amino acids and proteins.
- The rest is used in respiration to supply *energy*. Plants use energy for various purposes, such as growth (eg making cellulose from glucose, or proteins from amino acids). Energy is also used for absorbing minerals from the soil.

Not all food made in photosynthesis is used immediately – some is usually stored temporarily as starch. Over short periods, some starch is stored in the leaves, but at night this is broken down into sugar again to be *translocated* to other parts of the plant by the phloem. Many plants (such as taro and sweet potato) have large underground storage organs which store starch.

Photosynthesis and respiration

Photosynthesis is always going on faster than it seems, because respiration uses some of the O_2 produced from photosynthesis before O_2 can exit a leaf. In terms of gas exchange, respiration has the opposite effect of photosynthesis.

Example

Suppose a plant is making 12 units of oxygen per hour in photosynthesis, and using 4 units of oxygen per hour in respiration. Of the 12 produced, 4 are used, so only 8 reach the outside world. Hence only 8 units of oxygen are actually measured. This is the *net photosynthesis*. The true, or *gross photosynthesis* is actually 4 more than this.

Scientists measure the true rate of photosynthesis by first measuring the rate of respiration in the dark (when there is no photosynthesis). Then they measure the net photosynthesis in the light, and calculate the true photosynthesis by:

gross photosynthesis = net photosynthesis + respiration

Daytime photosynthesis *must* be faster than respiration – in order to grow, a plant must make more sugar in the daytime than it uses during an entire 24-hour period, or it couldn't grow.

Unit 11.2 Activity 1A: Leaves and photosynthesis

1. For each of the phrases **1–15**, write the letter **A–O** of the term to which it applies.

Phrase	Applied term
1. Alternative name for a vein in a leaf or young stem	A. Chloroplast
2. Diffusion of oxygen and CO_2 in opposite directions	B. Cuticle
3. Evaporation of water from leaves and stems of plants	C. Diffusion
4. Group of cells co-operating to carry out a more complex function	D. Epidermis
5. Group of tissues co-operating to carry out a more complex function	E. Gas exchange
6. Lower layer of mesophyll	F. Lignin
7. Outer layer of cells in a leaf	G. Organ
8. Pore through which gases enter and leave leaf	H. Palisade layer
9. Process by which CO_2 enters a leaf	I. Phloem
10. Site of photosynthesis in a plant cell	J. Spongy layer
11. Substance that stiffens cell walls	K. Stoma
12. Sugar-transporting tissue	L. Tissue
13. Upper layer of mesophyll	M. Transpiration
14. Water-transporting tissue	N. Vascular bundle
15. Waxy, water-resistant layer secreted by epidermis	O. Xylem

2. **a.** To produce glucose by the process of photosynthesis, a green leaf needs light energy, chlorophyll and two substances obtained from the environment. Name these two substances.

   b. The diagram shows a typical cross-section of a leaf.

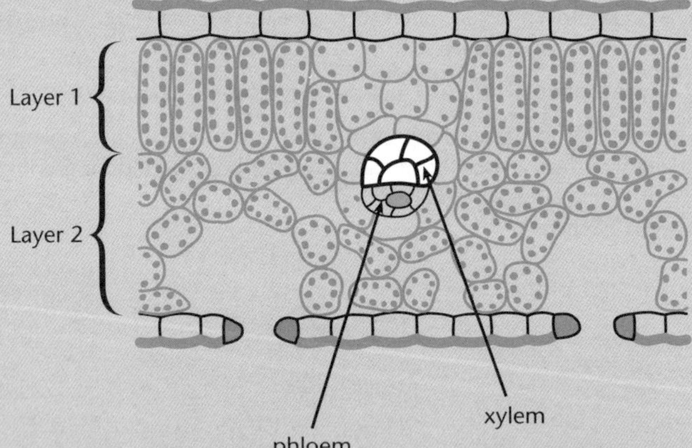

   Choose either Layer 1 or Layer 2, and describe one feature of the structure or arrangement of cells in that layer of the leaf that contributes to the efficiency of photosynthesis.

3. Stomata are microscopic pores or holes located in the epidermis layer of leaves.

   a. Describe the structure of the guard cells that control the opening and closing of these pores. You may use a labelled diagram to help your answer.

   b. Explain how the opening and closing of stomata change the rate of photosynthesis.

4. A student carried out an investigation to see where starch was produced in a variegated (white and green) geranium leaf that had a black cardboard letter E attached to it (Diagram A). The geranium plant had previously been kept in the dark for 12 hours. The leaf was exposed to normal light conditions for four hours. It was then removed from the plant and tested for the presence of starch. Diagram B shows the leaf after it had been tested for the presence of starch. The darkly shaded portions of the leaf contain starch.

Use this experiment to discuss the importance of two factors necessary for starch production in a leaf.

5. The diagram below shows a cross-section of a typical leaf from a dicotyledon flowering plant.

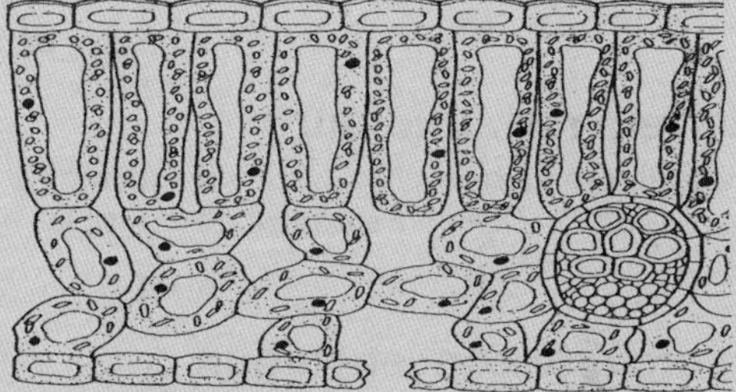

   a. Describe the appearance and position of the palisade mesophyll cells.
   b. Explain how CO_2 from the air reaches the chloroplasts of a palisade mesophyll cell.
   c. Although leaves of plants have many different shapes and sizes, the basic function of a leaf is to carry out photosynthesis as efficiently as possible.

      Discuss how the general shape, structure and position of a leaf can help make photosynthesis more efficient.

Effect of environmental factors on photosynthesis

The rate of photosynthesis can be expressed as the amount of oxygen produced per unit time, or the amount of CO_2 used per unit time, or the amount of organic matter produced per unit time.

The main environmental factors affecting photosynthesis are light intensity, carbon dioxide concentration, temperature, soil water content, and the availability of certain minerals.

Effect of light intensity

Brighter light does speed up photosynthesis – but only up to a point.

In very bright light (though less than the full midday sun), increasing the light supply makes no difference to the rate of photosynthesis. There comes a point beyond which other factors, such as the supply of CO_2, begin to hold up or *limit* the rate of photosynthesis. Even if there is adequate CO_2, the temperature may limit the rate at which enzymes can work.

At low light intensities the rate of photosynthesis appears to be zero – even though there is some light. This light intensity is the **compensation point** – respiration just balances photosynthesis (the plant is 'marking time').

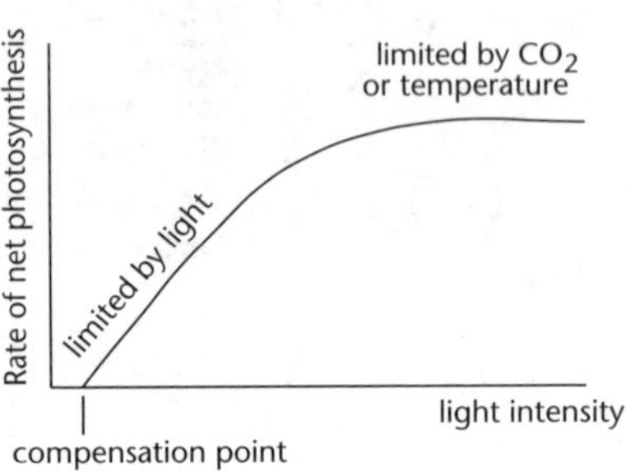

Effect of light intensity on rate of photosynthesis.

Effect of light wavelength ('colour')

The role of chlorophyll is to absorb light for photosynthesis. Chlorophyll appears green because it does not absorb green; red and blue light are absorbed. Red and blue light are thus by far the most effective in photosynthesis.

Effect of carbon dioxide concentration

As with light, the effects of more CO_2 eventually level off, because another factor begins to limit the rate of photosynthesis. Horticulturalists who grow plants under glass take advantage of the benefits of extra CO_2 by pumping the gas into the growth chamber.

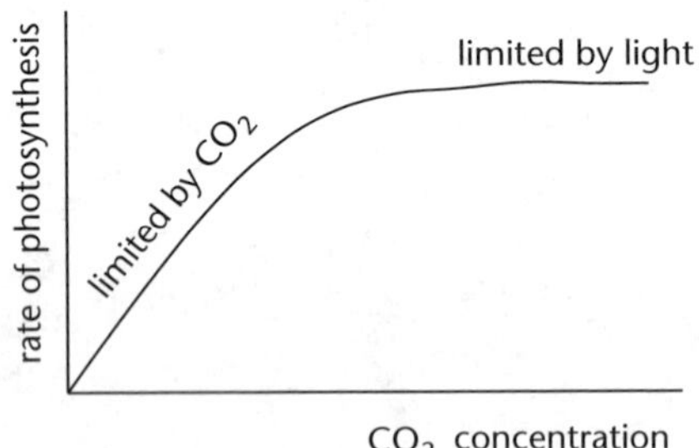

The effect of carbon dioxide concentration on photosynthesis.

Effect of temperature

The conversion of CO_2 and water to carbohydrate is a complex process involving many reactions catalysed by enzymes. Enzymes work faster at warmer temperatures. Provided the light is bright enough, photosynthesis speeds up if the plant is warmed – but only up to a point.

Above about 40°C, enzymes are inactivated (denatured), and photosynthesis is slower.

In dim light, gentle warming makes no difference to the rate of photosynthesis because there is not enough energy to drive the process any faster.

Effect of wind

Wind removes water vapour that collects outside the stomata, which creates a concentration gradient between the inside of the leaf and the outside. The concentration gradient, in turn, causes an increase in the loss of water from the stomata. As water is one of the raw materials for photosynthesis, decreasing the amount of water available to the photosynthetic cells decreases the rate of photosynthesis.

Effect of water (internal and external)

Anything affecting the availability of water to the photosynthetic cells will also affect the rate of photosynthesis. For example, if too much water is lost from the leaves by transpiration, then the rate of photosynthesis will drop (because water is a raw material for photosynthesis).

The effect of soil water shortage is indirect – in dry soils, a plant closes its stomata, thus shutting the leaf off from its supply of carbon dioxide, CO_2.

Unit 11.2 Activity 1B: How plants feed

1. A student made the statement, 'Light energy is needed for photosynthesis, but it is the colour and amount of light that is important.' Discuss the relationship between light energy and photosynthesis.
2. Canadian pond weed (*Elodea canadiensis*) is a common water plant that produces tiny bubbles of oxygen during photosynthesis. The number of bubbles produced per minute indicates the rate of photosynthesis. The graph shows how the rate of photosynthesis in Canadian pond weed relates to light intensity.

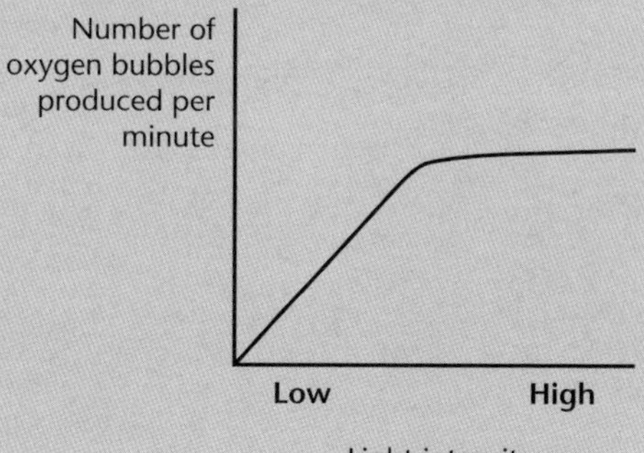

   a. Describe the effect of changing from low to high light intensity on the rate of photosynthesis.
   b. Give a reason why the photosynthesis rate does not increase any further at high light intensity.

3. The drawing shows a piece of lower leaf epidermis viewed through a microscope.

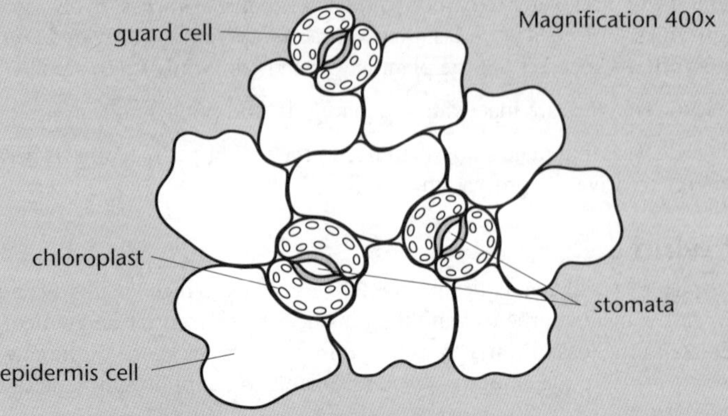

a. Describe one feature of guard cells that help them open and close the stomata.
b. Explain why stomata in the leaf epidermis are important for the process of photosynthesis.
c. Discuss how hot, dry, windy weather can affect the rate of photosynthesis.

Topic 2: Plant nutrition

This Topic reviews and consolidates the information presented in the previous Topic (p. 69) about the structure and function of plants. It deals with:

- The structures involved in plant nutrition – may be at cell level (eg structure and location of chloroplasts) or at organ level (eg structure and location of leaves).
- How the processes involved in plant nutrition are carried out (eg photosynthesis, parasitism).
- Reasons for the differences in structure and function between different plant groups – these could relate to habitat, size, life cycle, way of life.

It is recommended that this Topic be studied in conjunction with Topic 1 (p. 69) and also:

Unit 11.3 Transport Systems, Topic 1: Transport systems in plants (p. 121)

Unit 11.3 Transport Systems, Topic 2: Transpiration (p. 137)

The three processes are sufficiently linked to make their content part of a whole.

With the exception of parasitic plants, *all* plants are *autotrophs*, producing organic compounds from **photosynthesis**. Photosynthesis occurs in the **chloroplasts** of cells, mainly in the **leaves** (or **fronds** in ferns), where pigments such as **chlorophyll** catch solar energy to form *ATP* (adenosine triphosphate). Photosynthesis provides the chemical energy to bond CO_2 (from the air) and H_2O (from soil water) to form the sugar *glucose*. The gas O_2 (from water) is released as a waste product. Glucose is stored as *starch* (mainly in the roots), and is the starting point for the production of all the important organic compounds – carbohydrates, fats, proteins – needed by plants and all other life forms.

Plants are the *producers* for food webs in biological communities.

Photosynthesis

The equation for photosynthesis is:

Word equation

$$\text{carbon dioxide + water} \quad \frac{\text{solar energy*}}{\text{chlorophyll*}} \longrightarrow \text{glucose + oxygen}$$

*typically added with arrow, as are *essential* to process

Formula equation (simple)

$$6CO_2 + 6H_2O \quad \frac{\text{solar energy}}{\text{chlorophyll}} \longrightarrow C_6H_{12}O_6 + 6O_2$$

Inorganic nutrients

Plants require **inorganic nutrients** (**minerals**) – these are present in the soil as soluble ions, and are taken up by a plant's roots.

Nitrogen, N, phosphorus, P, and potassium, K, are the nutrients most commonly added in fertilisers. *Complete* fertilisers are those that contain these three elements.

Mineral	Form taken up by plants		Use(s) or role of nutrients in plants
Nitrogen	Nitrate ions	NO_3^-	Protein and co-enzyme structure, DNA and RNA
Phosphorus	Phosphate ions	PO_4^{3-}	Energy production (ATP) in photosynthesis and respiration, DNA and RNA
Potassium	Potassium ions	K^+	• Enzyme activator • Synthesis of carbohydrate and chlorophyll
Sulfur	Sulfate ions	SO_4^{2-}	Protein and co-enzyme structure
Calcium	Calcium ions	Ca^{2+}	• Forms part of cell wall • Regulates membrane permeability
Magnesium	Magnesium ions	Mg^{2+}	Part of chlorophyll molecule

Trace elements

Minerals required by plants in only *very limited* amounts are called **trace elements** or **micronutrients**.

Micronutrient	Role in plant metabolism
Iron	• Part of electron transport pathways that produce ATP from photosynthesis • Essential in synthesis of chlorophyll
Manganese	Enzyme activator in Krebs cycle
Copper	Part of electron transport pathway during ATP production in photosynthesis

Leaves and photosynthesis

Leaves are the site of photosynthesis. The structure and location of leaves show *adaptations* for photosynthesis. Leaves:

- Are *thin* so that light penetrates to all the photosynthetic cells.
- Have a *large surface area* to capture as much light as possible.
- Are arranged in *patterns* around the stem and branches, to *reduce overlap* and so allow *each* leaf to capture as much light as possible.

Leaves of the shrub rangiora show all the adaptations of being *thin*, having a *large surface area* and are arranged in *patterns* around the stem and branches.

The network of veins (xylem and phloem) is clearly visible.

Leaves of *Coprosma sp* showing their arrangement around the stem to reduce overlap.

Leaves in angiosperms have a characteristic layout of cell/tissues from top to bottom (dorsal to ventral).

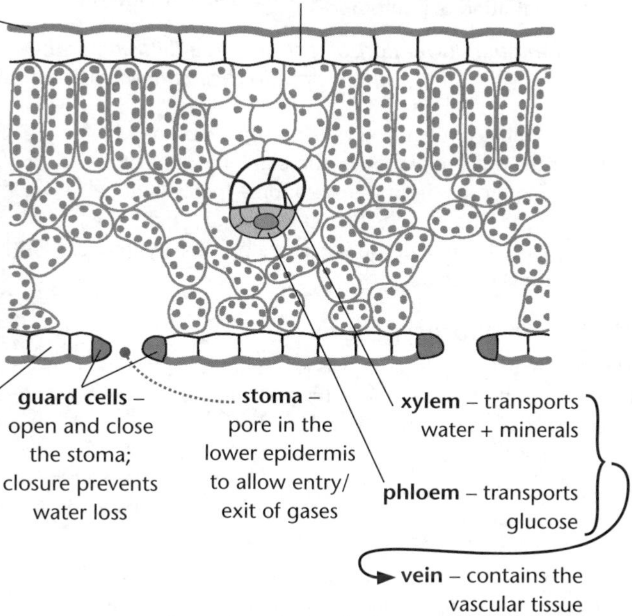

waxy cuticle – reduces water loss

upper epidermis – the 'skin' – clear to let sunlight through

palisade cell layer – cells are rich in chloroplasts for maximum photosynthesis

spongy cell layer – cells are loosely packed to give air spaces to facilitate diffusion of CO_2, O_2, H_2O vapour to/from cells

lower epidermis – the 'skin'

guard cells – open and close the stoma; closure prevents water loss

stoma – pore in the lower epidermis to allow entry/exit of gases

xylem – transports water + minerals

phloem – transports glucose

vein – contains the vascular tissue

Typical arrangement for a dicotyledon leaf.

The structure and arrangement of the cells *within* the leaf also show adaptations for photosynthesis.

The *palisade cell layer* (palisade mesophyll) is the main site of photosynthesis, with the cells containing the most chloroplasts.

Palisade cells are layered vertically under the upper epidermis, to catch maximum sunlight.

- The vertical rather than horizontal positioning of the cells reduces the amount of cell wall exposed to the light – the cell wall reflects and scatters light rays.
- The centre of each cell is a large vacuole, which stores the products of photosynthesis ('cell sap'); the chloroplasts are close to the cell membrane to catch light but also minimise the diffusion distance for CO_2 and H_2O across the cell membrane and into the chloroplasts (and O_2 out). CO_2 *diffuses* in (in response to its *concentration gradient*) from the surrounding air,

entering the leaf through the *stomata* (singular stoma) and travelling through the air spaces of the *spongy cell layer* (spongy mesophyll). O_2, produced in photosynthesis, diffuses out (reverse pathway) in response to its concentration gradient.

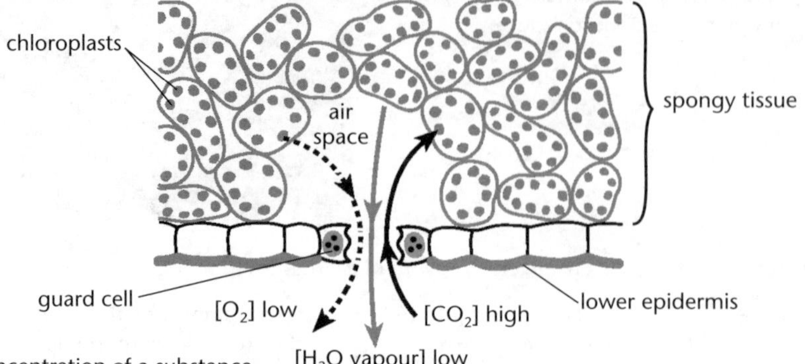

Section through lower part of leaf, showing diffusion pathways of CO_2, O_2 and H_2O vapour in response to their concentration gradients.

The membranes of plant leaf cells must be kept moist to allow for the diffusion of gases; therefore, the leaf needs to control water loss from transpiration – typically done by closing the stomata when temperatures are high. The spongy cells contain chloroplasts and also photosynthesise; the epidermis layers do not.

Water for photosynthesis

H_2O for photosynthesis (as well as for other plant processes) enters the plant from the soil via the root hairs, which are long extensions of the epidermal cells of the root.

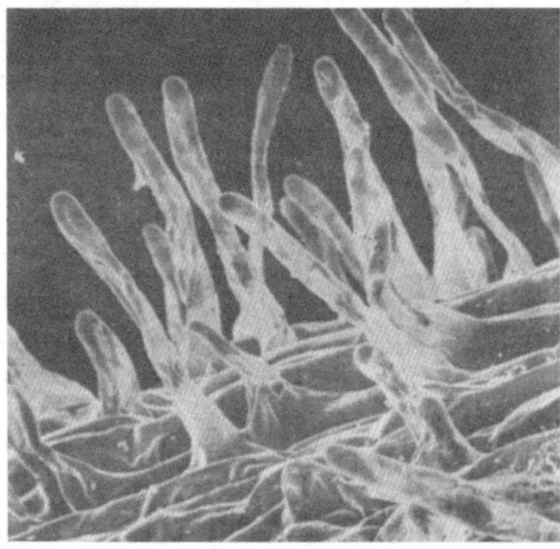

SEM of root hairs showing their elongated cell membrane (maximises surface area for entry of water).

The elongated shape of root hairs provides maximum SA (surface area) for the entry of water from *osmosis* (in response to the concentration gradient of water). The water molecules then diffuse from cell to cell across the cortex to the *xylem* vessels, which transport water up the roots and stem to the leaves. Transport is the result of the processes of capillarity, cohesion and transpiration pull. Xylem vessels are large, with thick, lignified walls to transport large volumes of water under pressure.

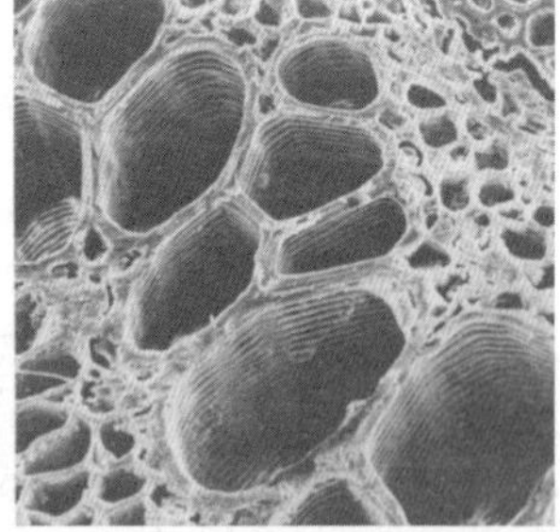

SEM of xylem vessels showing their large lumen and strengthening bands of lignin in the cell walls.

Water leaves the xylem from the veins in the leaf and enters the chloroplasts of the cells of the palisade and spongy layers.

Minerals for photosynthesis

Mineral ions dissolved in soil water *diffuse* into the root hairs in response to their concentration gradient, and are transported around the plant (in solution) in the xylem vessels. *Active uptake* of needed ions against a concentration gradient also occurs.

> In the *bryophytes* (mosses and liverworts), there is no vascular (transport) tissue as the plants are small and live in damp, shady areas. Therefore the transport of materials is by *simple diffusion* from cell to cell.

Glucose from photosynthesis

Glucose produced in photosynthesis is temporarily stored as *starch* in the leaf cells (glucose is soluble, starch is insoluble; if glucose was held in the cells, it would upset their osmotic balance). At night, the starch is converted back to glucose and diffuses into the *phloem* tubes of the veins. In the phloem tubes of veins, glucose is *translocated* to the roots (mainly) for long-term storage as **starch** (starch granules) within the cells of the cortex.

Chloroplasts

Chloroplasts are large, oval, cell organelles.

There may be up to 500 000 chloroplasts per square millimetre surface of leaf.

Inside each chloroplast are stacks of (thylakoid) membranes called the *grana*, between which is the liquid matrix of the organelle, the *stroma*.

- The light-trapping pigments (chlorophylls a and b absorb red and blue wavelengths; carotene and xanthophyll absorb yellow and orange wavelengths) are embedded in the grana.
- In the grana, the light-dependent chemical reactions of photosynthesis take place – the manufacture of ATP from solar energy; the splitting of water.

In the stroma, the light-independent (Calvin cycle) photosynthetic reactions take place – the bonding of CO_2 with H (from water), and their rearrangement to form glucose, $C_6H_{12}O_6$.

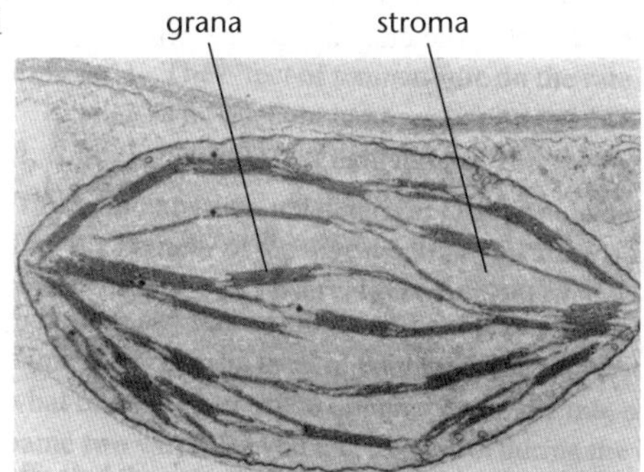

TEM of chloroplast showing grana and stroma.

Chloroplasts are also found in the outer layer of cells (cortex) in the stems of non-woody plants. Light is able to penetrate to these cells, allowing photosynthesis to occur to supplement that from the leaves. Pores (lenticels) in the epidermis of the stem allow CO_2 and O_2 to diffuse into/out of the cells.

Sun-adapted plants

Plants that live in areas exposed to high light intensities (eg canopy layer of forest community, deserts, high-altitude mountain slopes, grasslands), have to cope with light in excess of their photosynthetic capacities.

Adaptations to this can involve the leaf or the cells within the leaf.

Leaf adaptations

- Leaves tend to be *smaller*, reducing SA to catch sunlight.
- Leaves tend to be positioned *vertically*, reducing exposure to the intensity of the midday sun. The vertically positioned leaves (includes leaves that hang down – eg many *Eucalyptus* trees) are better placed to catch the less intense morning and afternoon sunlight.
- Leaves tend to be *thicker*, with *more cell layers*, as the more intense light has greater penetration.
- Some plants have the ability to suddenly collapse their leaves if light intensity increases (a photonastic response), preventing damage (eg *Oxalis*).
- The *waxy cuticle* may be increased, so more light is reflected from the leaf (as well as reducing water loss), eg mangroves.
- Leaves may be *hairy*, which both increases localised shade for the leaf and reduces water loss through transpiration – common in alpine plants that inhabit the higher altitudes of mountains.

Tissue adaptations

- Two or more *palisade cell layers* may be present.
- The *spongy cell layer* has few chloroplasts.
- Chloroplasts tend to be *smaller*, with fewer thylakoid membranes in the grana stacks, reducing the amount of light-absorbing pigments.
- Have *less* of the accessory pigments (chlorophyll b, carotene, xanthophyll) which absorb light from the blue end of the spectrum and pass it onto chlorophyll a, which absorbs light from the red end of the spectrum. The ratio of chlorophyll a to chlorophyll b is high.
- More of the enzyme *rubisco*, which catalyses the uptake of CO_2 in the Calvin cycle part of the photosynthetic reactions.
- Chloroplasts can move from the horizontal walls of the mesophyll cells to the vertical walls, reducing the amount of light absorbed by up to 20%.
- Occurrence of CAM photosynthesis in certain cacti and succulents (see Unit 11.3 Topic 2: Transpiration on p. 137).

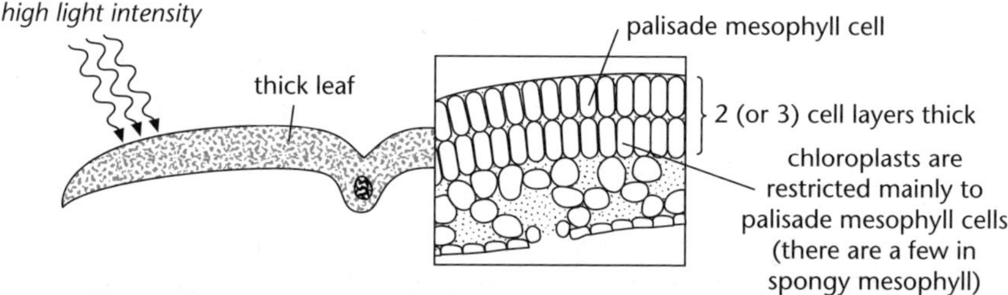

A sun-adapted leaf exposed to high light intensities absorbs much of the light available to the cells

Cross-section of typical sun-adapted plant leaf

The rate of photosynthesis tends to be fast with high light intensities, so the rate of growth in sun-adapted plants (eg grasses) tends to be higher than in shade-adapted plants. As high levels of light typically mean higher temperatures, sun-adapted plants also have adaptations to reduce water loss.

Shade-adapted plants

Plants that live in areas where they are exposed to low light intensities (eg the shrub layer and ground layer of a forest community) and those that exist below the surface of the water in ponds, lakes and rivers, have to maximise their ability to catch the available light and photosynthesise sufficiently to grow and survive.

Adaptations to this can involve the leaf or the cells within the leaf.

Leaf adaptations

- Leaves tend to be *large*, providing a larger SA to catch sunlight (eg rangiora). The size of the leaves may vary within a tree, being larger at the bottom and smaller at the top.
- Leaves tend to be positioned *horizontally*, so are exposed to as much light as possible; many plants can alter the *orientation* of their leaves to the sun as it changes its position (eg the 'prayer' plant moves its leaves from the vertical at night to the horizontal at daylight; geranium plants orientate their leaves to face the light from a window, and, if the plant is turned around, its leaves will slowly re-orientate to the light again).
- Leaves tend to be *thin* with *few cell layers* to allow light to penetrate to all the cell layers.

Tissue adaptations

- Only one *palisade cell layer* is present.
- The *spongy cell layer* has many chloroplasts.
- Chloroplasts tend to be *larger* and may have more thylakoid membranes in the stacks of grana – this will allow for more light-absorbing pigments.
- Tend to have *more* of the accessory pigments (chlorophyll b, carotene, xanthophyll), which absorb light from the blue end of the spectrum and pass it onto chlorophyll a, which absorbs light from the red end of the spectrum.

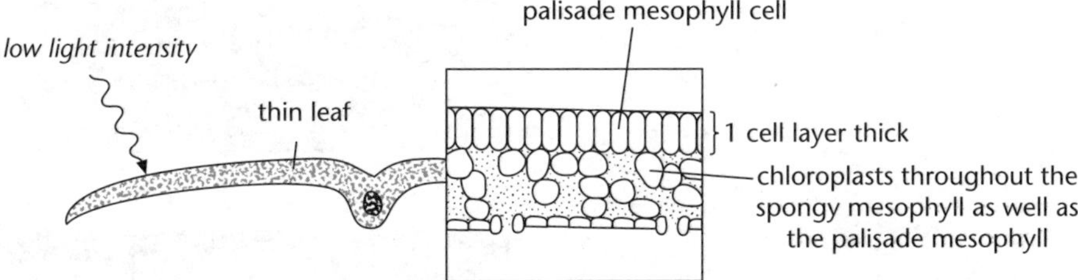

A shade-adapted plant leaf absorbs lower light intensities. If exposed to high light intensities, most of the light would pass through.

Cross-section of typical shade-adapted plant leaf

Lower light intensities reduce the rate of photosynthesis, therefore growth tends to be slower in shade-dwelling plants. However, water loss is typically not a problem, so they do not 'require' the adaptations (such as hairy leaves and thickened cuticles) that may be found in sun-adapted plants.

Carnivorous plants

There are about 550 species of carnivorous plants throughout the world. They inhabit 'bogs', ie swampy areas that are low in nutrients, especially nitrogen. To compensate for this, they trap insects (and spiders), then digest them to utilise the nutrients contained in their bodies. All carnivorous plants are photosynthetic and display the same features for photosynthesis as other plants, but their leaves have also become adapted to trapping insects, eg:

- Leaves are hinged down their midrib. The open leaf has sensitive hairs on the two inside halves. When an insect blunders into the hairs, the cells of the midrib rapidly lose water, so the spring-like hinge suddenly closes the leaf on the insect. The insect is prevented from escaping by the spines on the outer edge of the leaves, which form a cage when the leaf closes. Digestive enzymes secreted slowly break down the insect (eg Venus fly trap).

- Leaves have hairs that secrete sticky compounds which act like flypaper to trap insects. If enough hairs stick to the insect, it is unable to escape. The leaf encloses the insect and digestive enzymes are secreted to break it down (eg sundews, *Drosera*).

- Leaves that are modified to become a bowl or pitcher. The rim of the pitcher has cells that secrete sweet nectar to attract the insect. The insect topples over the rim, falling into the pitcher and can't climb up the slippery vertical walls to escape. The bottom of the pitcher contains a digestive solution with enzymes (especially those that digest protein) secreted by cells that line the inside wall of the lower half of the pitcher. The insect drowns then is slowly digested, eg *Nepenthes sp* (pitcher plant).

Parasitic plants

Parasitic plants may be true or partial parasites.

True parasites do not photosynthesise – they obtain food as well as water and minerals from the host plant, eg the PNG ficus tree (*Ficus septica*), dodder (*Cuscuta sp*).

Example

The lesser dodder

The lesser dodder (*Cuscuta epithymum*), which parasitises gorse, broom and heather, shows the following adaptations to its parasitic way of life:

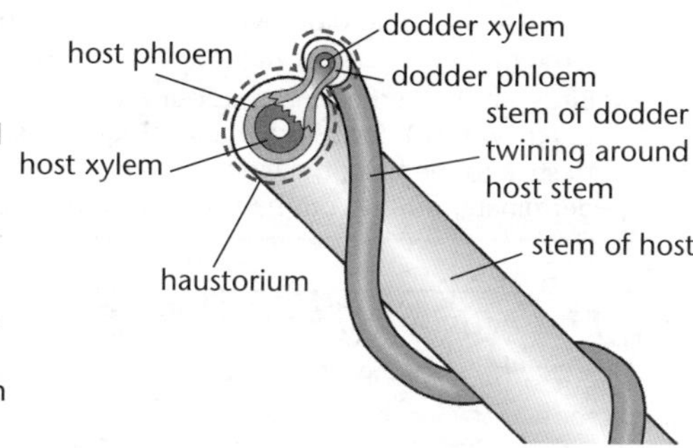

- Large numbers of seeds are produced to maximise the chances of finding a host plant, and each has a large endosperm food store.
- Seedlings grow in circles after germinating, to locate a host plant. A pad behind the growing tip enables attachment to a host. The stem is thin and twining, to find and encircle the host.
- Structures called **haustoria** (modified roots) grow into the host plant using digestive enzymes, and join to the host's xylem and phloem to allow the dodder to obtain water, minerals and food from the host.
- Leaves are very reduced, being small and scale-like without chlorophyll, as the plant no longer feeds itself from photosynthesis.
- Roots die away in the adult plant, as they are no longer needed to anchor the plant or absorb water and minerals.
- Vascular/transport tissue very reduced – there is little need for them.
- During winter, the body of the plant dies away except for small lumps of tissue around the haustoria. Dodder may live for two or more years.

Partial parasites or **hemiparasites** (like all plants) obtain their food through photosynthesis and have the same adaptations for photosynthesis as do 'normal' plants. However, hemiparasitic plants obtain their water and minerals from the host plant and show specific adaptations to this, eg New Zealand mistletoes (*Peraxilla tetrapetala, P. colensoi*). (Photo alongside.)

Example

Peraxilla

Peraxilla are hemiparasites on native beech trees (*Nothofagus sp*), but do not harm the host. These mistletoes can grow into bushy plants on the host tree, producing bright red flowers in summer. The flowers are unusual in that they can only be pollinated when native, nectar-sipping birds (tui, bellbird) twist off the top of the petals, causing the flower to burst open and fling out the pollen onto the bird's head. The pollen is transferred to the next flower that the bird opens.

Adaptations for the parasitic existence include:

- **Haustoria** – using digestive enzymes, these grow into, and join with, the *xylem* of the host – the mistletoe obtains water and minerals.
- Birds eat the berries and pass out the seed in their droppings onto a beech tree (dispersing to a new host). Seeds are made sticky by a substance called **viscin**, allowing the seeds to stick to the host; the viscin hardens, fixing the seed securely to the tree, where it will germinate, putting out haustoria into the host.

Unit 11.2 Activity 2A: Plant nutrition

1. Describe how the raw materials for photosynthesis are brought to the chloroplasts.
2. Describe how leaves are adapted for photosynthesis.
3. Explain how the cell layers in the typical angiosperm leaf are adapted to the process of photosynthesis.
4. Explain how the chloroplast is adapted to the process of photosynthesis.
5. Explain why carnivorous plants could be considered to be both autotrophs and heterotrophs.
6. Explain why parasitic plants:
   a. Produce large number of seeds.
   b. Have haustoria.
   c. Have very reduced or absent vascular tissue, leaves or roots.
7. Compare and contrast the leaves of sun-adapted and shade-adapted plants.
8. Discuss diversity in nutrition in three named groups of plants. In your answer:
   - Describe the structures associated with nutrition in each group.
   - Explain the ways the structures carry out the processes of nutrition.
   - Discuss why the systems in these three groups are different.

Unit 11.2 Nutrition

Topic 3: Cell processes – photosynthesis

The material covered in this Topic describes the plant structure that makes food by photosynthesis. The Topic deals with several of the bullet points listed on p.11 of the Biology Syllabus (2009):

- The process and stages of photosynthesis.
- Balanced word and chemical equations for photosynthesis.
- Use and storage of photosynthetic products such as oxygen and glucose in respiration.
- Factors that affect the rate of photosynthesis.

The Topic should be enriched by practical experiments to test for photosynthesis: light, carbon dioxide, chlorophyll and oxygen; and experiments to identify limiting factors that inhibit photosynthesis so that you come to a good understanding about the importance of photosynthesis.

Photosynthesis is:

'Undoubtedly the most important single metabolic innovation in the history of life on the planet.'

Margulis and Sagan
Microcosmos, *p. 78*

Photosynthesis occurs within the chloroplasts of plant cells, mainly those of the leaf, and only in daylight hours.

There are two main chemical pathways in which the raw materials – *water* and *carbon dioxide* – are joined to produce the sugar **glucose** (with oxygen being a waste product). The source of energy is solar (from the sun) and its entrapment requires the presence of pigment molecules such as chlorophyll. All chemical pathways in photosynthesis are controlled by enzymes.

First chemical pathway – light-dependent reaction

The **light-dependent reaction** takes place on the thylakoid membranes of the **grana** of chloroplasts.

- Chlorophyll (and other light-absorbing pigmented molecules such as carotene and xanthophyll) embedded in the thylakoid membranes have their electrons 'excited' by solar energy striking them. These high-energy electrons are passed along a series of carrier molecules and their energy used to make the 'energy molecule' ATP from ADP; the electrons are then returned to the chlorophyll.
- Water is split into its constituent elements – hydrogen, H, and oxygen, O. Oxygen is not required in the rest of the process of photosynthesis, so is released as oxygen gas, O_2. Hydrogen is picked up by a carrier molecule (NADP), and transferred to the second photosynthesis chemical pathway.

Second chemical pathway – light-independent reaction

The **light-independent reaction** is also known as the **Calvin cycle**. This occurs in the liquid matrix of the chloroplast, the **stroma**.

- CO_2 and H enter a complex biochemical cycle. As they are passed around the cycle, extensive rearrangement takes place and glucose, $C_6H_{12}O_6$, is formed as the final product. The molecules that form the cycle are *not* used up in the reaction; CO_2 and H are continuously fed into the cycle.
- ATP (produced in the light-dependent reaction) is used to run the cycle.

Word equation for photosynthesis

$$\text{carbon dioxide + water} \quad \frac{\text{solar energy*}}{\text{chlorophyll*}} \longrightarrow \text{glucose + Oxygen}$$

*typically added with arrow, as are *essential* to process

Formula equation (simple) for photosynthesis

$$6CO_2 + 6H_2O \quad \frac{\text{solar energy}}{\text{chlorophyll}} \longrightarrow C_6H_{12}O_6 + 6O_2$$

However, as the (waste) oxygen produced comes from the water only, it is more accurate to write the balanced equation as follows:

$$6CO_2 + 12H_2O \quad \frac{\text{solar energy}}{\text{chlorophyll}} \longrightarrow C_6H_{12}O_6 + 6O_2 + 6H_2O$$

The *glucose* produced is:

- Stored as insoluble *starch*, typically in the cells of the roots.
- Used in *respiration*.
- Used to make other needed *organic chemicals* – eg fats, amino acids (for proteins), cellulose.

Rate of photosynthesis

The *rate* of photosynthesis is determined by factors such as:

- *Temperature* – increasing the temperature increases the rate of photosynthesis up to an optimum temperature. When temperatures rise too far above the optimum temperature, the enzymes controlling the reaction **denature** and can no longer catalyse the reactions; photosynthesis ceases.

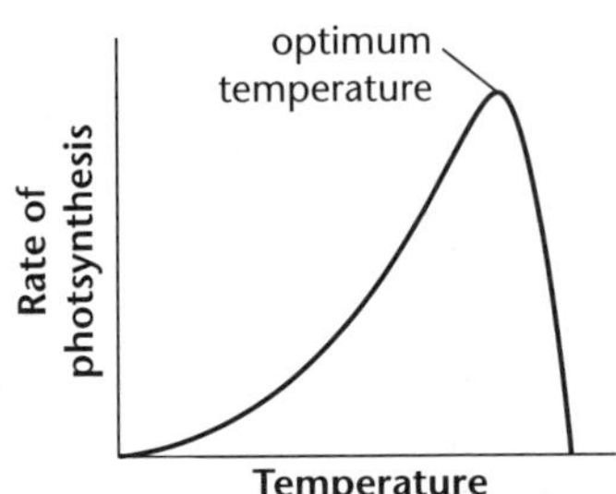

Effect of temperature on rate of photosynthesis

- *Light intensity* – increasing the light intensity increases the rate of photosynthesis up to a maximum. Above this maximum, further increases in light intensity have no further effect on the photosynthetic rate (because either the light-absorbing pigments are saturated and/or the concentration of CO_2 is limiting the Calvin cycle and/or temperature is low).

Rate of photsynthesis

Light intensity

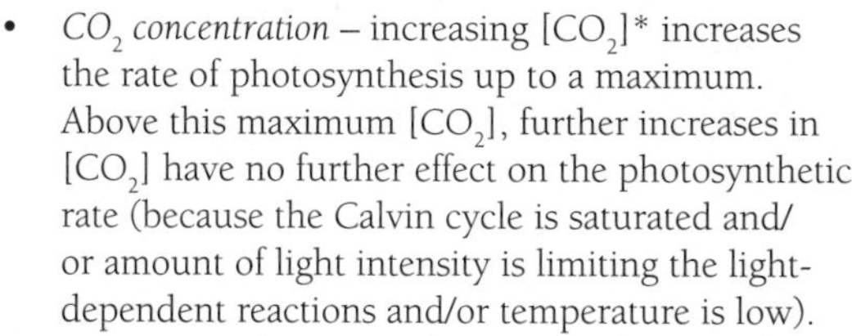

- *CO_2 concentration* – increasing $[CO_2]$* increases the rate of photosynthesis up to a maximum. Above this maximum $[CO_2]$, further increases in $[CO_2]$ have no further effect on the photosynthetic rate (because the Calvin cycle is saturated and/or amount of light intensity is limiting the light-dependent reactions and/or temperature is low).

*$[CO_2]$ – square brackets around a compound refer to its concentration.

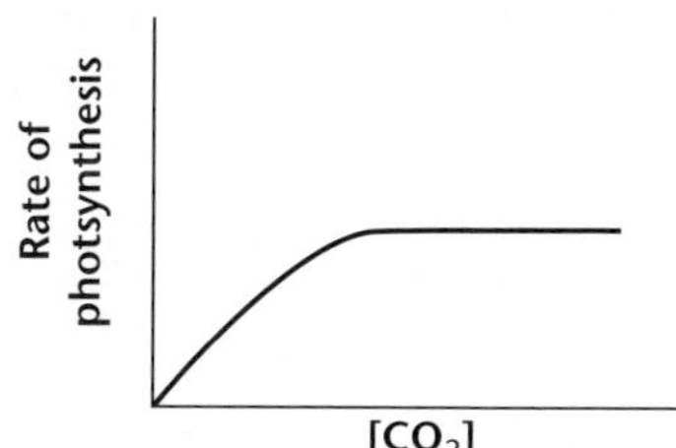

Because these three factors (temperature, light intensity and $[CO_2]$) combine to determine the rate of photosynthesis, maximum photosynthesis rates can alter if one of the other factors changes.

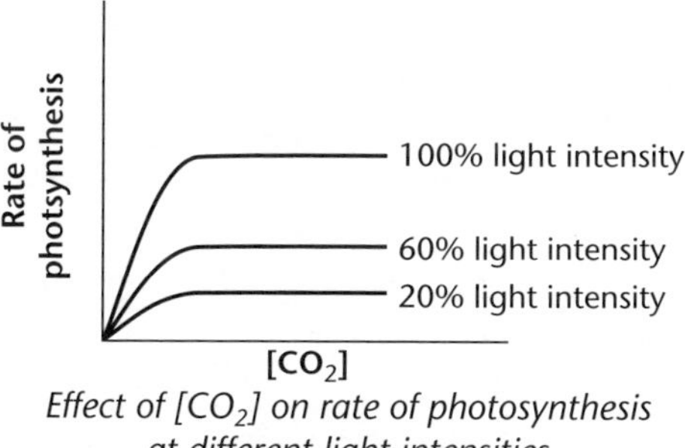

Effect of $[CO_2]$ on rate of photosynthesis at different light intensities

The rate of photosynthesis is typically measured as the amount of O_2 produced or the amount of CO_2 consumed in a given time.

> In the school lab, the easiest method to measure the rate of photosynthesis is to use aquatic plants such as oxygen weed (*Elodea*) and count the bubbles of O_2 released in a set time while varying environmental factors such as temperature or light intensity.

Other environmental factors that may influence photosynthesis are:

- *Water* – cells, even if dehydrated, contain quantities of water sufficient to allow photosynthesis; however, the closure of stomata to prevent water loss will also prevent entry of CO_2 which will limit photosynthesis.
- *Mineral ions* – certain mineral ions are essential for photosynthesis to occur (eg Fe is needed for the synthesis of chlorophyll, Mg is an essential part of the structure of the chlorophyll molecule (a deficiency of Mg causes yellowing of leaves)).
- *Light wavelength* – the light-absorbing pigments are most active at absorbing light in the blue and red wavelengths of the electromagnetic spectrum; wavelengths of green are not absorbed, but reflected (hence the green colour of plants).

Importance of photosynthesis

The importance of photosynthesis is that it converts *solar energy into chemical energy* as organic molecules (carbohydrates, proteins, fats).

Plants are called autotrophs ('self feeders) since they are the **producers** (of food) – the starting point of food chains. Organisms that feed on other organisms are called heterotrophs. The wastes and dead bodies of autotrophs and heterotrophs are broken down by decomposers (bacteria and fungi), so that the chemicals locked in the wastes and bodies can be released into the soil. These enter the roots of plants, allowing their growth and new supplies to food chains.

Photosynthesis by plants accounts for the high level (approx 20%) of *oxygen* in the atmosphere. This gas is essential for the process of *aerobic respiration*.

Without photosynthesis, animal life wouldn't exist.

Unit 11.2 Nutrition

Topic 4: How humans feed – heterotrophic nutrition

Do you know what our human digestive system looks like? The information presented in this Topic describes heterotrophic nutrition and the digestive system of humans (see p. 11 of the Biology Syllabus). In particular, this Topic looks at:

- Digestion – teeth, chewing, swallowing, the stomach and small intestine.
- Enzyme action.
- Absorption of glucose and amino acids, of fatty acids and glycerol, of water; energy demands; the large intestine.
- Problems with the digestive system – peptic ulcer, bowel cancer, constipation, diarrhoea.

The digestive system is concerned with processing food so that it can be made available to the cells of the body. The digestive system consists of the **alimentary canal** or **gut**. This is a long tube extending from the mouth to the anus, together with various glands that secrete juices into it.

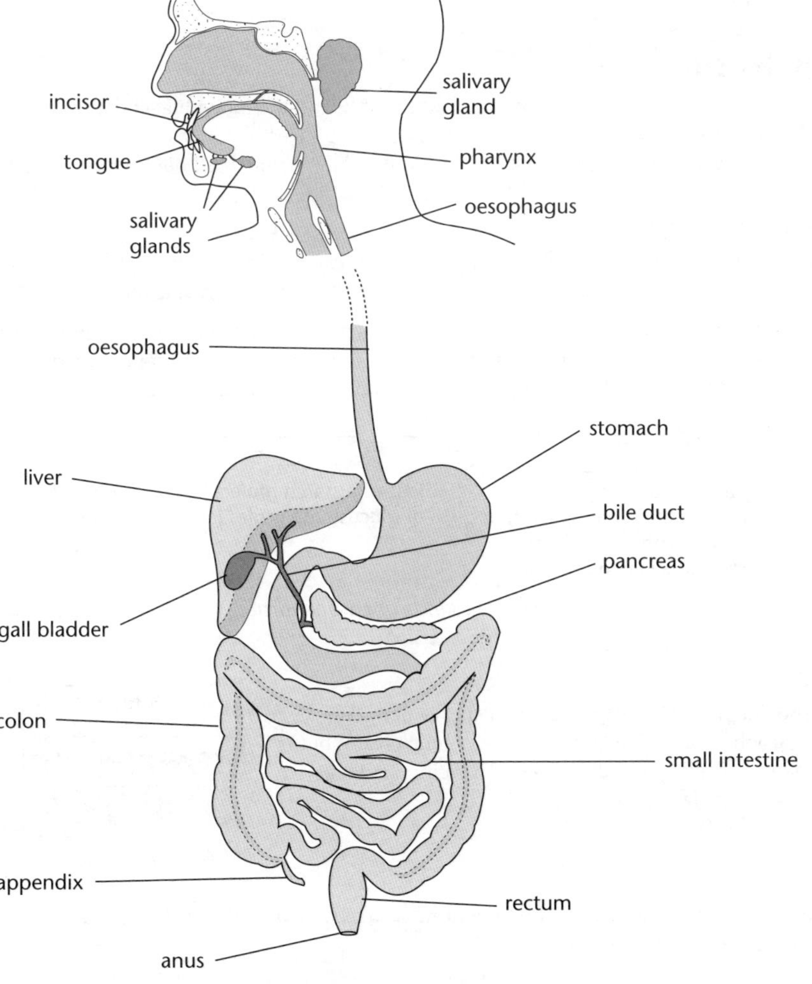

The alimentary canal.

Food-processing by the body involves five stages:

- **Ingestion** – taking food into the mouth.
- **Chewing** – chops up of the food into smaller pieces. This is sometimes called 'physical digestion'; it is fundamentally different from the next stage (digestion), which is a chemical process.
- **Digestion** – the breakdown of complex foods like starch and proteins into simpler ones like **sugars** and amino acids.
- **Absorption** – products of digestion are absorbed through the walls of the intestine into the blood.
- **Egestion** – the getting rid of undigested residue as faeces via the anus.

After it has been absorbed, the digested food is distributed round the body by the blood and absorbed by the target cells. The process by which food is absorbed and used by cells is called **assimilation**.

Nutrition

Nutrition is the relationship between food and health, and the individual chemicals in food are called **nutrients**. There are six main classes of nutrient – carbohydrates, fats, proteins, vitamins, minerals and water. Most of the first three need digestion before the body can use them.

Carbohydrates

Carbohydrates include sugars and starch and give us *energy*. Carbohydrate-rich foods include bread, spaghetti, cereals, peas, potatoes, sultanas, most biscuits, kumara, and rice.

The simplest carbohydrates are called simple sugars or *monosaccharides*, for example **glucose**. Glucose is the main fuel used by cells and is the form in which carbohydrate is carried in the blood. Simple sugars can be built up into more complex carbohydrates like starch.

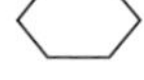

Glucose

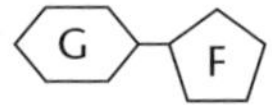

Sucrose (table sugar). This is a double sugar, consisting of a glucose and a fructose molecule joined together.

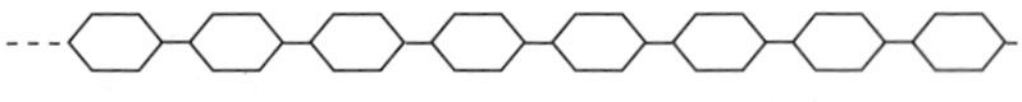

Part of a starch molecule, consisting of hundreds of glucose molecules joined together.

Different kinds of carbohydrate.

Fats

Fats provide more than twice as much energy per gram as do carbohydrates. Foods rich in fats include cream, butter and margarine, egg yolk, and cheese. Dripping, lard and fish oil are pure fat.

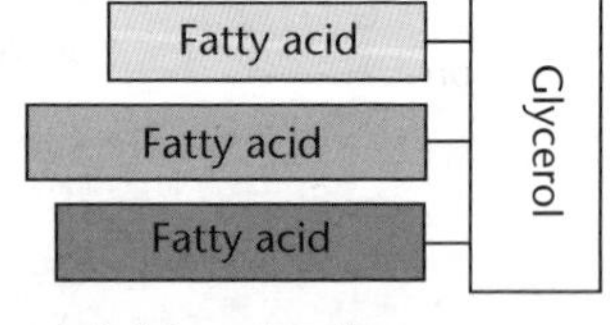

A fat molecule.

Fat molecules are made of three **fatty acid** molecules joined to **glycerol**. Fats that are liquid at room temperature are called *oils*.

Proteins

Proteins provide building materials for *growth*. Protein-rich foods include meat, egg white, fish, cheese, milk, and all seeds (nuts, beans, peas, and cereals).

Because proteins are constituents of all cells, they are important for increasing the number of cells, and thus for *growth* and *repair*.

A protein molecule consists of a long chain of up to a thousand or more smaller units called **amino acids**. There are 20 different kinds of amino acids commonly found in proteins, and in each protein they are arranged in a specific order or *sequence*. It is the sequence and type of amino acids that makes each protein different.

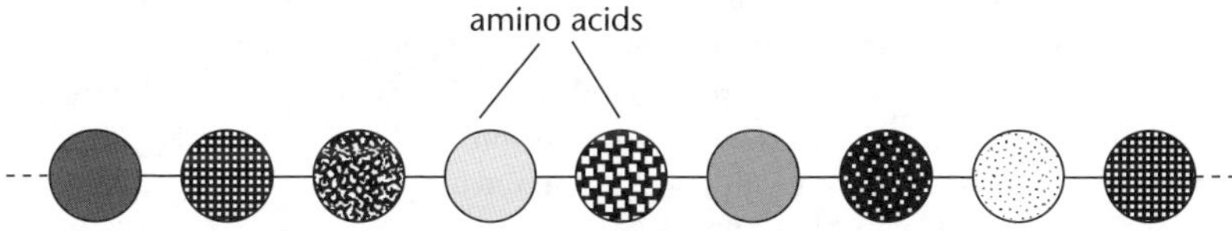

Part of a protein molecule.

Vitamins and minerals

These are needed in much smaller amounts than carbohydrates, fats or proteins, and do not provide energy. Vitamins are *organic* substances (contain carbon atoms joined to hydrogen), and are found only in living things. Minerals are *inorganic*; for example, calcium, potassium, phosphate, and numerous other ions or salts. Vitamins and some minerals play an important part in helping enzymes to work.

Digestion – how food gets into the body

Swallowed food isn't *actually* inside you – this is because the cavity of the gut is really a deep 'intucking' of the outside world. Before food can be used by the cells of the body, it therefore must pass through the lining of the gut 'into' the body.

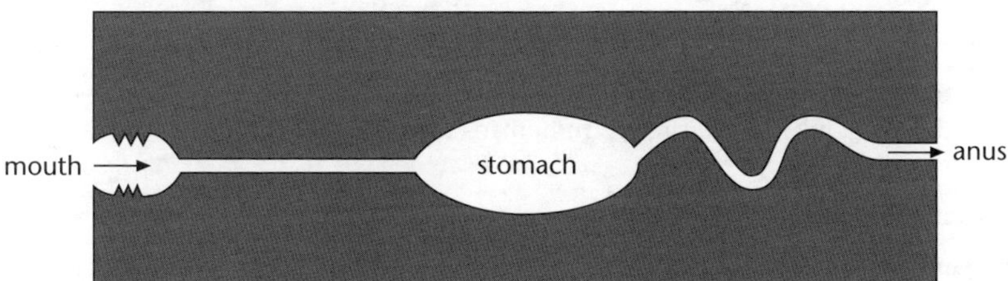

The gut is really an extension of the outside world. It would be possible to pass from mouth to anus without going through any living cells.

Simplified view of the gut.

Much of our food consists of molecules that are too large to pass through the wall of the gut, eg starch and proteins. The function of digestion is to break down these large molecules into molecules that are small enough to pass through the gut lining and into the blood.

Enzyme action

Digestion is a complex series of chemical reactions carried out by **enzymes** secreted by glands lining the gut. Most enzymes catalyse chemical reactions inside the cell that makes them, but digestive enzymes are secreted into the gut cavity.

Enzymes have the following important properties:

- They are *specific* – each enzyme can catalyse only one kind of reaction, eg the enzyme salivary amylase converts starch into maltose (a disaccharide), and the enzyme maltase converts maltose into glucose. Because each enzyme only acts on one chemical constituent of a meal, many different types of enzymes are needed to digest the whole meal.
- Each enzyme acts best at a particular pH – eg, amylase in saliva works best in neutral solution, whilst the enzyme pepsin in the stomach juices acts best in a very acid solution.
- They are very sensitive to temperature. Like other mammals, humans regulate their body temperature at about 37°C. This is about the *optimum* (best) temperature for enzyme action. At temperatures above 38°C, enzymes begin to **denature** (break down). At temperatures below 36°C, enzyme activity slows down.

Teeth

Before digestive enzymes can work they have to mix with the food. Solid foods are eaten in large lumps, which must first be broken up to increase the area of contact with the enzymes. This is the purpose of chewing, which is done by the teeth.

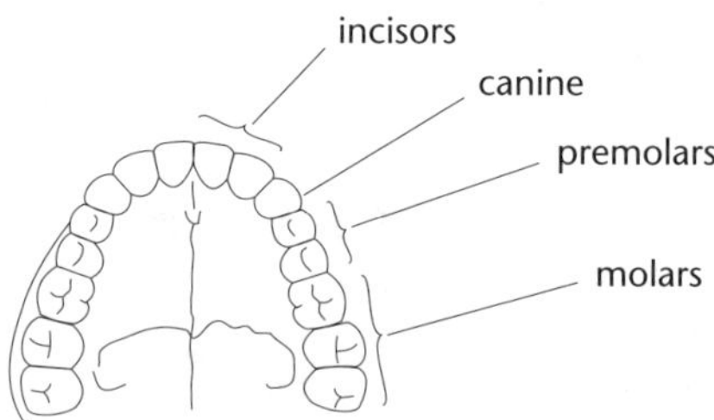

The upper jaw showing the different kinds of tooth.

Humans and most other mammals have two sets of teeth; a 'milk set' of *deciduous* teeth, which are replaced by an adult, *permanent*, set. In humans there are 20 **deciduous** teeth and 32 teeth in the adult set.

Humans and other mammals have different kinds of teeth, adapted for special tasks.

- **Incisors** are situated in the front of the mouth and have chisel-like edges for cutting and biting off pieces of food.
- **Canines** are next to the incisors. They are only slightly pointed in humans, but highly developed in carnivores (where they are adapted for piercing and holding onto struggling prey).
- **Premolars** have one root and two cusps (projections on the surface of the tooth). They are behind the canines and are used for crushing food into smaller pieces.
- **Molars** differ from the other adult teeth in that they do not replace earlier teeth. They have several roots and four or five cusps. They are larger than the premolars and help the premolars to crush the food.

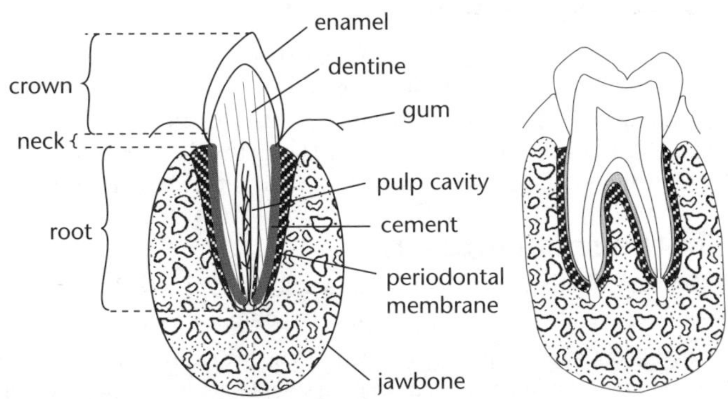

Vertical section of an incisor (left) and molar (right).

Teeth are anchored in sockets in the jawbone. Each tooth consists of a number of parts.

- **Enamel** is the hardest substance in the body and covers the crown of the tooth. It protects the tooth and forms a hard, biting surface. It consists almost entirely of calcium phosphate.
- **Dentine** is found just under the enamel and forms the bulk of the tooth. Though tough, it is not as hard as enamel or as resistant to decay. It is penetrated by thread-like extensions of cells in the pulp cavity. They make the dentine sensitive to touch, temperature and acids. Since it contains protein, dentine can be used as food by decay-causing bacteria.
- **Pulp cavity** is in the centre of the tooth and contains blood vessels which bring food and oxygen to the dentine and nerve endings sensitive to pain.
- **Cement** covers the root of the tooth and is similar to bone.

The tooth is anchored in the socket by a tough *periodontal membrane*. However, a tooth is still able to move slightly and this reduces the chances of it being broken during chewing.

Chewing

Solid food is normally chewed before swallowing. Aided by the secretion of saliva from three pairs of **salivary glands**, and manipulated by the tongue, chewing breaks up the food into a soggy lump or *bolus*. At the same time, the enzyme *salivary amylase* converts starch into the complex sugar *maltose*.

Swallowing

This is a complex process in which the bolus is propelled from mouth to stomach.

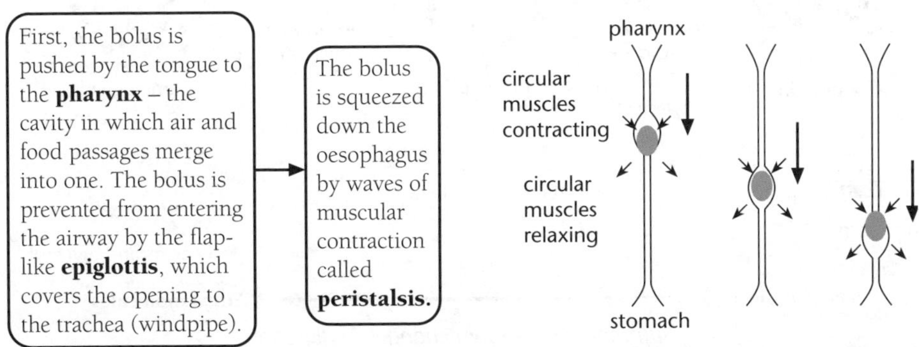

Peristalsis in swallowing.

Digestion in the stomach

Gastric juice is secreted onto the food, and at the same time it is mixed by the action of muscle in the stomach wall. An enzyme in gastric juice, **pepsin**, converts proteins into **polypeptides**. As a result, the food is slowly liquefied. Gastric juice contains hydrochloric acid, which not only enables pepsin to work, but also kills most bacteria. The exit to the stomach is guarded by a ring of muscle called a **sphincter**. This controls the rate at which food leaves the stomach.

Digestion in the small intestine

Slowly, the food is released into the first part of the small intestine, the **duodenum**. Two digestive juices – bile and pancreatic juice – are secreted onto the food.

- **Bile** is a yellow-green, alkaline liquid, secreted by the liver and temporarily stored in the **gall bladder**. Though it contains no enzymes, it contains salts which play a vital part in fat digestion, breaking up large fat globules into an **emulsion** of millions of tiny droplets. This greatly increases the surface area of the fat, speeding up the action of the fat-digesting enzyme **lipase**.

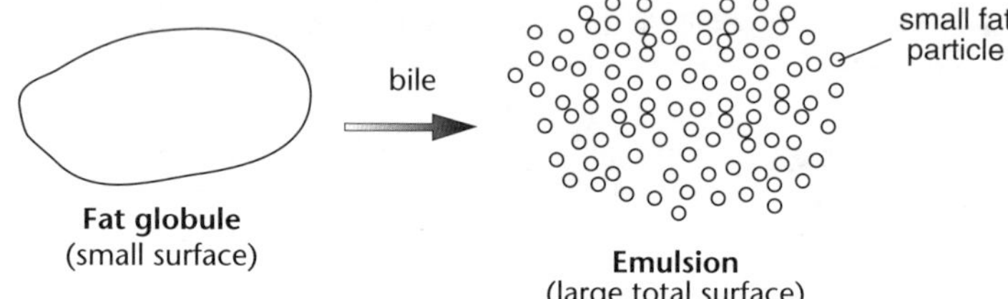

How bile salts help the digestion of fats.

- *Pancreatic juice* is secreted by the **pancreas** and contains enzymes that digest fat, proteins and starch.

Other enzymes in the small intestine complete the digestion of proteins and carbohydrates, producing a mixture of simple sugars, amino acids, fatty acids and glycerol.

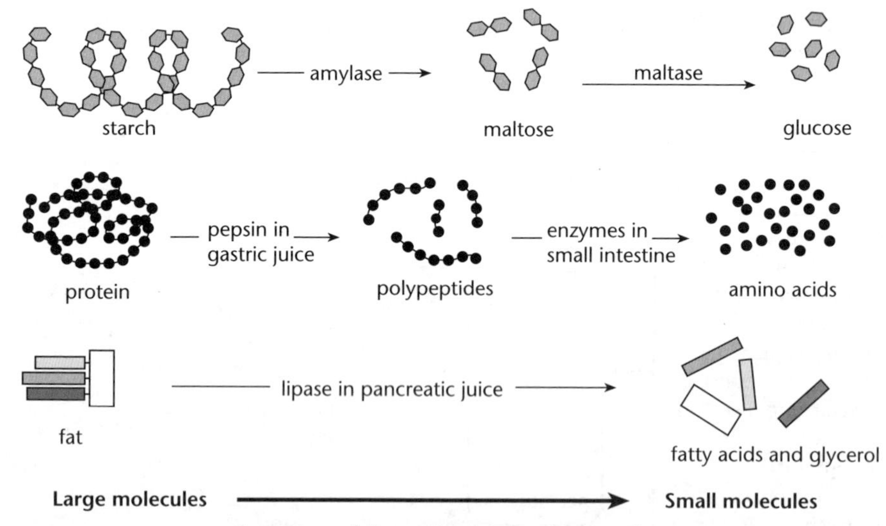

Summary of the main changes in digestion.

While the food is in the intestine, it is mixed up by muscular action and also propelled along by peristalsis.

Absorption of digested food

Absorption of glucose and amino acids

The glucose and amino acids produced in digestion have to pass through the lining of the small intestine. They are not absorbed by diffusion, but are 'pumped' through the lining by **active transport**, which requires energy. The area of the lining is greatly increased by vast numbers of finger-like **villi**. The surface area is further increased (to about 250 m^2) by millions of **microvilli** on the surface of the individual cells lining the intestine. Each villus contains a network of microscopic blood vessels called **capillaries**. These eventually join to form the **hepatic portal vein**, which takes blood from the gut to the liver.

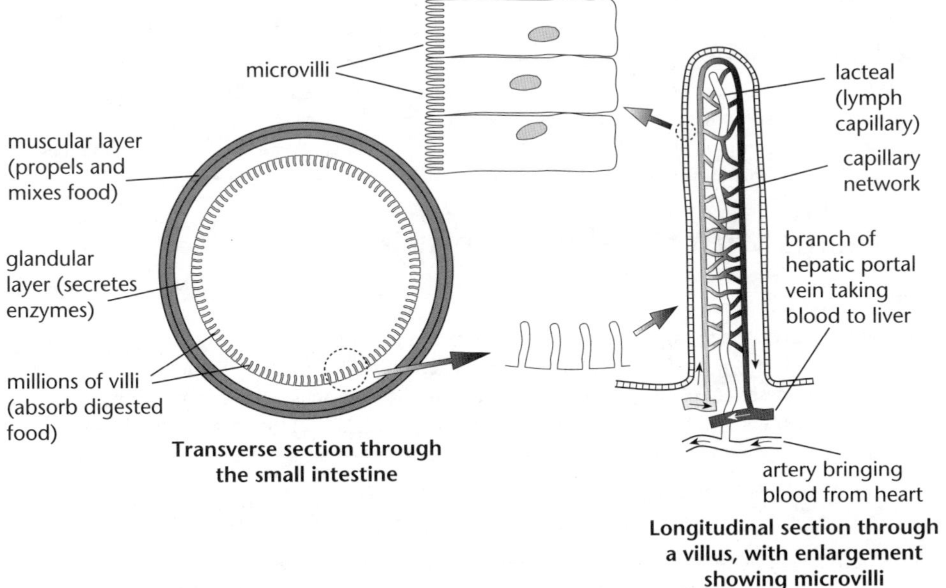

How the area of the lining of the small intestine is increased.

Absorption of fatty acids and glycerol

In crossing the lining of the intestine, fatty acids and glycerol recombine to form tiny fat droplets. These enter the tiny lymph capillary in each villus as an emulsion. This makes the lymph look milky (hence the name 'lacteal').

Absorption of water

The total volume of digestive juices secreted during 24 hours is about 8 litres. Water is too valuable to waste, and about 95% of the water in the digestive juices is reabsorbed from the small intestine. Most of the rest is reabsorbed in the colon.

Energy demands of food processing

The production of enzymes, the propulsion and mixing of the food and the absorption of the digested products all require energy. To provide this, the cells lining the gut have many *mitochondria*.

Example

The blood flow to the gut increases greatly when processing a large meal – try not to undertake vigorous exercise until a couple of hours after dinner.

What happens to the digested food?

The hepatic portal vein carries digested food to the liver. Some of the glucose is converted into glycogen and stored, the rest is used for energy in respiration. Some of the amino acids are used to make proteins; the rest are used for energy.

The large intestine

By the time the undigested food (mainly cellulose) has reached the large intestine, all that remains is cellulose (which cannot be digested), and some water. Most of the remaining water is absorbed from the colon. The millions of bacteria in the colon convert the material into **faeces**. These are temporarily stored in the rectum, prior to being *egested* via the anus.

Problems with the digestive system

Peptic ulcer

With a peptic ulcer, the lining of the stomach or duodenum is broken, exposing the sensitive deeper layers. Ulcers most commonly affect the stomach (gastric ulcer) and duodenum (duodenal ulcer) – both diseases are painful. An ulcer also decreases the efficiency of the stomach or duodenum because it affects the muscular contractions (peristalsis) of the organ walls. This in turn reduces the mixing and churning ability, and food is not digested as effectively.

The immediate cause seems to be that the stomach or duodenum ceases to be able to protect itself against gastric juice (ie, a small part of the gut begins to digest itself).

Ulcers occur most often in middle age, and seem to be linked to stress. A major factor that seems to be involved is infection by *Helicobacter pylori*. This is a bacterium that lives in the stomach wall of many people, and normally does no harm. In some people, the delicate balance between bacterium and human host becomes upset, and in some way triggers the processes leading to an ulcer. Antibiotics can be taken to treat the infection. Alcohol and smoking also seem to be factors contributing to the development of ulcers.

Bowel cancer

Bowel cancer is one of the more common forms of cancer leading to death in developed countries like New Zealand and Australia. The biggest risk factors are heredity (ie bowel cancer tends to run in families) and diets low in fibre and high in animal protein. The best way of avoiding this disease is to eat plenty of fresh fruit and vegetables containing fibre, avoid excessive alcohol, avoid smoking, reduce weight and take regular, moderate exercise.

Bowel cancer is growths that begin in the lining of the large intestine. These growths may be so large they block the colon or grow through the colon wall, resulting in its rupture. Bowel cancer symptoms include changes in bowel motions that don't get better (eg constipation, diarrhoea) and bleeding, which may result in anaemia. Treatment usually involves surgery, chemotherapy and/or radiotherapy.

Diarrhoea

Diarrhoea occurs when the food moves through the digestive system too quickly, and most of the water has not been reabsorbed. As a result the faeces are very runny. Diarrhoea can be caused by many things, including infections (eg the micro-organism *Giardia* found in some water sources, or bacteria that cause food poisoning); taking antibiotics; stress or anxiety; consuming too much water or other fluids. To avoid dehydration when suffering diarrhoea, it is important to replace the fluids lost. Purpose-made electrolyte solutions, dilute fruit juice with a tiny amount of salt dissolved in it or sweet drinks such as lemonade should be sipped slowly but regularly. In extreme cases, hospitalisation and a saline drip may be needed. Medications to stop the diarrhoea can be taken, but will not eliminate any infection or replace lost fluids.

Constipation

This condition occurs when there is insufficient movement of faecal material in the large intestine. Faeces become compacted and hard, and can be very difficult and painful to egest. This pain can make the problem worse as the person may not want to go to the toilet even if they have the urge to.

Constipation is usually caused by lack of fibre in the diet, which is easily resolved by eating higher-fibre foods such as peas and beans, wholemeal bread, fruits and vegetables. Constipation may also be caused by lifestyle changes such as pregnancy, travel or lack of exercise, aging, some medications, overuse of laxatives. Constipation can be treated by eating a high-fibre diet, drinking plenty of water, taking regular exercise, not unnecessarily using medications that cause constipation, and ensuring that sufficient time is taken to go to the toilet each day.

A selection of foods high in fibre.

Unit 11.2 Activity 4A: The digestive system

1. For each of the phrases **1–26**, write the letter **A–Z** of the term to which it applies.

1.	Absorbs water from undigested food	A.	amino acids
2.	Blind-ending part of large intestine	B.	appendix
3.	Carries food to stomach	C.	bile
4.	Converts starch to sugar	D.	cement
5.	Emulsifies fats	E.	colon
6.	End product of protein digestion	F.	dentine
7.	End product of starch digestion	G.	duodenum
8.	Fat-digesting enzyme	H.	emulsion
9.	Fine suspension of fat droplets	I.	enamel
10.	First part of small intestine	J.	epiglottis
11.	Hardest part of a tooth	K.	gall bladder
12.	Makes bile	L.	glucose
13.	Muscular waves propelling food along gut	M.	glycerol
14.	Part of a tooth attacked by bacteria	N.	hepatic portal vein
15.	Prevents food going down windpipe	O.	incisor
16.	Product of fat digestion	P.	lipase
17.	Ring of muscle	Q.	liver
18.	Secretes enzymes into duodenum	R.	oesophagus
19.	Secretes gastric juice	S.	pancreas
20.	Sticky paste formed on surface of teeth	T.	peristalsis
21.	Stores bile	U.	plaque
22.	Surrounds root of tooth	V.	rectum
23.	Takes digested food to liver	W.	salivary amylase
24.	Temporarily stores faeces	X.	sphincter
25.	Tiny finger-like projections lining small intestine	Y.	stomach
26.	Tooth at front of mouth	Z.	villi

2. a. The mechanical (physical) digestion of food begins in the mouth. Complete the table below to name and describe the function of the different types of tooth found in the human mouth.

Diagram of tooth	Name of tooth	Function

b. Explain why only starchy foods start to be chemically digested in the mouth.

c. Different parts of the digestive system have different pH levels. Discuss the importance of pH levels to the functioning of enzymes in the process of digestion.

3. a. Describe the purpose of absorption.

b. Explain how villi increase the effectiveness of absorption in the small intestine. You may use a diagram in your answer.

c. The small intestine is also a site of digestion. Discuss the roles of bile and pancreatic juice in the digestion of fat.

4. a. Each year thousands of people around the world are diagnosed as having bowel cancer. It is one of the most common cancers among both men and women. Describe the symptoms of bowel cancer.

b. Explain why diarrhoea, though usually less serious than bowel cancer, can sometimes still lead to death if it is untreated.

c. Discuss how normal stomach functioning is affected by ulcers.

5. a. Complete the following table to describe the role in digestion of the named digestive juices:

Digestive juice	Produced by	Role in digestion
Saliva	Salivary glands	i.
Gastric juice	Stomach	ii.
Bile	Liver	iii.

b. All of the digestive juices contain enzymes. Explain why more than one type of enzyme is needed in the human digestive system.

c. The pancreas produces pancreatic juice that contains enzymes that digest carbohydrates, proteins and fats. Gallstones in the liver or gallbladder cut down bile secretion, severely disrupting the digestive process and changing the pH significantly. Discuss how this change in pH level affects the digestive properties of pancreatic juice.

6. a. Describe the function of the colon.

b. Explain why the colon does not need villi.

7. Tina does not like to eat vegetables and fruit. Her mother is always telling her that she will get constipated. Tina looked up a medical website to find out if this is true. Some of the information she found is below:

> The most common cause of constipation is a diet low in fibre (found in vegetables, fruits and whole grains) and high in fats (found in cheese, eggs and meats). People who eat plenty of high-fibre foods are less likely to become constipated.
>
> Fibre – soluble and insoluble – is the part of fruits, vegetables and grains that the body cannot digest. Soluble fibre dissolves easily in water and takes on a soft, gel-like texture in the intestines. Insoluble fibre passes almost unchanged through the intestines.

a. What is constipation?

b. Explain how fibre reduces the chance of a person getting constipation.

8. The diagram below shows the organs in the human digestive system.

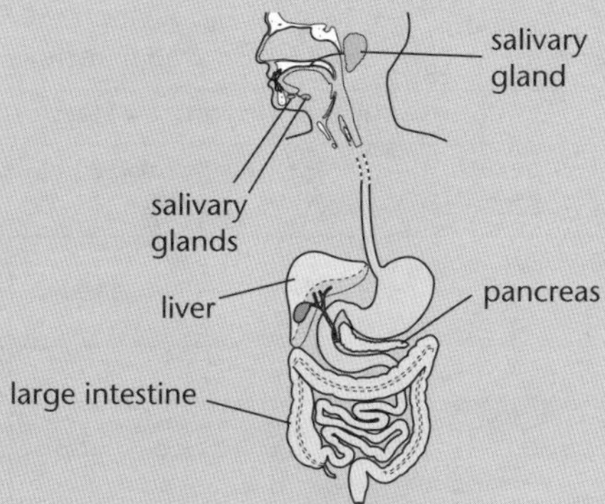

The human digestive system

a. Describe the purpose of the human digestive system.

b. Choose two organs of the human digestive system labelled on the diagram. Write their names in the first column of the table below. For each chosen organ, describe how it functions in the digestive process. An example has been done for you.

Organ of the digestive system	Description of how the organ functions in the digestive process
Oesophagus	*The oesophagus is a hollow tube that joins the mouth and the stomach. The muscles in the wall of the oesophagus contract to push balls of chewed food from the mouth down into the stomach.*
(1)	
(2)	

c. Each of the organs in the human digestive system has structural features that help it to carry out its function. For example, one of the features of the gall bladder is that the walls around it contain smooth muscle cells. When nerves cause these muscles to contract, bile is sent through a duct into the duodenum. This feature of the gall bladder allows our body to control when bile is released.

Select one of the following organs and discuss how the structure of the organ helps it to carry out its function in the human digestive system.

- Stomach
- Small intestine (duodenum and ileum)

9. Malili and Jessica were investigating the effect of temperature on the digestive enzyme salivary amylase. This enzyme changes starch to maltose sugar.

a. Describe the function of enzymes in digestion.

b. The table shows the results of the investigation. Explain why the different temperatures cause these results.

The action of the enzyme salivary amylase on starch at three different temperatures

Time (hours)	Maltose sugar formed from starch solution (%)		
	Temperature 22°C	Temperature 37°C	Temperature 50°C
0	0	0	0
0.5	0	2	0
1.0	8	12	3
1.5	20	34	9
2.0	41	65	16
2.5	62	77	21
3.0	74	92	27

c. Acid and bile are used to change the pH of food as it moves through the digestive system. Explain why the pH of the food needs to be different, in different organs of the digestive system.

Unit 11.2 Nutrition

Topic 5: How animals feed – heterotrophic nutrition

The previous Topic looked at how humans feed and in this Topic we take a look at nutrition and digestion in animals. The two Topics together cover the bullet points listed under 'Digestion' on p. 11 of the Syllabus. This Topic deals with:

- Structure of the digestive system.
- How the processes of ingestion, digestion, absorption, egestion are carried out.
- Mechanical (eg teeth) and chemical (eg enzymes) digestive processes.
- Reasons for the differences in structure and function between different groups – these could relate to diet, habitat, size, life cycle, way of life.

Animals are all **heterotrophs**, because they obtain their food by eating other organisms. In doing this, animals may be put into the functional groups of herbivores, carnivores, omnivores, parasites, filter feeders, scavengers, carrion feeders.

Food is used for:

- Fuel for respiration.
- Repair of tissues.
- Reproduction and growth.
- Maintaining a stable internal environment (homeostasis).

Most of the diet is made of the large organic compounds – carbohydrates (including sugars and starches), lipids (fats and oils), proteins. Smaller amounts are needed of the essential minerals (eg Ca, Fe, Se, Co) and vitamins (eg vitamin B complex). Water is needed in large amounts, as it makes up approx 70% of cells and tissues and is needed for many essential processes (including digestion).

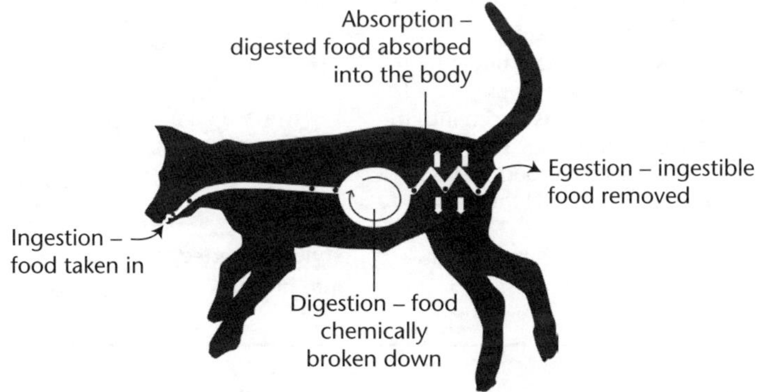

Processing of food

Ingestion

Animals **ingest** many different kinds of foods, either as small particles, fluids, or as larger particles of animal and plant material.

Small particle feeders

Fluid feeders

Some fluid feeders, like the hydatids tapeworm (an endoparsite), 'wallow' in their food, absorbing digested food directly through their skin. Other fluid feeders suck up their food using specialised mouthparts (eg butterflies feeding on the nectar of flowers use tube-like mouthparts). The mosquito is a highly-specialised fluid feeder, an ectoparasite feeding on blood. Like many ectoparasites, mosquitoes have piercing/sucking mouthparts.

Example

Feeding in the mosquito

The mouthparts of the mosquito suck blood by forming a piercing **proboscis** or tube. When a mosquito feeds, **mandibles** (jaws) inside the proboscis open up the wound, which is deepened by **maxillae**. Blood is sucked up in a tube formed from the mouthparts of the upper jaw. The lower jaw acts as a sheath for the proboscis.

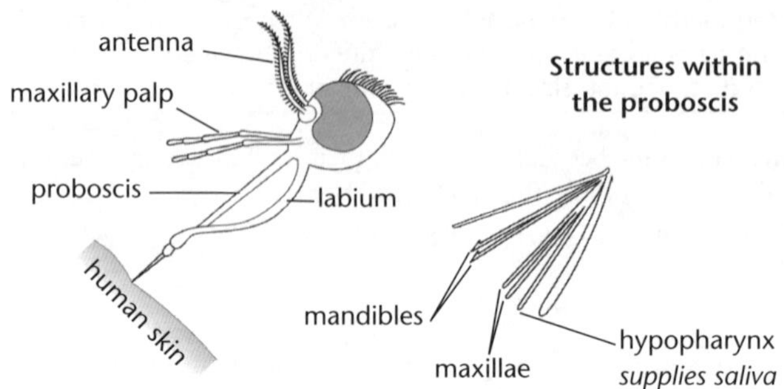

Specialised mouthparts of the mosquito.

Removal of blood is aided by saliva which contains an **anticoagulant** to prevent the victim's blood clotting. This is typical of other blood-feeding parasites (eg leeches).

Feeding in aphids

Sap in phloem vessels in plants is under high pressure; once a phloem vessel is punctured by the stylet of an aphid, sap is forced into the aphid's food canal. As they feed, aphids often transmit plant viruses to their food plants. These viruses can sometimes kill the plants.

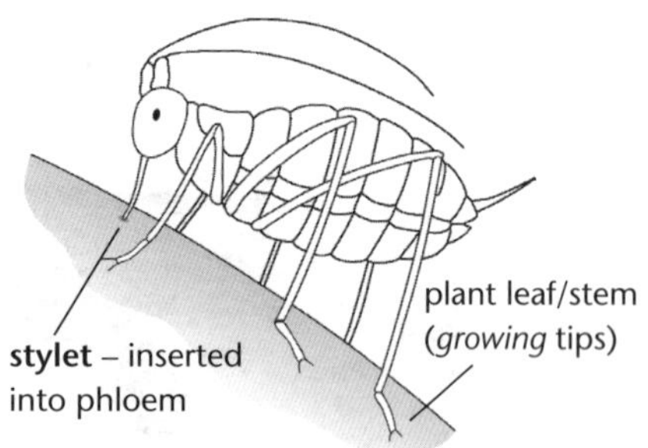

Filter feeders

Filter feeders sieve minute particles (plankton) from water. Filter feeding is a common feeding method found in many taxonomic groups, eg:

- Plankton-eating fish, such as herrings and mullet, trap food on filaments across the openings of their gill slits.
- Whales trap plankton in a thick, hairy sieve called baleen.
- Barnacles trap particles in a net formed from the fine bristles that cover their legs.
- Many shellfish such as pipi, toheroa and mussels are also filter feeders, using modified gills.

Example

Filter-feeding in the common mussel *(Perna canaliculus)*

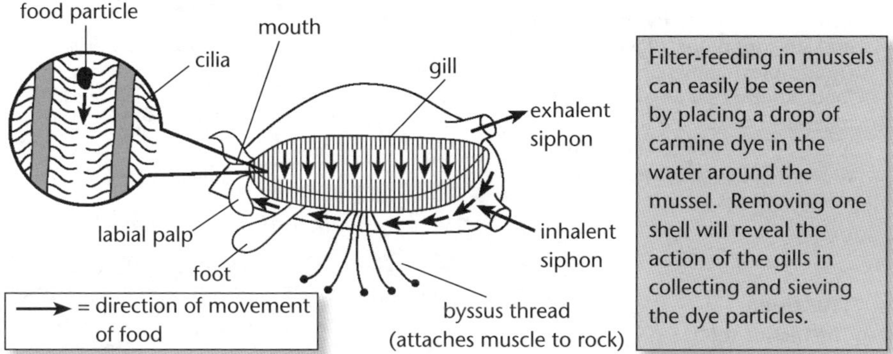

Filter-feeding in mussels can easily be seen by placing a drop of carmine dye in the water around the mussel. Removing one shell will reveal the action of the gills in collecting and sieving the dye particles.

Currents set up by the beating of cilia on the gills draw water into the *inhalent* siphon, then over the gills. The two gills are covered with sticky mucus, which traps particles. These particles are moved by cilia to the edge of the gill to form a sticky tube. Particles in the tube are sorted and directed to the mouth by the **labial palps**. Water leaves the mussel by the *exhalent* siphon.

Carnivores

Some carnivores, such as snakes, swallow their food whole, but most carnivores chew (or **masticate**) their food before it is swallowed.

Mammals have specialised teeth for eating. In carnivores, **incisor teeth** cut the food, **canine teeth** puncture, rip and shred it and **molars** grind the food up so it can be swallowed. Some carnivores such as dogs have **carnassial teeth** at the side of the jaw to crack bones and cut flesh (acting like scissors). Teeth break down the food into smaller chunks – this makes it easier to swallow and also increases the surface area, facilitating chemical digestion. Saliva mixed with the food in the mouth helps swallowing by acting as a lubricant (as well as starting chemical digestion).

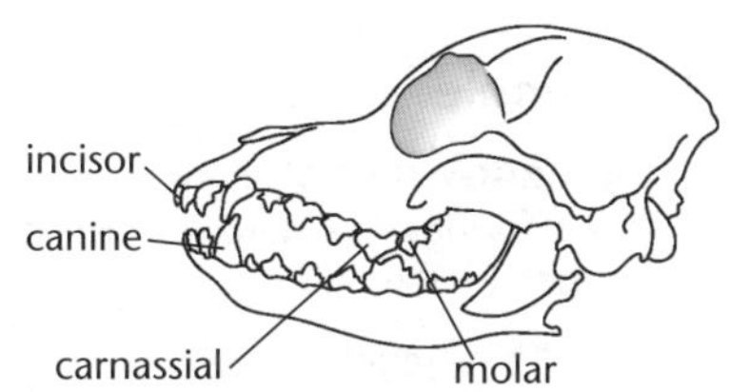

Skull of the domestic dog.

Herbivores

The cells that make up plants have a strong hard cellulose cell wall which must be broken up before the contents of the cells can be used. Herbivores have tough grinding mechanisms to crush plant cell walls.

Herbivorous molluscs, such as limpets, catseyes and chitons, rasp and shred algae using their file-like tongue, or **radula**. Insects and mammals are the only other groups of animals to have adapted to a plant diet.

Example

The cockroach

Foods are detected by the **antennae** and **sensory palps**. Powerful jaws **(mandibles)** gather and crush the food. Another set of jaws, called maxillae, manipulate the food so it can be swallowed. The **labium**, or lower lip, contains a tongue and salivary glands.

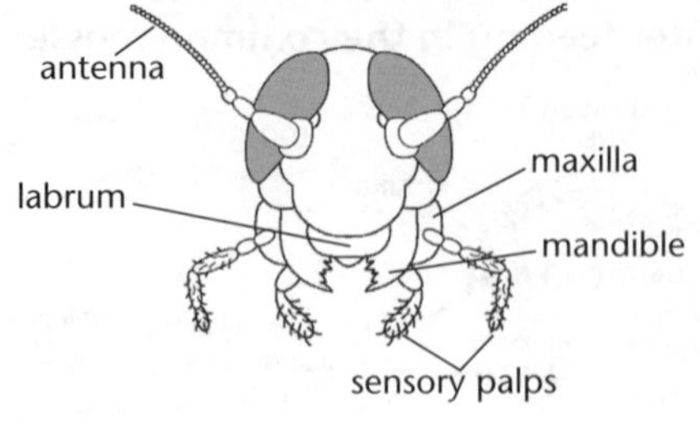

Mouthparts of the cockroach.

The deer

The **incisor teeth** nip off the vegetation against the bony pad. Food is temporarily stored in the gap behind the incisors (the *diastema*) before it is crushed and ground by the premolar teeth and **molar** teeth. Canine teeth are absent.

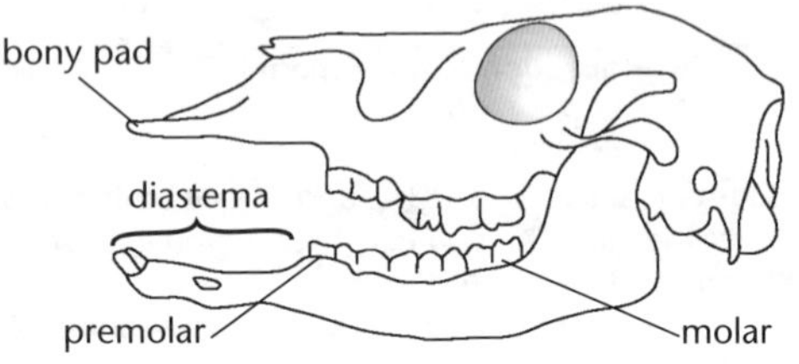

Skull of the deer showing teeth arrangement.

Digestion

Digestion involves the chemical breakdown of food. The main agents of chemical digestion are **enzymes**, which catalyse the breakdown of large molecules into smaller ones. After digestion, food particles are soluble and small enough to be easily transported around the body.

Example

Breakdown of starch

The breakdown of starch involves two enzymes *amylase* and *maltase*. In each reaction, water is needed ('hydrolysis').

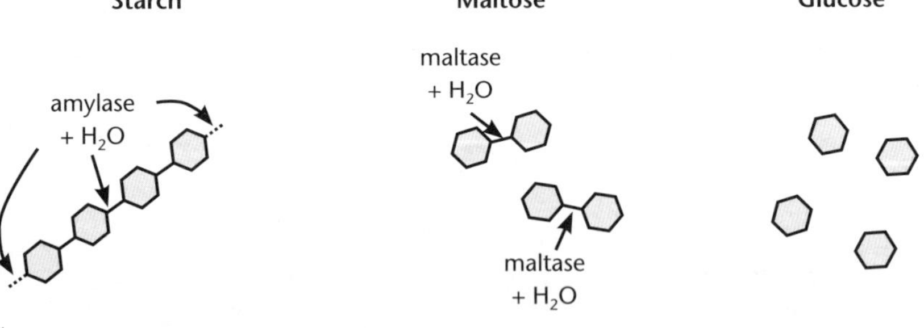

Digestion of starch using the enzymes amylase and maltase.

Sea anemones have a very simple sac-like digestive system.

Example

The digestive system of a sea anemone

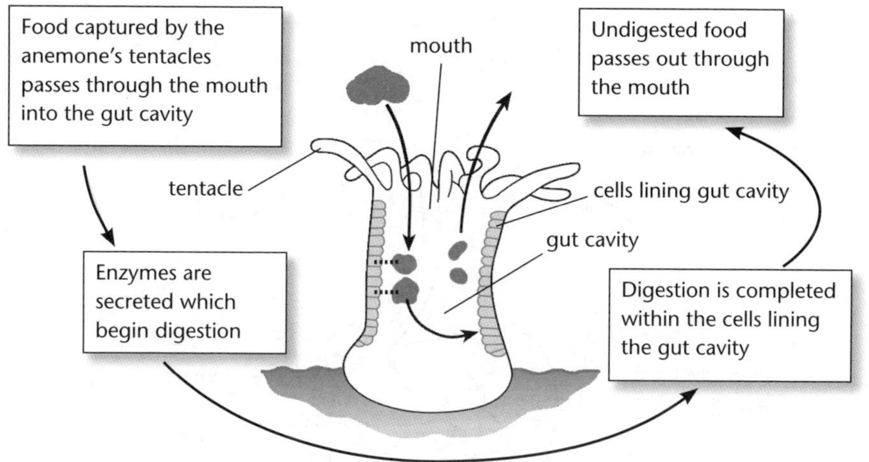

Ingestion, digestion and egestion in the sea anemone.

Digestion using a **sac-like gut** with a common entrance and exit, as in the sea anemone, is comparatively inefficient and only suitable for sedentary or slow-moving animals.
Larger and more complex animals have evolved a tube gut system or **alimentary canal**, which allows a one-way passage of food. Areas along a tube gut can become specialised for digestion, absorption and waste removal. Increased food processing efficiency also occurs by increasing the length and surface area of the gut by folding and coiling it inside the gut cavity. This provides a greater surface area for absorption.

Omnivores

Omnivores have a shorter gut system and less complex stomach than herbivores. A **caecum** for the breakdown of plant material may be developed, such as is found in rats.

Humans have no functioning caecum at all – what is left of the human caecum has been reduced over time through 'lack of use' to a small sac off the gut which forms the **appendix**.

Example

Alimentary (gut) system of humans

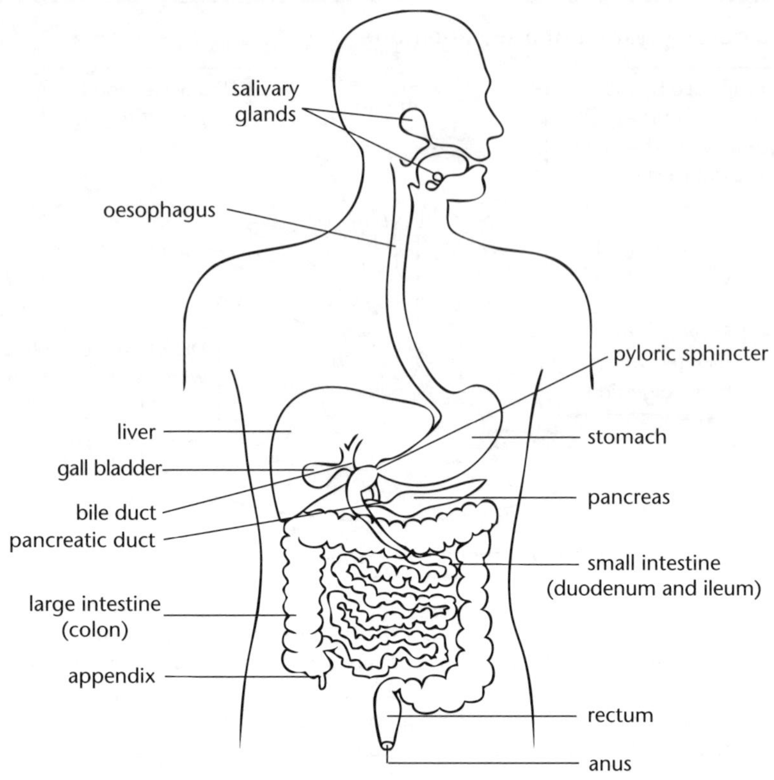

Human alimentary system.

The tongue mixes **saliva** with food and rolls the food into a ball which can be swallowed.

Salivary glands produce water and mucus to lubricate the passage of food, and the enzyme *amylase* to break down starch.

Food passes down the **oesophagus** by muscular action (swallowing) and by gravity. Movement of food throughout the rest of the gut system is by regular muscle movements called **peristalsis**.

The stomach stores and mixes food. Dilute hydrochloric acid kills most micro-organisms and chemically breaks down food into a 'soup' called **chyme**. The enzyme *pepsin* begins protein digestion.

Food takes about 3–4 hours to completely leave the stomach after a meal, a process which is regulated by the pyloric sphincter, a valve which frequently opens as pressure from the chyme increases. This means the small intestine works on a small but continuous supply of chyme from the stomach.

Carnivores

Animal tissue requires little digestion by the gut of a carnivore before it is absorbed, so carnivores have a simple, short gut compared with herbivores. Carnivorous mammals, such as the cat, have a digestive system similar to that of (omnivorous) humans, except a cat's small and large intestines are both comparatively shorter – meaning the overall digestive system is shorter.

Herbivores

Because plant tissue requires complex digestion before it can be assimilated, herbivores have a very long gut system compared with carnivores. Herbivorous mammals, such as the rabbit, have a digestive system similar to that of the (omnivorous) human, except a rabbit's small and large intestines are both comparatively longer, meaning the overall digestive system is longer. In herbivore gut systems, either the stomach (eg ruminants such as cows, sheep) or the caecum (eg rabbits, rodents) is well developed and houses the mutualistic bacteria essential for the digestion of the cellulose cell walls of plant material.

Example

Alimentary system of a grasshopper

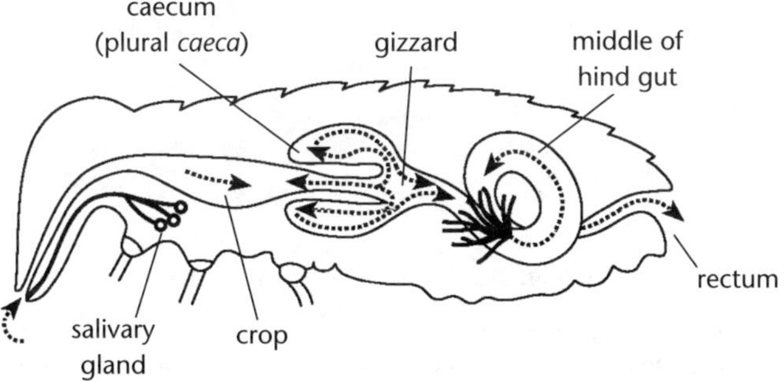

Complex gut of a herbivore, the grasshopper.

The **gizzard** is lined with 'teeth' which grind up large particles and return them to the **crop**, where most of the digestion occurs. **Caeca** contain bacteria and micro-organisms to break down cellulose.

Cows have a huge, four-chambered stomach for processing plant material. It takes up to 75% of the gut cavity and is up to 200 litres in volume. It contains cellulose-digesting bacteria.

Example

Gut of the domestic cow

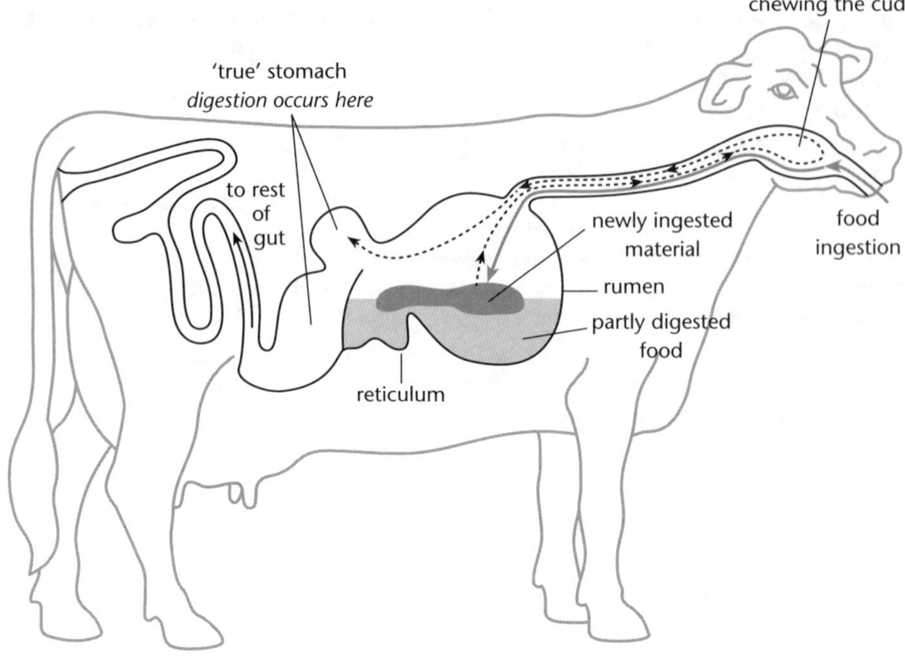

Stomach system of the domestic cow.

After ingestion, the food is partly digested by mutualistic bacteria in the **rumen** and **reticulum**. It is then returned to the mouth for further mechanical breakdown by the teeth – 'chewing the cud'. This cud re-enters the oesophagus but bypasses the rumen and reticulum to enter the 'true' stomach compartments, which contain acid and protein-digesting enzymes.

Digestion in humans

The duodenum is the first part of the small intestine. Its function is to digest foods.

- Protein digestion begun in the stomach is completed with enzymes from the **pancreas**.
- Carbohydrate is broken down by enzymes such as maltase, sucrase and amylase produced from the pancreas and the intestinal glands (cells lining the first part of the duodenum). Cellulose cannot be broken down so it passes along the small intestine as roughage.
- Fat is emulsified (broken into small droplets) by **bile** from the **gall bladder** in the **liver**. Lipase from the pancreas breaks down fat molecules into fatty acids and glycerol. Bile and alkaline secretions from the intestinal glands neutralise acid from the stomach in the small intestine.

The appendix lies between the **ileum** and **colon**. An infection of the appendix causes appendicitis. If the infection ruptures the appendix, peritonitis develops – the whole gut may become infected.

Absorption in humans

The ileum is the second and largest part of the small intestine. Its function is to absorb digested materials.

Efficient **absorption** of food molecules occurs in the ileum because of its very large surface area from infoldings of the wall of the ileum called **villi** (singular villus).

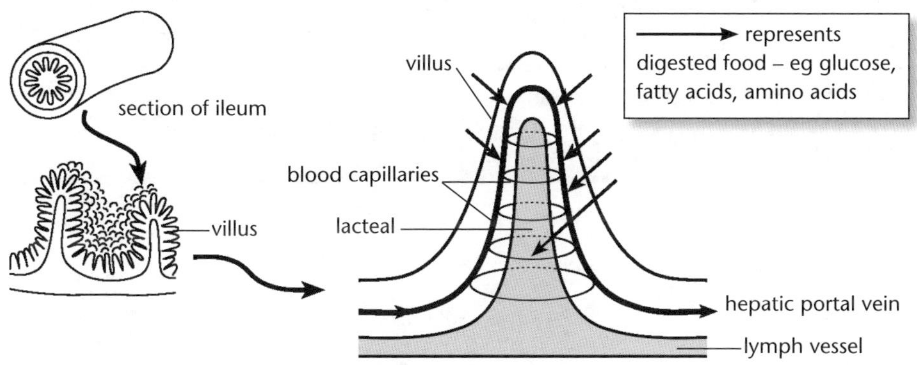

Absorption of food in the ileum.

Fatty acids pass into villi and along lacteal vessels within villi, which connect with vessels of the **lymph system**. All other molecules pass into blood capillaries within each villus. Blood is taken from villi by the **hepatic portal vein** to the liver, where food molecules are altered for storage, detoxified, or further broken down to be used by body cells.

Indigestible food that has not been absorbed passes into the large intestine as **faeces**.

Egestion in humans

The large intestine (or colon) forms faeces from indigestible wastes (eg fibrous plant material, including skins and husks). Useful materials (such as water, enzymes and minerals), diffuse out of the colon to surrounding blood vessels, making faeces quite solid.

Faeces are stored in the **rectum** and are egested through the **anus** when the **anal sphincter** opens.

Because of the high fibre content of plant material, herbivores produce large quantities of faecal material compared with omnivores, which produce more than carnivores (carnivore diet typically lacking in fibre).

Unit 11.2 Activity 5A: Nutrition in animals

1. Define the following terms:
   a. Ingestion.
   b. Digestion.
   c. Absorption.
   d. Egestion.
2. Distinguish between mechanical and chemical digestion.
3. Explain how ectoparasites are adapted to ingesting their food.
4. Explain the differences between the teeth of mammalian herbivores, carnivores and omnivores.

5. Explain the similarities in structure and function between:
   a. Teeth and a gizzard. b. Crop and a stomach.
6. Explain how filter feeders ingest food.
7. Explain how the surface area of ingested food is increased, and why this aids digestion.
8. Discuss the role of enzymes in the digestion of food.
9. Compare and contrast the digestive systems of herbivorous insects and herbivorous mammals.
10. Discuss diversity in nutrition in three named groups of animals. In your answer:
    - Describe the structure of the digestive system in each group.
    - Explain the ways in which the structures carry out the processes of digestion.
    - Discuss why the systems in these three groups are different.

Unit 11.3 Transport Systems

Topic 1: Transport systems in plants

We suggest that this Topic be studied in conjunction with Topics 1, 2 and 3 in Unit 11.2 Nutrition (see pp. 69, 83 and 93) and Topic 2 in this Unit (see p. 137) because the three processes are sufficiently linked to make their content part of a whole. This Topic deals with:

- The structures involved in plant transport – may be at cell level (eg structure and location of xylem and phloem, root hairs, stomata), or at organ level (eg structure of root, stem, leaves).
- How plant transport is carried out (eg osmosis, diffusion, transpiration pull).
- Reasons for the differences in structure and function between different groups – these could relate to habitat, size, life cycle, way of life (ref. Syllabus p. 14).

All plants are classified into two major groups (phyla):

- Bryophyta (non-vascular plants)
- Tracheophyta (vascular plants having transport tissues called xylem and phloem.

The structures involved in transport of materials in a plant are often also those involved in the *support* of the plant. As plants invaded the land (ie became terrestrial), support structures were needed, as plants could no longer depend on the buoyancy of water for support. Also, diffusion of gases and minerals from the surrounding water was no longer possible. Therefore, the evolution of transport and support systems was necessary to allow plants to colonise the land.

The transport of materials is largely to meet the needs of the photosynthetic process. Water is also required to keep the cell membranes moist, so that *gas exchange* can occur.

Water and minerals enter the plant from the soil into root or root-like structures, and need to be transported up the stem to the photosynthetic leaves or fronds; the sugars made in photosynthesis need to be transported down and around the plant. Transport occurs in the *vascular* tissue – *xylem* and *phloem*. Minerals and organic materials travel in separate tissues. Gases (needed/produced in both *photosynthesis* and *respiration*) diffuse into/out of the leaves and stems through pores in the epidermis.

Bryophyta

This group of plants is the *non-vascular* plants, ie they do not have specialised xylem for the transport of water and minerals or phloem for the transport of glucose. Their small size and simple structure allow them to use cell-to-cell diffusion (a slow process) for movement of water, minerals and glucose around the plant, restricting them to damp habitats.

- Water enters the root-like *rhizoids* by *osmosis*, and then from cell to cell up the plant to the leaf-like photosynthetic structures. This occurs as the *concentration gradient* of water is always lower higher up the plant, as water is used in photosynthesis or lost in transpiration. If the cells were to become saturated (eg during periods of heavy rain), then the *overall* movement of water would stop, as the concentration would be the same between cells (isotonic).
- Water can also move up the outside of the stem by *capillary action* in wet conditions.
- Minerals also enter the rhizoids, but by *diffusion* (or *facilitated* diffusion), then from cell to cell to where they are needed. The direction of movement is in response to the *concentration gradient*, though *active transport* across the cell membrane may also occur.

- The glucose produced in photosynthesis moves in a similar way to minerals.
- CO_2 and O_2 *diffuse* directly into and out of the plant cells from the air, in response to their *concentration gradients*.

Bryophytes, the *non-vascular* plants, live in damp areas and are sufficiently small plants that they have no vascular tissue. Two groups exist – mosses and liverworts (see pp. 58–59).

Tracheophyta

These are the *vascular* plants, ie they have transport tissue in the form of xylem and phloem. Vascular plants are typically much taller than non-vascular plants. Three major groups exist – ferns, conifers and angiosperms (see also p. 59).

Ferns

Ferns have specific transport tissue in the form of xylem and phloem. This allows the ferns to be much larger plants than mosses (eg mamaku tree fern can reach 15 m). In ground ferns, the *stem* is underground and horizontal, swollen with stored food (a *rhizome*). Adventitious fibrous roots come off the rhizome and penetrate the soil. *Fronds* (the photosynthetic 'leaves') arise into the air from the rhizome. In tree ferns, the stem is vertical and resembles a trunk, the fronds form at the top.

- Water enters the roots by osmosis, minerals by diffusion (as for mosses).
- Rhizomes have vascular bundles of xylem and phloem; xylem on the inside, phloem on the outside. Xylem tissue is simple pitted, lignified tracheids, not vessels. Phloem tissue is also simple, with sieve elements.
- Water and minerals travel in the xylem to and up the fronds, glucose travels down the fronds to the rhizome and roots in the phloem.

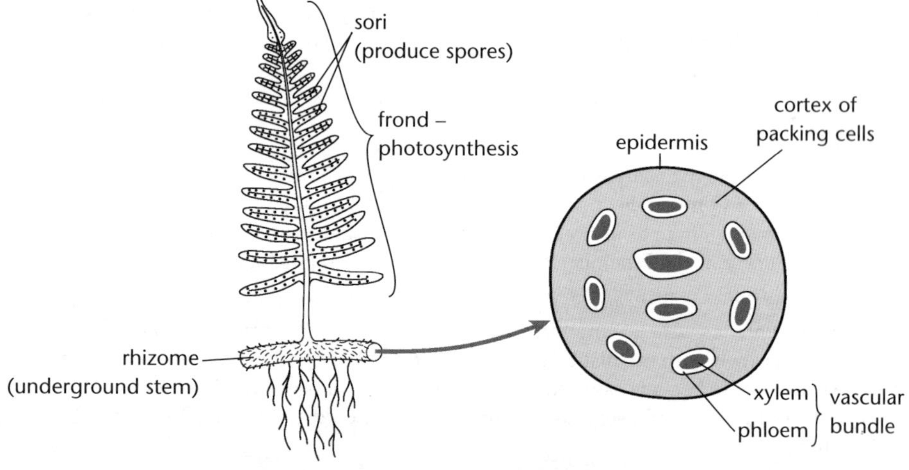

Ground fern – sporophyte generation. Fronds arise directly from underground rhizomes. Fronds bear the spore-producing sori on their underside.

Cross-section through a rhizome, showing the vascular bundles. Each bundle has a solid core of xylem surrounded by phloem; there is *no* cambium.

Angiosperms

Two groups exist – **monocotyledons** ('monocots') and **dicotyledons** ('dicots') (see p. 60).

Angiosperm	Leaves	Stems	Flowers
Monocotyledons Includes all the grasses, lilies, orchids.	Leaves are typically long and thin, with parallel veins (the transporting xylem and phloem) along their length.	Vascular bundles (of xylem and phloem) are scattered throughout the width of the stem, and there is rarely secondary growth of the stem.	Flowers typically have parts arranged in threes (or multiples of three). Seed has only one seed leaf or *cotyledon*.
Dicotyledons Most of the angiosperms.	Leaves are typically broad and have a network of veins.	Vascular tissue is arranged in a circle around the stem, and secondary growth of the stem occurs.	Flowers typically have their parts in fours or fives (or multiples thereof). Seed has two seed leaves or *cotyledons*.

The angiosperms continue the trend shown by the ferns to terrestrial living, and are completely adapted to a terrestrial existence (though some forms have become adapted again to an aquatic existence, eg water lilies). Angiosperms may be large and long lived. Angiosperms are broken up into two major divisions – the monocotyledons (monocots) and the dicotyledons (dicots). Though the two angiosperm groups show differences in the structure and location of tissues in the root and stem, transport in the two groups is essentially the same:

- Vascular tissue is continuous from roots to leaves.
- Water is transported from roots upwards to the leaves in the xylem vessels.
- Organic compounds are transported in the phloem both downwards (leaves to roots) and upwards (roots to areas of growth, eg buds).

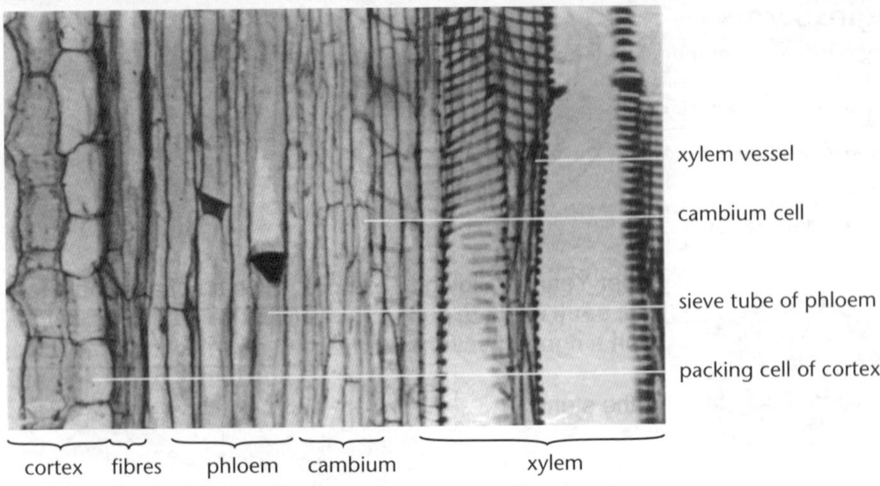

Photomicrograph of a longitudinal section (LS) of a stem through the vascular tissue.

Roots

Dicotyledon roots

Under the epidermis, *dicotyledon* roots have a wide area of tissue (the cortex) made of general 'packing' cells (parenchyma cells), where starch is stored. In the centre of the root is an area called the *stele*, which contains the xylem and phloem.

- The xylem begins as a central, star-shaped bundle, with clusters of phloem in the 'arms' of the star.
- Thickening of the stem results in a complete round centre of xylem, surrounded by a complete circle of phloem.
- The outer layer of the stele is the **endodermis**, a ring of cells which functions to control the entry of minerals from the soil into the xylem.

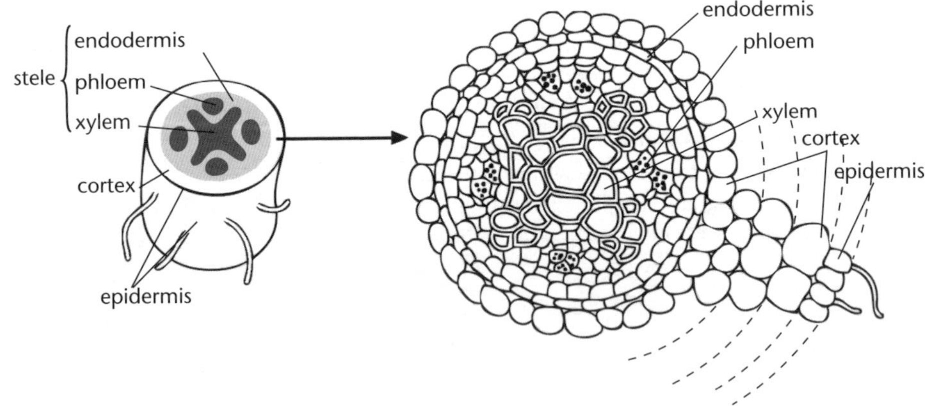

Transverse section of a dicot root.

Monocotyledon roots

Under the epidermis, *monocotyledon* roots also have a wide cortex for starch storage.

- The stele is centrally placed, with clusters of phloem and xylem alternating in a circle around the root.
- The endodermis is more thickened and prominent.
- Unlike dicots, there is a central band of packing cells present (the pith) inside the vascular tissue.

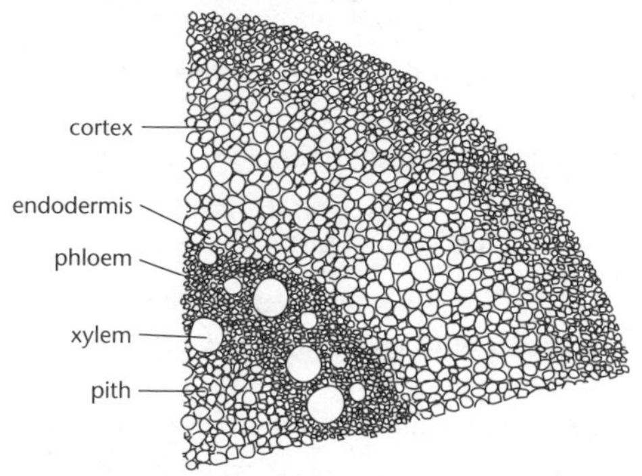

Transverse section of a monocot root.

Water and minerals enter the root from the soil via the *root hairs* (located just behind the growing root tip), and travel across the cells of the cortex to the xylem.
Sugars move from the leaf to the cortex of the root in the phloem.

Stems

The main difference between dicot stems and monocot stems is in the arrangement of the vascular tissue and the presence or absence of secondary growth.

Dicotyledon stems

In the stems of dicotyledons, the xylem and phloem form *vascular bundles* in the young plant.

Xylem is on the inside, and phloem on the outside of each vascular bundle, often with a cap of fibres (for support).

- Sandwiched between the xylem and the phloem is a cell layer called the **cambium**. Cambium cells are *meristem cells* (ie they are capable of undergoing *mitosis* to form new cells – this is how the stem is able to widen).
- Between the vascular bundles and the epidermis, is a layer of packing cells (the *cortex*); at the centre of the stem is another layer of packing cells (the *pith*).

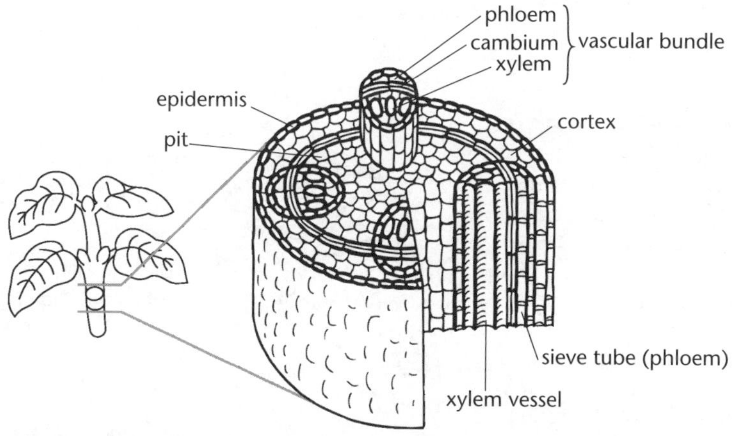

Section of a one-year-old stem of a dicot showing the arrangement of primary tissues.

As the young plant grows, the cambium cells divide to form a complete ring around the stem.

- As division continues, the cells made on the inside of the ring develop into xylem; the cells on the outside of the ring develop into phloem.
- Eventually, complete rings of both xylem and phloem form.

As the xylem is laid down on the inside, the cambium and tissues outside it are pushed outwards and the stem gets wider – this is *secondary growth*.

- The old xylem fills with deposits of resins and other materials, and is no longer functional in the transport of water and minerals; it becomes *wood*, which provides support in trees (and smaller woody plants).
- The old phloem and cortex get crushed as the stem widens.

The only functional xylem and phloem in the older dicot stem are those on either side of the cambium. Therefore, all the living tissue, including the transport tissue, is on the *outside* of the stem. This is why:

- Large, old trees, can become hollow and still survive.
- 'Ring barking' a tree will kill it.

In the outer cortex, a second meristem cell layer (the **cork cambium**) becomes functional, producing **cork** cells that combine with the dead phloem to form protective **bark** in place of the epidermis. The different rate of development and size of xylem from winter to spring/summer can often be seen as *growth* (or **annual**) **rings** across the trunk.

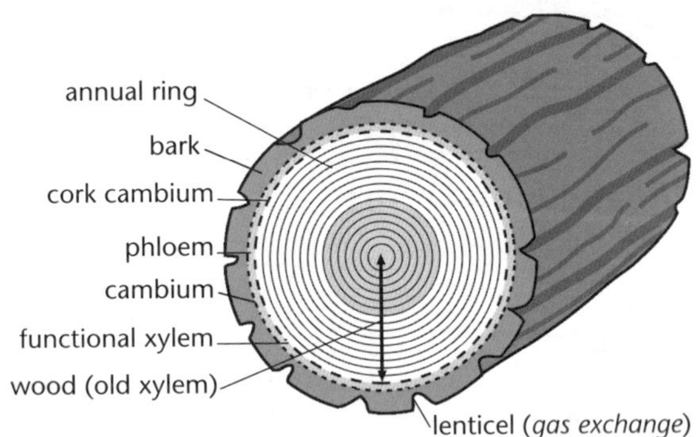

Some dicot stems remain green (herbaceous plants) and do not develop wood and bark. For their support, they are dependent on:

- The lignified xylem.
- Thick-walled fibres.
- Turgor pressure in the packing cells.

Monocotyledon stems

In monocot stems, the vascular bundles of xylem and phloem do not form a circle, but are scattered throughout the cortex and pith. The xylem is on the inside; a cap of fibres (for support) may be outside the phloem. There is no cambium between the xylem and phloem, so the monocot stem does not undergo the thickening and wood-forming process of the dicot stem.

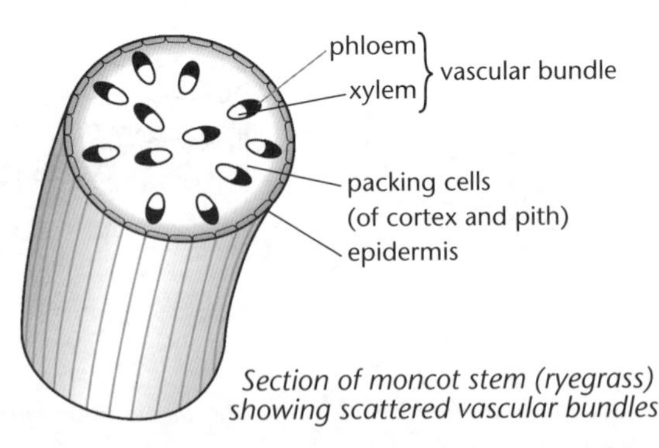

Section of moncot stem (ryegrass) showing scattered vascular bundles.

The monocot stem is therefore dependent on fibres and turgor pressure of cells for support (there are a few exceptions – such as bamboo, cabbage trees).

Leaves

Water and minerals travel up the xylem in the stem to the leaves by a combination of pushing (root pressure) and pulling (capillarity, cohesion, transpiration) processes.

Sugars travel both up and down the phloem by translocation.

Dicot leaves usually have a broad, flattened shape, with a central vein running along the length of the blade and a network of veins running towards the outer edges.

Monocot leaves are usually narrower and more spear-shaped, with parallel veins running the length of the leaf.

The *veins* contain the xylem and phloem, and are continuous with those of the stem. The veins pass through the photosynthetic mesophyll cells. Water passes from the xylem into the cells by osmosis, minerals by diffusion. The organic compounds from photosynthesis diffuse into the phloem and travel to where they are needed in the plant or to the storage sites. Gases enter the leaves through the *stomata*:

- In dicot leaves, stomata are usually confined to the lower epidermis (to reduce water loss).
- In monocots, stomata may be found on both the upper and lower epidermis.

Large *air spaces* occur in from the stomata and surround the mesophyll cells – this allows rapid diffusion of gases between the cells and the surrounding air. Water vapour also diffuses out through the stomata.

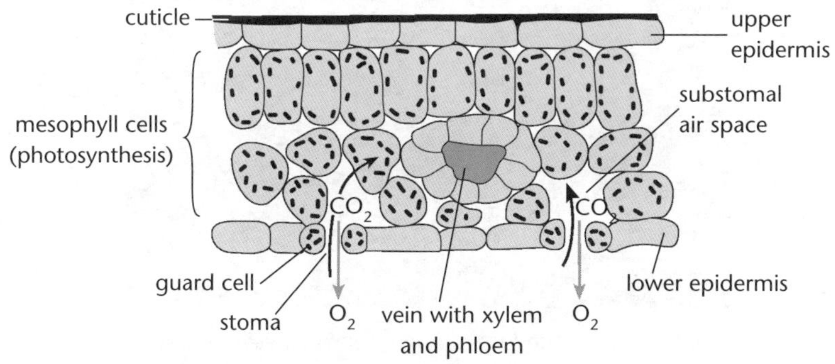

Cross-section of a dicot leaf.

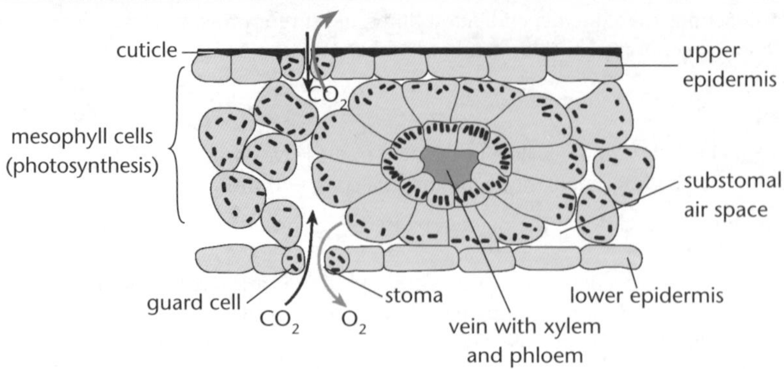

Cross-section of a monocot leaf.

Methods of transport

Minerals and water

Minerals and water are transported throughout plants in xylem tissue.

Cells of mature xylem are dead, as they lose their cell contents, their end walls disappear and the cells form into long, thin vessels, and can be several metres in length. These vessels are continuous from the roots up the stem to the veins of the leaves.

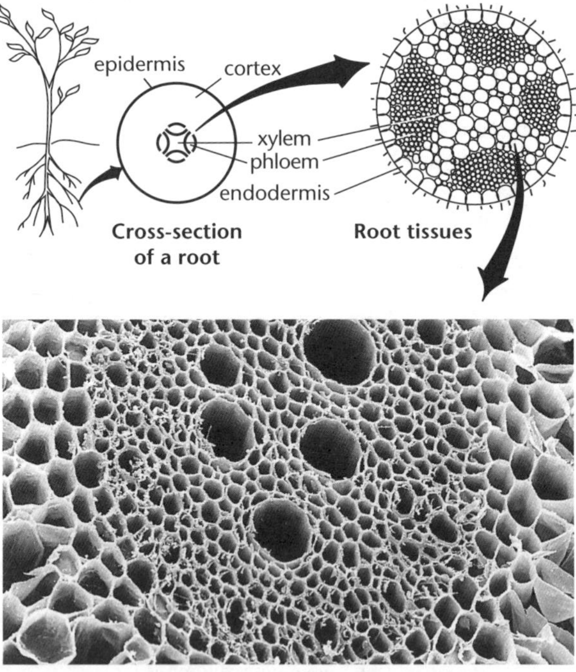

SEM photo – the large lumen cells are xylem vessels, the surrounding small lumen cells are fibres (for support) (magnification x 300).

Xylem vessels form the central core of plant roots. They have a large lumen and their walls are strengthened with **lignin**. Lignin provides support and helps the vessels withstand the suction pressure from long columns of water.

In ferns, the xylem does not develop into vessels, but remains as the more simple *tracheids*. These are long, thin cells with tapered ends. The ends dovetail with those of adjacent tracheids. The walls are lignified and the end walls pitted to allow the movement of water. Unlike xylem vessels, tracheids do not form long tubes and their lumen is smaller.

Water moves by *osmosis* from the soil into xylem, but minerals diffuse into xylem. If the needed mineral is in lower concentration in the soil than in the plant cells, then it may be taken in by active transport. Important minerals are K^+, Ca^{2+}, Mg^{2+}, Fe^{2+}, NO_3^-, SO_4^{2-}, PO_4^{3-}. Most of the absorption of water and minerals occurs through root hairs.

Root hairs, extensions of epidermal cells, give a large surface area for the absorption of materials. Root hairs are located behind the growing tips of roots, ensuring that, as roots grow, new sources of minerals and water are exploited.

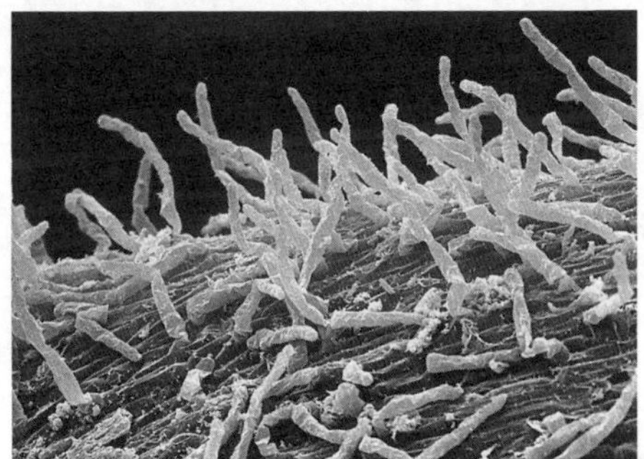

SEM photo of root hairs (magnification x 300).

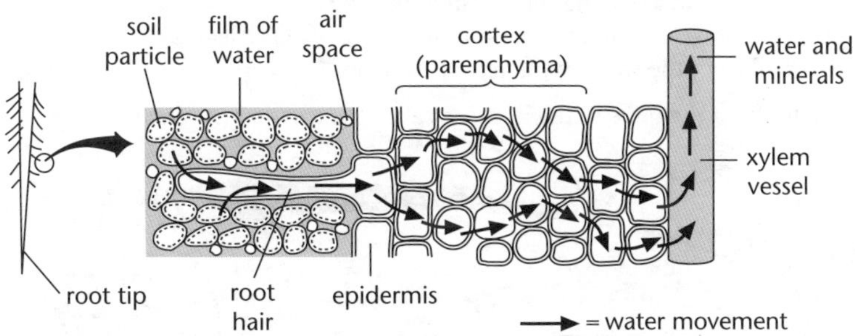

Movement of water and minerals through a root.

Root pressure *pushes* water up the plant. Other processes act to *pull* water upwards:

- **Capillary action** – water adheres to the sides of the microscopic xylem tubes (as it does in capillary tubing) and moves upwards.
- **Cohesion** – water molecules are attracted to adjacent water molecules by electrostatic forces (*hydrogen bonds*) – this gives the water column tensile strength, assisting upward pull.
- **Transpirational pull** – as water evaporates from the leaves, more water moves in to replace it, pulling water up the xylem.

Transpiration

Transpiration is the *loss of water from the leaves of a plant*. (See Unit 11.3 Topic 2, p.137.)

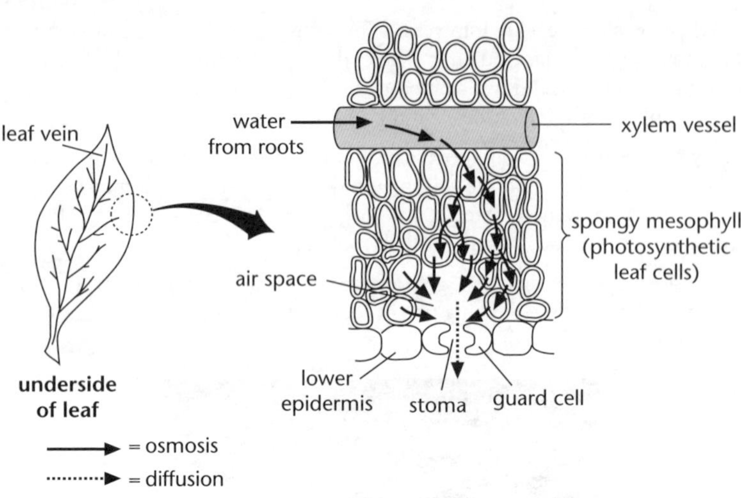

Role of leaf tissues, osmosis and diffusion in transpiration.

Water moves from xylem and from cell to cell via osmosis; water vapour diffuses through the air spaces and out of the stomata. The direction of movement is determined by the concentration gradient.

Water vapour lost from the leaf is replaced by water from xylem vessels. Strong attractive forces occur between water molecules, and, as a water molecule leaves the xylem, another water molecule is pulled along the xylem vessel to replace the water molecule that has been 'lost'. Water molecules pulled along the xylem in turn pull other water molecules. Since a continuous column of water exists in xylem from the roots to the leaves, water is moved by this transpirational pull from the roots to the rest of the plant.

The flow of water from the roots to the leaves is called the *transpiration stream*.

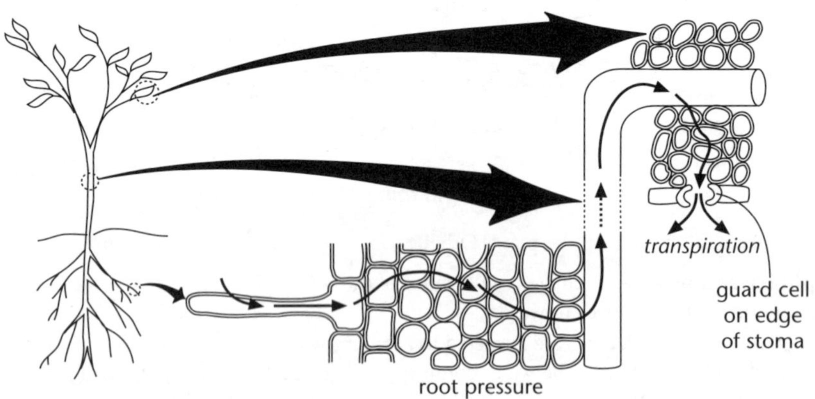

Relationship between roots, stem and leaves in the transpiration stream.

Besides moving water and minerals, transpiration also helps to cool plants because heat energy is lost as it is used to convert liquid water into water vapour, which then leaves through stomata.

Transport of organic compounds

The transport of organic materials ('phloem sap'), occurs in the phloem, and is called **translocation**.

Phloem occurs in vascular bundles, or as a ring of cells near the outside of a stem or branch, just under the bark.

Phloem tissue is made up of **sieve tube cells** and **companion cells**.

Sieve cells contain cytoplasm but no nucleus – their activity is probably controlled by the companion cells. Sieve cells form continuous tubes throughout the plant. The end wall of each sieve tube cell has become perforated (a sieve plate forms) to allow the flow of sap. The sieve tubes take organic nutrients and sap to where they are needed or to storage organs (such as the tubers of potatoes and the stem of sugar cane, the roots of carrots, parsnips and beetroot, and the fruits of bananas, pears, etc).

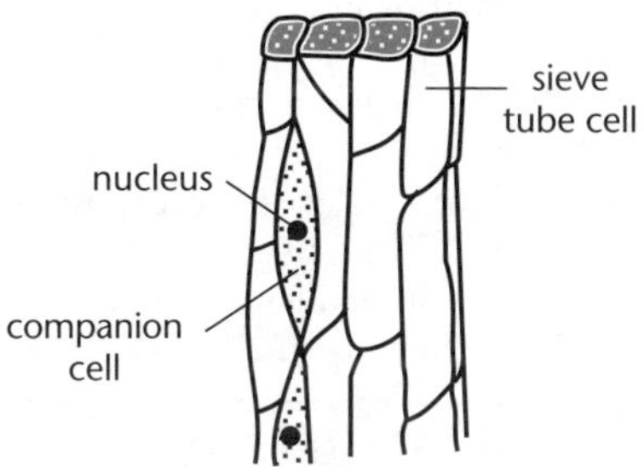

Longitudinal section of phloem tissue.

The largest component in the sap in phloem is the sugar sucrose (approx one-third); sucrose makes sap viscous (thick). The sugars are dissolved in water; hormones and amino acids may also be being transported in the sap. The sugar is stored as starch (which is insoluble). Translocation can be partially explained by the *pressure flow hypothesis*:

- *Pressure* – organic nutrients enter sieve cells by *diffusion* or *active transport* from photosynthesising cells, and water enters sieve cells by *osmosis*, because the sieve cells now have a low water concentration due to the presence of organic nutrients.
- *Flow* – water and organic nutrients move to the area of the plant where they are required. As the nutrients leave the sieve cells, they raise water concentration inside the sieve cells, causing water to leave the sieve cells as well by osmosis.

The movement of materials in phloem is still not completely understood. It is known that flow in phloem can be in either direction, though is always from a sugar source (eg roots from storage, leaves from photosynthesis) to a sugar sink, ie an area of need (eg developing flowers, fruit). Flow is too fast to be accounted for by simple diffusion. Flow needs energy – the transport is an active process. Differences in pressure between sugar sources (high) and sugar sinks (low) may account for transport of sugars from source to sink.

Transport of gases

Photosynthesis

Photosynthesis requires CO_2 and produces O_2 as a waste product.

- CO_2 *diffuses* from the air through the stomata into the leaves and through the air spaces that surround the mesophyll cells. It continues its diffusion across the moist membrane of these cells and into the chloroplasts that lie in the cytoplasm close to the membrane.

As CO_2 is used in photosynthesis, a concentration gradient is maintained from the air to the chloroplasts, so the direction of movement of CO_2 is inwards.

- O_2 produced in photosynthesis reverses this diffusion pathway, moving outwards as its concentration gradient is the reverse of that of CO_2.

In the green stems of herbaceous plants, the outer layer of cells in the cortex have chloroplasts (chlorenchyma cells), so CO_2 diffuses through the pores in the epidermis (*lenticels*) to reach the cells; the diffusion of O_2 is the reverse of that for CO_2.

Respiration

All the living cells in a plant need energy for their processes – this is released in (aerobic) respiration occurring in the mitochondria. Aerobic respiration requires O_2 and releases CO_2 as a waste product. Photosynthesis occurs only during daylight hours, respiration occurs all the time. Therefore, during daylight, much of the O_2 released in photosynthesis can supply respiration, and much of the CO_2 released in respiration can supply photosynthesis. During the night, photosynthesis does not occur; therefore, the concentration gradients of the two gases are reversed, so their diffusion pathways are also reversed. This results in O_2 diffusing into the plant through the stomata and lenticels to enter the cells and their mitochondria for respiration, while the CO_2 produced in respiration diffuses in the opposite direction, *out* of the plant.

Unit 11.3 Activity 1A: Transport of materials in plants

1. Describe the *tissues* involved in vascular plants in the transport of:
   a. Water and minerals.
   b. Sugars.
2. Describe the *processes* involved in the transport of the following in vascular plants:
   a. Water.
   b. Minerals.
   c. Sugars.
   d. Gases.
3. Describe how the angiosperm leaf is adapted to transport essential materials.
4. Explain why mosses are 'non-vascular' plants.
5. Explain the role of transpiration in the movement of water in plants.
6. Discuss the need for, and the processes involved in, the transport of O_2 and CO_2 in plants.
7. Discuss the need for, and the processes involved in, the transport of water in plants.
8. Discuss diversity in transport in three named groups of plants. In your answer:
   - Describe the structures associated with transport in each group.
   - Explain the ways the structures carry out the processes of transportation.
   - Discuss why the systems in these three groups are different.
9. The source of a plant's new cells is a type of plant tissue called its:
   A. Tracheid **B.** Protostem **C.** Cortex
   D. Mesophyl **E.** Meristem

10. The patch of actively dividing cells that is found near the tip of roots and shoots of plants is the:

A. Lateral meristem **B.** Apical meristem **C.** Intercalary meristem
D. Tracheid **E.** Vascular bundle

11. How does water move from soil to xylem?

A. diffusion **B.** active transport **C.** passive diffusion
D. osmosis **E.** facilitated diffusion

12. Answer True or False to the following questions:

a. The type of leaf that is divided into leaflets is the compound leaf.
True False

b. Fibrous root systems are good at preventing erosion and are found in most dicots.
True False

c. Grasses and other monocots usually have fibrous root systems.
True False

d. Herbaceous plants are those that contain little or no woody tissue.
True False

e. Leaves of monocots usually have netlike venation.
True False

f. The vascular cambium produces xylem in woody plants and the cork cambium produces phloem in woody plants.
True False

Unit 11.3 Activity 1B: Short answer questions

1. What is the function of root hair cells?
2. Explain what causes stomata to close when a plant wilts.
3. Explain how adhesion helps to move water through a plant.
4. What organic substance is responsible for strengthening and preventing the xylem vessels from collapsing under tension?
5. What are the functions of xylem and phloem vessels?
6. Why do flowering plants transport organic compound (sugar) including minerals and amino acids to all their parts?
7. Explain the structural difference between dicot and monocot leaves.

Unit 11.3 Activity 1C: Crossword puzzle

Across

1. Movement of dissolved food substances both up and down plants

3. Columns of living cells in phloem

5. A factor that affects rate of transpiration

6. A non-vascular plant

7. Substances travel though this, in a one-way direction

10. Xylem and phloem tissues together form

12. Process that do not require energy

13. Opens to let in carbon dioxide for photosynthesis

15. A shorten term for all monocotyledonous plants

16. Plant nutrients that travel through xylem

Down

2. Water molecules are attracted to each other

4. Made of columns of living cells

8. Process by which organic nutrients enter sieve cells

9. An opposite process of photosynthesis

11. Through which water enters into the plant

14. Singular form of 13 Across

Unit 11.3 Activity 1D: True or false?

1. Xylem vessels are columns of dead cells and phloem vessels of living cells.
2. Stomata are openings on leave surface that permits gas exchange.
3. Vascular bundles of dicots form circle whilst monocots are scattered throughout the cortex and pith.
4. Minerals and water are transported throughout the plant in phloem vessels.
5. Transpiration is the process of water loss from plant leaves.
6. Respiration and photosynthesis releases O2 and CO2 as a waste product respectively during gas exchange.

Unit 11.3 Activity 1E: Multiple choice

1. The movement of water from roots to leaves is called
   A. transpiration
   B. transpiration stream
   C. translocation
   D. water potential gradient
   E. evaporation
2. Water moves up a tall tree against the force of gravity. Which of the following mechanism(s) moves the water upwards?
   A. cohesion
   B. tension
   C. adhesion
   D. all above mechanisms
   E. mechanisms A and C only
3. Which of the following is/are the most important cell type(s) in phloem tissue for transport?
   A. sieve tube elements
   B. companion cells
   C. phloem fibres
   D. phloem parenchyma
   E. both A and B are the most important
4. Flowering plants have distinct parts which enable them to adapt and survive in their growing environments. What are these distinct parts?
   A. roots
   B. stems
   C. leaves
   D. flowers
   E. all of the above

5. Flowering plants are divided into two groups – dicotyledgon (dicots) and monocotyledon (monocots). The dicotyledon and monocotyledon are different in their structure and location of tissues in certain parts. What is the major difference between dicot and monocot stems?
   A. organic compound (sugar) is only transported from leaves to the roots through phloem vessels in stems of monocots
   B. xylem and phloem vessels in dicots form vascular bundles in mature plants
   C. arrangement of xylem and phloem vessels and presence or absence of secondary growth
   D. vascular bundles of monocots form circle whilst those of dicots are scattered
   E. none of the above
6. The process involving transportation of organic materials in phloem vessels is called
   A. transpiration
   B. diffusion
   C. respiration
   D. translocation
   E. osmosis
7. Water and minerals travel along the stem to the leaves by a combination of
   A. osmosis and diffusion processes
   B. pushing and pulling processes
   C. transpiration and photosynthesis processes
   D. respiration and translocation processes
   E. none of the above

Unit 11.3 Transport Systems

Topic 2: Transpiration

Transpiration is a process like sweating whereby plants lose water vapour. It is part of the water cycle. It is recommended that students conduct simple experiments to explore the differences in water loss from a range of leaf surfaces (see Syllabus p. 14). This Topic deals with:

- The structures involved in transpiration – may be at cell level (eg structure and location of stomata, xylem, root hairs) or at organ level (eg structure of roots, stem, leaves).
- How and why the processes involved in transpiration are carried out; how water loss is controlled.
- Reasons for the differences in structure and function between different groups – these could relate to habitat, size, life cycle, way of life.

We suggest that this Topic be studied in conjunction with Topics 1, 2 and 3 in Unit 11.2 Nutrition (see pp. 69, 83 and 93) and Topic 1 in this Unit (see p. 121) because the processes are sufficiently linked to make their content part of a whole.

Transpiration *is the loss of water from the leaves of the plant.* It is a natural (ie passive) process, in that water will vaporise with increasing temperature and water vapour will diffuse in response to a concentration gradient. Therefore, water entering the leaf from the xylem will not only enter the mesophyll cells for photosynthesis, but will vaporise in the air spaces and diffuse out of the *stomata* in transpiration (the rate of water loss will increase with increasing temperature). As a plant needs water for photosynthesis and to keep cell membranes moist for gas exchange (for photosynthesis and respiration), mechanisms are needed to reduce the loss of water and so keep these essential life processes going.

Transpiration assists the plant by:

- Being the main cause of the movement of water from the roots to the leaves (*transpiration pull*).
- Keeping the plant cool in hot weather, as water draws heat from leaves to vaporise.

Factors affecting transpiration

Since transpiration depends upon the *concentration gradient of water* between the air spaces in the leaves and the atmosphere, any factor that *increases this gradient will increase the rate of transpiration*, and so increase the rate at which a plant loses water.

- *Wind speed* – as wind speed increases, water vapour is blown away from the stomata faster, increasing the concentration gradient between the inside and the outside of the leaf, increasing the rate of transpiration.
- *Humidity* – the more humid the atmosphere, the more water vapour it contains. Therefore, increasing the humidity decreases the concentration gradient between the inside and the outside of the leaf, so decreasing transpiration.
- *Temperature* – increasing temperature increases the rate at which water molecules move, increasing the rate of diffusion. Therefore, increasing the temperature increases the rate of transpiration.
- *Light intensity* – increasing light intensity is associated with increasing temperature (light and heat are forms of energy), therefore increasing the rate of transpiration.
- *Soil moisture* – the more water in the soil, the lower the concentration gradient between the soil water and the root hairs, so there is less net movement in of water. Less water

moving up to the leaves means less water available for transpiration, so the rate of transpiration is reduced. Conversely, if soil moisture is too low, no water will be available for movement up the stem, and no transpiration will occur, and water will move out of the cells resulting in loss of turgor and the plant wilts. (It is thus essential to get the water balance right in horticulture!)

Role of stomata in transpiration

Stomata are the pores in the epidermis layer of the leaf. They are formed by two cells, called **guard cells**, which are able to open and close the stoma.

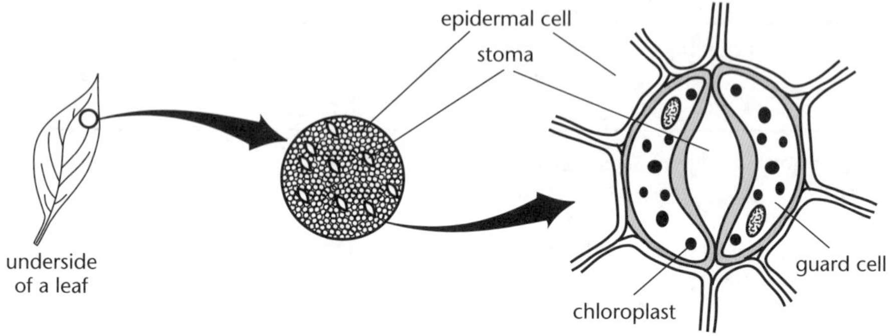

Structure of stoma in a dicotyledon.

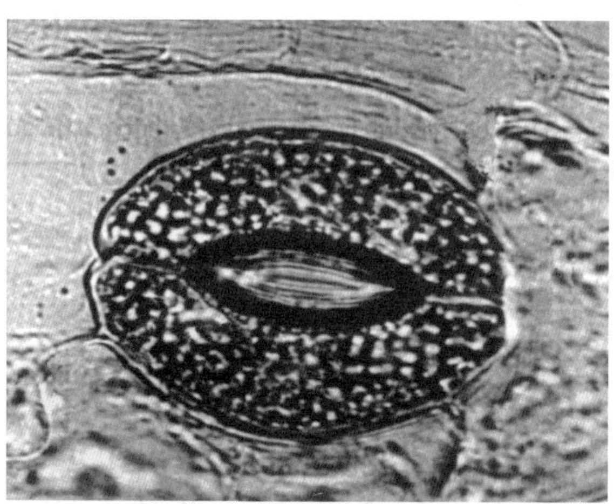

Photomicrograph of stoma and guard cells in the epidermis of a leaf (x 400).

The opening and closing of a stoma is controlled by the *turgor pressure* of the guard cells. Water enters and leaves guard cells in response to its concentration gradient (eg during the daytime, the chloroplasts in the guard cells photosynthesise, increasing the solute levels in the cytoplasm, which draws water into the cells in osmosis). Because the inner walls of the guard cells are thickened, the increase in turgor pressure from the water in the vacuoles causes the cells to bend and form the stoma between them.

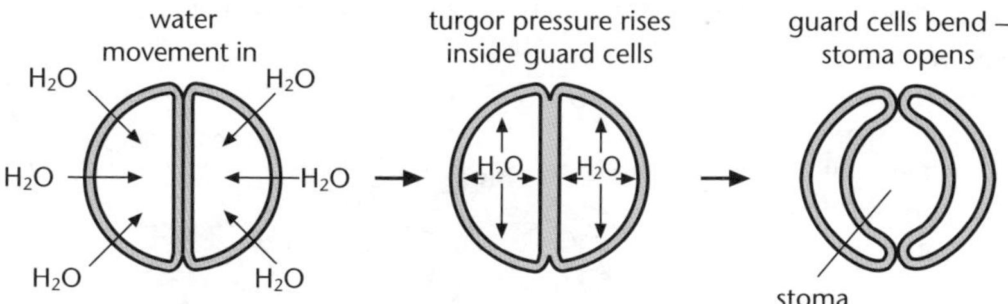

When the concentration gradient between guard cells and the surrounds (ie cells/air spaces) is reversed (eg at night when the chloroplasts no longer photosynthesise and the solute levels in the guard cells fall, or when the water level in the leaf falls with increasing transpiration on hot days), water leaves guard cells and they 'collapse', closing a stoma.

When the weather is hot and /or dry and /or windy, there is a conflict between the need to keep the stomata open to allow entry of CO_2 for photosynthesis, and the need to close them to avoid excessive water loss. Plants display various adaptations to reduce this conflict (ie reducing water loss while trying to maintain photosynthesis rate).

Mesophytes

Mesophytes are plants adapted to living in *temperate* zones such as New Zealand, where the climate does not usually have extremes of heat and/or dryness. Therefore, water loss is typically only a problem in times of high temperatures (such as midday in summer), or high daytime winds. Mesophyte adaptations may include:

- *Waxy cuticle* on the upper leaf epidermis and sometimes the lower epidermis to *reduce* water loss (up to 40% of leaf transpiration can occur through the cuticle if it is very thin).
- Stomata confined to the cooler, *lower epidermis*.
- *Stomata closed* to reduce water loss in periods of extreme heat and/or wind.
- Deciduous trees *shed their leaves* in winter, as water frozen in the soil cannot enter a plant and so no transpirational pull could occur.

Xerophytes

Xerophytes are plants adapted to living in dry, arid places, such as sand dunes, deserts, alpine regions.

Xerophyte adaptations include:

- Desert shrubs have short trunks, numerous branches and small thick leaves (small in stature).
- Many plants in desert conditions have spines or volatile chemical compounds that are used as a defence against water-seeking herbivores.
- Long roots that penetrate deep in the soil/sand to maximise the chances of obtaining water (eg marram grass).
- Water-storage tissue in their stems, eg succulents.

Specific adaptations to reduce water loss and /or store water are concentrated in the leaves, eg:

- *Stomata are sunken* in the lower epidermis – this traps a pocket of still, moist air, reducing transpiration.
- Leaves are *rolled* – trapping a layer of still, moist air around the stomata that are on the lower epidermis, reducing transpiration.
- Leaves may be covered in *hairs*, especially around the stomata – reduces transpiration by reducing the concentration gradient for diffusion by reducing wind movement.

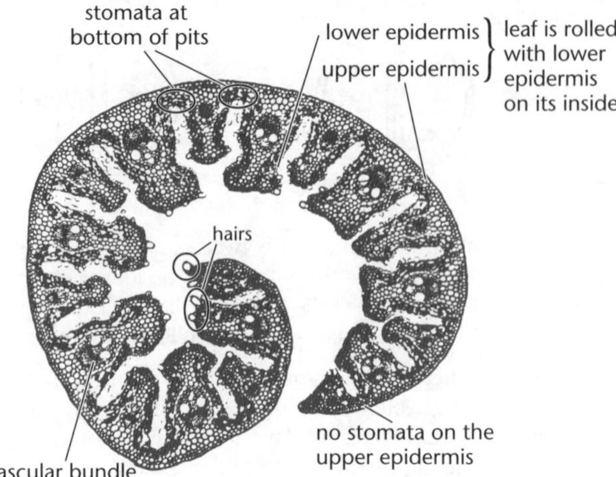

Cross-section of a leaf of marram grass, a sand-dune plant, showing adaptations for reducing water loss – sunken stomata, rolled leaves and hairs.

- A *very thick waxy cuticle* on the epidermis of the leaves, which greatly reduces water loss.
- *Woody support* may be present in leaves (the cell walls become lignified), which stops them collapsing/wilting; a collapsed leaf is exposed to less light than a horizontal leaf.
- Leaves have become *very small* to reduce the SA available for transpiration, eg the needle-like leaves of conifers (photo alongside) and the spines of cacti.

Cacti

In cacti, the leaves have become spines and no longer photosynthesise (they protect against herbivores). It is the stem which photosynthesises. The stem has many packing cells in the cortex that can store water, and the epidermis has a thick waxy cuticle to reduce water loss. Cacti may close stomata /lenticels during the day to reduce water loss, and open them at night to let in CO_2; the CO_2 combines with an organic acid to be stored until daytime when it is released and enters the chloroplasts to make glucose in the Calvin cycle phase of photosynthesis (using the solar energy trapped in the light-dependent reactions of photosynthesis). This is known as *CAM photosynthesis*. However, this reduces the overall extent of photosynthesis, which is why cacti are very slow-growing plants.

The root system of cacti tends to be shallow rather than long; this is to take advantage of any water entering the top layers of the soil /sand from overnight condensation.

Hydrophytes

Hydrophytes are plants adapted to living in an aquatic environment, and so have to cope with an excess (*not* a shortage) of water. Plants may be completely submerged (eg *Elodea*), or have leaves that float on the surface (eg water lilies). (Photo alongside.)

- Minerals, and CO_2 and O_2, are all dissolved in water, but CO_2 and O_2 are in lesser amounts than they are present in an equal volume of air. The minerals and gases can all enter the plant by *direct diffusion*, but this is slower than from air. Therefore, the tissues (both stems and leaves) of aquatic plants tend to *have very large air spaces* around the cells, especially those of the palisade and spongy mesophyll, to facilitate diffusion of needed substances and to aid buoyancy.
- Hydrophytes with floating leaves (such as water lilies), have *stomata* only on the upper epidermis, which allows gas exchange with the air. As water vapour can also diffuse from the upper epidermis into the air, a *waxy cuticle* is present to reduce loss of water vapour (as occurs on hot days).
- The leaves of submerged plants do not have a cuticle or stomata. Their leaves are typically *thin and ribbon-like, to* increase the SA for diffusion into the leaf and to all the cells (eg *Elodea*).
- *Vascular and support tissues tend to be reduced*, as water provides buoyancy and water can enter the plant and reach many of the cells by osmosis (especially if the plant and/or leaves are long and thin).

- *Roots also tend to be reduced* (even absent), as minerals are dissolved in the water that surrounds the plant, and can enter by direct diffusion to reach most of the cells; anchorage of the plant is only important if the plant is exposed to strong currents (eg the weed eel grass has an extensive root system to withstand wave and tidal action).
- *Flowers* usually open above the water (eg water lilies), to facilitate pollination.

Unit 11.3 Activity 2A: Transpiration

1. Define transpiration, and explain how it occurs.
2. Explain why plants need water.
3. Explain why transpiration helps plants.
4. Explain how stomata are able to be opened and closed.
5. **a.** Name the factors that increase transpiration, and explain how each increase is caused.
   b. Name the factors that decrease transpiration, and explain how each decrease is caused.
6. Explain the significance of CAM photosynthesis.
7. Compare and contrast the adaptations of mesophytes and xerophytes.
8. Compare and contrast the adaptations of mesophytes and hydrophytes.
9. Discuss diversity in transpiration in three named groups of plants. In your answer:
   - Describe the structures associated with transpiration in each group.
   - Explain the ways the structures carry out the processes of transpiration.
   - Discuss why the systems in these three groups are different.
10. From the graph below, answer the questions:

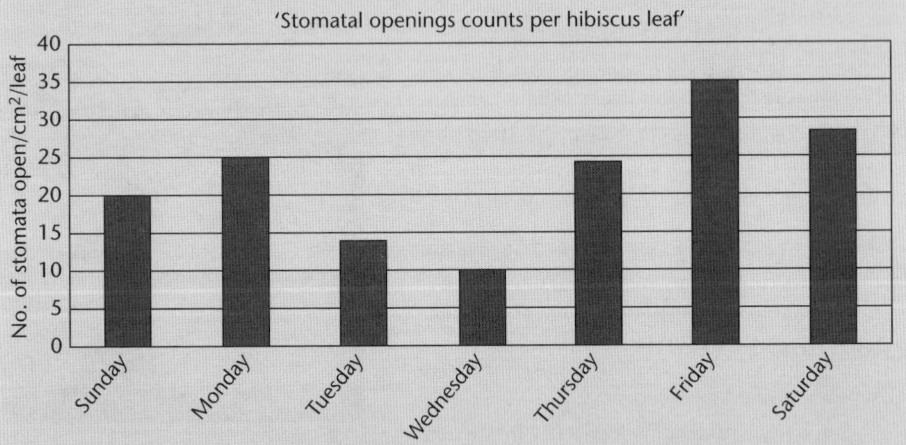

   a. Which day had the highest number of stomatal openings?
   b. If all the leaves have an area of 30cm^2, how many stomata were open on one leaf on Friday?
   c. How many stomata were open on 20 leaves on Wednesday?
   d. How many stomata were open on 20 leaves on Monday?
   e. Name the factors that may have contributed to the readings on Wednesday. Explain why.

Unit 11.3 Transport Systems

Topic 3: Transport systems in animals

This Topic describes the ways in which food, water, blood and oxygen are circulated in animals, as outlined in the content set out on p. 14 of the Grade 11 Biology Syllabus.
The Topic deals with:

- The structure and function of the parts of internal transport systems.
- The process of internal transportation – eg transport medium, pump, vessels, valves, muscles, blood cells, oxygen-carrying pigments, diffusion, circulation.
- Reasons for the differences in structure and function of internal transport systems between different groups – these could relate to mobility and energy needs, size, habitat, life cycle, way of life.

When animals reach a certain size and complexity, they can no longer rely on simple diffusion of substances from the environment into cells and from cell to cell to meet their needs. A *transport system* is required, to:

- Supply all cells with the nutrients (eg glucose, mineral ions, amino acids), water and oxygen, that they require.
- Remove the waste products from cell processes – eg carbon dioxide (from respiration), urea (from deamination of amino acids from protein in the diet).
- Remove heat energy (from cellular respiration).
- Transport hormones from source to target organs (eg the pituitary gland produces follicle-stimulating hormone, which is transported to the ovaries where it controls the production of eggs).
- Carry lymphocytes, phagocytes and antibodies to sites of infection to fight pathogens.

Organisms with no internal transport systems

The simple, two-cell layer of the body of cnidaria (eg sea anemone) encloses a sac-like gut (the 'gastrovascular cavity'); the animals' relative immobility means that *simple diffusion* can meet their needs – therefore they have no transport system.

The flattened platyhelminthes (eg *Planaria*) have a sufficiently large SA:V (Surface Area to Volume) ratio that their cells can obtain and remove gases by simple diffusion with the outside environment, and from cell to cell within the body. A branching gut ensures that all nutrients can be directly supplied to the cells.

Vascular systems

Larger, specialised, active organisms have evolved a **vascular system**. Such systems circulate the transport medium (the 'blood') using a pump (the '**heart**') around the body – hence the term **circulatory system**.

Substances enter the blood from the source organ (eg O_2 from the lungs or gills) and are transported directly to the cells. Needed substances enter the cells (eg O_2, glucose) while waste substances (eg CO_2, urea) leave the cells (via diffusion, facilitated diffusion, active transport).

Respiratory pigments are complex proteins containing either copper, Cu (*haemocyanin*), or iron, Fe (**haemoglobin**), to bind to the oxygen molecules. Respiratory pigments are commonly found in blood and can transport a much greater quantity of oxygen than can a similar volume of water.

- The pigment haemocyanin (found in molluscs, arachnids, crustaceans) becomes a bluish colour when oxygenated. Haemocyanin is able to pick up oxygen under low pressure, so is an adaptation that allows animals from these groups to live in damp or aquatic environments with low levels of oxygen.
- **Haemoglobin** is found in all chordates (in the red blood cells) and is able to carry 60 times more O_2 than an equivalent volume of H_2O. The packaging of the pigment in the red blood cells (RBCs) reduces the viscosity of the blood, allowing for easier flow (compare with earthworms, where haemoglobin is carried in the plasma and their circulation is more sluggish – which reduces their level of activity).

There are two main types of circulatory system:

- *Open* – the blood circulates in a body cavity called the **haemocoel** (eg insects).
- *Closed* – the blood circulates within blood *vessels* (most animals).

Open circulatory systems

Open circulatory systems are found in molluscs (except the highly active squid and octopus) and arthropods. These are small animals, and typically not very active (active insects do not rely on the blood system for the transport of gases), as the blood flow in open systems is more sluggish than that in closed systems. In open systems, the organs are bathed in the blood, so all receive the same amount of blood flow.

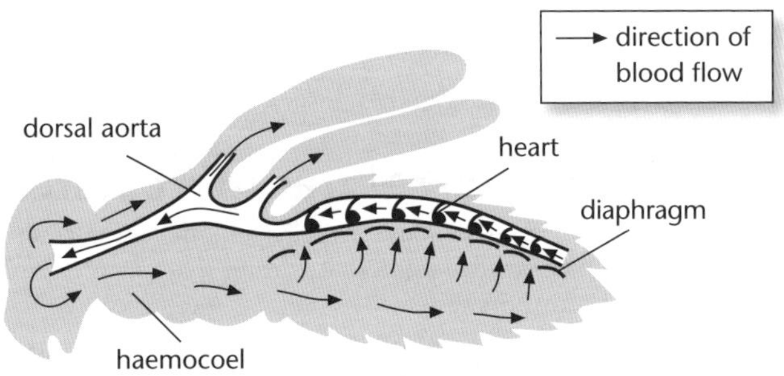

The open circulatory system of an insect.

In insects, the heart pumps blood into the main blood vessel, the **aorta**, which empties into the body cavity. Blood supports and bathes all the internal organs in fluid, returning to the hearts through holes in a diaphragm.

Insect blood also acts as a store of water, and assists in moulting by altering body pressure in different parts of the insect.

The circulatory system of insects does *not* transport O_2 and CO_2, as the tracheal gas exchange system is separate from the blood. The blood of insects, or *haemolymph*, is thick and a whitish colour; it moves sluggishly.

The haemolymph is a mixture of blood and **interstitial fluids**.

Nutrients and wastes are exchanged between the haemolymph and the cells of the body which gets re-circulated back to the heart.

Closed circulatory systems

Closed circulatory systems are found in earthworms and chordates.

Closed systems have the advantage of being able to adjust the flow of blood to individual organs in response to need – eg increased flow to the muscles during exercise, reduced flow to the skin for heat retention in cold conditions (mammals).

In closed circulatory systems, blood and interstitial fluid are physically separated by blood vessel walls and they differ in their components and chemical composition.

All closed circulatory systems share the following features:

- Blood is contained in tube-like vessels that transport it around the body.
- Blood also contains dissolved salts that are transported around the body.
- Blood contains disease-fighting cells and molecules of the immune system.
- Blood is pumped under pressure by a muscular, chambered heart.
- The activity of the system can be adjusted to meet the animal's metabolic demands.
- The system can heal itself from minor injuries by forming clots at the sites.
- The system grows in size as the animal grows.

Earthworms

Earthworms have a simple closed system. The blood transports O_2 and CO_2, and has the oxygen-carrying pigment haemoglobin.

Example

Earthworm closed circulatory system

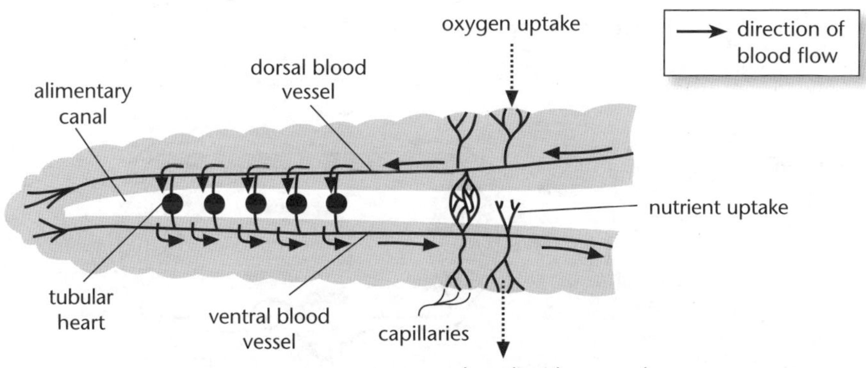

The closed circulatory system of the common earthworm.

Blood is pumped along the **ventral** (*'bottom side'*) **vessel** to the body by five simple tubular hearts. The ventral vessel branches out to supply blood to each body segment. The smallest blood vessels, called **capillaries,** provide intimate contact between blood and body cells, enabling gas exchange, nutrient uptake, etc, to occur.

Blood returns to the hearts along the *dorsal vessel,* moving along by *peristaltic* contractions of its walls. Blood is prevented from flowing backwards by **valves.**

To increase the oxygen-carrying capacity of the blood, earthworm blood contains **haemoglobin,** a red blood pigment. Haemoglobin enables blood to carry many times more oxygen than the same volume of water. O_2 enters the body by diffusing through the moist skin (from air spaces in the soil) and into the capillaries – CO_2 diffuses from the capillaries out through the skin.

The earthworm circulatory system consists of one loop. Oxygenated blood mixes with deoxygenated blood – this is less efficient than double circulatory systems (where oxygenated and deoxygenated blood are kept separate).

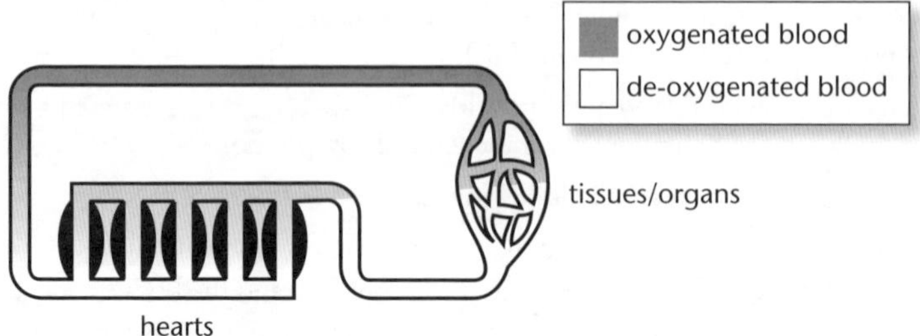

Single-loop circulatory system of the earthworm.

Vertebrates

Circulation in vertebrates has evolved from a single system where blood is pumped by a two-chambered heart (eg fish) to a double system with a three-chambered heart (eg amphibians and reptiles) and a four-chambered heart (eg birds and mammals).

Fish

Fish have a single-loop circulatory system, like earthworms. The two-chambered heart consists of:

- A thin-walled **atrium** to collect blood from the body.
- A thick-walled **ventricle** to pump blood through the gills to the body.

Blood from the ventricle passes to the gills where it is *oxygenated*. From the gills, blood moves to the rest of the body.

Blood loses pressure because of friction between the blood and the blood vessels as it passes through the gills and body, so blood returning to the heart is of lower pressure than blood leaving the heart. Blood flow around the body to the heart is aided by the back muscles, which alternately contract and expand **sinuses** (blood cavities), forcing blood along the blood vessels as the fish swims.

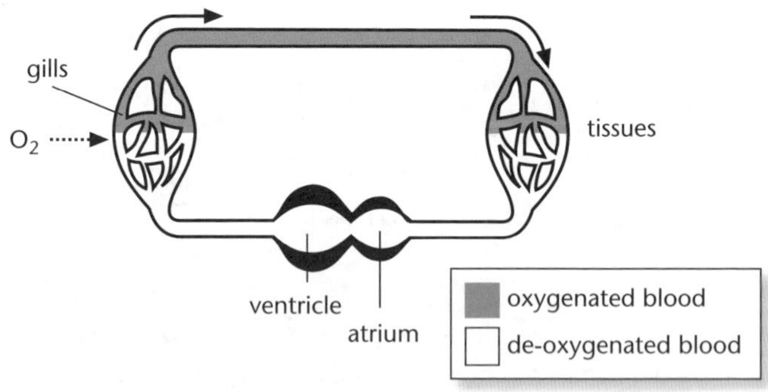

Single-loop, twin-chambered heart circulatory system of fish.

The fish's circulatory system would be inadequate in land vertebrates, because gravity would prevent blood flowing back to the heart. Land vertebrates have two circulatory loops, so blood is pressurised *twice*.

Reptiles and amphibians

Reptiles and amphibians have two circulatory loops – one to the lungs or gills (where blood is oxygenated), the other loop to repressurise this blood to take it to the rest of the body. The three-chambered heart has only one ventricle, so some mixing of oxygenated and de-oxygenated blood occurs. In amphibians, gas exchange may also occur through the skin, which further causes mixing of oxygenated and de-oxygenated blood. Blood to the body is not fully oxygenated, reducing the rate of metabolism and activity.

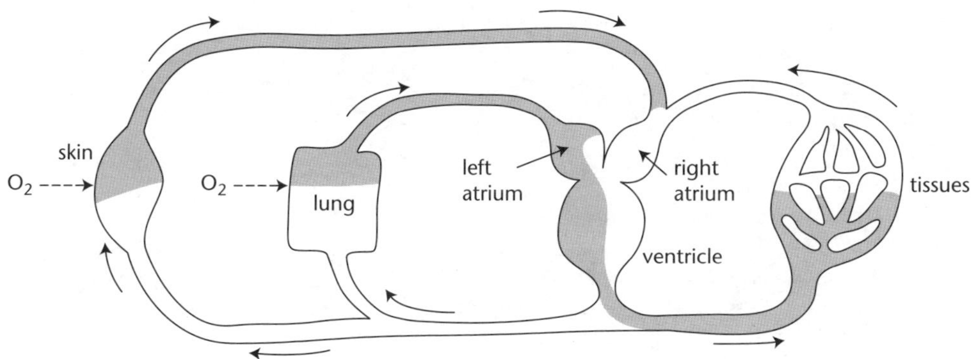

Double-loop, three-chambered heart circulatory system of a frog

Mammals and birds

The circulatory loops of mammals and birds are totally separate. No mixing of oxygenated and de-oxygenated blood occurs – this supports the high metabolic rate demanded by the active lifestyle of these two groups. The four-chambered heart (two **atria** and two **ventricles**) acts as two pumps, with the right-hand side pumping blood to the lungs, the left-hand side pumping blood to the rest of the body.

> In diagrams of the heart, the left atrium and ventricle are drawn as they would appear looking from the body 'outwards', so in diagrams the left side of the heart is drawn on the right and vice versa.

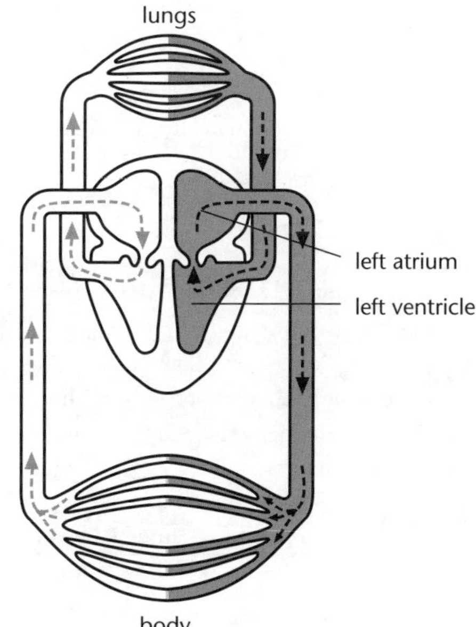

Double-loop, four-chambered heart circulatory system of mammals and birds.

Unit 11.3 Activity 3A: Transport systems in animals

1. One of these statements does not best describe the requirements of the transport system in animals. Which one is it?
   A. to supply most cells with the nutrients, water and oxygen they need
   B. to remove all waste products from cellular metabolism
   C. to carry white blood cells and antibodies to sites of infection
   D. to remove heat energy produced from cellular respiration
   E. to transport hormones from source to target organs

2. Select the organism below that has a need for an internal transport system.
   A. sea anemone
   B. hydra
   C. stingray
   D. jellyfish
   E. corals

3. Which of these statements is not true of respiratory pigments?
   A. they are complex proteins containing copper and iron
   B. the protein complex is known as haemocyanin
   C. the haemoglobin binds oxygen molecules
   D. they can be found in blood
   E. haemoglobin proteins are found in chordates only

4. Which one of these animals does not have an open circulatory system?
   A. snail
   B. squid
   C. oyster
   D. clam
   E. sea slug

5. Two-loop closed circulatory systems are typical of which of these groups of organisms?
   A. fish
   B. birds
   C. mammals
   D. worms
   E. amphibians

6. One of these systems describes the circulatory system that the crowned pigeon (guria) possesses. Which one is it?
   A. double-loop, three-chambered heart
   B. single-loop, twin-chambered heart
   C. single-loop, double-chambered heart
   D. double-loop, four-chambered heart
   E. single-loop, three-chambered heart

Unit 11.3 Transport Systems

Topic 4: The human circulatory system

Authors: Martin Hanson with Beatrix Marjen-Waiin

This Topic explores the circulatory and lymphatic systems in the human body, their functions and structures; and it enables students to identify lifestyle diseases related to the human circulatory system (see 'Knowledge', Syllabus, p. 13). The Topic looks at:

- The blood system – plasma, corpuscles, platelets.
- Human circulation – heart, arteries, veins, capillaries, blood clots, blood groups, giving blood.
- The lymphatic system.
- Diseases of the blood system.

It is recommended that students also read Topic 2: Human responses to micro-organisms (p. 341) in the Supplementary Unit.

The blood vascular system is the body's *transport system*. It carries food (eg glucose and amino acids), oxygen, wastes (carbon dioxide and urea), hormones, defence proteins and defence cells, and heat.

Parts of the blood system

The human blood system consists of five components:

- **Blood** – the material in which substances are transported.
- **Heart** – pumps the blood round the body.
- **Arteries** – carry blood *away from* the heart and distribute it around the body.
- **Veins** – return blood from all parts of the body *towards* the heart.
- **Capillaries** – microscopic vessels connecting the smallest arteries and the smallest veins. Their function is to allow substances to enter and leave the blood.

What blood is made of

Blood consists of blood cells, sometimes called **corpuscles** (about 45% of the blood volume), suspended in a pale straw-coloured liquid called **plasma**.

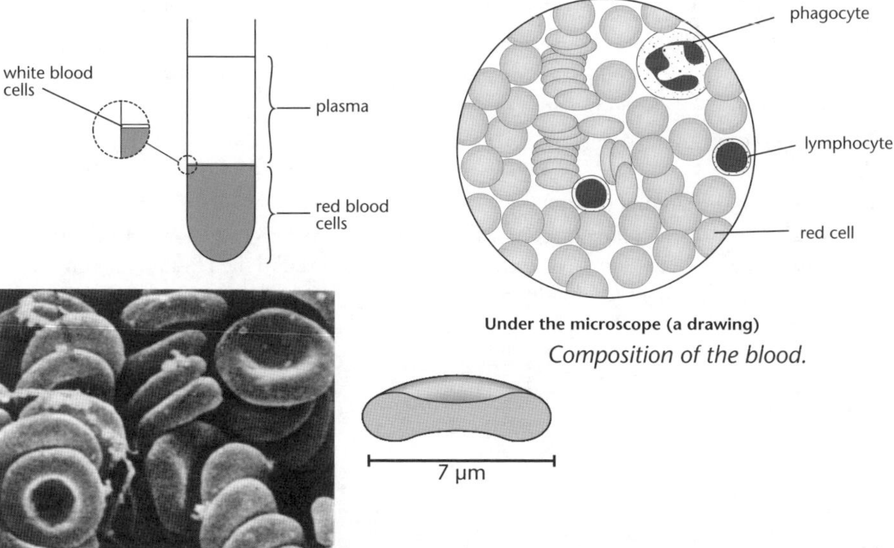

Under the microscope (a drawing)

Composition of the blood.

Under the microscope (a photograph)

A 70 kg adult has about 5.5 L of blood.

Blood is made up of:

- Plasma (fluid) – 55% of the total blood volume.
- Blood cells – 45% of the total blood volume.

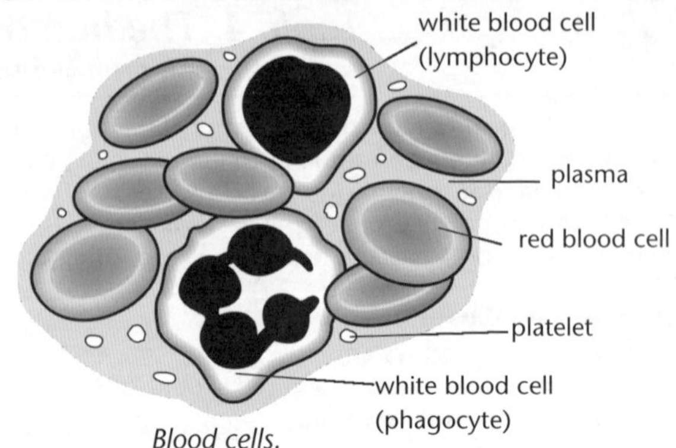

Blood cells.

The most common blood cells are **red blood cells** (5 million per mL of blood). The other two blood cells are **white blood cells** (7 500 per mL) and **platelets** (7 000 per mL).

Blood plasma

Plasma is a straw coloured fluid, made up of:

- Water – 91%.
- Protein – 7.5%.
- Ions, lipids, enzymes, hormones, gases, vitamins, nutrients – 1.5%.

The major blood proteins are:

- Albumin – maintains water levels of the blood.
- Globulin – contains antibodies.
- Fibrinogen – involved in blood clotting.

The composition of plasma constantly changes, as materials enter and leave the capillaries from the surrounding cells.

Plasma is a complex liquid consisting of about 91% water and a wide variety of solutes:

- Inorganic ions – eg sodium, magnesium, calcium, chloride, and hydrogen carbonate.
- Nutrients – eg glucose, amino acids, vitamins.
- Waste products – eg CO_2 and urea.
- Hormones – eg insulin, adrenaline.
- Antibodies.
- Other proteins (eg clotting factors).
- Carbon dioxide – carbon dioxide is transported from the body cells (where it is produced) to the heart and then the lungs (where it is exhaled) dissolved in the plasma. Carbon dioxide dissolves in water to form hydrogen carbonate ions, and most of the CO_2 is carried in the body in this form.

$$CO_2 + H_2O \underset{\text{lungs}}{\overset{\text{rest of body}}{\rightleftharpoons}} HCO_3^-$$

hydrogen carbonate ions

Corpuscles

These are all produced in the bone marrow and are of two general types – red cells and white cells.

Red blood cells (erythrocytes)

These carry oxygen from the lungs to all other parts of the body. In mammals they are biconcave discs and have no nucleus. The absence of a nucleus allows them to contain more **haemoglobin**, a red protein that carries oxygen.

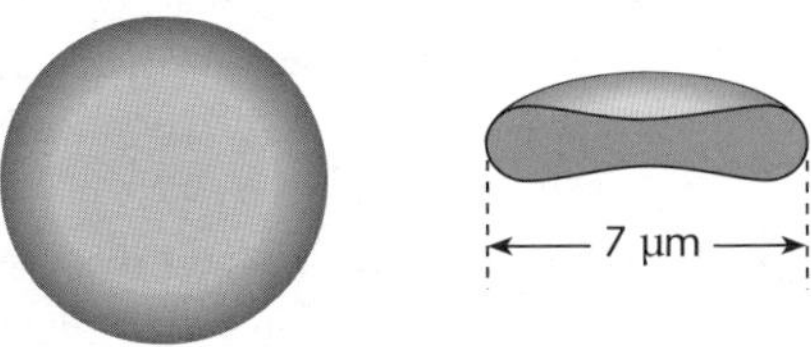

A red corpuscle.

At the high oxygen concentrations in the lung capillaries, haemoglobin combines with oxygen to form *oxyhaemoglobin*. At the same time, the blood changes from dark red to scarlet. At the low oxygen concentrations that occur in all other capillaries, oxyhaemoglobin splits up into haemoglobin and oxygen and the blood becomes dark red again:

$$\text{haemoglobin} + \text{oxygen} \underset{\text{rest of body}}{\overset{\text{lungs}}{\rightleftharpoons}} \text{oxyhaemoglobin}$$

Most organs remove less than half the oxygen from the blood flowing through them, so 'deoxygenated' blood still contains a reserve of oxygen.

Haemoglobin contains iron – one of the main reasons why iron is needed in the diet.

The biconcave shape of the red cells gives them a high surface area compared with their volume. It also enables them to take up a 'bell shape' as they pass along the capillaries, assisting their movement.

> The biconcave shape of red blood cells does *not* enable them to swell without bursting as they absorb oxygen (the increase in volume is minute).

Because they have no nucleus, red cells cannot reproduce or repair themselves, and only live about four months. New red cells are constantly being produced by cell division in the bone marrow.

Worn-out red cells are destroyed in the liver.

In the breakdown of haemoglobin, green-yellow **bile pigments** are produced. These are excreted in the bile and give it its characteristic colour. The iron is not excreted but is stored in the liver and 'recycled' to make haemoglobin in new red cells.

When tissues are bruised, the blood that escapes from damaged blood vessels is broken down at the damage site. The bile pigments that are produced may give a bruise its yellow-green colour.

White blood cells (leucocytes)

These *defend* the body against **pathogens**, or disease-causing organisms. They are produced in the bone marrow and are of two main kinds – **lymphocytes** and **phagocytes**. Both lymphocytes and phagocytes play essential roles in defence against disease. Phagocytes 'eat up' bacteria. One kind of lymphocyte can develop into **plasma cells**, which make defence proteins called **antibodies**.

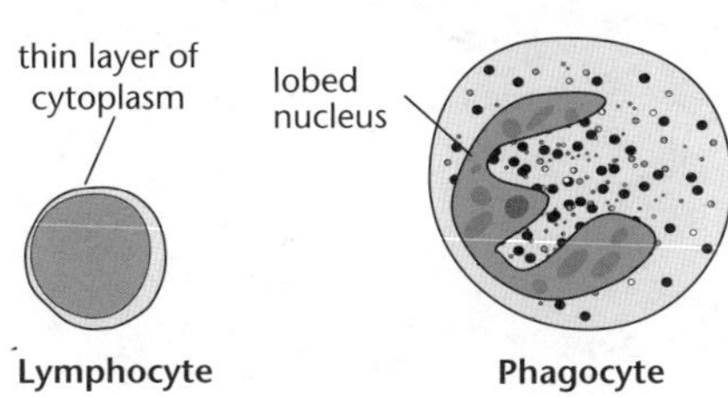

Lymphocyte and phagocyte.

Leucocytes are made in the bone marrow and contain a nucleus and protect the body against disease. *Lymphocytes* are white blood cells that produce **antibodies**. Antibodies are proteins that 'tag' or mark foreign objects in the body (such as *Streptococcus* bacteria that cause sore throats and the pollens that cause hayfever). Foreign objects, which are tagged with antibodies, can be recognised and destroyed by other leucocytes, called **phagocytes**. Phagocytes can alter their shape and move to sites of infection where they engulf pathogens – this is achieved by cytoplasmic streaming and phagocytosis.

The specific region on a foreign object that the antibody recognises is called an **antigen**. Antigens are usually proteins. Each different antigen stimulates the production of a specific antibody that will only recognise that particular antigen (eg the antibody that 'tags' a *Streptococcus* bacterium will not 'tag' any other species of bacteria). White blood cells are not confined to blood vessels. They can leave and travel through the lymph and tissue fluid around cells, seeking out pathogens.

When leucocytes fail to mature properly, a person suffers from **leukaemia**.

Platelets (thrombocytes)

These are not cells, but tiny cell fragments (not much larger than bacteria) that break away from certain cells in the bone marrow. Thrombocytes are involved in blood **clotting**. When a blood vessel is injured, platelets gather at the injury, forming a temporary clot. These platelets release a chemical that converts the soluble blood protein **fibrinogen** into insoluble **fibrin**. Fibrin forms a tangle of fibres at the clot, trapping further blood cells to form a permanent clot.

Unit 11.3 Activity 4A: Blood composition

1. For each of the phrases **1–10**, write the letter **A–J** of the term to which it applies.

Phrase	Applied term
1. Blood cell that can develop into an antibody-producing cell	A. Anaemia
2. Cell fragment important in blood clotting	B. Antibody
3. Cell that eats bacteria	C. Bile pigments
4. Dietary shortage of iron may cause this	D. Haemoglobin
5. Form in which carbon dioxide is carried in the blood	E. Hydrogen carbonate
6. Liquid part of the blood	F. Liver
7. Organ where old red cells are broken down	G. Lymphocyte
8. Oxygen-carrying protein	H. Phagocyte
9. Protein that helps destroy bacteria and viruses	I. Plasma
10. Waste substances produced in breakdown of haemoglobin	J. Platelet

2. The diagram shows components of blood.

Describe the function of each of these components of blood.

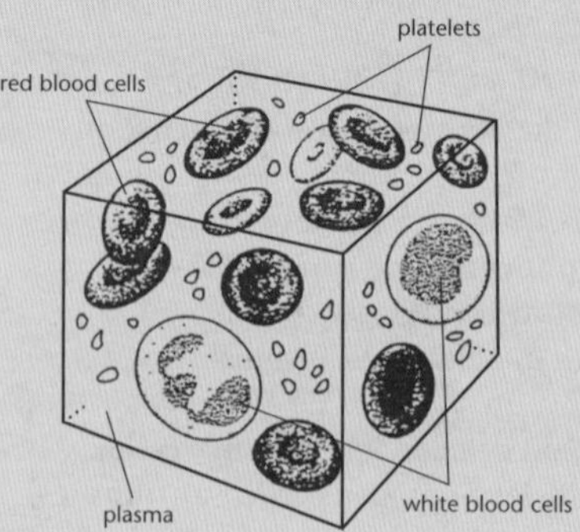

Human circulation

The blood system of a mammal is called a *double circulation* arrangement because blood passes through the heart *twice* in each complete circuit. The right side of the heart pumps deoxygenated blood through the lungs, and the left side of the heart pumps oxygenated blood to the rest of the body. The mammalian heart is therefore a double pump.

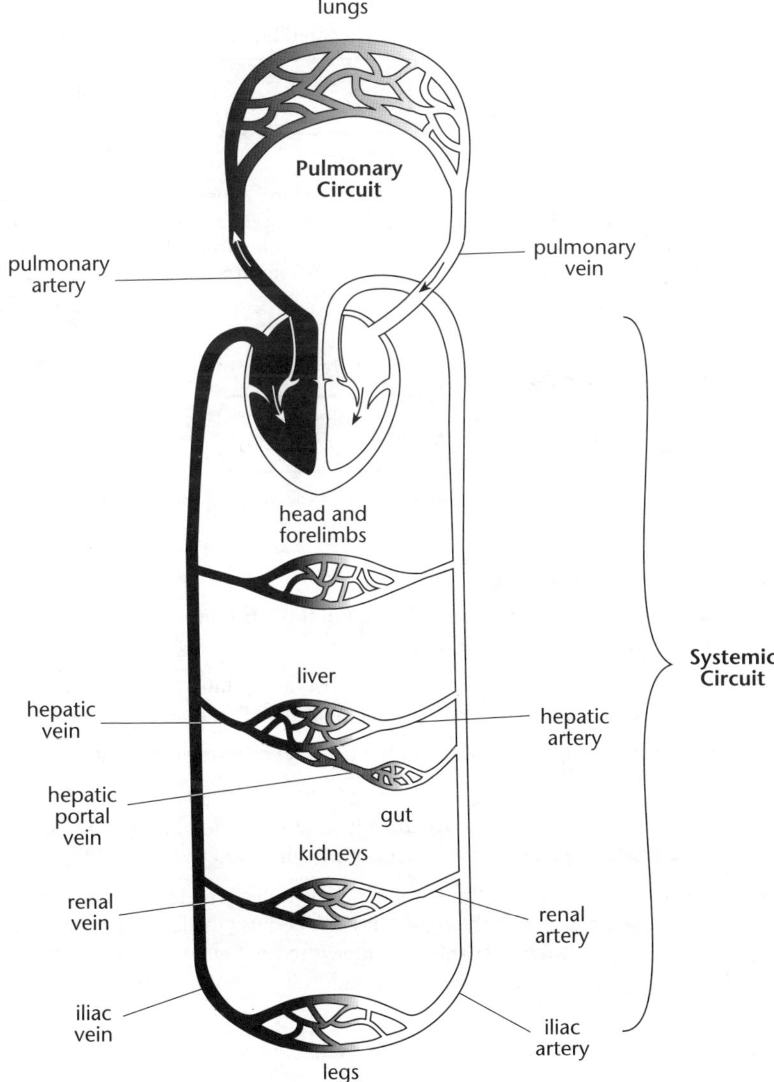

General plan of the human circulation.

Transport of blood gases

As blood passes through the lungs, oxygen from the lungs binds to haemoglobin (Hb) in red blood cells to form **oxy-haemoglobin** (oxy-Hb). Oxy-Hb releases oxygen in regions of the body low in oxygen. Hb is a large complex protein, containing a few iron, Fe, atoms.

Carbon dioxide is transported from tissues to the lungs as bicarbonate ions, HCO_3^-, in the blood plasma.

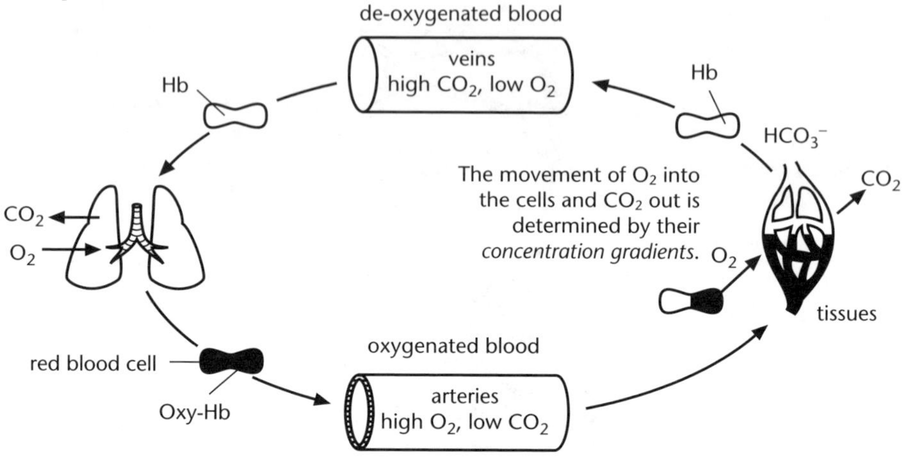

Transport of oxygen and carbon dioxide in the blood system.

The human heart

The human heart is a two-sided muscular pump. It is surrounded by a thin layer of lubricating fluid, contained in a tough bag, the **pericardium**.

The human heart is about the size of a person's fist, and pumps about 5 litres of blood per minute in a resting person. The heart is made of muscle that does *not* become fatigued.

The septum separates the right side from the left side, ensuring oxygenated and deoxygenated blood do not mix.

Each side consists of a thin-walled **atrium** which receives blood from *veins*, and a much more muscular **ventricle**, which pumps blood out through an artery.

The right ventricle forces blood through the nearby lungs. The muscle of the left ventricle is three times thicker than that of the right ventricle, because the left ventricle must raise the blood pressure high enough to enable the kidneys to filter efficiently and well. Blood pumped to the lungs from the right ventricle must be at lower pressure, so that alveoli are not damaged/destroyed.

Valves occur between the atria and ventricles, and between the ventricles and arteries, to ensure directional flow of blood through the heart.

The *left side* pumps oxygenated blood through the aorta to the rest of the body – the blood loses some of its oxygen and then returns to the right side of the heart through two huge veins called **venae cavae**.

This is the *systemic circuit.*

The *right side* pumps deoxygenated blood through the **pulmonary artery** to the lungs – the blood picks up oxygen and loses some of its CO_2. The blood then returns to the left side via the **pulmonary veins**.

This is the *pulmonary circuit.*

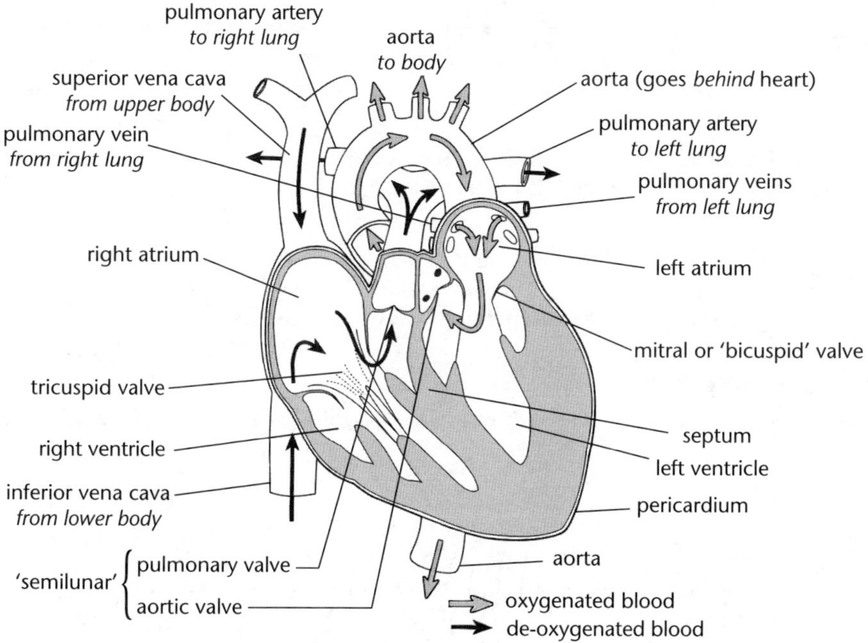

Blood flow in the human heart.

All the blood that leaves the left side of the heart comes back to the right side, so the two sides of the heart are actually 'in series' rather than 'in parallel'. The two halves must therefore pump *equal volumes* of blood. However, the left ventricle has to exert a much higher pressure than the right (and is much more muscular), because the kidneys (which are part of the systemic circuit) need high pressure blood to filter effectively. It is not true that the left ventricle needs to exert a higher pressure because it has to pump blood further around the body.

Each chamber of the heart alternately contracts and relaxes. When the two ventricles are contracting, the two atria are relaxing, and vice versa.

The blood is prevented from flowing backwards by four valves, of two kinds.

- Between each atrium and ventricle is an **atrioventricular valve** (A-V valve). Its flaps are prevented from turning 'inside out' by tendons attached to the ventricle wall. The A-V valve on the right side is called the **tricuspid valve**, because it has *three* 'tri' flaps. The valve on the left side is called the **bicuspid valve** (with *two* 'bi' flaps).
- At the base of the aorta is a **semilunar valve**, consisting of three half-moon shaped flaps. Another semilunar valve lies at the base of the pulmonary artery.

The heart valves have no power of their own – each is pushed open or closed by blood pressure during a cycle of events in the heart.

The **pacemaker** for the heart is located in the right atrium. It produces an electrical impulse which spreads out across the heart and causes the muscles that make up the atria, then those that make up the ventricles, to contract.

The pacemaker is controlled by nerves from the brain, and responds to the higher carbon dioxide content of the blood during exercise by making the heart beat faster. The pacemaker is also stimulated by the hormone **adrenalin**, produced during stress or excitement.

The cardiac cycle

The contractions of the chambers of the heart follow a regular pattern, called the **cardiac cycle**.

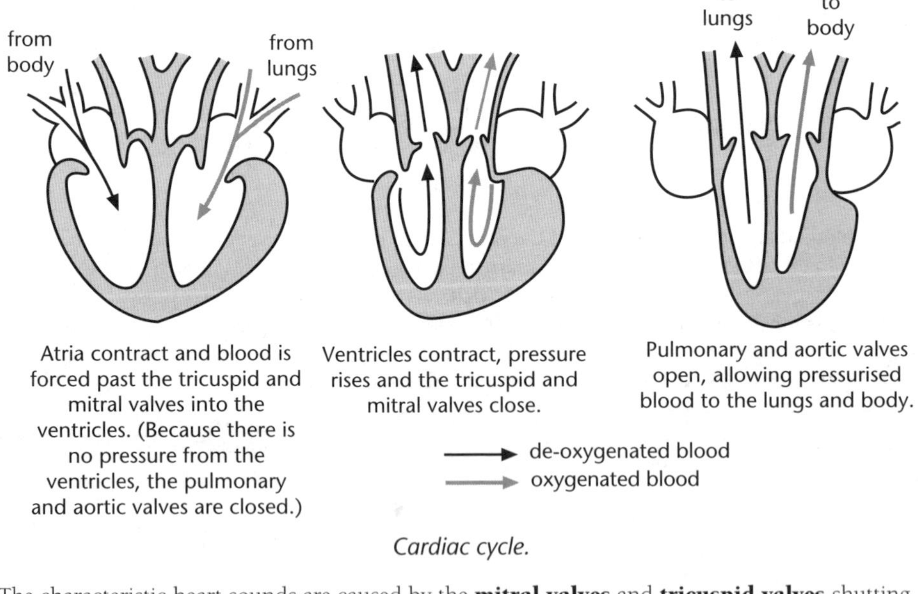

Cardiac cycle.

The characteristic heart sounds are caused by the **mitral valves** and **tricuspid valves** shutting (the first 'thump'), followed by the **aortic valves** and **pulmonary valves** (the second, louder 'thump'), which can be heard on a stethoscope.

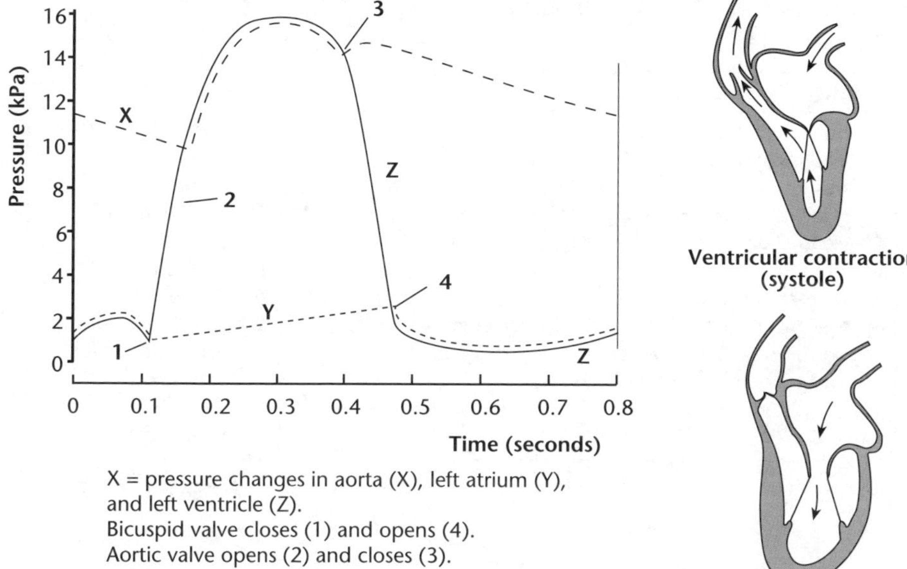

Events of the cardiac cycle (left) and the left ventricle at two phases of its cycle.

The walls of the heart consist of **cardiac muscle**, which differs from skeletal muscle in two important ways:

- It contracts spontaneously (without outside stimulation).
- After each contraction it has to relax, so it cannot make a sustained contraction and cannot get cramp or **fatigue**.

Although the heart is full of blood, the muscle has its own internal blood supply – the **coronary arteries** and the **cardiac veins**. The two coronary arteries are the first branches of the aorta.

The heart of an average adult beats about 70 times a minute at rest, and pumps about 5 litres per minute (both values vary considerably). During exercise, the heart output increases, because it beats faster and also pumps more out each beat.

Heart output is regulated partly by the hormone **adrenaline** (which speeds the heartbeat up), and partly by the two sets of autonomic nerves that supply the heart. The sympathetic nerves have the same effect as adrenaline, whilst the parasympathetic nerves act to slow the heart down.

Heart sounds

The clapping shut of the heart valves gives rise to the heart sounds which can be heard on a stethoscope. Closure of the A-V valves produce a soft sound ('lub'), with the semilunar valves causing a sharper sound ('dup').

Unit 11.3 Activity 4B: Human heart

1. For each of the phrases **1–12**, write the letter **A–L** of the term to which it applies.

Phrase	Applied term
1. Brings deoxygenated blood to the heart	A. Adrenaline
2. Brings oxygenated blood to the heart	B. Aorta
3. Carries blood away from the left side of the heart	C. Atrium
4. Heart chamber that pumps blood into an artery	D. Cardiac muscle
5. Heart chamber that receives blood from veins	E. Coronary artery
6. Hormone that stimulates the heart	F. Pericardium
7. Membrane around heart	G. Pulmonary artery
8. Prevents back-flow of blood from the aorta	H. Pulmonary vein
9. Prevents back-flow of blood from the right ventricle	I. Semilunar valve
10. Takes blood to the heart muscle	J. Tricuspid valve
11. Takes deoxygenated blood away from the heart	K. Vena cava
12. The main tissue of the heart	L. Ventricle

2. The diagram shows some of the main structures of the human heart. Copy the diagram.

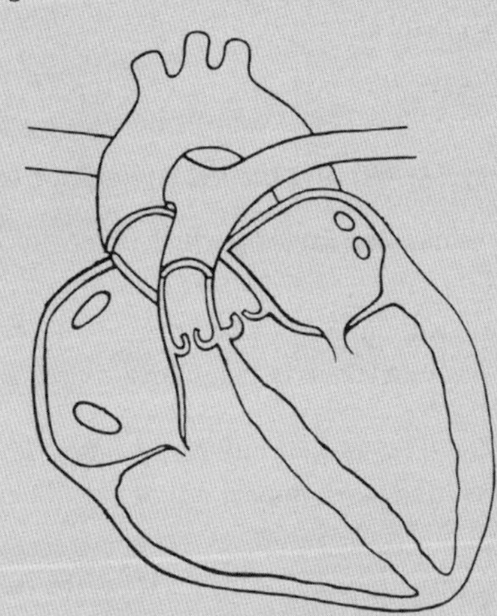

a. i. On the diagram, circle clearly the places in the heart where you would find valves.
   ii. Describe the function of these valves.

b. On the diagram, draw a vena cava and a pulmonary vein to show where they enter the heart. Label these two blood vessels.

Blood vessels

Arteries carry blood *away* from the heart. All arteries (*except* the pulmonary artery) carry oxygenated blood. To resist the high pressure, the artery walls have to be *thick*.

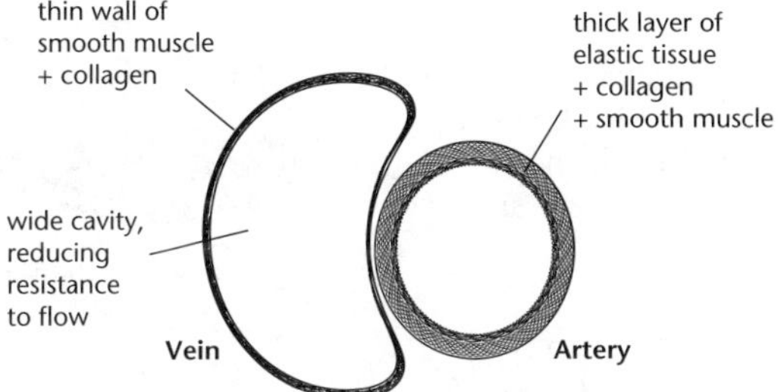

Transverse sections of an artery and vein.

The larger arteries have very *elastic* walls. This helps to smooth out the surge of a pulse, so by the time the blood reaches the smallest arteries the flow is steady.

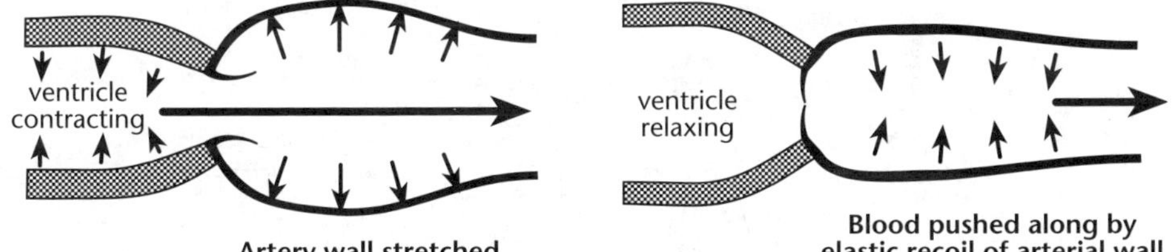

How elastic arteries smooth out the pulse.

The smallest arteries are called **arterioles**, and their walls have large amounts of smooth muscle running round them. Although this muscle is supplied by nerves, it is not under control of the will. Arterial muscle does *not propel* the blood, but regulates its *distribution*.

Example

During exercise, blood is diverted to the limb muscles from other parts. The muscle in the small arteries supplying the limb muscles relaxes, so blood pressure forces the arteries to get *wider*. This allows more blood to get through. At the same time, the arteries supplying the gut become *narrower*, because the muscle in their walls contracts, allowing less blood to pass through.

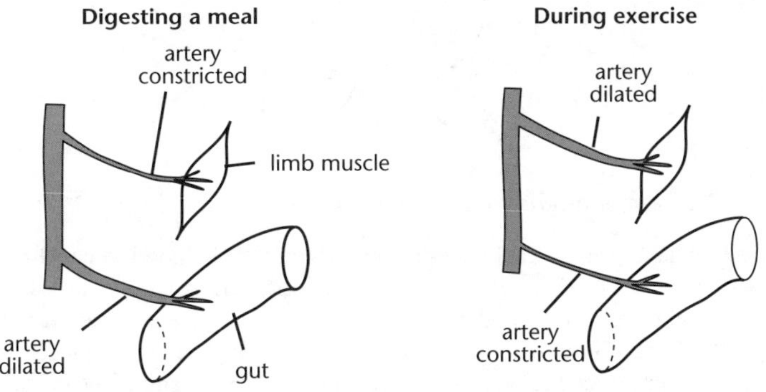

How arteries regulate the distribution of blood.

Arteries take blood away from the heart. Artery walls are thickened with connective tissue to withstand blood pressure, yet elastic enough to expand and contract to even out pressure from each heartbeat.

When arteries lose their elasticity (because of bad diet or from disease), high blood pressure develops, which puts a strain on the heart, affects kidney function, and may even cause arteries to burst.

Section through an artery.

The human arterial system

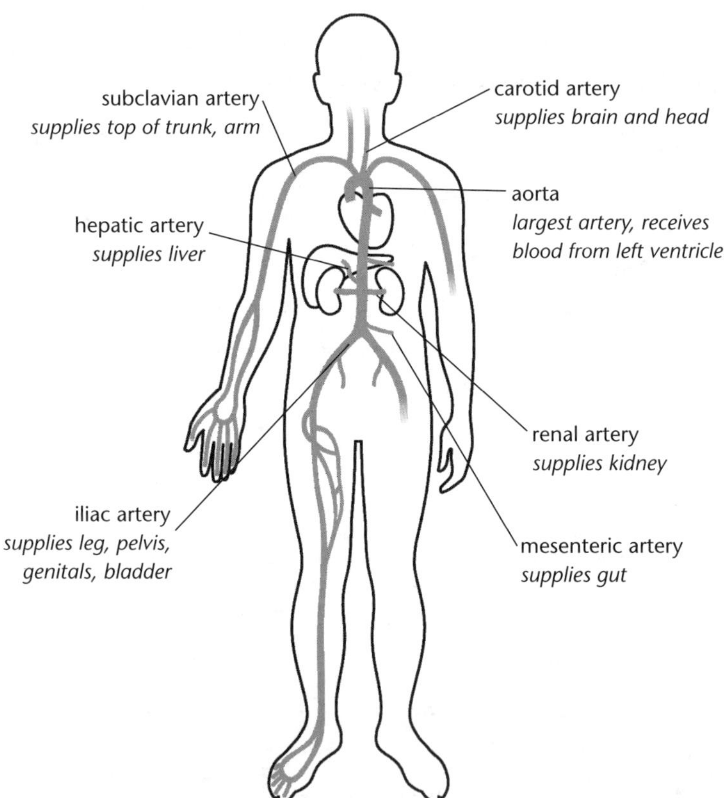

Major arteries of the human circulatory system.

Arteries divide into smaller vessels called **arterioles**, which supply blood to **capillaries**. Capillaries are the smallest blood vessels. Their walls are only one cell thick, allowing rapid exchange of materials between the blood and individual cells. Blood enters capillaries under pressure, forcing fluid and useful materials out of the capillaries to surrounding cells. Fluid containing waste products or cell secretions is drawn into the *venule* end of a capillary by *osmosis* and *diffusion*.

Veins

These carry blood *towards* the heart and have much thinner walls than arteries. All veins (*except* the pulmonary vein) carry deoxygenated blood. By the time the blood has reached the veins, there is not much pressure 'to push it along'. Return of blood to the heart is helped by valves and the wide diameter of veins.

- Most veins have *valves*. During exercise, veins are squeezed by contraction of neighbouring limb muscles. Active limb muscles therefore pump blood back to the heart, and this is one reason why blood flow is sluggish if you stand still for too long. Blood is also pumped back to the heart by the action of breathing. When you are breathing out, the pressure in the chest increases, squeezing the venae cavae. The valves ensure that the blood is pushed towards the heart.

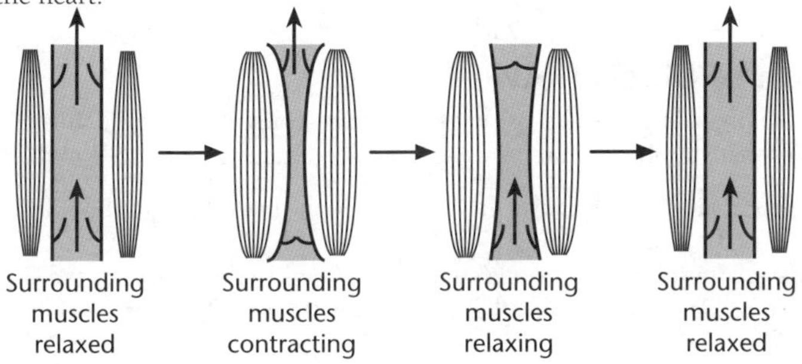

How blood is pumped back to the heart during exercise.

- Veins are wider than arteries, and there are more of them (eg there are many veins in the forearm but only two main arteries). Just as a river flows more slowly where it enters a stretch that is wider or deeper, so too does the blood. Slower flow reduces the resistance to flow.

Blood moves through veins towards the heart by:

- The squeezing of muscles against the walls of veins.
- Gravity (blood from the neck and head).
- Pressure in the chest cavity during inspiration 'pulls' blood into the heart.
- Some (but very low) blood pressure helps push blood through veins.

Blood is prevented from flowing backwards through veins by valves:

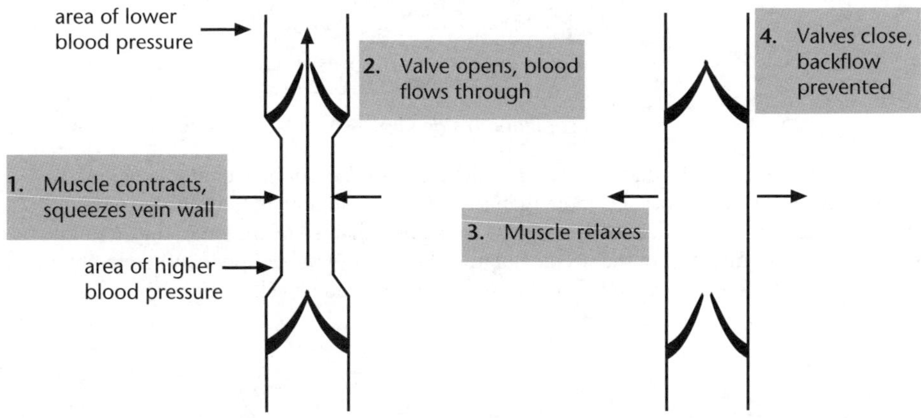

Valve action stops venous blood flowing 'backwards'.

If the valves in veins stop functioning, blood may flow backwards along the veins, putting excess pressure on them. These veins may enlarge and become painful, forming **varicose** veins in the legs, or piles (**haemorrhoids**) if the veins are in the rectum.

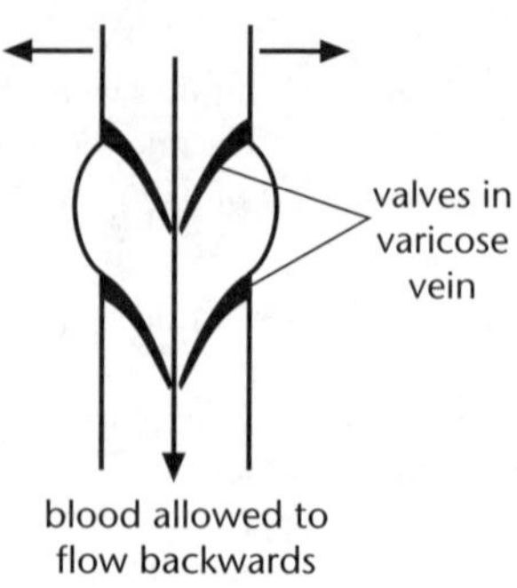

Varicose vein.

Regulation of blood flow

Blood flow to an organ is prevented or slowed down by constriction of the muscles in the walls of the arterioles that supply blood to an organ (ie the arterioles are reduced in diameter).

Relaxation of arteriolar muscles allows greater blood flow. Blood flow to the skin is controlled in this manner.

- On hot days, arteriolar muscles relax and blood flows to the surface of the skin (which will now appear redder in colour), so that heat energy from the blood is quickly lost and the person is kept cooler.
- On cold days, blood flow to the skin is reduced and the person's complexion therefore appears paler.

Capillaries

It is in these microscopic vessels that the composition of the blood changes. Capillaries have a number of features which enable substances to enter and leave the blood:

- Their walls are extremely *thin* (about 0.2 µm, or about $\frac{1}{5}$ the width of a bacterium!), consisting of a layer of very flat cells.
- They are extremely small and numerous, so they have a *large total surface area* – over 6 000 m^2 in an average adult – larger than the area of a hockey pitch!

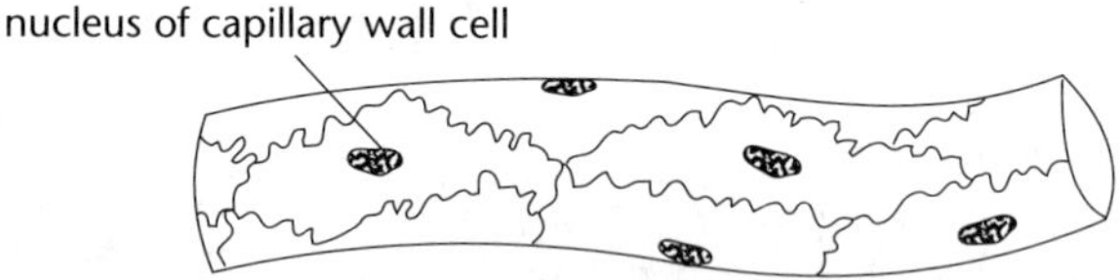

A capillary in side view.

Most capillaries are about the diameter of a red blood cell, and about half a millimetre long. Because there are so many, no cell is more than a fraction of a millimetre from the nearest capillary. As red blood cells squeeze through the capillaries, their shape becomes distorted into a bell shape. As a result, most of the haemoglobin is displaced to the periphery of the red blood cells, reducing the distance oxygen has to diffuse to the surrounding cells.

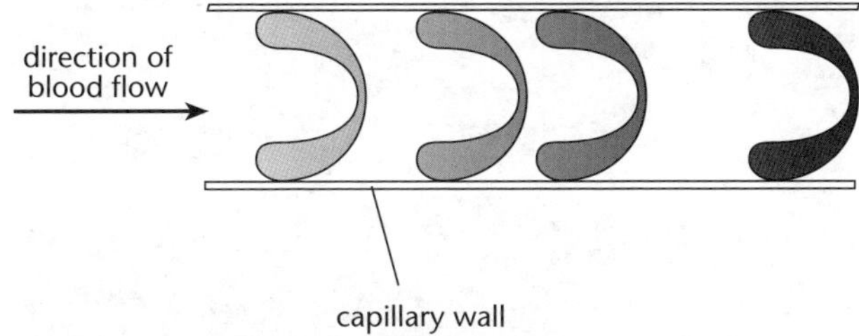

Red cells squeezing along a capillary.

How substances enter and leave capillaries

Substances enter and leave capillaries in two ways:

- By *diffusion* – eg, oxygen and CO_2. These molecules are small enough to pass through the semipermeable membranes of the capillary wall cells.
- Through tiny pores between capillary wall cells – these pores are wide enough for glucose, amino acids and salts to pass through, but too small for the plasma proteins. Capillary walls therefore act as *filters*.

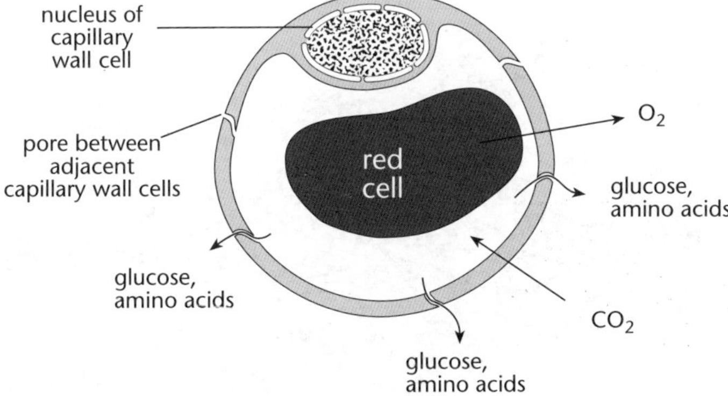

A capillary in cross-section, showing how substances enter and leave the blood.

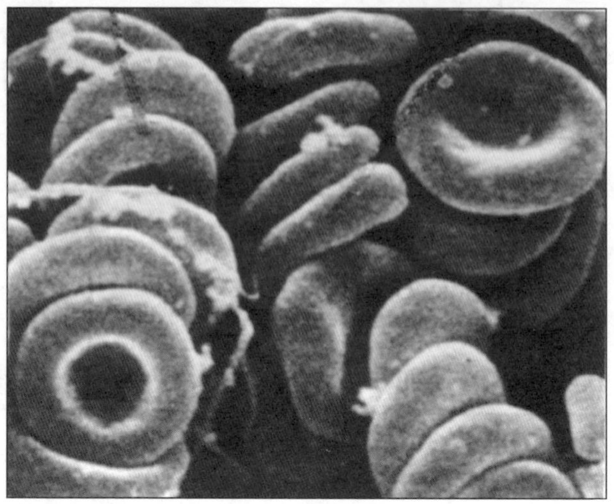

3D photo of red blood cells. A capillary is little more than the diameter of one of these cells, so the cells have to pass along a capillary in single file.

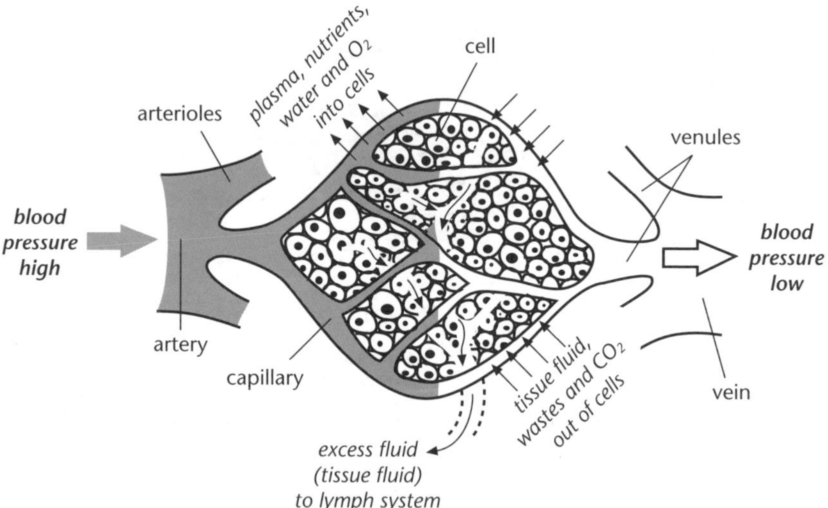

Blood movement through, and fluid movement in and out of, a capillary system.

Speed of blood flow

Blood takes an average of about 20 seconds to travel from any given part of the body back to the same place. As it travels along the arterial 'tree' it slows down, and by the time it reaches the capillaries it is moving at about 1 mm s^{-1}. It slows down because every time an artery branches, the total width of the branches is greater than that of the 'trunk'. In the veins the reverse is true, so the blood speeds up as it travels along the venous system.

Extension

Transport over short distances – a closer look at diffusion

Over very short distances, such as into and out of individual cells, oxygen and carbon dioxide move by a process called **diffusion**. This is the movement of a substance from where it is more concentrated to where it is less concentrated, by random movement of its particles.

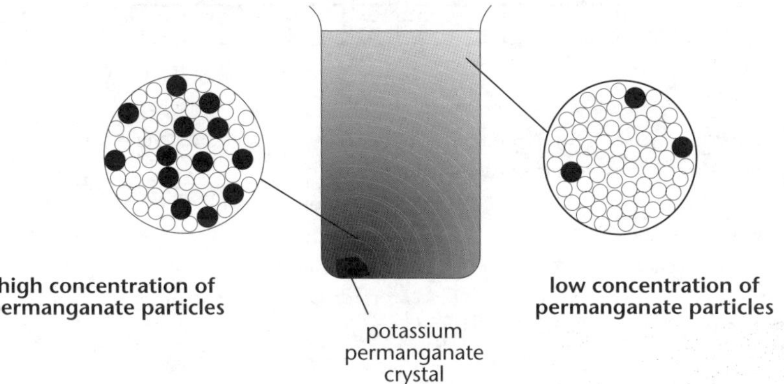

Diffusion of potassium permanganate in water.

Inside a cell, oxygen is being used up and so it is less concentrated than outside the cell. Oxygen therefore diffuses *into* the cell. At the same time, CO_2 is being produced inside the cell, so it diffuses *out*.

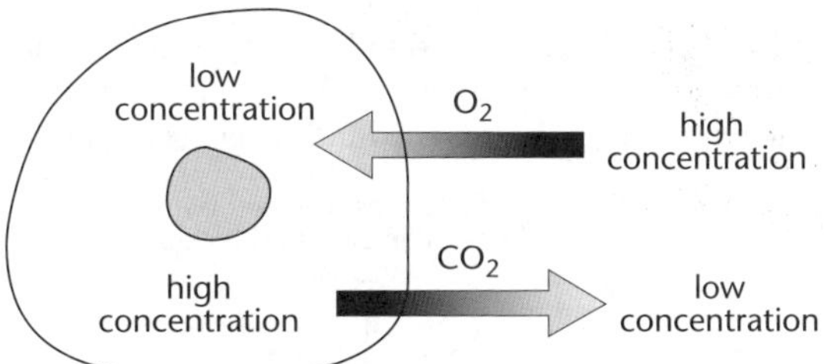

Diffusion into and out of cells.

Diffusion is far too slow to carry materials round the body – this is why a transport system is needed.

Unit 11.3 Activity 4C: Human circulation

1. As blood flows along capillaries, substances enter and leave it.
   The table shows changes in the composition of the blood as it passes along capillaries in seven parts of the body – bone marrow, liver after a meal, lungs, active skeletal muscle, the small intestine after a meal, the kidneys, and brain. Each of these organs is represented by one of the letters **A–G**. In each cell of the table, a '+' sign indicates a substance entering the blood, a '–' sign indicates a substance leaving the blood, and a '0' sign indicates no change. Identify each of the organs **A–G**.

	A	B	C	D	E	F	G
Oxygen	–	+	–	–	–	–	–
CO_2	+	–	+	+	+	+	+
Glucose	–	–	+	–	–	–	–
Urea	0	0	0	+	–	0	0
Lactic acid	0	0	0	–	0	+	0
Iron	0	0	+	0	0	0	–

2. The human circulatory system transports blood to every cell of the body. It consists of arteries, veins, capillaries and the heart.
   a. Describe two structural differences between the aorta and the vena cava.
   b. Explain the importance of each of the two differences.
   c. Describe the main difference in the composition of the blood between the pulmonary artery and the pulmonary vein.
   d. Discuss the function of the capillaries in the circulatory system.
3. Discuss how substances are transported around the body. Your answer should refer to at least four substances that are transported and include the role of the:
   - Heart.
   - Plasma.
   - Red blood cells.
4. Discuss how veins and arteries differ in their structure, and how this affects their function.

Blood clots

When blood vessel walls are cut, the blood clots, helping to seal the wound. This not only helps stop blood loss, but also makes it harder for harmful bacteria to get in. The trigger for clotting is the release of substances from injured blood vessel walls, and from blood platelets. As a result, the soluble plasma protein **fibrinogen** is converted into a network of insoluble **fibrin**.

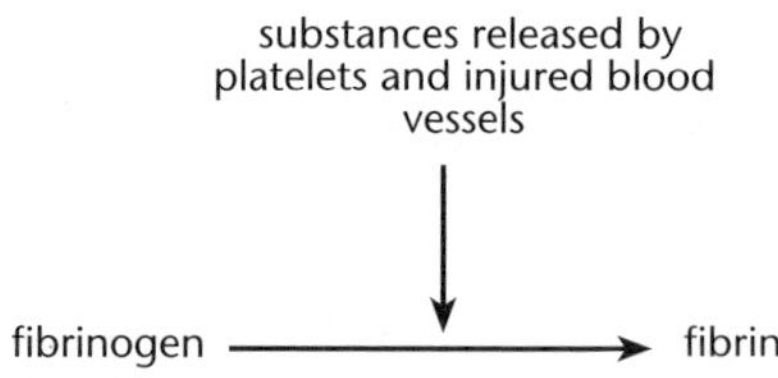

Simplified scheme of the blood clotting mechanism.

Red cells become trapped in the fibrin, forming a *clot*. The liquid remaining after blood has clotted is called *serum*.

The conversion of fibrinogen to fibrin is triggered by substances released by the injured blood vessel walls and blood platelets. The process is complicated, depending on over a dozen proteins. **Haemophilia** is the hereditary inability to make one of these proteins (Factor VIII).

Though the clotting mechanism works extremely well, a clot occasionally forms without injury to blood vessels. Such a clot is called a **thrombosis**, and usually forms in a vein. It becomes dangerous if it breaks free and is carried along until it blocks an artery. Thromboses are more common in smokers, and can also develop as a result of long periods of inactivity, such as may occur on long plane flights.

Clotting also requires the presence of calcium ions. When blood is donated, the receiving bottle contains a chemical that removes calcium ions, thus preventing clotting.

Blood groups

Blood can only be *transfused* from one person to another if the donor and patient are of compatible *blood groups*. When wrong blood groups are mixed, the red cells of the donor clump together or *agglutinate* (this is different from clotting – in clotting, the chemical change occurs in the *plasma*). Agglutination is dangerous, because the capillaries can become blocked.

Everyone belongs to one of four major blood groups, depending upon which antigens are present on their red cells.

Group A people have antigen A; Group B people have antigen B; Group AB people have both A and B antigens, and Group O people have neither. People of Group A have anti-B in their plasma (antibody to antigen B), and people of Group B have anti-A. Group O people have both anti-A and anti-B, and Group AB people have neither antibody.

If Group A blood is given to a Group O patient, the anti-A antibody in the patient's plasma combines with the A antigen on the donor's red cells, causing them to clump together. If Group O blood is given to a Group A patient, the anti-A antibody in the donor's blood is so diluted in the patient's blood that it 'can be ignored'.

		Donor			
		A	B	AB	O
Patient	A	✓	×	×	✓
	B	×	✓	×	✓
	AB	✓	✓	✓	✓
	O	×	×	×	✓

Safe (✓) and unsafe (×) transfusions.

Safe and dangerous transfusions.

The ABO blood groups form only one of a number of blood group systems. Another important one is the **Rhesus system**. Besides being of group A, B, AB or O, every person is either *Rhesus positive* or *Rhesus negative*. The Rhesus blood group can be important in pregnancy if a Rhesus positive baby is carried by a Rhesus negative mother.

Giving blood

Unless a person is involved in top-level 'aerobic' sports, he or she can give 600 cm^3 of blood without much noticeable effect. The very slight fall in blood pressure is minimised by contraction of the walls of the blood vessels (mainly the veins), and the loss of fluid is quickly made up by a slight reduction in urine output, aided by a drink provided by the transfusion service.

In the longer term, the production of red cells by the bone marrow is increased under the influence of a hormone called **erythropoetin**, or **EPO**, made by the kidneys. EPO production increases in people living at high altitude, so their red cell count is higher. Some athletes take EPO artificially to boost their red cell count, but it is 'cheating' and can be fatal.

The lymphatic system

Capillary walls have minute pores so they are very leaky, allowing all substances (except most of the plasma protein) to pass through. As a result, colourless **tissue fluid** seeps out from the arterial end of capillaries (where the pressure is higher) into the surrounding tissues. Tissue fluid is like plasma but with less protein. The cells of the body are thus bathed in an 'internal environment' containing dissolved glucose, amino acids, salts, carbon dioxide and oxygen.

Tissue fluid contains less protein than does plasma, so it has a slightly higher water concentration and there is a tendency for water to re-enter the capillaries by osmosis. At the arterial end of a capillary, the pressure of blood prevents this. Near the venous end, the blood pressure has fallen and is not high enough to prevent some tissue fluid from re-entering the blood, but slightly less fluid re-enters the capillary than leaves it. Tissue fluid would accumulate in the tissues if it were not for the **lymphatic system**. This is a kind of 'overflow' system, which returns surplus tissue fluid to the blood.

Like the arterial and venous systems, the lymphatic system extends to all parts of the body. It differs from the blood system in that its finest branches or **lymphatic capillaries** are *blind-ending*. They eventually join to form two main lymphatic vessels, which open into the large veins near the base of the neck. The liquid in the lymphatic vessels is called **lymph**.

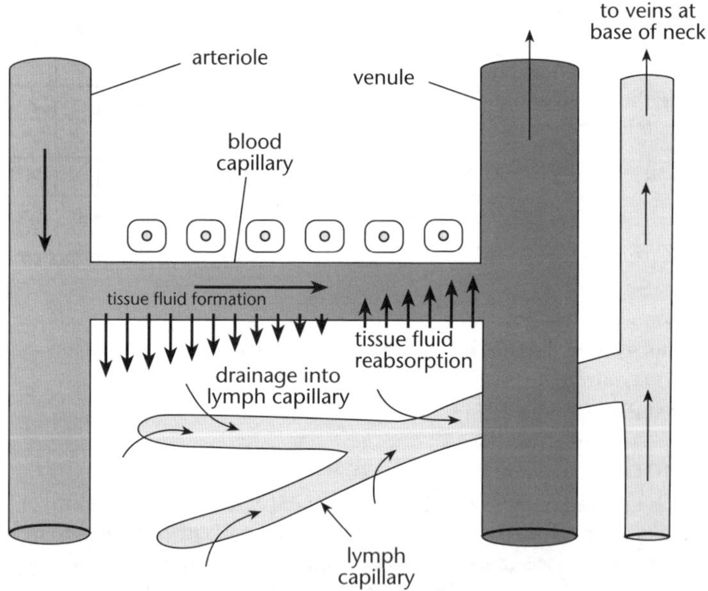

How surplus tissue fluid is drained by the lymphatic system.

Lymph flow is sluggish and is greatly helped by the action of muscles during exercise. As in veins, one-way flow is ensured by *valves*.

At intervals along lymphatic vessels are **lymph nodes**. These are part of the defence system of the body and are packed with white blood corpuscles. During infection, the number of white corpuscles increases, causing the nodes to become enlarged and tender. This condition is often called 'swollen glands'.

The **lymph system** works like a drainage system. Lymph capillaries and lymph vessels, similar to veins, drain excess tissue fluid from around cells. Lymph fluid is similar to blood, except that it contains no red blood cells and little protein.

Lymph returns to the blood system via two large lymph vessels which empty into the superior **vena cava**, the large vein that connects to the right auricle of the heart.

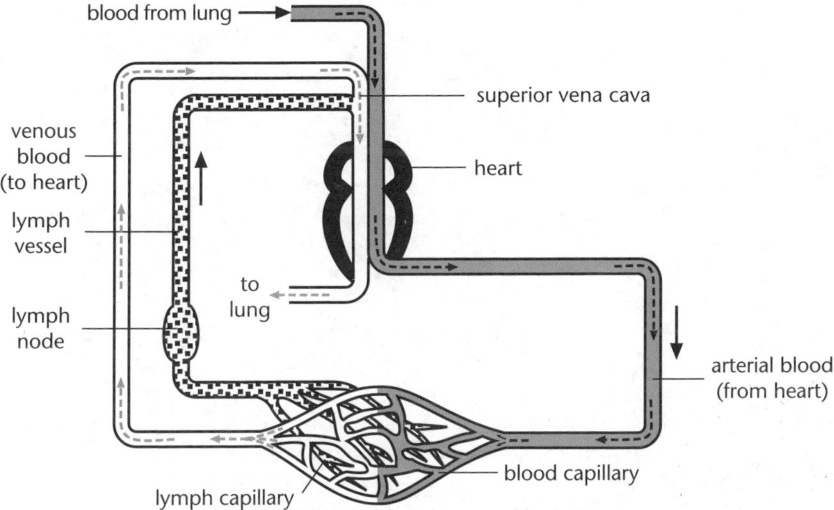

Relationship between the lymph and blood systems.

Movement of lymph occurs as a result of muscles squeezing against the walls of lymph vessels. Lymph vessels also have one-way valves.

Unit 11.3 Activity 4D: Internal transport in animals

1. Distinguish between open and closed transport systems.
2. Explain why cnidaria, although multicellular, have no special transport system, while a transport system is essential to the survival of insects.
3. Describe the composition of blood plasma.
4. Explain why oxygen-carrying pigments are common in animal internal transport systems.
5. Explain how the function of the following blood cells is related to their structure:
   a. Red blood cells.
   b. Phagocytes.
6. Compare and contrast the structure of arteries and veins.

7. Using the diagram following to assist you, explain how the structure of the heart is related to its function.

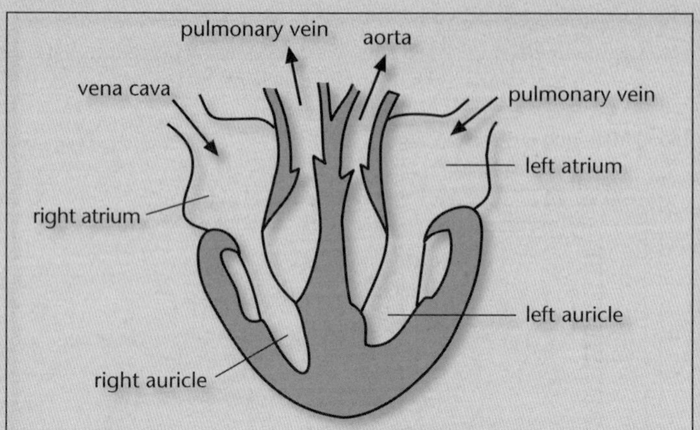

8. Explain why fish live successfully with a single-loop circulatory system, while a double system is needed by birds and mammals.
9. The metabolic processes of cells require a constant supply of nutrients and oxygen and constant removal of metabolic wastes. Discuss the ways in which the animals of three named taxonomic or functional groups have solved the problem of transporting these substances from one part of their body to another.
10. Circulatory systems may be classified into 'open or closed' systems, 'single or double' systems. Most animals fall into one or more of these categories. Select three different taxonomic or functional groups, and, using named animals as examples, discuss the structure of their circulatory systems and the reasons for their differences. In your answer:
    - Describe the system for each of the named animals.
    - Explain how each of these systems operates.
    - Explain the differences in the systems in relation to the different ways of life of the three animals.

Extension

Things that affect diffusion rate

The rate at which the concentration of a substance changes with distance is expressed by the idea of a *concentration*.

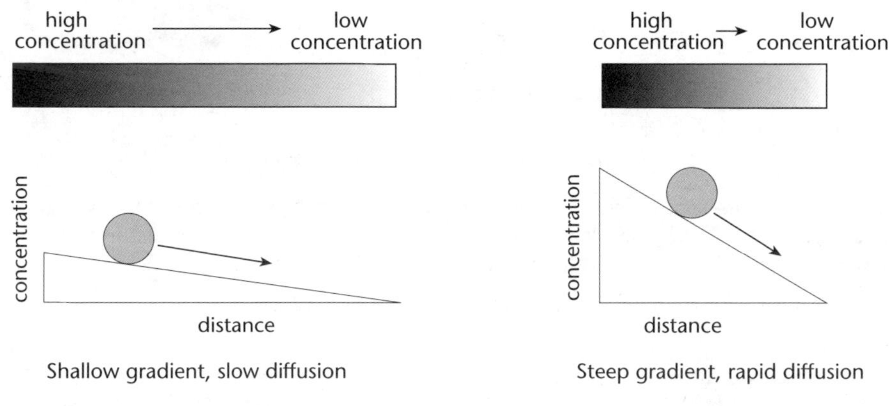

How the steepness of a concentration gradient affects diffusion rate.

A concentration gradient can be compared to a hill. Just as a ball rolls more quickly down a steep hill than down a gentle one, a substance diffuses more quickly along a steep concentration gradient than along a shallow one.

For a given concentration difference between two areas, the closer the areas are together the steeper the gradient. This is why diffusion into and out of adjacent cells can be rapid.

In the human body, especially in the lungs, the rate of diffusion is affected by several things:

- It is faster along a steep concentration gradient.
- It is about 10 000 times faster in gases than in liquids.
- Small molecules diffuse faster than large ones.
- It is faster over a larger area than over a small one.

Diseases of the blood system

Atherosclerosis

This is one of the greatest causes of early death in industrialised countries. It is the narrowing of arteries by the deposition of fatty material in the walls, reducing the flow of blood. It usually begins many years before symptoms develop.

The trigger that starts atherosclerosis is damage to the lining of an artery, caused by factors such as carbon monoxide (from cigarette smoke), high blood cholesterol level, high blood pressure, and diabetes.

The damage to the lining of an artery causes the development of an **atherosclerotic plaque**, a mound of cholesterol and other fatty material covered by a cap of fibrous tissue. If the fibrous cap is thin it may rupture, exposing the fatty material below and triggering the formation of a clot. This may block the artery completely. If this occurs in the heart, the affected heart cells die, causing a **cardiac infarction** or 'heart attack'. If it occurs in the brain, it causes a 'stroke'.

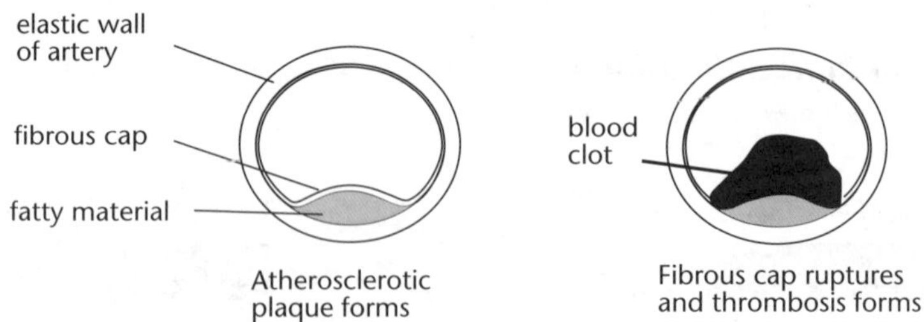

Formation of an atherosclerotic plaque.

Plaques that contain more fibrous tissue generally do not rupture but continue to grow slowly, producing a localised prolonged starvation of the tissues. In the heart, this condition is called **ischaemia**. It may produce pain called **angina** during exertion, which may subside after rest.

Atherosclerosis is known to be associated with a number of *risk factors*:

- Smoking.
- Diet (avoid **saturated fats** and cholesterol).
- Obesity (to avoid obesity, keep your energy intake to that needed to balance your energy output).
- High blood pressure.
- Diabetes.
- Hereditary factors.

Atherosclerotic plaque consists largely of cholesterol, so it is not surprising the disease is associated with a high blood cholesterol level. This is not necessarily due to too much cholesterol in the diet, because the body makes cholesterol from dietary fat. Even on a low-cholesterol diet, a high proportion of blood cholesterol is produced in the body from saturated fat. High blood cholesterol can have two causes:

- The diet may contain too much cholesterol, or too much animal fat.
- Some people have genes that make it difficult for them to remove cholesterol made in the body. (Cholesterol is normally removed from the blood as fast as it is produced, so even on a normal diet, such affected people's cholesterol levels can be too high.)

Prevention and treatment of atherosclerosis

Atherosclerosis can be prevented by reducing or eliminating risk factors. Medications can be taken by people with a genetic predisposition to atherosclerosis to prevent it occurring. Treatments are usually surgical and involve:

- Inserting a **stent** into the artery at the site of the plaque. A stent is a wire mesh tube that keeps the artery open, improving blood flow. The stent is usually inserted at the same time as angioplasty is done.
- **Angioplasty** involves inserting a balloon into a blocked artery and inflating it at the site of the blockage. The inflated balloon pushes (squashes) the plaque against the artery wall, restoring blood flow. Often, a stent is placed over the balloon, so when the balloon inflates, the stent also expands to the same diameter as the artery. The balloon is removed, and the stent stays in the artery permanently.

- Coronary bypass is major surgery required in cases where the blockages are too severe for angioplasty or stents. The blocked artery (or arteries) is removed and replaced with blood vessels that have been removed from other places in the body, often from the leg.

High blood pressure (hypertension)

The pressure of blood in the arteries rises and falls with the pulse.

- The pressure *peaks* correspond to ejection of blood from the left ventricle upon its contraction (systole) – the **systolic pressure**.
- The *troughs* between the surges correspond to the relaxation (diastole) of the left ventricle – the **diastolic pressure**.

Blood pressure is measured by an inflatable cuff placed round the arm and is expressed as the height (in mm) to which the pressure can push up a column of mercury.

In healthy young adults, systolic and diastolic pressures average 120 and 80 mm mercury respectively, and such a blood pressure is referred to as '120/80' or '120 over 80'. A person with **hypertension** (high blood pressure) has a blood pressure of 140/90 or greater. However, for people with conditions such as diabetes or kidney disease, it may be dangerous if the blood pressure is 130/80.

Hypertension results from excessive and sustained contraction of the arterial muscle, and is a major risk factor in life expectancy.

Hypertension may contribute to the rupture of a blood vessel. If such a rupture occurs in the brain, there is a loss of blood supply to the affected part, leading to a 'stroke'. Hypertension can also lead to a heart attack or heart failure.

It may be difficult to determine the cause of hypertension in an individual. There are several contributing factors that may lead to hypertension, including:

- Genetics.
- Some hormones, eg insulin.
- A diet with moderate to high levels of salt.
- Certain diseases, eg kidney disease, adrenal gland abnormalities.
- Certain medications, eg steroids.
- Age – the artery walls lose elasticity with age.
- Pregnancy – hypertension during pregnancy can occur for some women, leading to complications that could threaten the foetus.

Hypertension can be controlled by lifestyle changes, by medications or a combination of both. Lifestyle changes include:

- Reducing bodyweight and doing regular aerobic exercise.
- Reducing the amount of salt consumed.
- Having a diet rich in fruit, vegetables, nuts and low-fat dairy products. Calcium in these dairy products has been linked to a decrease in hypertension.
- Stopping smoking.
- Stopping the drinking of alcohol.
- Making time to relax each day, eg through meditation.

Hypotension

Hypotension is the opposite of hypertension. The blood pressure is so low that blood may not be able to be pumped to the extremities of the body, including the head. A person may feel faint or light-headed, particularly when standing. Hypotension is caused by loss of blood volume or by blood vessels that have too large a diameter. It will disappear when blood lost is replaced, and can be managed by medications that cause blood vessel walls to contract, decreasing their diameter.

Stroke

A stroke is a sudden interruption in the blood flow through the brain. It can be caused by a clot or by a ruptured blood vessel. The part of the brain affected dies, but how serious this is depends on the size of the blood vessel and the part of the brain it serves.

If the part of the brain controlling muscular activity is involved, the affected side of the body is that opposite to the affected part of the brain (because each side of the brain controls the opposite side of the body). Thus, paralysis of muscles on the left side of the body may result from a stroke on the right side of the brain.

Varicose veins

This is a very common condition in the legs. In a person who has been standing for a while, blood tends to accumulate towards the bottom of the legs. As a result of a build-up of pressure, the walls of the veins bulge and become permanently stretched and the valves become non-functional. It is particularly common in women and in people who have to stand over long periods. When a person is standing, the veins are not 'massaged' by contraction of surrounding muscles and the return of blood to the heart is reduced, causing blood pressure in the veins to increase. Another factor is pregnancy, when the distended uterus presses on the abdominal veins.

Anaemia

Anaemia is a reduction of the oxygen-carrying capacity of the blood, due either to fewer red cells, or lower haemoglobin content, or both.

- In moderately severe anaemia, the person is pale and may be breathless after quite mild exertion.
- In very severe cases, the kidneys, heart or liver are affected.

Tests for anaemia may involve measuring the amount of haemoglobin in a given volume of blood, or taking a red cell count, or measuring the volume of red cells in each 100 cm^3 of blood. Even in healthy people, these figures vary – the red cell count tends to be higher in males (4.3–5.9 million per mm^3) than in females (3.5–5.0 million per mm^3).

Though there are well over a dozen causes of anaemia, they fall into three major groups.

- Loss of blood. This may be external, or it may be internal and slow, perhaps associated with a tumour. When bleeding is internal, the iron can be recycled after the haemoglobin has been broken down.
- Reduced rate of red cell and/or haemoglobin production. The best known and most common causes result from shortage of iron (a constituent of haemoglobin). These may be from iron deficiency in the diet, particularly in those whose dietary needs are greater, such as children and women (women lose blood in menstruation). Another cause of reduced production is *pernicious anaemia*, due to lack of vitamin B_{12}. This vitamin is entirely absent from plant food, so extreme vegetarians (vegans) are at risk.
- Increased rate of red cell destruction.

Leukaemia

The various kinds of leukaemia are characterised by uncontrolled division of the cells that produce white blood cells. The result is an abnormally high white cell count. The abnormally large number of white blood cells in the bone marrow means other types of blood cells (erythrocytes and platelets) are not made in sufficient numbers. This deficit leads to anaemia, and blood that does not clot easily, resulting in easy bruising and bleeding. The causes of leukaemia are not known, but several factors increase its likelihood, including:

- Radiation.
- Certain chemicals, eg benzene.
- Certain viruses.
- Genetics.

Some kinds of leukaemia can now be cured, but others are still difficult to treat. Treatments involve chemotherapy and/or radiotherapy to destroy the excess white blood cells, although these treatments make the person highly susceptible to bacterial infection. A bone marrow transplant from a suitable donor involves removing the affected bone marrow and replacing it with normal bone marrow in the hope it will produce new, normal white blood cells.

Thrombosis

This is the formation of a blood clot without injury to a blood vessel. The clot, or **thrombus**, may form in an artery, a vein, or in a heart chamber. Long periods of inactivity (such as occur on intercontinental flights) may result in *deep vein thrombosis* (*DVT*). A clot formed in a vein may break free and be carried to the lungs, blocking a pulmonary artery. If a clot blocks a coronary artery, it may cause a 'heart attack'. If a clot blocks an artery in the brain, it causes a 'stroke'. Risk factors include smoking, a diet too rich in saturated fats, and heredity.

Toxaemia of pregnancy

This condition affects about 6% of pregnant women, usually in the later stages of pregnancy. The cause seems to be linked to the failure of the blood supply in the placenta to develop properly. As a result, the patient may develop high blood pressure and large numbers of minute clots in the blood. The mother's kidneys, brain, liver and other organs may be adversely affected. Mild toxaemia can be treated. In severe cases, birth may be induced early (the condition disappears once the baby has been born).

Unit 11.3 Activity 4E: Blood diseases and the lymphatic system

1. For each of the phrases **1–8**, write the letter **A–H** of the term to which it applies.

Phrase	Applied term
1. A sudden interruption of blood supply to the brain	A. Anaemia
2. High blood pressure	B. Angina
3. Inadequate blood supply to the placenta in pregnancy	C. Atherosclerosis
4. Non-functional valves in a blood vessel	D. Hypertension
5. Pain in the heart caused by reduced blood supply	E. Leukaemia
6. Partial blockage of an artery by a fatty deposit	F. Stroke
7. Reduced oxygen-carrying capacity of the blood	G. Toxaemia
8. Uncontrolled production of white blood cells	H. Varicose veins

2. True or false?

a. All arteries carry oxygenated blood.

b. Blood slows down as it travels along the venous system.

c. Tissue fluid is the same as blood minus the corpuscles.

d. Blood entering the right side of the heart contains no oxygen.

e. Blood spurts out of a cut artery but flows steadily out of a cut vein.

f. Capillary walls are only one cell thick.

g. Red cells are biconcave so that they can swell up without bursting when they take up oxygen.

h. When blood of incompatible groups is mixed, it clots.

i. The function of the muscle in the walls of arteries is to pump the blood along.

j. If, during vigorous exercise, the blood flow through muscles were to double, the flow through the lungs would also double.

3. The diagram shows the coronary arteries on the outside of the heart.

a. Describe the function of the coronary arteries.

b. Identify two lifestyle factors other than smoking that may lead to coronary heart disease.

c. Explain how one of the factors you identified in part **b.** may cause coronary heart disease.

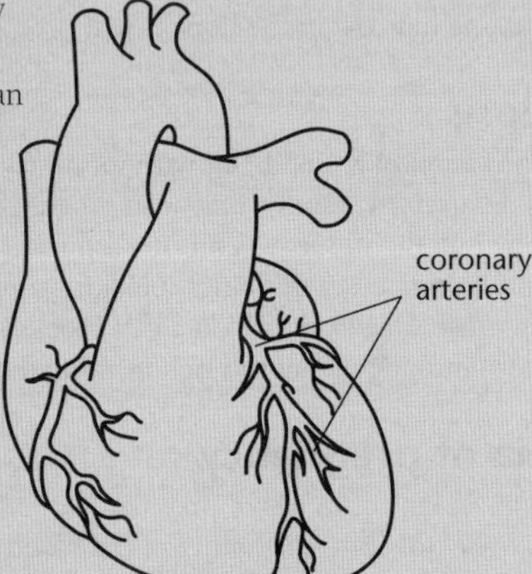

4. a. A person's blood pressure is $\frac{110}{70}$. Explain what each of these figures means, and explain the significance of these figures to the person.

b. Describe hypertension.

c. Describe what could happen to the blood vessels or heart to cause hypertension.

5. Lack of iron in the diet may cause iron-deficiency anaemia. When this happens, new red blood cells are made with insufficient haemoglobin. Discuss why iron deficiency anaemia causes a person to feel exhausted after completing a 5-km run.

6. Which of these actions prevent blood from flowing backwards?

A. muscular contraction of cell wall

B. opening of vein valves under low blood pressure

C. closing of vein valves after muscular relaxation

D. closing of vein valves before muscular relaxation

E. opening of vein valves under high blood pressure

7. a. What is deoxygenated blood?

b. The diagram shows how a blood clot forms when the skin is wounded. Use the diagram to explain how a blood clot forms to stop bleeding.

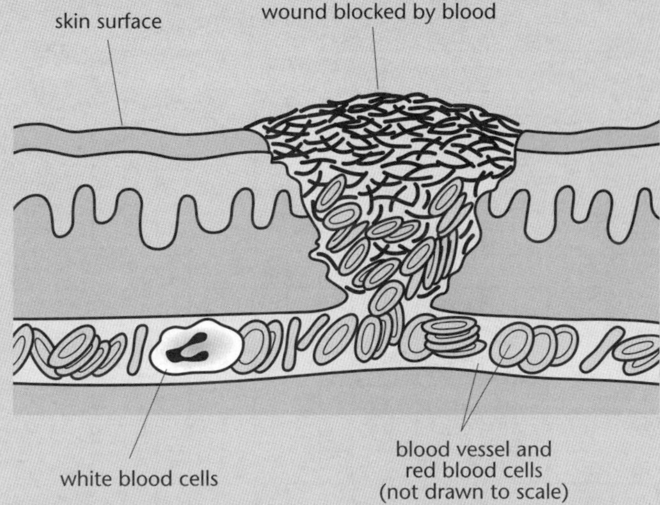

8. Select the statement that best describes 'varicose veins'.

A. limited number of veins

B. dead veins with waste

C. build up of air pressure in veins, hence the swelling

D. excessive build up of blood in veins due to non-functional vein valves

E. bad blood caught up in veins

9. In which part of the human circulatory system does the exchange of gases take place?

A. main artery

B. veins

C. aorta

D. arterioles

E. capillaries

10. The characteristic heart sounds are caused by which of these combinations of operations?

A. shutting of mitral valves and tricuspid valves

B. shutting of aortic and pulmonary valves

C. shutting of aortic and pulmonary valves followed by shutting of mitral and tricuspid valves

D. shutting of mitral and tricuspid valves followed by shutting of the aortic and pulmonary valves

E. none of the above

11. Which of the following statements about the lymphatic system is false?

A. lymph fluid is similar to blood in that it has red blood cells

B. the lymphatic system is responsible for draining tissue fluid

C. lymphatic capillaries are blind-ending

D. the lymphatic system has intervals along the lymphatic vessels packed with white blood cells

E. one-way flow is ensured in lymph vessels by valves

12. Complete the crossword puzzle on the human circulatory system.

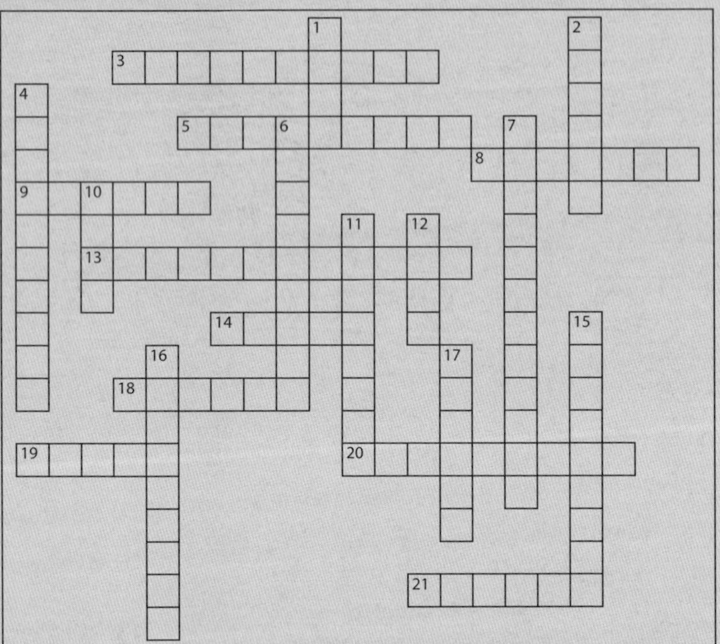

Across

3. Have a three-chambered heart

5. A system that helps fight infection

8. Have a four-chambered heart

9. Blood flow is restricted to vessels

13. Carry oxygen in blood

14. The "pump" of the circulatory system

18. Big vessel that carries blood away from the heart

19. Blood cells that fight disease

20. Tiny vessels involved in material exchange

21. Small vein

Down

1. Have a two-chambered heart

2. Liquid part of blood

4. Heart muscle

6. Blood flow through gills or lungs

7. Single beat of the heart (two words)

10. Blood circulates through body cavities

12. Big vessel that carries blood to the heart

15. Delivers blood to the body or lungs

16. Small artery

17. Delivers blood to the ventricles

Crossword puzzle from 'Instructor Resources/Manual' accompanying *Biology: Concepts and Investigations* by Mariëlle Hoefnagels (McGraw–Hill).

Unit 11.3 Transport Systems

Topic 5: Excretory systems in animals and humans

Authors: Martin Hanson with Beatrix Marjen-Waiin

This Topic continues the exploration of the diversity in the structure and function of animals and in that context it relates back to Unit 11.1 Living Things.
Specifically, the Topic deals with:

- The structure and function of the parts of excretory systems in animals.
- A description of the excretory products, explanations of where they come from, why they need to be excreted (toxicity), how they are excreted (role of water).
- Reasons for the differences in structure and function of excretory systems between different groups – these could relate to mobility and energy needs, size, habitat, life cycle, way of life.

This Topic further explores the functioning of the human circulatory, respiratory and excretory systems by looking at:

- Excretion – getting rid of waste substances.
- Structure of the human excretory system.
- Urine production as a two-stage process, involving filtration and reabsorption.
- Kidney diseases.

In some respects, animals are like car engines – they take in food ('fuel'), and break it down into simple waste products with the release of energy which they use to drive essential processes. These waste products are to some degree toxic, and must be got rid of.

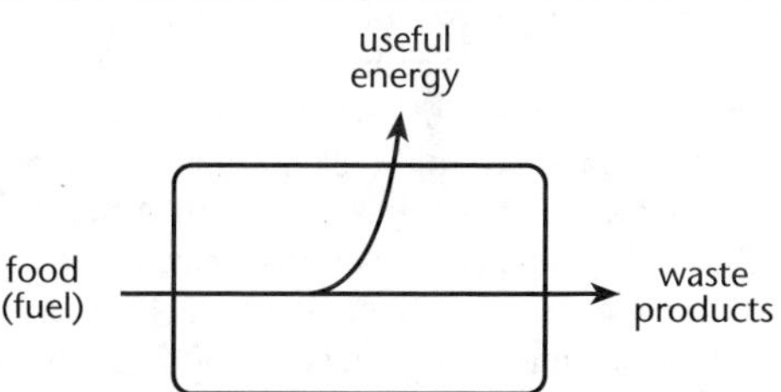

The body compared to a petrol engine.

Getting rid of the waste substances produced in chemical processes (metabolism) in cells is called **excretion**. The two most important processes that make waste products are:

- **Respiration** – produces *carbon dioxide*.
- **Deamination** of amino acids in the liver – produces *urea*.

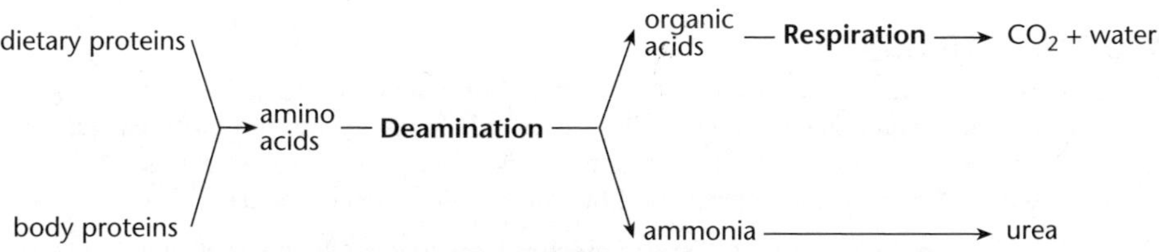

Deamination, and the fate of the ammonia produced.

Carbon dioxide is excreted via the lungs.

In deamination, the nitrogen-containing part of an amino acid is removed as ammonia and the rest is used in respiration to supply energy. Ammonia is extremely poisonous and is immediately converted into urea, which is much less toxic.

Excretion is quite different from *egestion*, which is the getting rid of undigested food as faeces. (Faeces are not produced in metabolism but form in the gut cavity which, strictly speaking, is an extension of the outside world – ie *outside* the body.)

Excretion removes **metabolic wastes** from the body, including substances in *excess* of the body's needs.

- CO_2 from respiration is removed by the gas exchange organs (eg lungs, gills).
- *Excess* H_2O and *mineral ions* from the diet and cell processes are removed by the excretory organs (eg kidneys) as well as the skin in some animals (eg sweating in humans).
- *Nitrogen compounds* (eg urea) from the breakdown of excess amino acids in the diet are removed by the excretory organs (eg kidneys). Nitrogen, N, compounds are typically toxic to the body.

Proteins are a large part of the diet of most animals. During digestion, proteins are broken down into their constituent amino acids, which in turn are used to make the specific proteins of the animal. Amino acids in excess of need may be converted into fatty acids or carbohydrates, or used in respiration. In these conversions, the N-containing parts of the amino acid are not used, and need to be excreted because of their toxicity.
These N-containing compounds are converted to one of the following:

- **Ammonia** – a toxic compound; N wastes can only be excreted in this form if the ammonia can be rapidly diluted, so ammonia excretion occurs only in aquatic organisms (eg some fish).
- **Urea** – less toxic than ammonia, so can be tolerated in the body temporarily before excretion; need quantities of water to dissolve and dilute the urea (eg excretion in mammals).
- **Uric acid** – virtually non-toxic and excreted as a solid by insects, birds and reptiles (allows for water conservation). As it is virtually non-toxic, it is a safe excretory product for the embryos of birds and reptiles (which are contained in shelled eggs).

Homeostasis

Excretion is an essential process for the maintaining of **homeostasis** – the ability of the body to maintain a constant internal environment. Body cells only function efficiently within a narrow range of conditions – eg pH, salinity, temperature, CO_2 concentration (CO_2 makes solutions acidic, eg blood plasma), osmotic pressure (from the concentration of water).

The body has a complex set of **homeostatic mechanisms** to keep the internal environment constant. **Hormones** are important in homeostasis:

- *Insulin* secreted by pancreatic cells is responsible for maintaining the blood *glucose* levels within a narrow range.
- *ADH* secreted by the pituitary is responsible for maintaining blood water levels.

Feedback loops are an important part of homeostatic mechanisms.

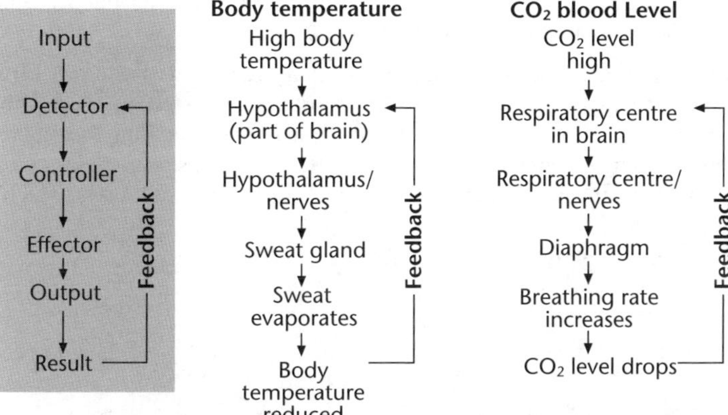

An animal's excretory system maintains the delicate balance of its chemical environment.

Excretion in earthworms

Earthworms belong to the phylum Annelida, the segmented worms. The bodies of all annelids are divided into many distinct body segments.

The earthworm has an excretory system that deals with water as well as nitrogenous wastes. Each segment (except the first three and the last one) has a pair of coiled excretory tubes called **nephridia**. The nephridia are open at both ends.

- The inner opening is a funnel-shaped, ciliated **nephrostome**, which opens into the internal body cavity or coelom (which contains the coelomic fluid).
- The outer opening or **nephridiopore** opens to the exterior. The opening of the nephridiopore is controlled by a sphincter muscle.
- The coiled tube is surrounded by blood capillaries. The tube is long to allow reabsorption of substances into the blood.

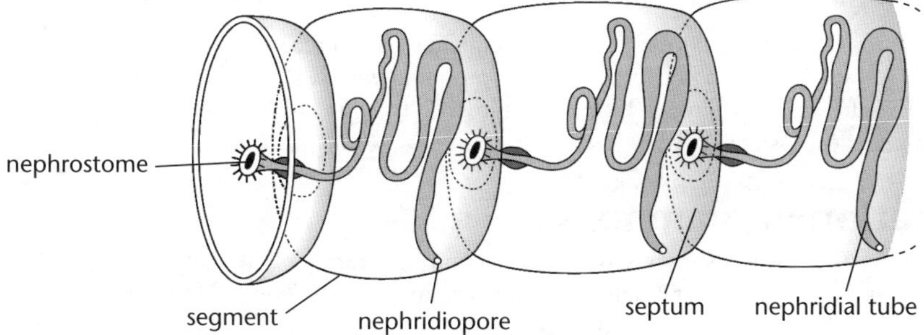

Earthworm nephridia in consecutive segments (one of each pair shown).

Each nephridium actually occupies two segments, because it opens externally on the ventral ('bottom') side of one segment and internally into the coelom of the previous segment, perforating the septum (membrane between segments).

The coelomic fluid contains both waste substances (such as ammonia and uric acid) and useful material. The coelomic fluid is sucked into the nephrostome by the action of cilia. The fluid passes down the narrow ciliated tubule. The blood capillaries surrounding the tubule reabsorb water and useful substances (eg mineral ions), while the waste substances move on through the nephridial tube towards the nephridiopore. At intervals, the sphincter muscle, which closes the pore, relaxes and waste is eliminated.

CO_2 is excreted over the whole skin surface, diffusing out from the blood.

Excretion in insects

The excretory organs of insects are called **Malpighian tubules** and number from 2 to 200 or more in different insects.

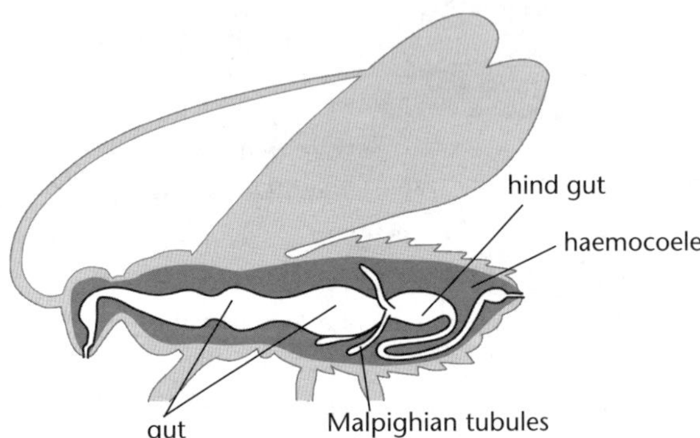

Malpighian tubules are elongated blind sacs, attached to the gut, that lie loosely inside the body cavity (or coelom), extracting nitrogenous wastes from the surrounding blood and converting them into **uric acid**. Water and dissolved ions that have entered the tubules are absorbed back into the blood in the coelomic cavity, so water is conserved.

Uric acid is virtually non-toxic and not very soluble. As uric acid moves along the tubules, it loses water and forms crystals, which are mixed with indigestible food wastes to form faeces. The egested wastes are virtually free of water. This ability to conserve water enables insects to inhabit the driest habitats on Earth.

Birds and reptiles also produce uric acid. Most mammals produce urea – this is a mildly toxic nitrogenous waste that can be stored only when diluted with water in a **bladder**. Having a bladder would make flying very difficult for birds! CO_2 in insects is excreted directly into the air from the tracheal system.

Excretion in mammals

The excretory organs in mammals are the very complex **kidneys**. The most important nitrogenous waste is **urea**, $CO(NH_2)_2$, which is produced in the **liver** from the breakdown of amino acids in a process known as **deamination** (a continuous process, as only very limited amounts of amino acids can be stored in the body). Urea is mildly toxic, so must be diluted with water prior to excretion by the kidneys. The kidneys also excrete excess water and mineral ions/salts (some excretion of water and salts also occurs with sweating). Water is also lost from the lungs in gas exchange; CO_2 is excreted from the lungs.

Excretion in humans

Urea is carried from the liver to the kidneys, where it is excreted together with other dissolved substances as **urine**. Excretion of urea is only one of several vital functions of the kidneys. Other functions include:

- Osmoregulation (the regulation of body water content).
- The regulation of blood pressure.
- The regulation of the pH of the blood.

The kidney

The kidney is one of the most important organs involved in homeostasis (ie the maintenance of a stable internal environment), because the kidneys:

- Remove toxic wastes and excess dissolved ions.
- Maintain the body's water balance and metabolic balance (solute and mineral levels in the blood and in the body's cells).

The kidneys do this by filtering blood (about 120 mL per minute) supplied by the **renal arteries**. From this filtered blood, useful materials are reabsorbed back into the bloodstream, but excess minerals, urea and other toxic wastes are held in the kidney and diluted with water to form **urine**. On average, about 1.5 litres of urine are produced per day (though this varies according to circumstances).

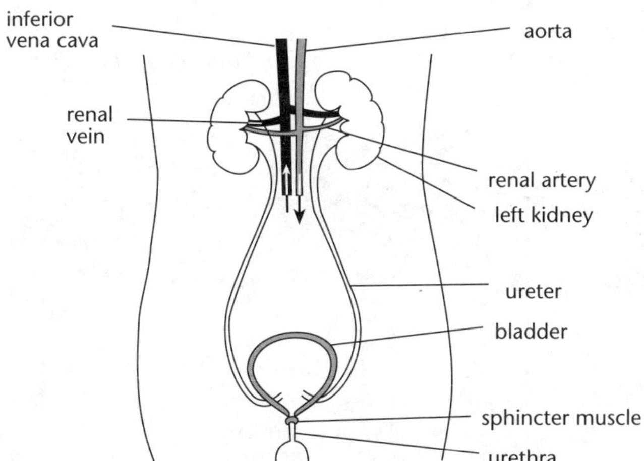

The human excretory system.

Some idea of the importance of the kidneys can be shown from the fact that even though their combined mass is only about 0.5% of the body, they take about 20% of the resting heart output of blood (ie, each gram of kidney takes about $\frac{20}{0.5}$ = 40 times as much as its 'fair share' of blood.)

Blood reaches the kidneys via the **renal arteries** and leaves via the **renal veins**.

Urine is carried away from the kidneys via the **ureters**. These propel the urine along by muscular waves called **peristalsis**. The ureters open into the **bladder**, where urine is temporarily stored. The opening of the bladder is guarded by a ring-like **sphincter muscle**. When the bladder walls are stretched, signals are sent to the brain via nerves, producing the desire to urinate – the sphincter muscle relaxes and urine is expelled by contraction of the smooth muscle in the bladder wall. Urine passes to the exterior through a tube called the **urethra**.

Structure of the kidney

A kidney has two layers – an outer, brown **cortex**, and an inner, pink **medulla**.

The inner part of the medulla forms bulges called **pyramids**. On the surface of each pyramid are the openings of hundreds of tiny **collecting ducts**. These carry urine into a cavity inside the kidney called the **pelvis**. Urine is carried from the pelvis of the kidney by the ureter.

Each collecting duct is the 'trunk' of a tree-like system of about a million blind-ending tubes called **nephrons**.

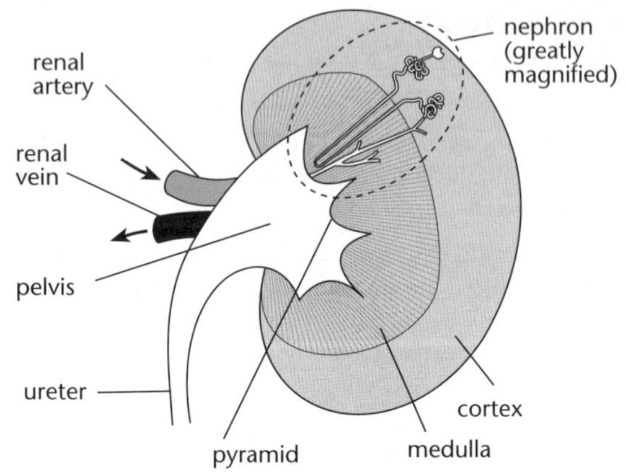

nephron (greatly magnified)
renal artery
renal vein
pelvis
ureter
cortex
pyramid
medulla

Structure of the kidney.

The filtering of the blood in the kidney occurs in millions of minute filters called **nephrons**. Nephrons are situated in the **cortex** and **medulla** of the kidney. Urine produced from the nephrons enters the kidney **pelvis**.

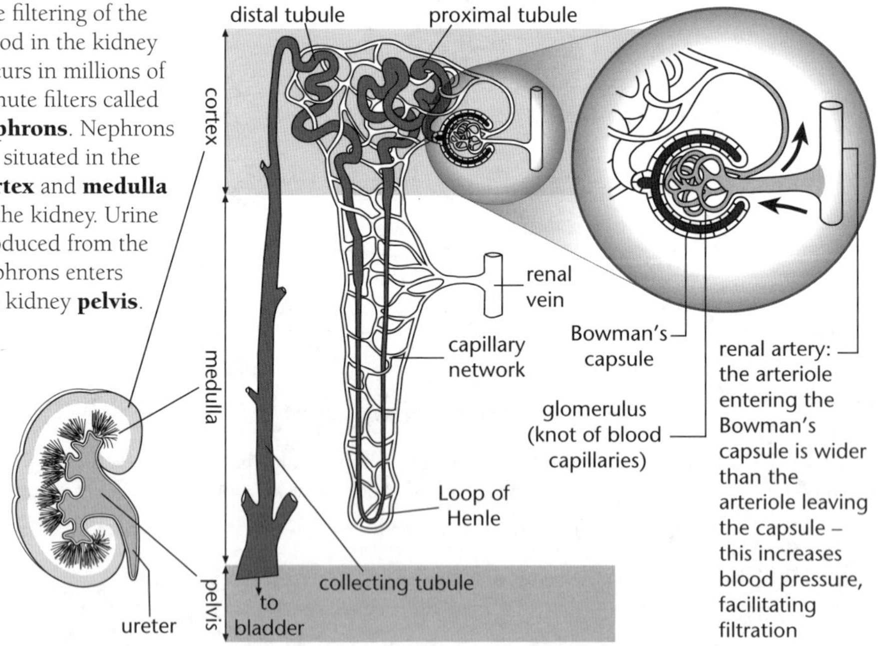

distal tubule
proximal tubule
cortex
medulla
pelvis
renal vein
capillary network
Bowman's capsule
glomerulus (knot of blood capillaries)
renal artery: the arteriole entering the Bowman's capsule is wider than the arteriole leaving the capsule – this increases blood pressure, facilitating filtration
Loop of Henle
collecting tubule
to bladder
ureter

Section through a kidney and expanded section showing a nephron

A nephron can be divided into two parts:

- **Bowman's capsule** – the cup-shaped, blind-ending of the nephron.
- A long **renal tubule** – about 4 cm long and consists of two *twisted* parts, separated by a hairpin **loop of Henle**.

The Bowman's capsules and the twisted parts of the renal tubules are all situated in the cortex. The loops of Henle and collecting ducts are in the medulla.

Dipping into each Bowman's capsule is a cluster of capillaries called a **glomerulus**. Blood reaches each glomerulus by a branch of the renal artery called an **afferent arteriole**; blood leaves the glomerulus via an **efferent arteriole** ('afferent' means going towards, 'efferent' means going away from). The efferent arteriole then splits up into a network of capillaries round the renal tubule. These capillaries eventually join to form branches of the renal vein.

Urine production

Urine is produced in two stages – filtration and reabsorption.

Filtration

Each glomerulus and Bowman's capsule acts as a tiny pressure filter, in which fluid is forced from the blood through the capillary walls. The energy for filtration comes from the blood pressure generated by the heart muscle.

Filtration is rapid because the blood pressure in the glomerular capillaries is about twice that in capillaries in other parts of the body. One reason for this is that the efferent arteriole is narrower than the afferent arteriole, forming a 'bottleneck'.

The glomerular capillaries allow all substances in the plasma except proteins to leak out. The liquid filtering through into the Bowman's capsule is therefore the same as plasma minus its proteins, and is called the **glomerular filtrate**. It contains many useful substances, such as glucose, amino acids and other foods, besides harmful substances, such as urea. The useful substances are reabsorbed back into the blood during reabsorption.

Reabsorption

Each renal tubule is surrounded by a network of capillaries. As liquid moves along the first twisted tubule (or proximal tubule), all the glucose and amino acids, together with most of the salts, are reabsorbed into the blood by *active transport*. The energy for this process comes from respiration in the cells of the renal tubule. Because dissolved substances are being removed from the liquid in the tubule, its *water concentration* rises and becomes higher than that in the blood. This causes water to pass by osmosis from the tubule into the blood. As a result, most of the water in the filtrate is also reabsorbed from the first twisted tubule back into the blood. The urea is thus much more concentrated, and by the time the liquid has reached the end of the collecting ducts it is called **urine**.

Secretion of hydrogen ions

Hydrogen ions formed from digestion are actively transported from the capillary network around the nephron into the distal tubule to form part of the urine. This is a homeostatic mechanism, as these hydrogen ions are removed from the blood to maintain correct acidity (pH) levels.

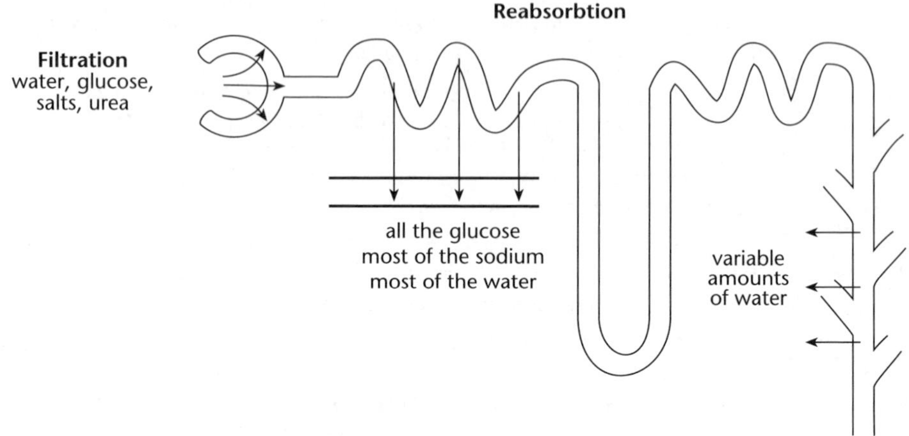

How urine is produced.

The average daily urine output is about 1.5 litres – less than 1% of the 180 litres filtered.

Urine composition

The figures below give the composition of blood plasma and urine. The *composition of urine varies a lot* according to a person's diet, activity, intake of liquid and body temperature.

Component	Blood plasma	Urine
Water, H_2O (%)	90–93	95
Urea, $(NH_2)_2CO$ (g)	0.03	2
Uric acid, $C_5H_4N_4O_3$ (g)	0.003	0.05
Ammonia, NH_3 (g)	0.0001	0.05
Sodium ions, Na^+ (g)	0.3	0.6
Potassium ions, K^+ (g)	0.2	0.15
Chloride ions, Cl^- (g)	0.37	0.6
Phosphate ions, PO_4^{3-} (g)	0.003	0.12

Urine composition in humans (representative).

Urine output varies considerably:

- When the body is dehydrated, less urine is produced.
- If the blood is slightly diluted by drinking a lot of water, urine output increases.

The regulation of the water concentration of the blood (osmoregulation) is an example of *homeostasis*.

Besides excretion and osmoregulation, the kidneys have several other functions, eg they help regulate blood pressure, blood pH, and the number of red blood cells.

Regulation of body water levels

Receptors in the hypothalamus of the brain respond to low water concentration in the blood by sending nerve impulses to the pituitary gland. The pituitary releases **ADH** (**antidiuretic hormone**), which increases the permeability of nephron tubules to water. The extra water reabsorbed restores the water levels of the blood and the pituitary stops producing ADH. Water regulation by the body is a typical homeostatic mechanism.

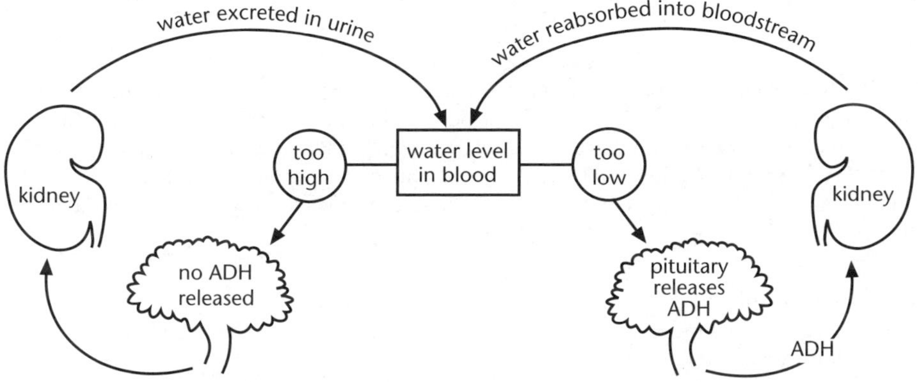

Homeostatic water regulation system.

Unit 11.3 Activity 5A: Excretion in animals

1. Distinguish between excretion and egestion.
2. Explain the role of excretion in animals.
3. Explain the role of excretion in osmoregulation.
4. Explain the role of kidneys in maintaining homeostasis.
5. Explain why excretory systems are closely associated with circulatory systems.
6. Explain why most multicellular animals need an excretory system.

Kidney diseases

Kidney stones

Kidney stones are lumps of mineral material – usually calcium phosphate or calcium oxalate – precipitated in the pelvis of the kidney. They are usually extremely painful. Previously they were removed by surgery, but it is now possible to use high-frequency vibrations to break kidney stones into fragments small enough to pass out with the urine. If they don't pass out or aren't removed, kidney stones can block urine flow and may result in infection and kidney damage. Kidney stones can be prevented by drinking plenty of water each day, and having a diet low in protein, nitrogen, salt and foods containing oxalates (found in many commonly eaten fruits, vegetables and cereals).

Nephritis

Nephritis is inflammation of one or both kidneys – often as a result of a bacterial infection – that affects the glomeruli. Often there are no symptoms, but when there are, they usually involve production of a smaller volume of dark urine that contains blood and protein. Losing too much protein from the body is a serious condition since proteins are essential to many life processes, eg enzymes, blood clotting and transport of substances (including oxygen) around the body. Treatment with antibiotics will kill the bacteria but, to avoid damage to the kidneys, food and fluids high in salt and protein are limited, and the volume of fluids consumed is also reduced.

Reflux

Usually, urine entering the bladder through the ureter is prevented from flowing back up the ureter by a valve that shuts. However, in some people, this valve doesn't function correctly and urine can reflux back into the kidney. Any bacteria present in the bladder can then enter the kidneys, causing infection and kidney damage. One of the symptoms of reflux is recurring kidney infections. These can be treated using antibiotics, but surgery may be needed to repair the defective valve.

Blood pressure

The kidneys play an important role in regulating blood pressure – in turn, high blood pressure can result in damage to the kidneys.

The kidneys regulate the amount of fluid in the blood. Too much fluid in the blood as a result of damage to, or disease in, the kidney may result in high blood pressure. High blood pressure over a long period of time results in damage to blood vessels. If blood vessels supplying the kidney are damaged, then substances may not be:

- Filtered from the blood vessel into the glomerulus, or
- Reabsorbed into the blood vessels from the loop of Henle.

A kidney disease contributing to high blood pressure needs to be identified and treated, while high blood pressure can be managed through changes in lifestyle and medications (*see* Chapter 8).

Unit 11.3 Activity 5B: Excretion in humans

1. For each of the phrases **1–10**, write the letter **A–J** of the term to which it applies.

Phrase	Applied term
1. Blind-ending, first part of nephron	A. Bladder
2. Brings blood to kidney	B. Bowman's capsule
3. Carries urine away from bladder	C. Cortex
4. Carries urine away from kidney	D. Glomerulus
5. Cluster of capillaries in which filtration occurs	E. Loop of Henle
6. Hairpin part of nephron	F. Medulla
7. Inner region of kidney	G. Renal artery
8. Muscular chamber that temporarily stores urine	H. Urea
9. Nitrogenous waste, produced in liver	I. Ureter
10. Outer region of kidney	J. Urethra

2.

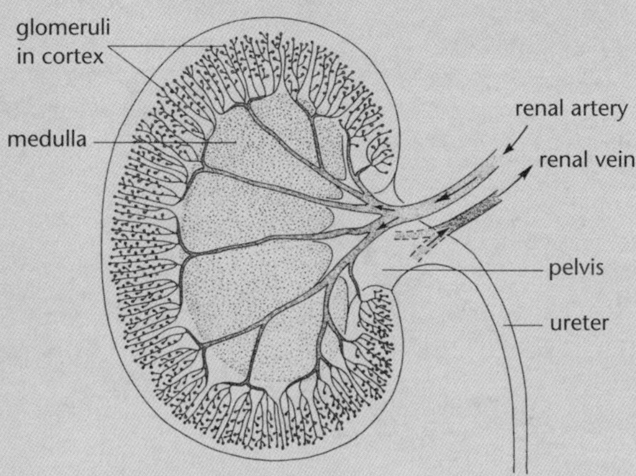

a. Describe the function of the renal artery.
b. Describe one function of the kidney.

3. Each kidney is made up of many nephrons. The diagram shows part of a nephron. Explain what is happening in Area A in the diagram.

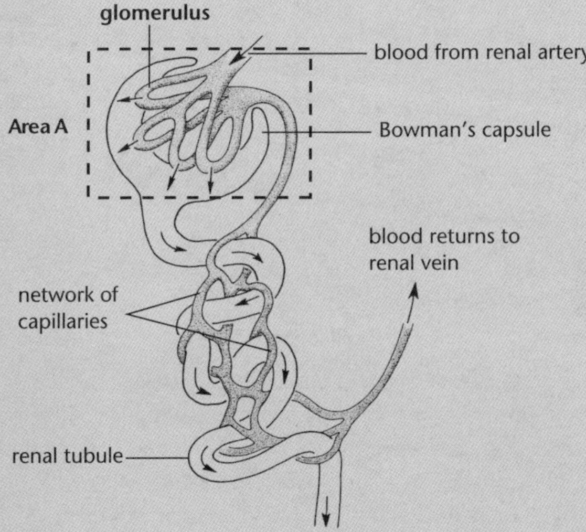

4. When there is a decrease in the amount of urine produced over a period of time, substances such as salts and minerals in the urine may stick to the walls of the kidney to form kidney stones.
a. Explain why a person who has had a kidney stone should drink large amounts of water to prevent another kidney stone forming.
b. Describe a possible outcome of a kidney stone forming in the ureter.
c. Describe a difference between urea and urine.

5. Discuss reasons for the similarities and difference between the composition of blood in the glomerulus and the network of capillaries around the renal tubule.
6. Each human kidney is made up of millions of kidney tubules. The diagram shows one kidney tubule.

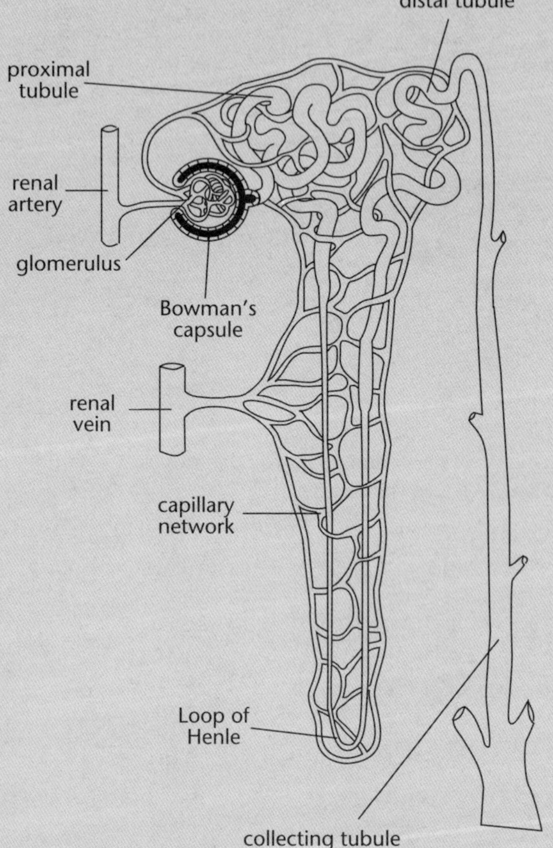

   a. Name three substances that pass from blood into the kidney tubule.
   b. Explain how the substances in part **a.** pass from blood into the kidney tubule.
   c. Some substances or components of blood should never enter the kidney tubule.
      i. Name two of these substances or blood components.
      ii. Explain why these two substances or blood components cannot enter the kidney tubule.
7. In hot weather, when a person sweats a lot, they urinate less often and the urine is a dark colour. In cold weather, a person sweats little; they urinate more and the urine is pale in colour. Explain these observations referring to kidney function.
8. Discuss how a kidney allows the waste products to be filtered from the blood. You need to consider the composition of the blood in the renal artery and the composition of the blood in the renal vein.

9. The diagram shows the renal system.

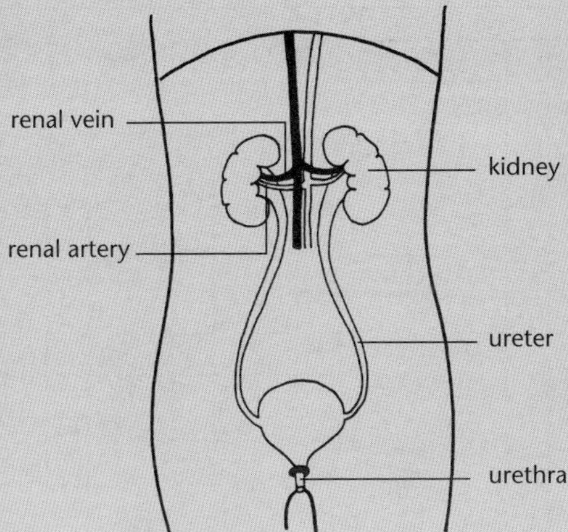

Describe the function of the following parts:

a. Ureter.

b. Urethra.

c. Kidney.

10. Describe the pathway of blood through the kidney from the aorta to the vena cava.

11. Nephritis causes inflammation of the nephrons in the kidney, and the glomeruli cannot work properly. Explain how the damaged glomeruli prevent the kidney from working properly, and any possible consequences.

12. Which one of these statements is *not* typical of kidney function?

A. excretion of urea
B. regulation of body water content
C. regulation of body temperature
D. regulation of blood pressure
E. regulation of blood pH

13. Urine is formed in the kidneys and is carried to the bladder via the:

A. urethra
B. sphincter muscle
C. collecting duct
D. ureter
E. nephrons

14. Select the substance below that is *not* a main component of urine.

A. water
B. sodium
C. urea
D. glucose
E. salts

15. Which one of these statements does *not* describe re-absorption processes in kidneys?

A. glucose, amino acids and salts are reabsorbed
B. re-absorption occurs by passive means
C. re-absorption takes place in the proximal tubules
D. re-absorption into blood occurs by active transport
E. the remaining urea is concentrated

16. Complete the crossword puzzle on the excretory system:

Across

1. The control of body fluids
3. Endocrine gland in brain that assists osmoregulation
7. This type of waste is handled by kidney
9. Body temperature determined by environment
10. This system regulates body salts and fluids
11. Organ that handles nitrogenous wastes
13. Secretes aldosterone for osmoregulation (two words)
14. The control of body temperature

Down

2. The recovering of critical salts by the kidney
4. Functional unit of the kidney
5. Capillary knot in the kidney
6. Blood flow that assists with thermoregulaton
8. A common type of nitrogenous waste in terrestrial animals
12. Controls body temperature using anatomical features

Crossword puzzle from 'Instructor Resources/Manual' accompanying *Biology: Concepts and Investigations* by Mariëlle Hoefnagels (McGraw–Hill).

Unit 11.4 Respiration and Gas Exchange

Topic 1: Respiration and gas exchange in plants

Authors: Martin Hanson with Beatrix Marjen-Waiin

The first three Topics in this Unit look at the principles of gaseous exchange in plants and animals, and the fourth Topic covers the principles of movement of substances in living cells.

This material is relevant to Unit 11.4 of the learning outcomes of the Biology Syllabus pp. 16–18 and the Biology Teacher's Guide pp. 31–32.

Related Topics in this book:

Unit 11.1 Living Things, Topic 7: Diversity of plants (p. 57)
Unit 11.3 Transport Systems, Topic 1: Transport systems in plants (p. 121) and Topic 2: Transpiration (p. 137).

Respiration and gas exchange in micro-organisms is summarised on p. 195 but it is recommended that students read Topic 2: Human responses to micro-organisms (p. 341) in the Supplementary Unit.

Plants take in CO_2 for photosynthesis and release O_2 as a by-product of that process. The mesophyll layer of the leaf contains most of the cells that carry out photosynthesis in the presence of water and carbon dioxide. Carbon dioxide enters the leaf via stomata openings on the underside of the leaf and the oxygen produced exits the leaf via the same route. Other atmospheric gases may also enter. The opening and closing of the stomata are regulated by specialised epidermal cells, called guard cells.

Photosynthesis primarily occurs in the palisade layer of the mesophyll, but can take place in the spongy mesophyll as well. The spongy mesophyll cells are loosely arranged in comparison to the palisade layer and this arrangement allows for a lot of air space between cells. Water required for photosynthesis is supplied by the xylem vessels to moisten the mesophyll.

The large air space and film of water on cell surfaces allow for the rapid diffusion of gases across the cell membranes of the mesophyll. Carbon dioxide diffuses into the mesophyll cells and oxygen diffuses out. Gaseous exchange always takes place by simple diffusion across a cell membrane. This process occurs efficiently if the gases are dissolved in a film of water found coating the cell.

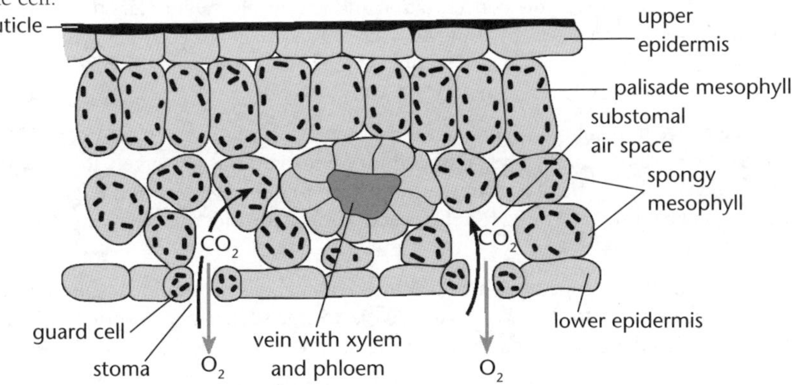

Cross-section showing the layers of a leaf.

photosynthetic equation:

$$\underset{gas}{CO_2} + 2H_2O \xrightarrow[\text{in leaves}]{} CH_2O + H_2O + \underset{gas}{O_2}$$

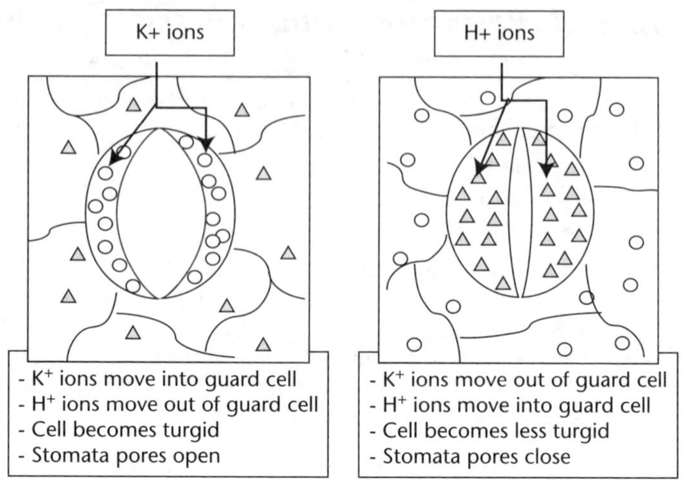

This diagram shows an open stomata (left) and a closed stomata (right).

Stomata open during the day for carbon dioxide to enter and carry out photosynthesis. They close at night or in the dark, and photosynthesis stops. The opening and closing of the stomata is controlled by the changing of the shape of its guard cell. The changes in the shape of the guard cell can be due to:

- Effect of light and darkness.
- Effect of temperature/dehydration.
- Effect of CO_2 concentration.
- Hormonal control.
- Biological rhythms – circadian rhythm.

Studies have shown that the opening and closing of the guard cells is due to the movement of H^+ and K^+ ions across the plasma membrane of the guard cells. The active diffusion of K^+ ions by the guard cell increases solute concentration and as a result water enters the guard cell from the surrounding epidermal cells by osmosis. This causes the guard cell to become turgid, and the stomata open. The stomata close when the K^+ ions are pumped out of the guard cell and the guard cell loses its turgidity.

Gaseous exchange in plants commonly takes place in the leaves of the plants via the stomata. This occurs because the components required for photosynthesis are largely found within the leaves. However, other parts of a plant that have been found to be involved in gas exchange include young green stems, young root hairs and associated epidermal cells, and the lenticels of older stems and roots.

The sugars made in photosynthesis are used by the plant cells for aerobic respiration, which releases chemical energy for other cellular activities. Some sugars are converted into organic compounds which the plant needs, such as starch and cellulose.

Photosynthesis by plants accounts for the high level (approximately 20%) of oxygen in the atmosphere. This gas is essential for the process of aerobic respiration.

Without photosynthesis, animals would not exist.

Respiration and gas exchange in micro-organisms

Micro-organisms are tiny single-celled organisms that do not have any specialised breathing organs. Oxygen is carried across the plasma membranes and cell walls.

Some species like single-celled green algae and photosynthetic bacteria behave like plants and carry out photosynthesis with their very own photosynthetic pigments. These photoautotrophic micro-organisms are responsible for production of over half of the earth's atmospheric oxygen. The remainder is produced by green plants.

Most other species of micro-organisms do not use oxygen but use alternative sources to complete their metabolic processes. These microbes utilise anaerobic respiration or fermentation. They are largely called chemoheterotrophs.

Anaerobic respiration and fermentation is used by micro-organisms in waterlogged or very deep soil depths, stagnant ponds or animal intestines. Inorganic substances like nitrate or sulfate are used by the microbe as a substitute for molecular oxygen as the terminal acceptor in the electron transport chain.

The fermentation process does not need the electron transport chain step. Certain bacteria and some fungi utilise this system to yield enough energy to sustain their metabolic needs.

Note: micro-organisms are covered in much more detail in the Supplementary Unit (p. 329).

Unit 11.4 Activity 1A: Respiration and gas exchange in plants

1. Which of these statements is true of respiration in plants?

- **A.** the effect of respiration is typically seen at night time
- **B.** the effect of respiration is typically seen at day time
- **C.** plants consume more CO_2 at night
- **D.** plants produce more O_2
- **E.** the effect of respiration is typically masked by photosynthesis at night

2. Name three features of a leaf that allow it to effectively take up CO_2.

3. Complete the word equation (A) and formula equation (B).

(A) glucose + ________ → __________ + water + ATP + ______

(B) _______ + $6O_2$ → $6CO_2$ + ________ + 38ATP + _______

4. In which part of the leaf would you find the most chloroplasts?

- **A.** guard cells
- **B.** epidermis
- **C.** mesophyll
- **D.** palisade
- **E.** all of the above

5. An experimental setup was designed to investigate the source of water loss in leaves. Plant stems were placed in water in cylinder tubes and the leaves were painted with clear nail polish in three different ways.

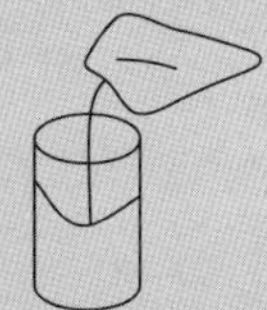
A. Leaf top painted

B. Underside painted

C. Both sides painted

D. No sides painted

The level of water was equal (50mL) in all cylinders at the start of the experiment. A layer of oil was added to the top of the water to prevent water loss from evaporation. At the end of the week, the results of water loss were as shown in the table.

Table 1. A week's observation and readings

A	B	C	D
45	50	50	45

a. Explain why A lost more water than B and C.
b. What do the results mean? Write one sentence to summarise the results.
c. Why did B and C not lose water?
d. What do these experimental results tell us about stomata?

6. The evaporation of water from the leaf of a plant is:
A. transpiration
B. totally prevented by the leaf's cuticle
C. hydrolysis
D. condensation
E. sublimation

7. Which of the following will *not* increase the rate of transpiration in a plant?
A. low humidity
B. high wind speeds
C. high humidity
D. high temperature
E. none of the above are correct

8. The pressure placed on the inside of a cell membrane of a plant by water is called:
A. adhesion
B. cohesion
C. osmosis
D. turgor
E. hydrolysis

9. If water is abundant a plant's guard cells will ______________ and the stomata will ______________.
A. swell, close
B. collapse, close
C. collapse, open
D. swell, open
E. none of the above are correct

10. Answer True or False to the following:
a. The part of a plant's roots that forces water to enter cells and move into the xylem is the Casparian trip.
True False
b. The cohesion–tension theory explains how water moves within a plant.
True False
c. Water is pulled from a plant's roots to its leaves by transpiration.
True False

Unit 11.4 Respiration and Gas Exchange

Topic 2: Respiration, energy and gas exchange

Authors: Martin Hanson with Beatrix Marjen-Waiin

This Topic covers aspects of respiration (p. 9 Biology Syllabus) by looking at:

- Respiration and fermentaton as energy-yielding chemical processes in cells.
- Aerobic and anaerobic respiration.

Cells and energy

Living cells are continuously active. They use energy for purposes such as:

- Building large molecules from small ones.
- Movement – eg, when muscle cells change their shape, or when chromosomes are moved during cell division.
- Active transport – eg, absorbing digested food from the intestine, or reabsorbing glucose in the kidneys.

Cells can extract energy (ATP) not just from carbohydrates but also from fats (lipids) and proteins.

All animals respire aerobically (except for endoparasites, which are parasites, like tapeworms, that live inside their hosts). During the process of aerobic respiration, nutrient compounds are broken down by catabolic reactions, which are chemical reactions that break down large molecules into smaller molecules. Over 90% of energy produced by a cell depends on the presence of oxygen which completes the process.

Animals that are very active have high metabolic rates and consume a lot of energy. They require more oxygen per gram of body weight than sedentary animals with low metabolic rates.

An organism's metabolic rates are also tied to the temperature ranges with which they are tolerant. Therefore, endothermic animals like mammals which have a high internal body temperature will need more oxygen per gram of body weight than ectothermic animals like fish or amphibians living in colder temperatures. Animals have evolved different strategies to meet their unique individual oxygen needs.

Human cells get their energy from 'fuels' such as glucose as a result of the process of cellular **respiration**. Respiration in living cells can occur aerobically (in the presence of oxygen) or anaerobically (in the absence of oxygen).

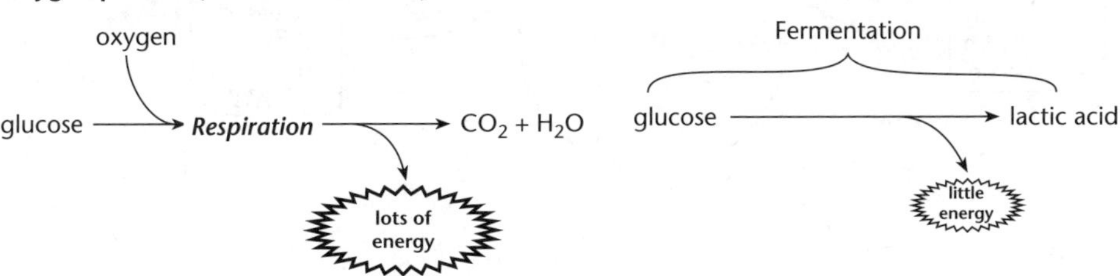

The two energy-giving processes in animal cells.

Sometimes, muscles need to work so hard that the blood system cannot supply oxygen fast enough. In such situations, the muscles are able to get energy from glucose without oxygen, or **anaerobically**. This process is called **fermentation** and produces **lactic acid**.

Lactic acid is slightly toxic and its accumulation in the muscles is one reason they cannot work 'flat out' for very long. During recovery, some of the lactic acid is oxidised in respiration. This provides enough energy to convert the rest back into glucose.

Respiration

Respiration occurs in the mitochondria of the cells in *all* organisms *all* the time (both day and night), ie 24/7. Respiration is the process by which the cell breaks down glucose to produce **ATP (adenosine triphosphate)**; heat energy is a by-product. ATP is the so-called 'energy molecule' in all organisms and is used to fuel all the chemical reactions of the cell, eg:

- Active transport of substances across membranes.
- Synthesis of molecules – eg proteins from amino acids.
- Movement – eg phagocytosis, action of cilia and flagella, action of actin and myosin in muscle contraction.
- Bioluminescence (light production) in cells of such animals as glow-worms and fireflies.

The breakdown of glucose may be *aerobic* or *anaerobic*; aerobic produces much larger amounts of ATP per glucose molecule than anaerobic.

ATP is constantly made in cells from **ADP (adenosine diphosphate)**. The energy from glucose metabolism adds a high-energy phosphate bond to ADP to make ATP:

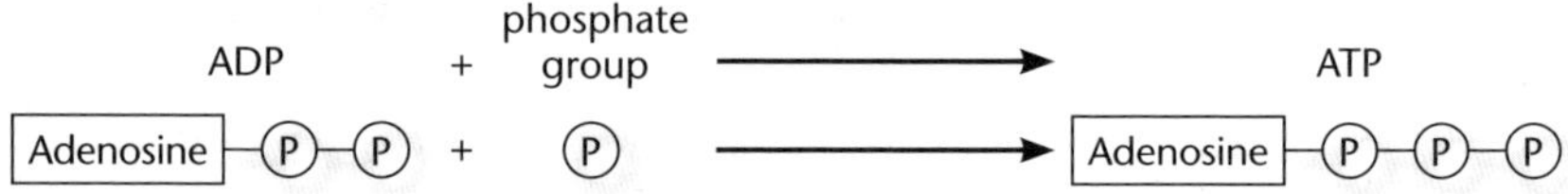

When a cell needs energy, the high-energy phosphate bond is broken and ATP returns back to ADP (ADP could be thought of as a 'flat battery' and ATP as a 'charged battery'). The process is very rapid and ongoing.

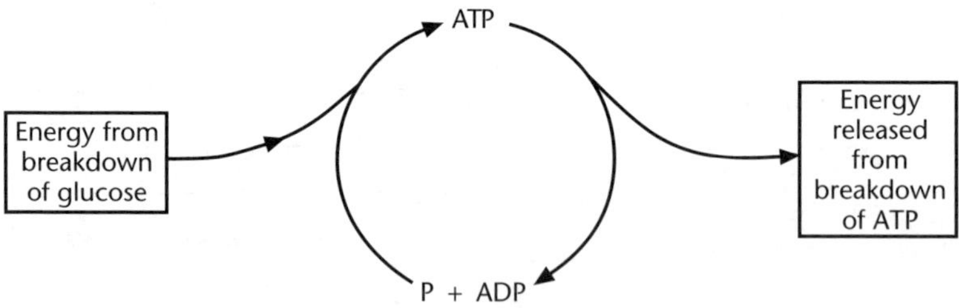

ADP/ATP cycle in cells.

The heat energy produced as a by-product of respiration (aerobic respiration is approximately 45% efficient) is used in maintaining body temperature in homeotherms ('warm-blooded' animals – birds and mammals).

Aerobic respiration

Aerobic respiration needs *oxygen* for the complete breakdown of glucose into carbon dioxide and water; energy is released in the form of ATP and heat. There are three main enzyme-controlled chemical pathways in aerobic respiration – glycolysis, Krebs cycle and the electron transfer chain.

Glycolysis, the first chemical pathway, occurs in the **cytoplasm** of the cell:

- Each glucose molecule is broken down into two pyruvate molecules.
- Two molecules of ATP are produced in the breakdown.

The **Krebs cycle** (also known as the *Citric acid* cycle), is the second chemical pathway of aerobic respiration, and it occurs in the matrix of the mitochondria:

- Pyruvate (in the form of acetyl CoA) is fed into a complex biochemical cycle. As the pyruvate passes around the cycle, extensive rearrangement occurs, with CO_2 molecules and H atoms being produced. The molecules that make up the cycle are not used up in the reaction.
- CO_2 is the waste product, and diffuses out of the mitochondria and the cell.
- H atoms are picked up by a carrier molecule (NAD, a co-enzyme) and taken to the third chemical pathway of aerobic respiration.

The **electron transfer chain** (also known as the *Hydrogen transfer chain* or the *Respiratory chain*) is the final chemical pathway of aerobic respiration, and it occurs on the **cristae** of the mitochondria:

- H atoms are ionised and their high-energy electrons are passed along a series of acceptor molecules (cytochromes) attached to the cristae.
- As the electrons are 'bounced' along the electron transfer chain, their energy is used to form ATP from ADP.
- At the end of the electron transfer chain, the electrons are returned to the H ions, which become atoms again and combine with O_2 to form H_2O.

It is this third stage of aerobic respiration which produces the most ATP. One molecule of glucose that enters glycolysis yields 38 molecules of ATP by the end of the electron transfer chain.

The equation for aerobic respiration is:

Word equation

glucose + oxygen ⟶ carbon dioxide + water + ATP + heat

Formula equation

$$C_6H_{12}O_6 + 6O_2 \longrightarrow 6CO_2 + 6H_2O + 38ATP + heat$$

Mitochondria

Mitochondria (singular = mitochondrion) are commonly known as the 'powerhouses of the cell', as they are the site of **aerobic respiration**, in which glucose, $C_6H_{12}O_6$, is broken down in a *series of enzyme-controlled chemical reactions* to become CO_2 and H_2O, along with the production of the 'energy molecule' ATP.

Mitochondria are elongated ovals in shape (like sausages) and their inner membrane is thrown up into folds called **cristae** (singular = crista). Cristae provide a large surface area for the respiratory chemical reaction known as the *hydrogen transfer chain* to take place. The space between the cristae is known as the *matrix* and is the site of the respiratory chemical reaction known as the *Krebs cycle*.

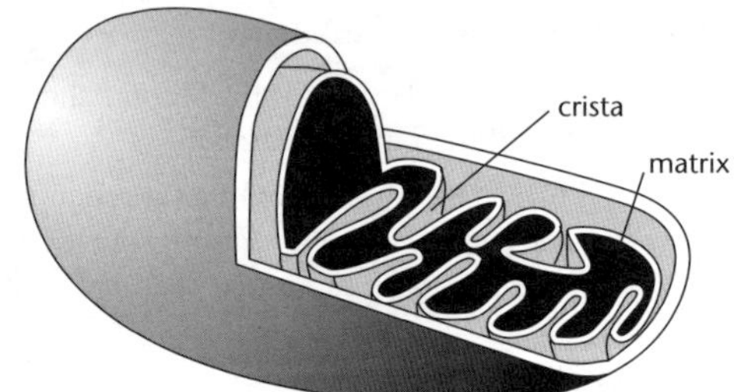

Cut-away view of a mitchondrion.

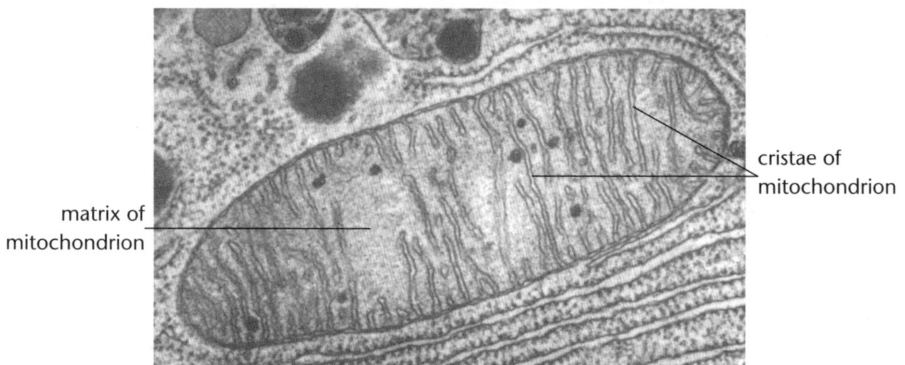

TEM photomicrograph of a mitochondrion in cytoplasm.

Mitochondria are more common in animal than in plant cells (as the energy demands of animal cells are typically higher), and are especially common in cells with high energy demands (eg sperm, muscle) – the most active of these cells may have close to 100 mitochondria. Cells lining the kidney tubules have large numbers of mitochondria, as they are very active in re-absorbing substances by *active transport*.

Mitochondria have their own DNA and protein-making machinery (ribosomes, RNA), and it is believed that they may once have been free-living organisms (possibly bacteria) that were 'captured' by other cells and then retained to become an organelle. Mitochondria can reproduce themselves.

Anaerobic respiration

Anaerobic respiration occurs in the *absence* of oxygen. *Only* glycolysis takes place, therefore *only* 2 molecules of ATP are produced from each glucose molecule.

Anaerobic respiration in animals

Anaerobic respiration in animals occurs when oxygen is in short supply; this typically occurs during prolonged exercise (only animals such as endoparasites will respire anaerobically as a way of life). The pyruvate formed in glycolysis is broken down into **lactic acid**. The build up of lactic acid causes *muscle fatigue*. The muscles need to stop working to allow replenished supplies of oxygen to continue the breakdown of the lactic acid into water and carbon dioxide.

Rate of respiration

The *rate* of respiration is determined by factors such as:

- *Temperature* – increasing the temperature increases the rate of respiration up to an optimum temperature. As temperatures rise too high above the optimum temperature, the enzymes controlling the reaction denature and can no longer catalyse reactions; respiration ceases (and the organism dies). In homeotherms, the optimum temperature is core body temperature (eg 37°C in humans).
- *Body's energy demands* – the rate of respiration will increase up to a maximum as the demand by the cells of tissues increases (eg muscle cells of legs when a human sprints). The amount of O_2 needed and the amount of CO_2 produced increase, and breathing rate increases to compensate. The build-up of CO_2 will slow down the rate of respiration.

The rate of respiration is typically measured by the amount of O_2 consumed or the amount of CO_2 produced in a given time.

The full name for respiration is **cellular respiration** – as it occurs in the cells to release energy, a chemical process. It is *not* the same process as **breathing** (inhaling and exhaling of gases into the body, a physical process) or **gas exchange** (exchange of O_2 and CO_2 across a membrane by diffusion, a passive process).

Comparing photosynthesis and respiration

Photosynthesis	Respiration
• Converts solar energy into chemical energy in the form of glucose (and starch).	• Converts chemical energy in the form of glucose into chemical energy in the form of ATP and heat energy.
• Uses CO_2 and H_2O as raw materials to make glucose and release O_2 as a waste product.	• Aerobic respiration uses glucose and O_2 as the raw materials and releases H_2O and CO_2 (CO_2 is a waste product).
• Occurs in plant cells (chloroplasts) only during daylight hours.	• Occurs in the cells (mitochondria) of all organisms all the time.

In plants, the effect of respiration in the daytime is typically masked by photosynthesis, with the *net* result being the consumption of CO_2 and the production of O_2. Therefore, the effects of respiration are typically seen only at night in plants.

Unit 11.4 Activity 2A: Photosynthesis and respiration

1. Distinguish between the following pairs of terms:
   a. Mitochondria and chloroplasts.
   b. Grana and stroma.
   c. Light-dependent and light-independent reactions.
   d. ATP and ADP.
   e. Aerobic and anaerobic respiration.
   f. Glycolysis and Krebs cycle
   g. Krebs cycle and electron transfer chain.
   h. Calvin cycle and Krebs cycle.
2. Explain the structure of mitochondria.
3. Explain why so much more energy is produced in aerobic respiration compared with anaerobic respiration.
4. Respiration is sometimes said to be the reverse of photosynthesis. Explain why.
5. The diagram alongside shows a cell from the palisade layer of a leaf. Explain how the rate of photosynthesis is affected by the location of the chloroplasts in a palisade layer cell.

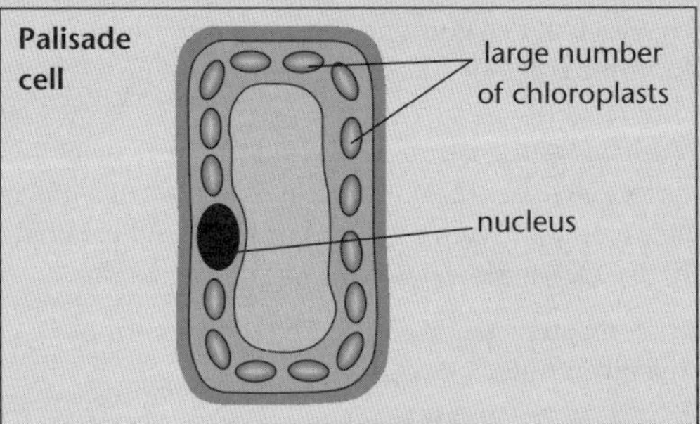

6. Use the information in the following graphs to discuss the factors affecting the rate of photosynthesis:

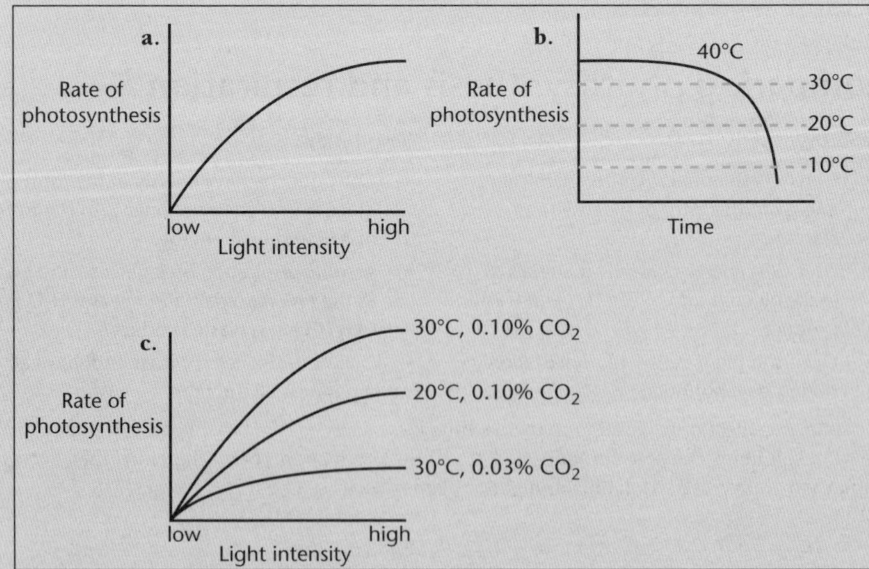

Unit 11.4 Respiration and Gas Exchange

Topic 3: Gas exchange in animals

Authors: Martin Hanson with Beatrix Marjen-Waiin

This Topic explores the different ways in which different animals respire and exchange gases for survival. It looks specifically at gas exchange in different creatures: earthworms, insects, amphibians, fish, birds, mammals and the human breathing system (Biology Syllabus pp. 16–17). The Topic deals with:

- The requirements of a gas exchange system – thin, moist, large surface area (SA), diffusion, concentration gradients.
- The structure and function of the parts of gas exchange systems.
- How the gas exchange system carries out gas exchange – includes breathing, transport.
- Reasons for the differences in structure and function between different groups – these could relate to mobility and energy needs, size, habitat, life cycle, way of life.
- The human breathing system.
- Disorders of the breathing system.

All animals respire aerobically (except for endoparasites) to release the energy needed for their cell processes. Aerobic respiration needs the gas *oxygen*; it produces the gas *carbon dioxide*. These two gases need to be *exchanged across membranes of gas exchange surfaces*. This exchange occurs across cell membranes:

- Of individual cells of the *skin* (eg earthworms, some amphibians).
- Through specialised organs such as *gills* (eg fish, some crustacea) and *lungs* (eg reptiles, birds, mammals).

Gas exchange across membranes occurs by **diffusion** (for O_2, this is likely to be by *facilitated diffusion*). Therefore, all gas exchange surfaces must be:

- *Moist* – O_2 must dissolve before it can diffuse.
- *Thin* – reduces the diffusion distance, making diffusion quicker.
- *Large in surface area* – increases quantity of gases diffused per unit time.

Animals may need to **breathe** (also known as *ventilation*) to *facilitate* gas exchange. Breathing is a *physical process*, involving muscular contractions to move air or water over the gas exchange surfaces (eg filaments of gills, alveoli of lungs). More rapid gas exchange allows for a higher rate of respiration. This allows for increased metabolism and increased activity levels for the animal.

Transport of gases to/from gas exchange surfaces uses the **circulatory system** in most animals; therefore, the gas exchange and transport systems are in close contact. (Insects are a notable exception, as their tracheal gas exchange system is separate from the blood system.)

The following distinctions need to be noted:

- **Respiration** – a *chemical process* that occurs in cells to *release energy*.
- **Gas exchange** – a *physical process* (**diffusion**) in which the gas O_2 (needed for respiration) is exchanged for CO_2 (produced in respiration) across a *membranous surface*.
- **Breathing** – a *physical process* involving *muscular* movements to take O_2 and CO_2 to and from gas exchange surfaces.

Gas exchange in earthworms

Earthworms have an external gas exchange system, as the entire skin of the earthworm is the gas exchange surface. The skin is kept moist by body processes (ie *physiologically*) through mucus glands secreting a moistening fluid, and *behaviourally* by the worm inhabiting damp areas, avoiding sunlight and feeding at night.

The skin surface is a suitable exchange surface in such a relatively 'large' animal as the earthworm, because earthworms have:

- A low metabolic rate, which does not use up oxygen rapidly.
- A *moist*, highly *vascularised* ('full of blood vessels') skin that allows extensive gas exchange between the air and blood.
- A circulatory system, which contains haemoglobin in the blood.
- A long, thin body shape, giving a large SA (surface area) for gas exchange.
- Blood vessels in the skin to ensure a short diffusion pathway for gases.

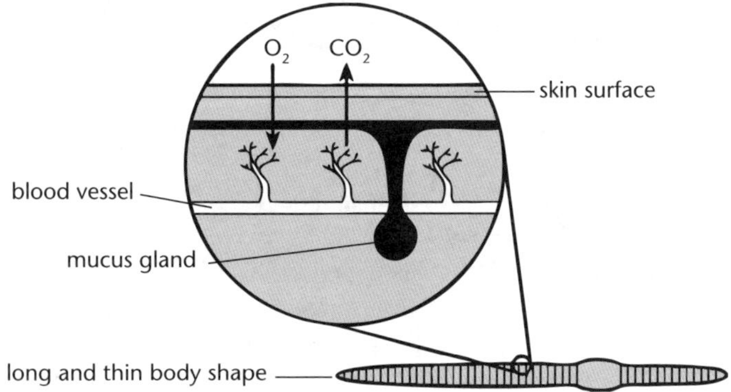

Gas exchange across the skin surface of the earthworm.

Gas exchange in insects

The internal gas exchange system of insects is unusual in that the blood of the transport system is *not* used to carry O_2 and CO_2 between the air and the body tissues. Insect have a **tracheal system** for gas exchange. Openings with valves (the **spiracles** – can be closed to conserve water) occur along each side of the abdomen. **Trachea**, bound with **chitin** to keep them open, lead from the spiracles and branch as they penetrate the body. The finest branches (**tracheoles**) lack chitin and pass between individual body cells. Oxygen dissolves in the moist environment surrounding the cells, and diffuses into the cells; carbon dioxide diffuses out. **Air sacs** may be present to increase the efficiency of the trachea. Muscular pumping of the abdomen (**breathing**) assists the movement of the gases into/out of the body and along the trachea.

Insects have an *open* transport system, in which blood circulates sluggishly and transport of gases is slow (transport is much faster in the *closed* transport systems of most other animals). It is the tracheal system that allows for more rapid transport of gases that means insects have the higher metabolic rates needed for their active lifestyles, especially flying. The tracheal system, together with the exoskeleton, act to limit the size of insects.

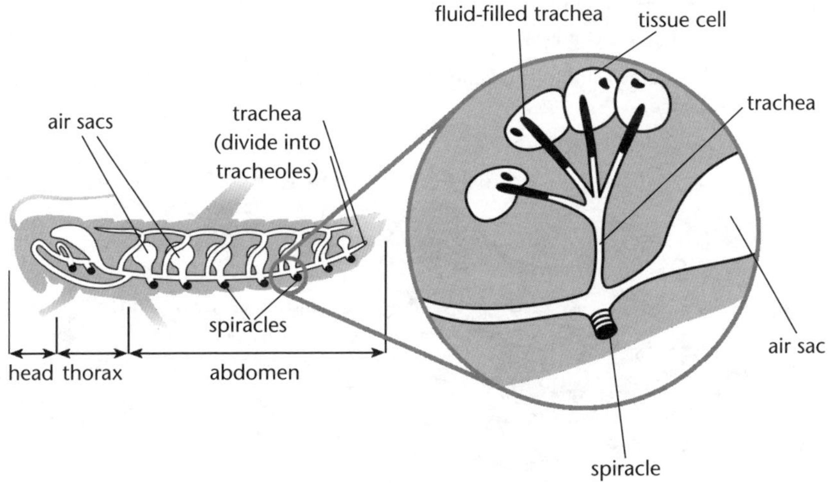

Gas exchange in the grasshopper.

Air taken in during **inhaling** enters the grasshopper through small holes in the thorax called **spiracles**.

Exhaling also occurs through the spiracles in the abdomen. Closing of spiracles and protective hairs around the spiracles reduces water loss.

Movements of the abdomen, thorax and **air sacs** assist air movement.

The circulatory system is not involved in gas exchange at all.

Gas exchange in amphibians

Amphibians are dependent on an aquatic or moist environment for both reproduction and gas exchange. Many amphibians have an aquatic larval stage when they are young (eg tadpoles of frogs), which metamorphoses into the mainly land-dwelling adult.

The aquatic young have **gills** for gas exchange, but these would be unsuitable for the terrestrial living adults, as gills:

- Cannot be supported by air.
- Cannot be kept moist.
- Cannot be easily protected.

Adults have a simple, balloon-like internal **lung** for gas exchange in the terrestrial environment. Many also rely on their *moist skin* for gas exchange. Oxygen diffusing in enters the blood capillaries for transport. The two gas exchange areas (internal lung and moist skin) means oxygenated and deoxygenated blood are mixed in the blood vessels. This reduces the rate of gas diffusion into and out of the cells (as the concentration gradient is reduced), limiting the rate of respiration and metabolism and therefore the activity levels of the animal.

Frogs gulp to facilitate gas exchange.

Reptiles, birds and mammals are fully adapted to terrestrial life, so have a complex internal lung system for gas exchange.

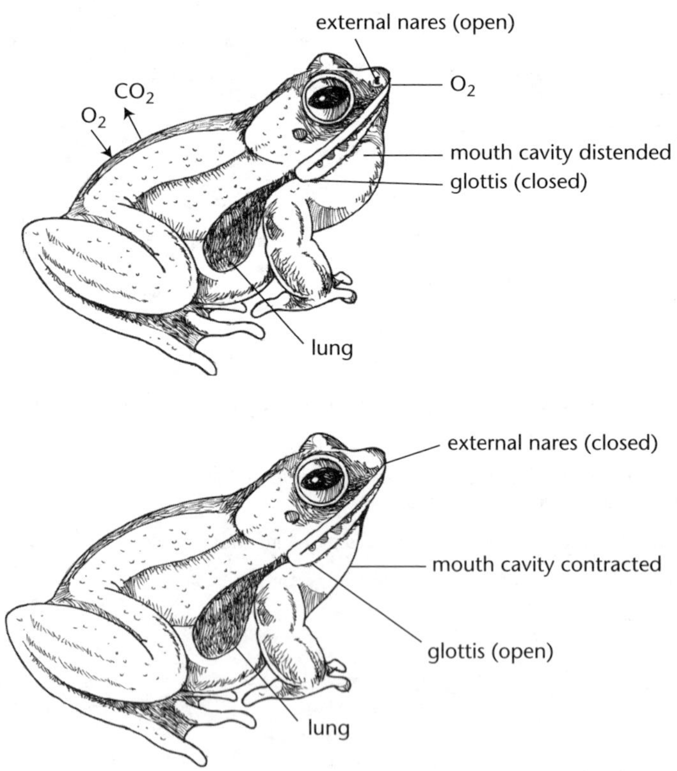

Gas exchange in fish

The gas exchange surface between the water and the blood in fish is the **gills**. The gills lie free in the water and are composed of many thin **filaments**, further divided into thin foldings called **lamellae**. Lamellae serve to greatly increase the SA for gas exchange – water contains much less dissolved oxygen than air. To further enhance gas exchange, fish gulp in water (breathing), and, by closing their mouths, force the water to exit over the gills.

- In the bony fish (Osteichthyes), the gills are protected by a bony cover called the **operculum**; their gill filaments are supported by bony structures.
- In the cartilaginous sharks and rays, the gills open to the sea via several slits; water is forced over the gills by the movement of the fish through the water.

Gills are dependent on the buoyancy of water for their support; in air the gills collapse and the fish suffocates. Gills are rich in blood capillaries. The direction of the blood flow in the gills is in the opposite direction to the flow of water over the gills – this is a *counter-current exchange* and serves to maximise the *concentration gradient*, facilitating the diffusion of O_2 from the water into the blood and CO_2 from the blood into the water.

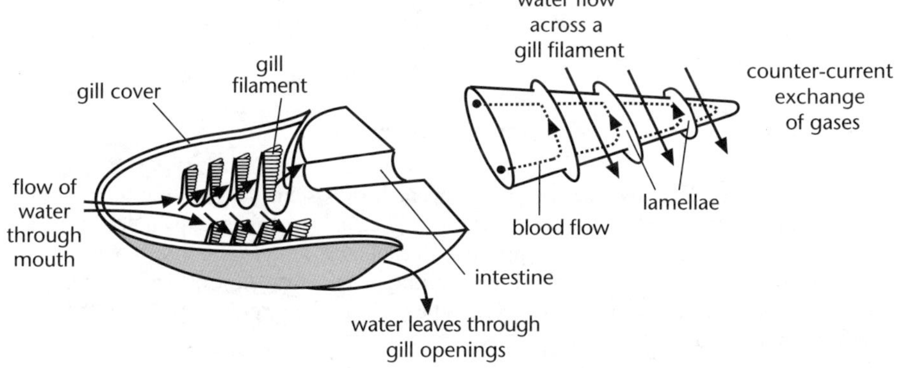

Gas exchange across the gill surface of a fish.

Most multicellular aquatic animals with gills have what is known as an evaginated gas exchange structure (gills). A few aquatic animals, such as the sea cucumber, use an invaginated gas exchange system. The counter-current exchange system is important to these aquatic organisms' existence in their habitats.

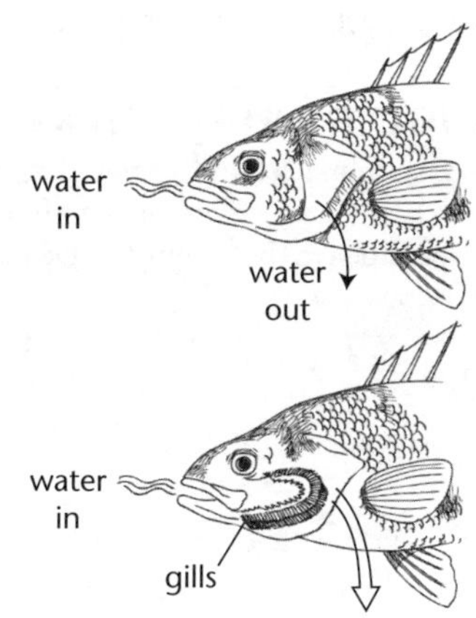

Flow of water through a fish.

Gas exchange in birds

Reptiles, birds and mammals are fully adapted to terrestrial life. They have developed a complex internal lung system for gas exchange.

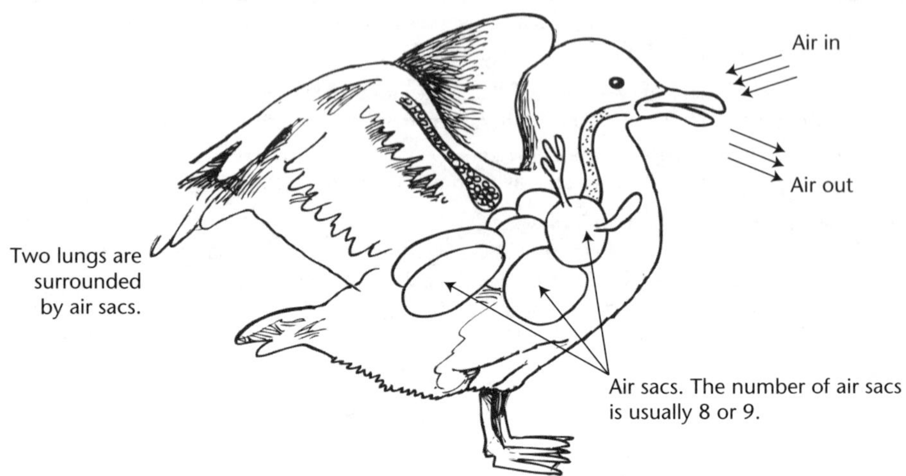

Respiratory system in a bird.

Birds have a different pattern of air flow in their respiratory system. This is because, in addition to paired lungs, they have about eight to nine thin-walled air sacs that occupy much of their body cavity. These air sacs are poorly supplied with capillaries and do not absorb oxygen or release carbon dioxide. The air sacs allow for a continuous one-way flow of air. Air enters into posterior air sacs and then into the lungs. From there, the exchanged gases pass into the anterior air sacs and out the trachea.

Air flows in one direction in bird lungs while the blood in the capillaries at the exchange surface flows in the opposite direction. This counter-current system makes birds far more efficient than mammals in extracting oxygen from air.

Evolution of lungs in vertebrates

The lungs of vertebrates (animals with backbones) show a progressive increase in surface area. The simplest types are sacs supplied with blood. The larger they evolved, the more partitioning they developed to give them a greater exchange surface area to accommodate their gas exchange needs.

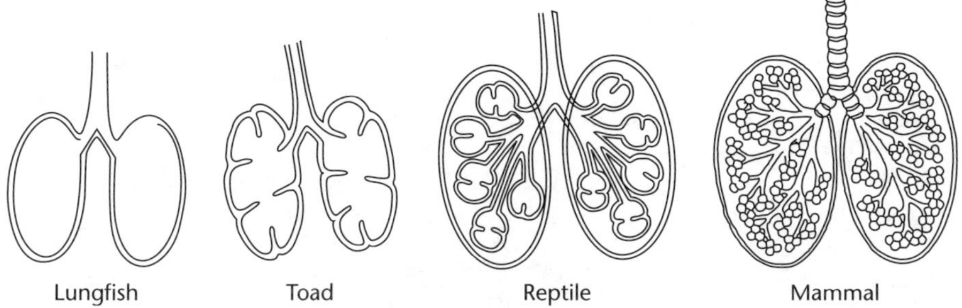

The evolution of lungs in vertebrates.

As mammals evolved, the partitioning of their lungs evolved and they also developed alveoli which provided a much greater surface area for gas exchange.

Gas exchange in mammals

Mammals are very active homeotherms ('warm blooded'), so have a high oxygen demand, met by complex internal **lungs** for gas exchange which closely connect with a closed, double-transport system.

Air passes first through the **nasal cavity**, which is moist so the moisture content of the air will be high enough to prevent the surfaces of the rest of the system from becoming too dry. The air is also *warmed* as it passes through the nasal cavity. Any large dust particles are trapped by hairs in the nose before entering the **trachea**.

Food or liquid is directed to the stomach and prevented from entering the lungs by a flap of tissue called the **epiglottis**, which covers the trachea during swallowing.

The trachea is strengthened and supported by rings of **cartilage** (similar to chitin in insects); the cartilage is flexible enough to allow the neck to bend, but strong enough not to collapse and shut off the trachea as the neck moves.

The trachea is covered in a film of mucus, which traps small particles which are brought up from the trachea to the nasal cavity by the beating of **cilia**. The mucus and other waste products are swallowed or ejected when you 'blow your nose'.

The trachea divides into two **bronchi** (singular bronchus), one to each lung. The bronchi divide into **bronchioles** (which do not have cartilage bonds); bronchioles divide further still, ending in minute air sacs called **alveoli** (singular alveolus), which provide a very large moist surface area (about 100 m^2) for gas exchange to occur. Oxygen diffuses from the alveoli into the blood capillaries and into the red blood cells, where it is picked up by the haemoglobin for transport to the body cells. Carbon dioxide diffuses out of the cells into the blood plasma for transport to the alveoli.

Lungs are classified as an invaginated gas exchange organ. An invaginated gas exchange organ is limited to a region of the body of the animal and must be richly supplied by a blood transport system.

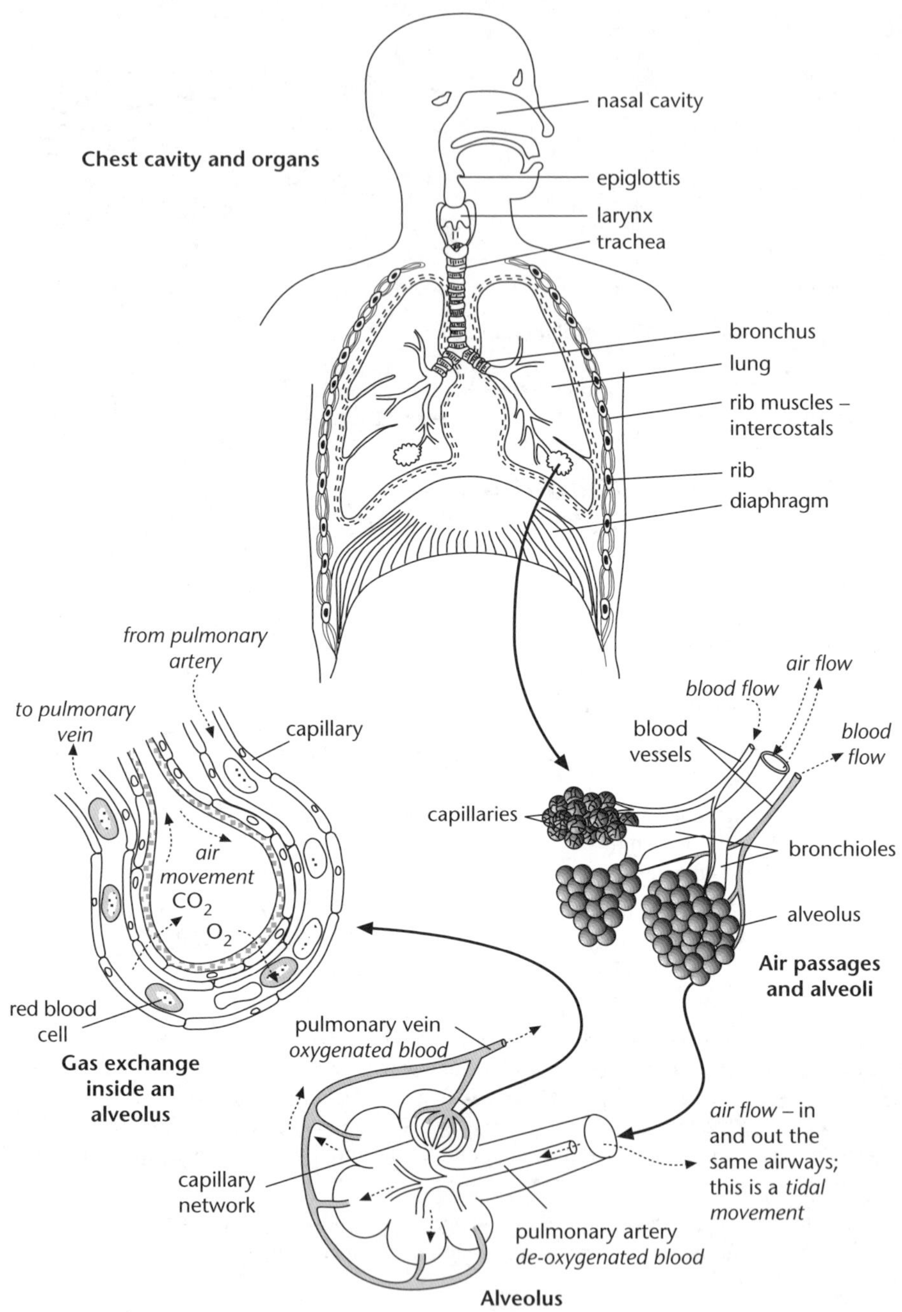

Gas exchange within the human lung.

Breathing

During inhalation, air is sucked into the lungs as the chest cavity enlarges and air pressure inside the lungs falls below outside air pressure. The **diaphragm** is the main muscle involved, but the shoulder and rib muscles (intercostals) also assist.

Exhalation is passive – the diaphragm and rib muscles relax, reducing the space inside the chest cavity, so air, now higher in pressure than air outside, is pushed out of the lungs and the lungs deflate.

The rate at which breathing occurs is controlled by the **respiratory centre** at the base of the brain. Signals from this centre automatically contract the diaphragm and rib muscles to cause inhalation. These signals are triggered by the level of carbon dioxide, CO_2, in the blood: the higher the level of CO_2, the more rapid the rate of breathing.

During normal relaxed breathing, about 500 mL of air is inhaled per breath. A resting adult takes about 12 breaths per minute, making 12 × 500 mL = 6 litres of air inhaled per minute.

During strenuous exercise, the breathing rate and volume of air taken in during each breath rise dramatically, and inhalation rates as high as 100 litres of air/minute are achieved.

During exercise, toxic **lactic acid** accumulates in muscle tissues. Buildup of lactic acid is known as **oxygen debt**. Oxygen is needed to change this lactic acid back to glucose. Rapid breathing after exercise has finished removes lactic acid.

Breathing is not a very efficient process, as only one quarter of the oxygen inhaled ends up in the pulmonary bloodstream, because:

- 30% of the air which enters the lung is in 'dead space' in the bronchi or bronchioles.
- Oxygen must diffuse from the lung to the bloodstream, and diffusion is a relatively slow process. Only some of the inhaled oxygen will have diffused into the bloodstream by the time exhalation occurs.

Unit 11.4 Activity 3A: Gas exchange in animals

1. Describe the requirements for any gas exchange surfaces in animals.

2. For animals, distinguish between the following:

  a. Respiration and gas exchange.

  b. Breathing and gas exchange.

3. Describe the location of gas exchange surfaces in animals.

4. Explain the need for maintaining a high concentration gradient for O_2 and CO_2 in gas exchange surfaces in animals.

5. Explain why most gas exchange surfaces in animals are closely associated with circulatory systems.

6. Explain why most multicellular animals need a gas exchange system.

7. Compare and contrast the tracheal system of insects with the lung system of mammals.

8. Compare and contrast the lung system of mammals with the gill system of fish.

The human breathing system

The *lungs* are two spongy organs on either side of the heart. Each lies in a **pleural cavity** between two **pleural membranes**. These membranes (inner and outer) are really the same membrane doubled back on itself. The pleural cavities are very narrow and contain just enough fluid (a few cm^3) to lubricate the movements of the lungs during breathing. *Pleurisy* is an infection of the pleural membranes and is described near the end of this chapter.

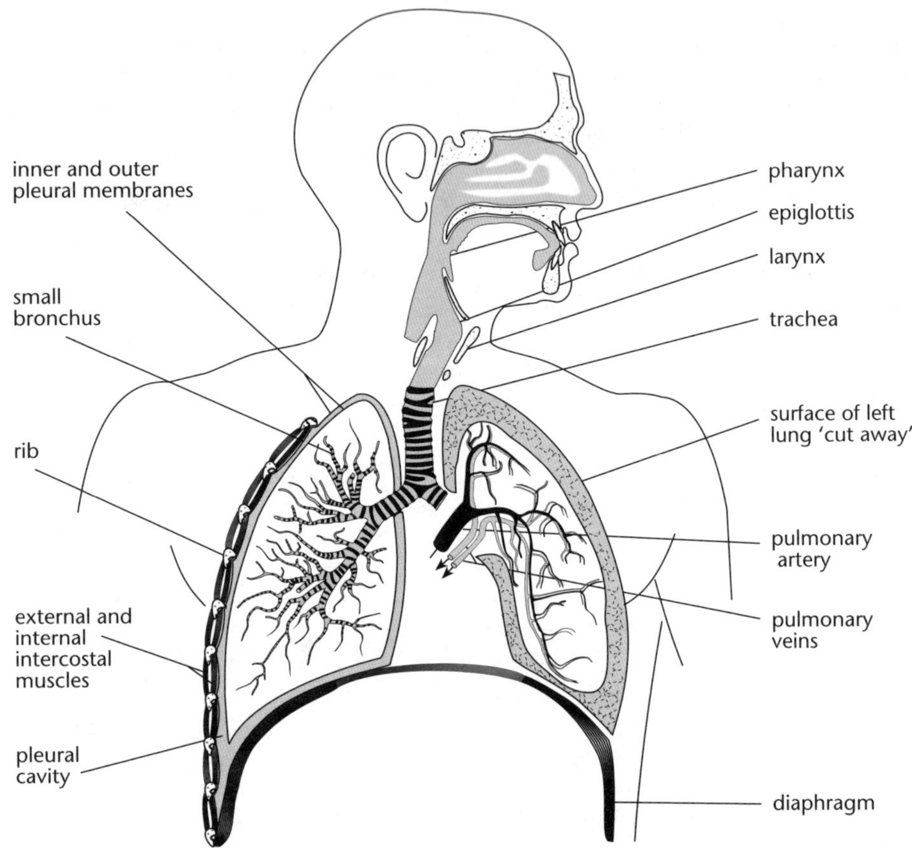

Human breathing system.

Air is carried to and from the lungs via a system of airways called the **bronchial tree**.

The 'trunk' of the tree is the **trachea** or windpipe.

- At the top of the trachea is the **larynx** (voice-box), which contains the sound-producing vocal cords.
- Above the larynx is the **pharynx**, through which both food and air pass. During swallowing, food is prevented from entering the larynx and trachea by the flap-like **epiglottis**. The walls of the trachea are supported by 'C'-shaped pieces of cartilage.

- The trachea divides into two main **bronchi** (singular, bronchus), which divide into smaller and smaller bronchi. The very small branches of the bronchial tree (less than 1 mm diameter) have no cartilage and are called **bronchioles**.
- The finest bronchioles open into clusters of air sacs or **alveoli** (singular **alveolus**). There are about 300 million alveoli in the average human.

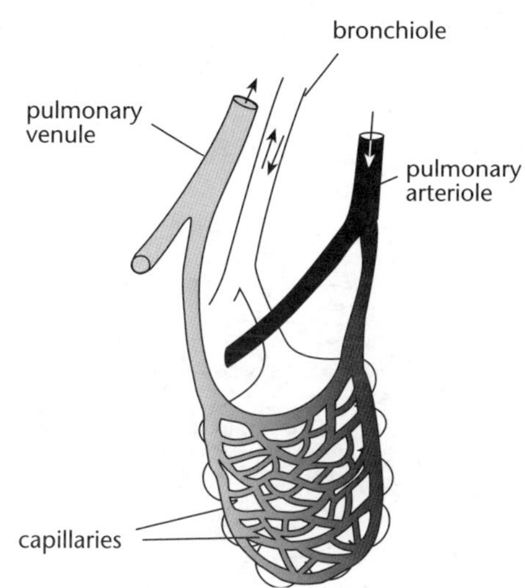

An alveolus and its capillaries.

The finest bronchioles open into clusters of air sacs or **alveoli** (singular **alveolus**). In the average human there are about 300 million alveoli.

- Each alveolus is covered by a dense network of capillaries supplied by a branch of the **pulmonary artery** and drained by a branch of the **pulmonary vein**. The alveoli and their capillaries are the *gas exchange surface*.
- Each alveolus is surrounded by blood in capillaries and the blood is surrounded by air in the alveoli.

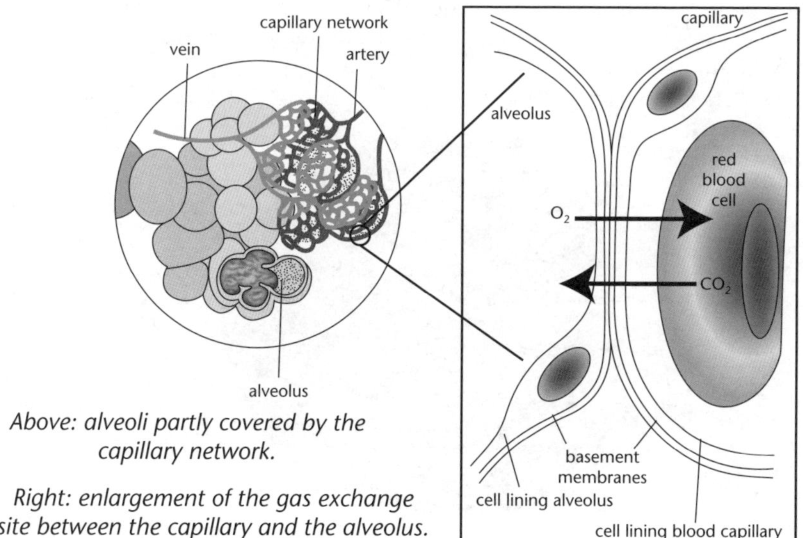

Above: alveoli partly covered by the capillary network.

Right: enlargement of the gas exchange site between the capillary and the alveolus.

When we inhale, the air we take in reaches the alveoli. Gas exchange occurs across the membrane of the alveolus and the capillary network.

Gas exchange

Gas exchange occurs because the concentration of oxygen in the air in the alveoli is higher than in the blood, and the CO_2 concentration is higher in the blood than in the alveoli – thus oxygen diffuses from air to blood, and CO_2 diffuses in the opposite direction. At the same time the blood changes from dark red to scarlet.

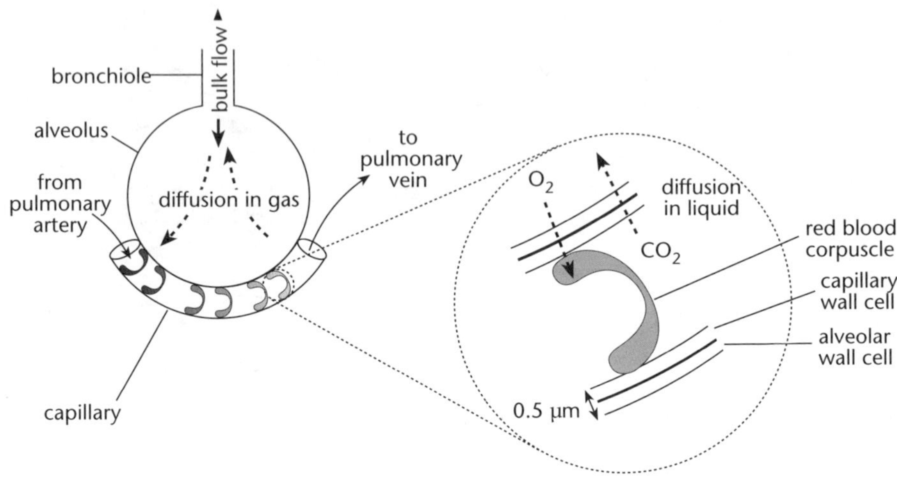

Diffusion of gases between air and blood.

Two features of the lungs help rapid gas exchange:

- The distance between the air and the blood is extremely short because the walls of the alveoli and capillaries are extremely thin (less than 1 µm). The oxygen and CO_2 concentration gradients are thus very steep. (This is important because diffusion in liquids is 10 000 times slower than in gases.)
- The dense network of alveolar capillaries has a *very large surface area* (about 125 square metres).

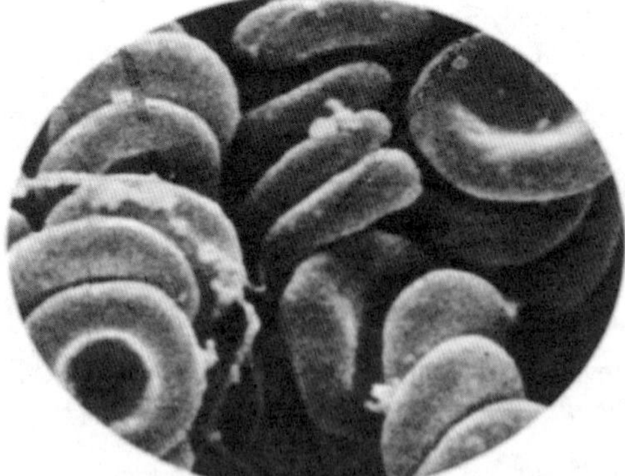

Photomicrograph of red blood cells taken with a scanning electron microscope. Each blood cell just fits into a capillary. The shape of the cells increases their surface area so allowing rapid uptake of oxygen entering from the alveoli.

Gas exchange in the alveoli

During exhalation, the first air to be breathed out is the air that had occupied the trachea and bronchi, and is called *bronchial air*.

The last part of an exhaled breath is *alveolar air*, and consists of air that had been in the alveoli.

When bronchial and alveolar air are analysed, it is found that only the alveolar air changes as a result of being in the body, *so the alveoli must be where gas exchange occurs*.

	Oxygen and carbon dioxide content of dried air (%)			
	Atmospheric air	Bronchial air	Alveolar air	Whole exhaled breath
Oxygen	20.96	20.96	13.8	16.4
CO_2	0.04	0.04	5.5	4.1

The figures are for *dried* air because the percentage water vapour in inhaled air varies. Exhaled air is saturated with water vapour which, at body temperature, accounts for about 6% of the air.

The 'whole exhaled breath' is a mixture of alveolar and bronchial air.

Analysis of bronchial and alveolar air.

Note that nitrogen gas makes up 78% of both inhaled air and exhaled air. Nitrogen is not exchanged in the alveoli, so all inhaled nitrogen is exhaled again. While the body does need nitrogen, it cannot be absorbed in the gaseous state.

Cleaning the airways

With their thin, moist walls, the alveoli would be a very good place for bacteria to invade the body.

One of the functions of the air passages is to remove bacteria and dust from the air *before* it reaches the air sacs. Some of the cells lining these airways secrete mucus or slime, which traps dust and bacteria. Other cells have many hair-like **cilia** that beat and slowly move the mucus and trapped particles towards the pharynx – backwards in the nose and upwards in the trachea and bronchi. As it reaches the pharynx, the mucus and trapped particles it are swallowed. Most of the bacteria arriving in the stomach are then killed by acid in the gastric juice.

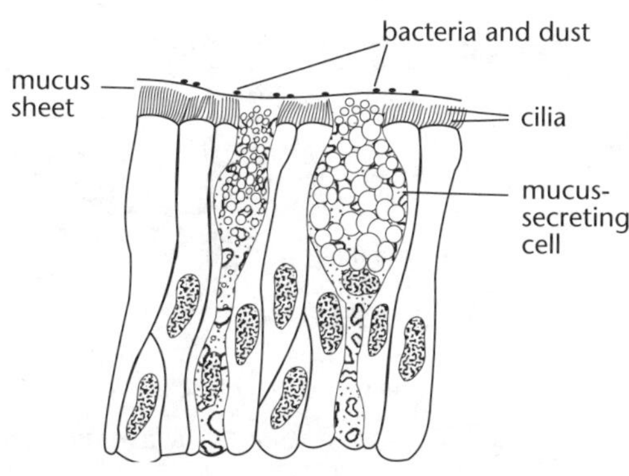

Cells that clean the airways.

This cleaning mechanism is harmed by smoking. Cigarette smoke temporarily paralyses the cilia, so that the lungs are more likely to become infected.

Unit 11.4 Activity 3B: Gas exchange

1. Describe two disadvantages of breathing through the mouth rather than through the nose.
2. Give the function of the cartilage that surrounds the trachea and bronchi.
3. Explain how breathing keeps a person alive.
4. Explain the purpose of respiration.
5. The table shows the concentration of gases in inhaled and exhaled air. Discuss the factors that may cause the differences between inhaled and exhaled air, for each of the named gases.

Gas	Inhaled air	Exhaled air
Oxygen	21%	16%
Carbon dioxide	0.03%	3%
Nitrogen	78%	78%
Water vapour	Small amount	Very moist

The breathing mechanism

The lungs are moved by action of the floor and sides of the chest. Unlike the flow of blood, the air movement is *tidal*, flowing alternately into and out of the lungs. There are thus two parts to the breathing cycle – *inspiration* (breathing in) and *expiration* (breathing out).

Inspiration

In inspiration, the volume of the chest is increased, causing the pressure of air in it to decrease. Air therefore flows from the higher pressure outside to the lower pressure inside the lungs. Air is therefore not *sucked* in – it is *pushed in* from outside by atmospheric pressure.

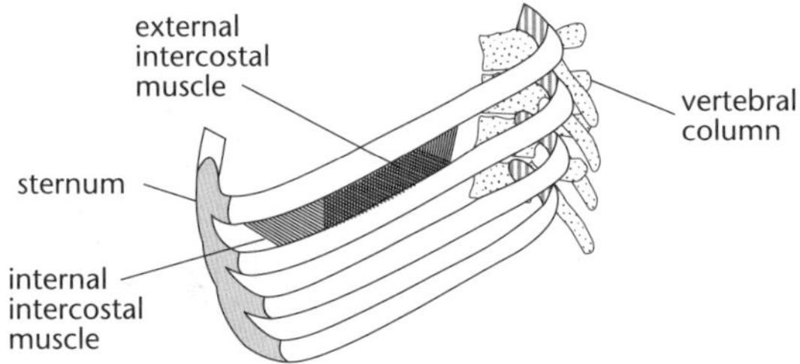

The intercostal muscles are shown only between two of the ribs.

The intercostal muscles.

Two sets of skeletal muscles increase the volume of the chest:

- The rib (intercostal) muscles – these form two layers. In breathing when at rest, only the *external* layer is used. When these contract the ribs move *up and outwards*. This increases the volume of the chest, and at the same time stretches the lungs and the cartilage of the rib cage.
- The diaphragm – this is a domed sheet of muscle forming the floor of the chest. When the diaphragm contracts, the floor of the chest is *lowered*. This pushes the abdominal organs down and outwards, stretching the muscular wall of the abdomen.

Although these movements are automatic, the diaphragm and intercostal muscles are also under voluntary control.

Expiration

Whilst inspiration is always active (because it involves muscle contraction), expiration in a resting person is *passive*. The main forces causing lungs to decrease in volume are due to the *elasticity* of the tissues that had been stretched during the previous inspiration. The most important of these are the lungs, the cartilage of the rib cage, and the muscles of the abdominal wall.

In forced breathing (during exercise, when blowing up a balloon, etc), the internal intercostal muscles are used to lower the ribs.

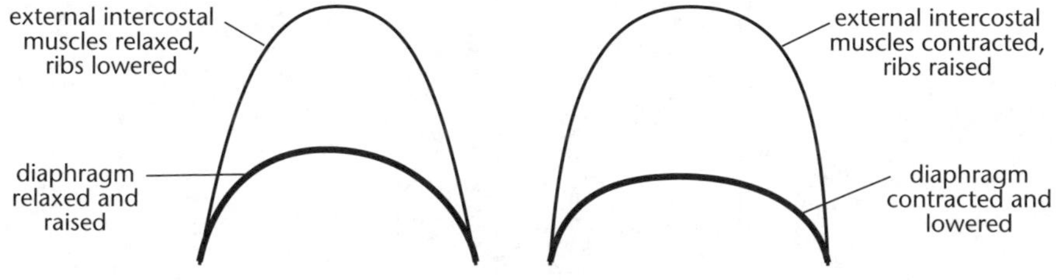

View is 'from the front'.

Changes in ribs and diaphragm during breathing.

During *forced* breathing (eg when panting or blowing up a balloon), expiration is *active*, and involves the contraction of the inner intercostal muscles. Also, the muscles of the abdominal wall contract and push the intestines and liver up against the diaphragm, forcing it upwards.

Lung volume during breathing movements

The total lung capacity in an average-sized adult is about 6 litres, but the maximum volume of a single breath – the **vital capacity** – is only about 4.5 litres. The remaining 1.5 litres is the **residual volume**, and cannot be exhaled because the lungs 'cannot be completely deflated'. The volume of air breathed in and out during natural breathing movements is the **tidal volume**.

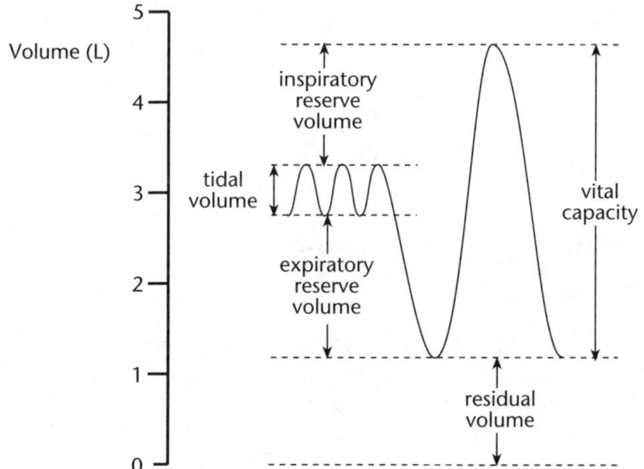

The trace shows normal breathing only uses about 10% of the lung capacity. An important result of this is that each inhalation only increases the volume of the alveoli by a small proportion, so the composition of the alveolar air (and thus the rate of gas exchange) changes little during the breathing cycle.

Trace showing the various lung volumes.

Recording lung volume during breathing movements

A simple way of measuring the volume of air exhaled involves an inverted bell jar.

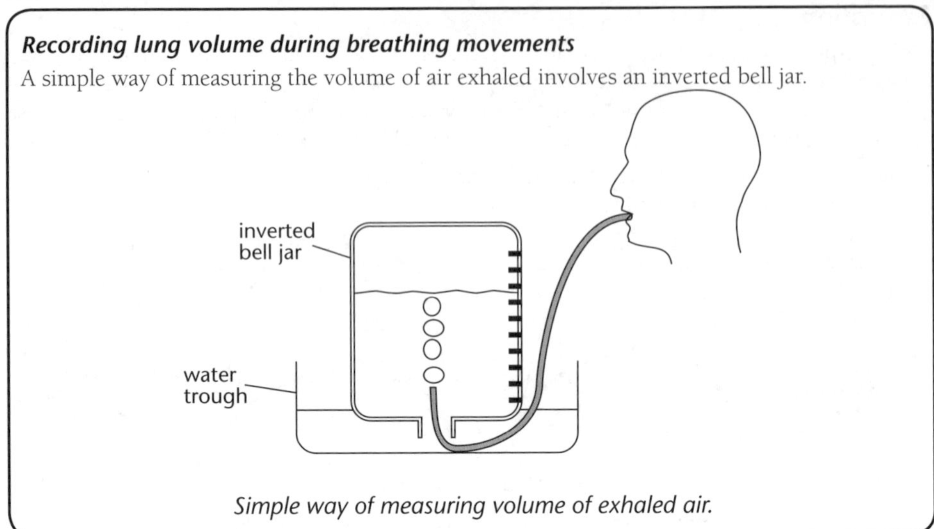

Simple way of measuring volume of exhaled air.

Breathing regulation

During exercise, the rate of transport of oxygen and CO_2 between lungs and muscles increases. This involves an increase in the rate at which air and blood are pumped through the lungs and muscles.

The rate and depth of breathing are controlled by a part of the brain called the *breathing centre*. This connects with the diaphragm and intercostal muscles by nerve fibres. During exercise, the carbon dioxide level in the blood rises slightly and the oxygen level falls slightly.

- The increase in carbon dioxide in the blood causes the breathing centre to increase the frequency of nerve impulses to the breathing muscles, causing them to work harder.
- The higher the CO_2 level, the harder the breathing muscles work and the greater the rate of excretion of CO_2.

Thus, the more the blood CO_2 level rises, the greater the tendency to correct it. This self-correcting tendency is called **negative feedback**, and is an essential feature of homeostatic mechanisms.

Disorders of the breathing system

Emphysema

This condition is one of the many harmful effects of smoking. The alveolar walls break down, causing them to coalesce into a smaller number of larger spaces. This greatly reduces the gas exchange surface, with a resulting chronic shortage of oxygen and the accumulation of carbon dioxide, so the person is permanently short of breath.

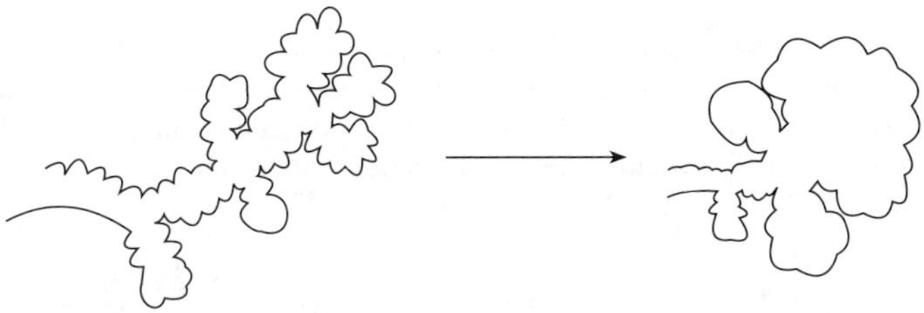

Normal alveoli

Alveoli coalesce in emphysema

Changes in the alveoli in emphysema.

The reduced number of alveolar capillaries increases the resistance to pulmonary blood flow, causing the right side of the heart to enlarge. The increased pressure in the pulmonary capillaries causes fluid to escape from them and to accumulate in the lungs. With less elastic tissue in the lung to hold them open, the bronchioles tend to become narrower, making it harder to breathe. The disease may last 20 years or more before death 'ends the suffering'.

Bronchitis

This is inflammation of the bronchial passages, with a persistent cough. In chronic bronchitis, the cough persists over long periods, and in severe cases, never gets better. The cough brings up lots of mucus, which can make the cough sound very 'wheezy'.

The main cause of chronic bronchitis is smoking (bronchitis is 4–10 times more common in smokers than in non-smokers). Tobacco smoke paralyses and eventually kills the cilia cells in the bronchi. This results in an increased risk of bacteria getting into the airway and lungs and causing infection. Often, people with chronic bronchitis also have respiratory infections. The excess mucus produced, and increasing damage to the lung tissue, result in shortness of breath and a sore or tight feeling inside the chest. People with chronic bronchitis must give up smoking or avoid whatever pollutant is causing their bronchitis for the breathing system to recover, but recovery is slow and the lungs and airways may never fully or even partially recover.

Acute bronchitis is caused by an infection (bacterial or viral) and lasts only a short time. If the cause is a bacterium, the bronchitis can be treated by antibiotics. Otherwise, the infection will clear up over time if the person doesn't overexert him- or herself, stays warm and avoids airborne pollutants.

Asthma

In asthma, the smooth muscle in the bronchi and bronchioles contracts too much, causing the bronchi and bronchioles to become narrower. At the same time, the mucus-secreting cells lining the bronchial tubes become overactive, so a thick layer of mucus develops. Both these effects make breathing difficult.

Asthma attacks usually occur in response to **allergens**, foreign substances (eg faeces of dust mites) that trigger a too-powerful immune response. There is also a genetic contribution, because asthma tends to run in families.

In a very severe asthma attack, the person's airways are so constricted that little or no air is reaching the lungs. This may result in a heart attack and death if not treated immediately. There is no cure for asthma, but the condition can be managed by avoiding triggers that may bring on an attack. Medications, usually inhaled, called bronchodilators can be used to relax the airway muscles, restoring airflow.

Pleurisy

Pleurisy is inflammation of the pleural membranes that surround the lungs and line the ribcage. Pleurisy can make breathing very painful because the two surfaces of the membrane rub together as they move with inhalation and exhalation. The condition is most commonly caused by infection (viral or bacterial), but can also result from injury to the chest area.

Pneumonia

Pneumonia is an infection of the lungs, caused mainly by bacteria (any one of half a dozen types). It mainly affects the very young and very old. Pneumonia results in the alveoli becoming inflamed and filled with fluid. This means breathing and gas exchange are very inefficient, and a person with pneumonia may have to be attached to a ventilator machine to help with breathing. If the pneumonia is bacterial, it can be treated with antibiotics. Otherwise, the person needs to rest and stay warm.

Lung cancer

Lung cancer is the uncontrolled growth of tissue within the lung, replacing healthy lung tissue and resulting in shortness of breath, coughing up of blood and other symptoms. The disease is one of the most common forms of cancer, and one of the most easily preventable, since the main cause is smoking. About 90% of lung cancer deaths are caused by smoking, but it is also linked to exposure to materials such as asbestos, to radioactive substances (eg radon gas), and to certain viruses. The dangers of smoking have been known for over half a century. The smoking habit is taken up mostly by young people – the age group least likely to think of long-term health consequences. By the time most people realise the dangers of smoking, the habit may have become so addictive they cannot give up. Lung cancer is most commonly treated by surgery (to remove affected tissue, but only up to a point), chemotherapy and radiotherapy (usually together).

Unit 11.4 Activity 3C: Respiration, breathing and gas exchange

1. For each of the phrases **1–15**, write the letter **A–N** of the term to which it applies.

Phrase	Applied term
1. Air channel in lung supported by cartilage in walls	A. Alveoli
2. Air channel in lungs without cartilage in walls	B. Bronchiole
3. Breathing in	C. Bronchus
4. Breathing out	D. Cartilage
5. Carries air to and from lungs	E. Cilia
6. Contains lubricating fluid around a lung	F. Diaphragm
7. Contains the vocal cords	G. Epiglottis
8. Flap covering opening to larynx	H. Expiration
9. Move mucus along bronchi and trachea	I. Inspiration
10. Move ribs	J. Intercostal muscles
11. Muscle separating chest from abdomen	K. Larynx
12. Site of gas exchange	L. Pleural cavity
13. Supports the walls of the trachea and bronchi	M. Pulmonary artery
14. Takes blood from lungs	N. Pulmonary vein
15. Takes blood to lungs	O. Trachea

2. Describe how each part below works to draw air into the lungs.
    - Diaphragm.
    - Intercostal muscles.
3. The diagram shows the structure of the mucus membrane which lines the air passages.

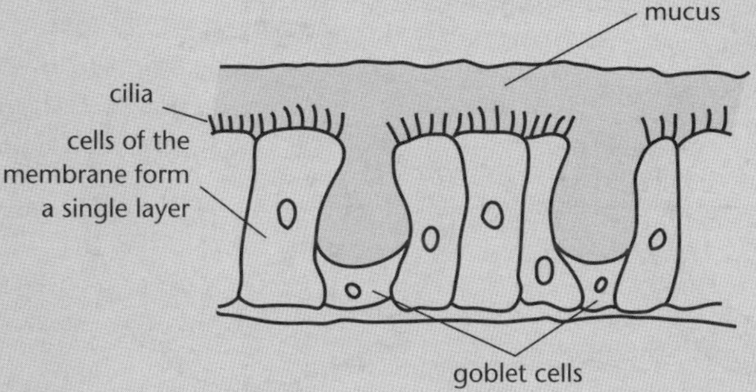

    a. Describe the function of the goblet cell.
    b. Explain the function of the cilia that line the air passages.
    c. Explain one effect smoking has on the lining of the air passages.
    d. Describe the function of the alveoli (air sacs).
4. The diagram shows a normal alveolus, and one damaged by many years of cigarette smoking.

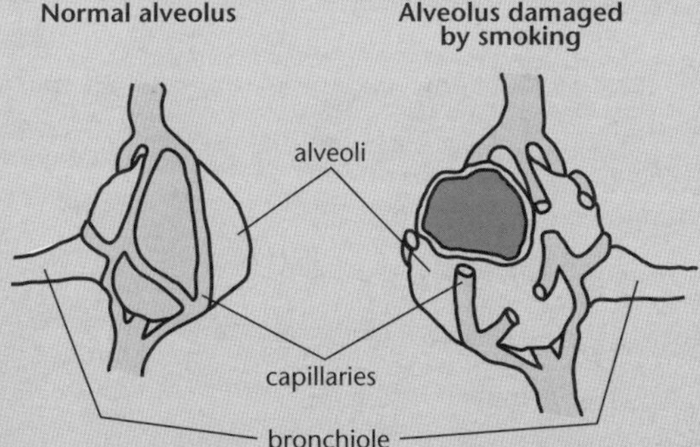

    Identify and discuss two ways this damage affects the functioning of the respiratory system.
5. The diagram shows a cross-section of an alveolus (air sac).

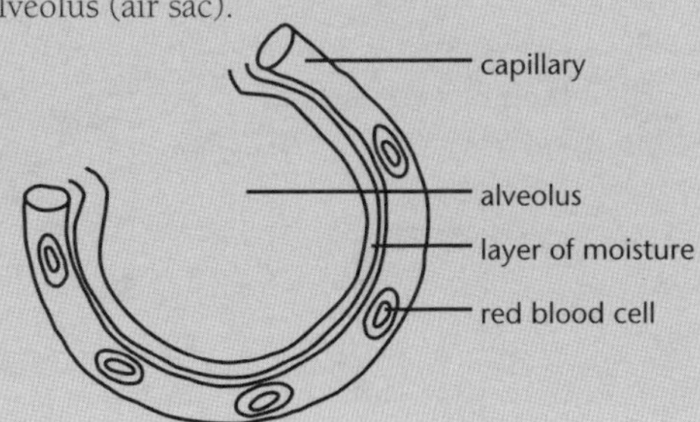

    a. Describe the function of the following parts:
        i. Alveolus.
        ii. Red blood cell.
        iii. Layer of moisture.
    b. Certain characteristics allow gas exchange to occur more quickly. Give two characteristics of an efficient gas exchange surface.

6. The diagram shows the difference between the bronchiole of a normal person, and that of one who is affected by asthma during exercise. Use the diagram to explain how asthma affects the function of the bronchioles.

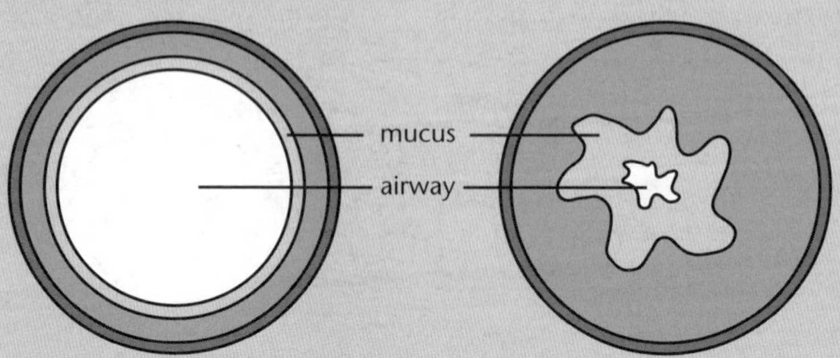

7. CO_2 is considered a harmful by-product of cellular respiration because it:
   a. lowers pH of blood
   b. leads to damage of plasma membrane
   c. competes with O_2 for transport in the blood
   d. poisons the blood
   e. a and b only
8. How is the inhaled air prepared by the nose and mouth before it reaches the lungs?
   a. it is not prepared
   b. it is humidified
   c. it is warmed, humidified and filtered
   d. it is filtered
   e. it is warmed
9. Intercostal muscles ________ during inhalation to ________ chest volume. The diaphragm ________ during inhalation.
   a. contract, decrease, lowers
   b. contract, increase, lowers
   c. contract, increase, rises
   d. relaxes, increase, lowers
   e. relaxes, decreases, rises

Unit 11.5 Response to Stimuli

Topic 1: Growth and response in plants

Authors: Martin Hanson with Takis Solulu

All living organisms are sensitive. But what about plants? How do they respond to stimuli? Does the whole plant respond or only certain parts of the plant?
This Topic explains plant tropisms and the factors that cause them (pp. 19–20 Biology Syllabus) by looking at:

- External features of plants.
- Stem growth.
- Sensitivity in plants – tropisms.
- Hormones and plant responses.

Growth in plants

Growth is a characteristic of all living things. Growth means the production of new cells, usually accompanied by an increase in body size. Plants grow rather differently from animals.

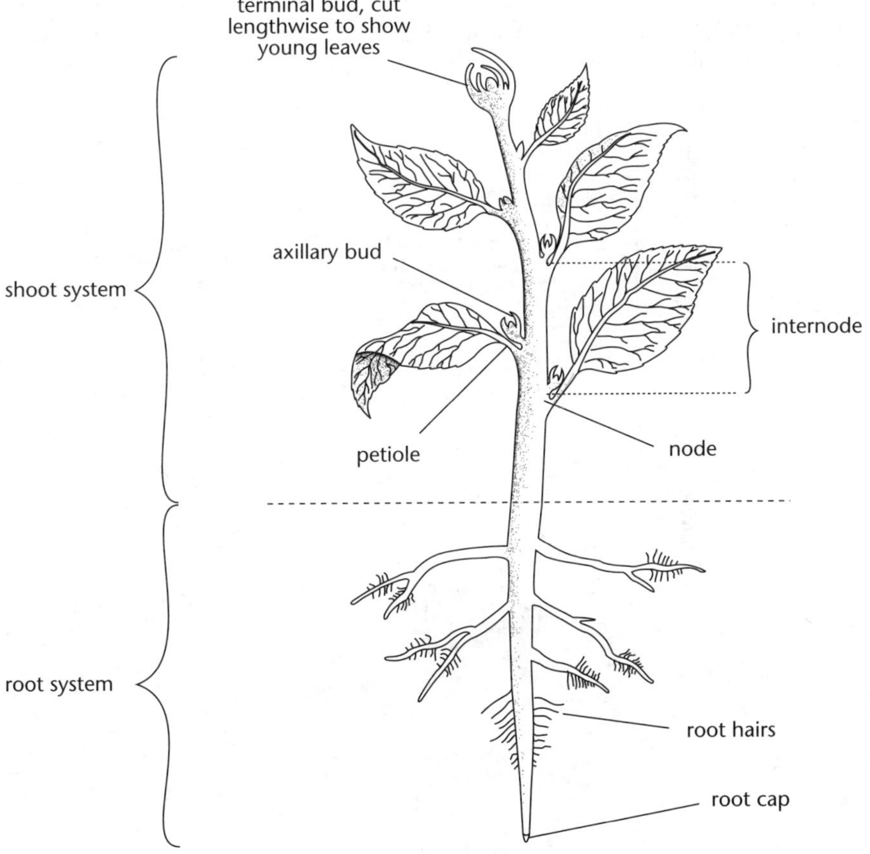

Basic form of a flowering plant.

A typical land plant consists of two parts – a *root system* that absorbs water and minerals, and a *shoot system* that uses light energy to convert CO_2 and water into carbohydrate by photosynthesis. The roots and shoots are interdependent – the shoots depend on the roots for water and minerals, the roots depend on the shoots for energy in the form of carbohydrate.

A fully-developed shoot system consists of a stem-bearing leaves. The leaves arise from points called *nodes*; the region between two nodes is an **internode**.

At the tip of the shoot the young leaves are very small and clustered together to form a **terminal bud**. The stalk of a leaf is called the **petiole**, and angle between a petiole and the stem is called the leaf **axil**. In each axil, there is an **axillary bud**, which may grow out into a side shoot.

Groups of plants

Flowering plants can be divided into two large groups, according to whether they have one or two cotyledons in the embryo.

- **Monocotyledons** include plants like grasses, palms, and lilies, and usually have parallel veins in the leaves.
- **Dicotyledons** include clover, buttercups, daisies, and most flowering trees, and have leaves whose leaf veins form a network.

How plants grow

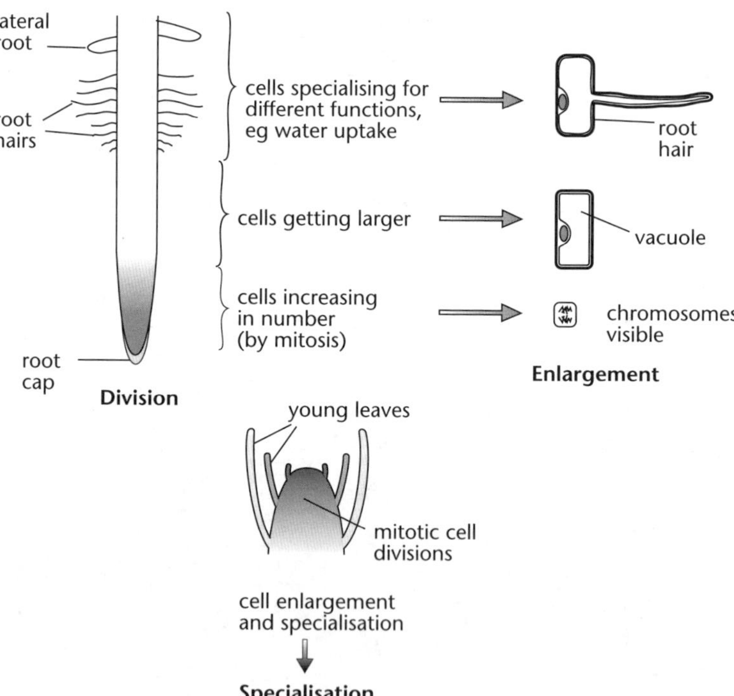

The three cell processes in growth – division, enlargement and specialisation – are highlighted.

Growth regions in a root tip.

Primary growth

Growth in length of a stem or root is called **primary growth**.

Plants grow at the tips of their shoots and roots, so these areas are always the youngest parts. Both roots and shoots increase in length by cell *division* followed by cell *enlargement*. These processes take place in different regions.

- Cell division occurs in the **apical meristems** at the tips of the shoot and root. Cells increase in number by mitosis.
- Behind the tip, cells begin to enlarge.
- Further back, cells begin to **differentiate**, or specialise for particular functions, eg in the root, the outermost layer of cells develop long outgrowths called **root hairs**, specialised for absorbing water. In the shoot, cells on the surface specialise for water conservation, by developing a waxy cuticle. Other cells specialise in different ways, for functions such as transport, support, or storage.

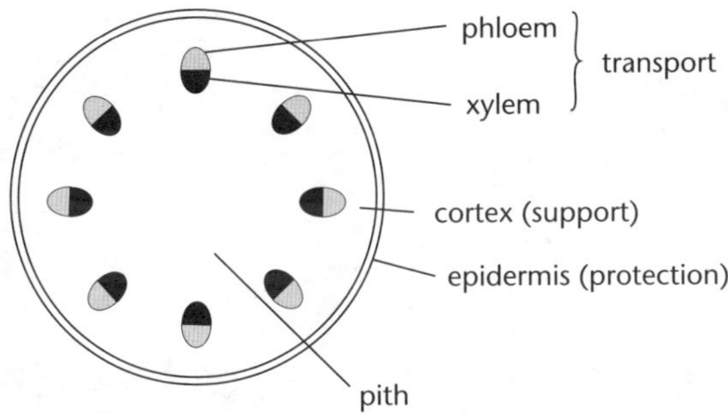

Transverse section through a young stem of a dicotyledon.

Some of the details of plant growth differ between monocotyledons and dicotyledons. *Much of what follows refers to dicotyledons.*

Secondary growth

Once growth in length (primary growth) has ceased, many plants begin **secondary growth**, or an increase in thickness. The details of secondary growth are best described in a stem.

The rings on this fossilised tree are the result of secondary growth that took place millions of years ago.

Increase in width in a young stem

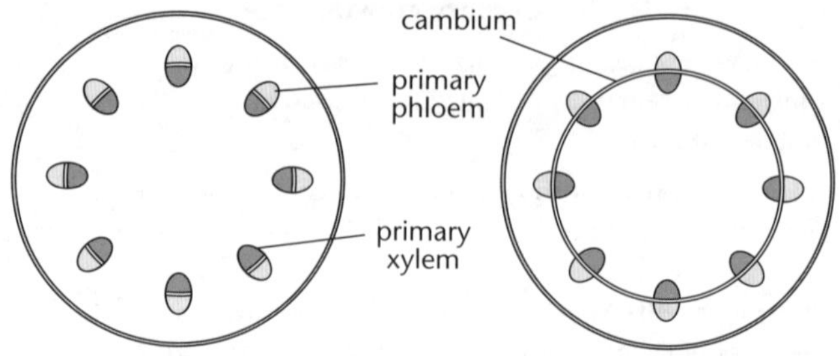

1. Ring of vascular bundles.

2. Cambium extends between bundles, forming a cylinder.

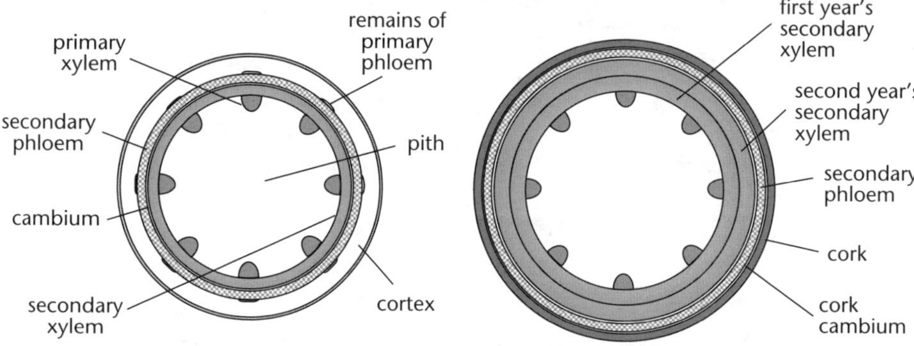

3. One-year-old twig. Cambium produces cylinder of secondary xylem and secondary phloem.

4. Two-year-old twig. Cambium produces another cylinder of secondary xylem and phloem; cork cambium develops and produces cork, replacing the epidermis and cortex.

Development shown is that of secondary vascular tissues over the first two years in a woody dicotyledonous stem.

Development of secondary vascular tissues.

In the middle of each vascular bundle is a narrow strip of cells called the **cambium**. The cells between vascular bundles become converted into cambial cells, so that the cambium comes to form a cylinder. Cambial cells divide repeatedly, producing new cells:

- Cells produced towards the inside become **secondary xylem** cells.
- Cells produced towards the outside become **secondary phloem** cells.

In shrubs and trees, the stem survives year after year. Each year the cambium produces more secondary xylem and secondary phloem. In climates where growth stops or slows down for part of the year, the secondary xylem forms a series of **growth rings**. Because one ring is formed in each growing season, the number of annual rings gives the age of a tree.

Growth rings do not form in secondary phloem – phloem is a soft tissue and eventually becomes crushed rather than accumulating year after year like secondary xylem.

Bark

As a result of secondary thickening, the epidermis and cortex are stretched and eventually torn apart. Stems and roots that grow in thickness therefore need a protective layer that can continue to get thicker – this is the *bark*.

Bark consists largely of a tissue called **cork** which is continuously produced by a layer of cells called the **cork cambium**. Cork cells have greasy walls and fit very close together, forming a waterproof layer which also gives protection against animals and fungi.

Plant sensitivity – tropisms

Like all living things, plants respond to **stimuli** (singular, stimulus), or changes in the surroundings.

A **tropism** is the growth of a plant part in a direction that depends on the direction of the stimulus.

- In *positive* tropisms, growth is *towards* the stimulus.
- In *negative* tropisms, growth is *away* from the stimulus.

Tropisms are particularly obvious in germination.

Most stems show *positive phototropism* – growth *towards* the light.

The plumules of seedlings show *negative gravitropism* – growth away from gravity.

Radicles show *positive gravitropism* – growth towards gravity.

The tendrils of climbing plants show *thigmotropism* – they bend towards, and coil round, anything they touch.

As **pollen tubes** approach the ovule, they show *positive chemotropism*, since they grow towards chemicals (produced by the female gamete).

Roots show *positive hydrotropism* – growth towards water.

How tropisms occur

Over a century ago, Charles Darwin did some simple experiments on phototropism using oat seedlings.

As in all grasses, the first leaves are protected as they grow up through the soil by a spear-like sheath called a **coleoptile**.

When Darwin covered the tips of coleoptiles they did not bend towards the light. When he covered the base with black sand, they did.

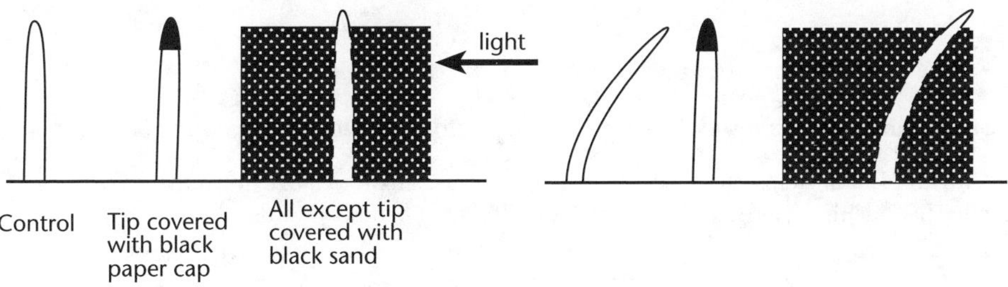

Darwin's phototropism experiment.

Darwin noted that the part of the coleoptile that bent towards the light was well behind the tip. He drew two conclusions:

- The stimulus must be detected by the tip.
- Some form of message or signal must pass from the tip to the region lower down where the bending occurred.

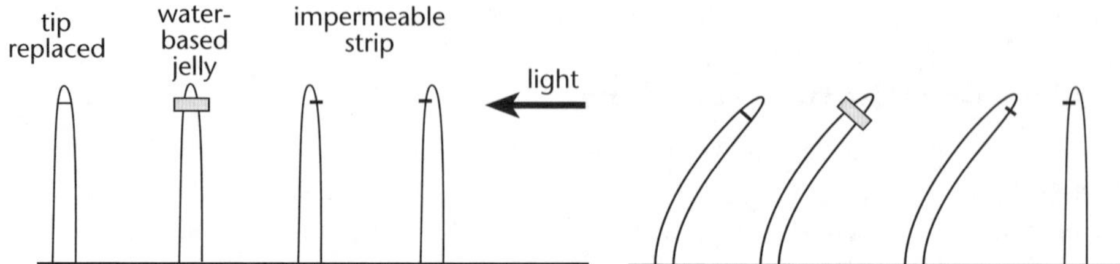

Experiments showing how the part of the coleoptile that detects light communicates with the part that responds.

Later experiments by other scientists showed what kind of message it was.

- After cutting off the tip and replacing it, a coleoptile could still respond to light. *This showed that the message could not be a nervous one* (nerves would have been cut).
- When the tip was separated from the base by a water-based jelly, the coleoptile curved normally. When a thin strip of impermeable material was inserted into the shaded side, the coleoptile did not respond to light, but when it was inserted into the illuminated side of the coleoptile, it did. This showed that *the message must travel down the shaded side of the coleoptile.*

Later experiments showed the message is a chemical called **auxin**. Auxin stimulates elongation of cells. When a shoot is illuminated from one side, auxin is transported down the shaded side. This shaded side then elongates more rapidly than the illuminated side, causing the shoot to bend *towards* the light.

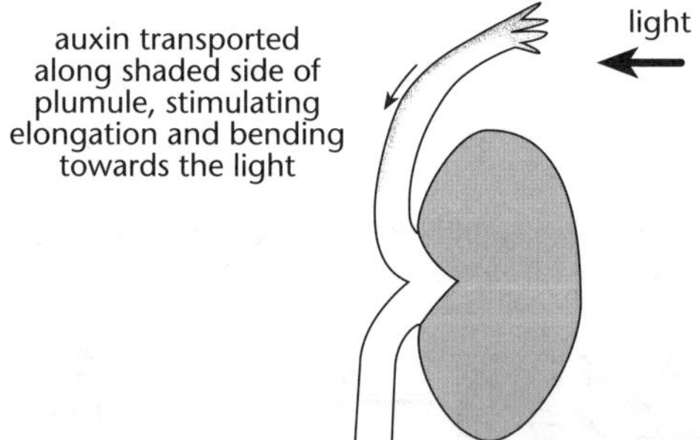

Auxin causes a shoot to bend during phototropism in a germinating bean.

In gravitropism in stems, the effect of gravity is to cause auxin to move down the *lower* side of a stem. This side then elongates more quickly, so the stem bends upwards.

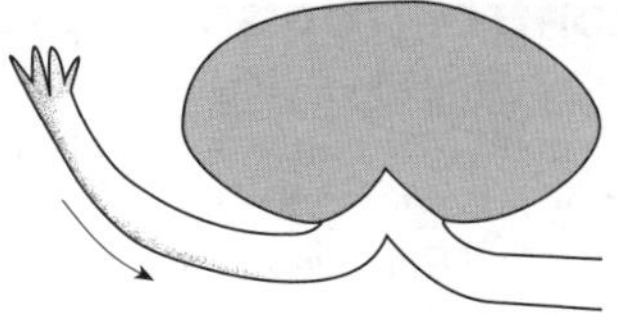

Auxin causes a shoot to bend during gravitropism in a germinating bean.

Plant hormones

Plant hormones are organic compounds (chemicals) produced by plants in small amounts in one part of the plant and transported (through vascular tissues) to another part where needed. Plant hormones are also termed *growth factors* as these regulate plant growth and development (flower, fruit development, senescence or aging), and control plant response to a variety of environmental stimuli that include both **abiotic** (light, gravity, water) and **biotic** environment (herbivores eg insects; disease-causing micro-organisms eg fungi, bacteria, virus). The table below lists the five major groups of plant hormones.

Major plant hormones: site of production, selected functions and practical application

Hormone	Where produced or found in plants	Selected function/description	Application/ usefulness
Auxin	Growing tips of shoot; young leaves	Indoleacetic acid (IAA) is the most common naturally occurring form involved in phototropism, gravitropism, thigomotropism. The chemical **auxin** (p. 228) is a hormone.	
Gibberellins	Growing tips; young shoot; seeds	Promote stem elongation between nodes, roots, shoots, young leaves; break seed and bud dormancy and influence germination of seeds.	
Cytokinins	Growing tips of roots	Promote cell division; prevent senescence (aging); along with auxin, promote differentiation leading to roots, shoots, leaves, or floral shoots.	*Tissue culture application*: Technique of growing plants from plant cells and tissues.
Abscisic acid	Seeds; roots	Initiates and maintains seed and bud dormancy; promotes formation of buds; promotes closure of stomata	
Ethylene	Aging parts of plant; ripe fruit	*A gaseous hormone* Promotes abscission (drop-off) of leaf, fruit, flower drop; ripening of fruit; involved in thigmotropism	*Commercial application*: Ethylene applied to fruits to ensure even-ripening at same time, such as banana, tomatoes, pineapples.

Hormones control tropism in plants

Unlike animals, plants cannot hide, bite or flee. Instead they must adjust to the environment to survive. Plants respond to the environment primarily by altering their growth patterns (or tropism), mediated by hormones. Besides *phototropism* (light), *gravitropism/geotropism* (gravity), *hydrotropism* (water) and *thigmotropism* (touch), other known tropisms include:

- Chemotropism – response to *chemicals*.
- Traumotropism – response to *trauma*.
- Skototropism – response to *dark*.
- Aerotropism – response to *oxygen*.

Plant hormones: practical applications

The farming industry (fruit industry) uses different plant hormones to control how different fruits develop.

- **Ethylene causes the ripening of fruits.**
  Ethylene stimulates enzymes that break down cell walls, break down chlorophyll and convert starch to sugars. This makes the fruit soft, ripe, sweet and ready to eat. For example, bananas are harvested and transported before they are ripe because they are less likely to be damaged this way. They are then exposed to ethylene (ethene) on arrival so they all ripen at the same time.
- **Auxins and gibberellins make fruit develop.**
  Auxins and gibberellins are sprayed onto unpollinated flowers, which makes the fruit develop without fertilisation. For example, seedless fruits (grapes, oranges, tomatoes) can be produced using auxins and gibberrellins.
- **Auxins can prevent or trigger fruit drop.**
  Applying a low concentration of auxins in the early stages of fruit production prevents the fruit from dropping off the plant. But applying a high concentration of auxins at a later stage of fruit production triggers the fruit to drop. For example, apples can be made to drop off the tree at exactly the right time.
- **Auxins are also used commercially by farmers.**
  Auxins are used in selective chemical weed-killers (herbicides) – auxins make weeds produce long stems instead of lots of leaves. This makes the weeds grow too fast, and they can't get enough water or nutrients, so they die.
- **Auxins are used as rooting hormones.**
  Auxins make a cutting (part of the plant, eg a stem cutting) grow roots. The cuttings can then be planted and grown into a new plant. Many cuttings can be taken from just one original plant and treated with rooting hormones, so lots of the same plant can be grown quickly and cheaply from just one plant.

Unit 11.5 Activity 1A: Hormones and plant responses

1. Auxin causes bending in stems and roots by:
- **A.** causing cell elongation.
- **B.** preventing cell elongation.
- **C.** breaking down cell walls.
- **D.** causing cell shrinkage.
- **E.** inhibiting lateral bud growth.

2. Which of the following is *not* an effect of gibberellins?
- **A.** stem elongation between nodes
- **B.** cell elongation
- **C.** breaking of dormancy
- **D.** ripening of fruit
- **E.** excessive stem elongation

3. A tomato grower in National Capital District plans to sell her produce to Stop N Stop supermarkets in Port Moresby, and wants all her tomatoes to ripen at the same time, just before she sells them.
- **a.** Name a plant hormone she could use to make the tomatoes ripen.
- **b.** Explain how the plant hormone named in part (a) makes tomatoes ripen.
- **c.** Suggest a commercial advantage of being able to pick and transport tomatoes before they are ripe.

Unit 11.5 Responses to Stimuli

Topic 2: Sensitivity and coordination in animals

Sensitivity in animals is studied under coordination, which looks at the way all organs and systems of the body are made to work efficiently together (pp. 19–20 Biology Syllabus). This Topic deals with:

- The structure and function of the nervous system, senses, endocrine (hormone) system.
- A description of how the nervous system functions, nerve impulses, pathways for sensory and motor signals, feedback control of co-ordination.
- Reasons for the differences in structure and function of sensitivity and co-ordination systems between different groups – these could relate to mobility and energy needs, size, habitat, life cycle, way of life.

Animals sense external stimuli such as light, sound, chemicals, heat and touch, through sensory receptors.

Animals respond to external stimuli and co-ordinate their internal body processes (gas exchange, digestion, excretion, growth, etc), through the **nervous system** and **endocrine system**. These systems control **effector organs** (muscles or glands).

Nervous systems respond *quickly*, because information is transmitted by electrical impulses.

Endocrine systems usually respond slowly by releasing **hormones** (eg ADH, insulin) which control body processes. Hormones travel in the blood.

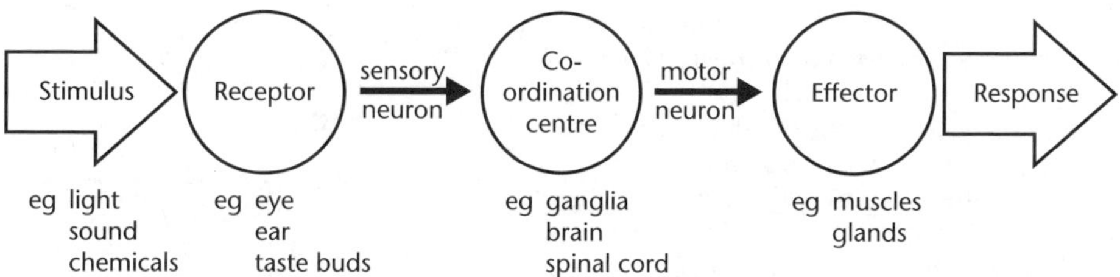

Nerve cells are called **neurons**. Neurons have three distinct parts:

- The *cell body* – contains the nucleus, cytoplasm and organelles such as mitochondria.
- *Dendrites* – short, branching structures, which collect stimuli/nerve impulses.
- *Axon* – a long extension from the cell body, which transmits the nerve impulse.

The position of the cell body varies between *sensory neurons* (which collect the information from sensory cells or organs), and *motor neurons* (transmit the nerve impulse to the effector organs – muscles, glands). In vertebrate neurons, the axon may have a *myelin sheath* of fatty cells which insulate it – the nerve impulse is transmitted faster in a sheathed axon than in an unsheathed.

A tiny gap, about 25 nm (0.000025 mm) between the axon of one neuron and the dendrites of another neuron separates adjacent neurons – this is the **synapse**.

Structure of a typical sensory neuron

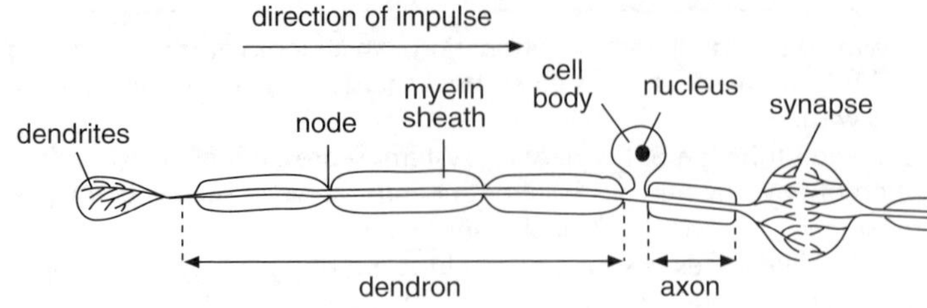

Structure of a typical motor neuron

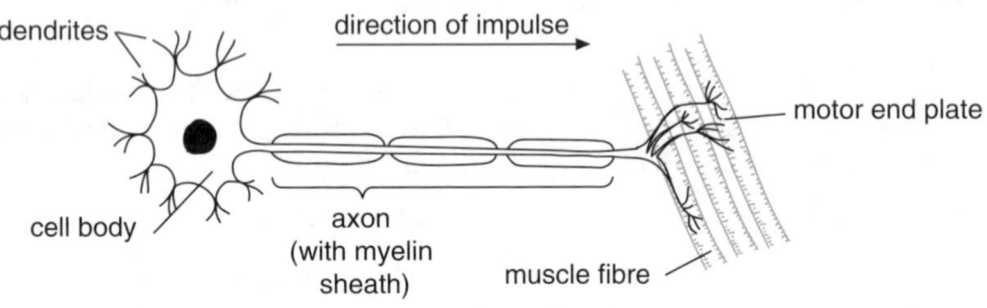

In animals with a central nervous system, nerves are *bundles* of neurons. Transmission of information is via the **nerve impulse**, which is an *electrical change* that travels along the axon. Axons are negatively charged inside. A stimulus received by the dendrites causes sodium ions, Na^+, outside the axon to rapidly diffuse into the axon, making it positive. The Na^+ ions are then rapidly pumped out again (active transport), and the negative charge inside is restored (takes about a millisecond). This change in charge flows along the axon – this is the nerve impulse which can travel at speeds of up to 120 m s^{-1} (in mammals).

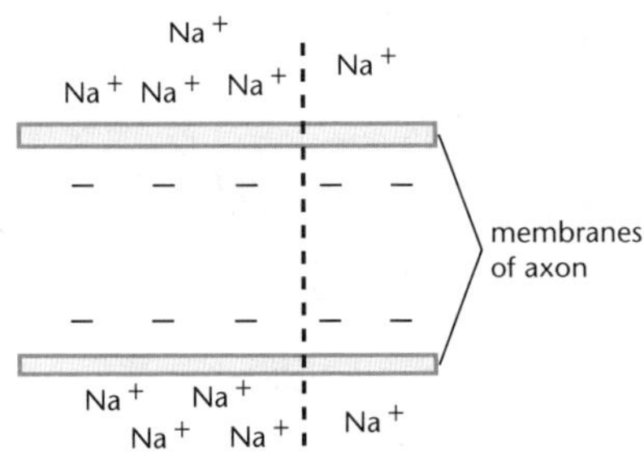

Neuron before stimulation

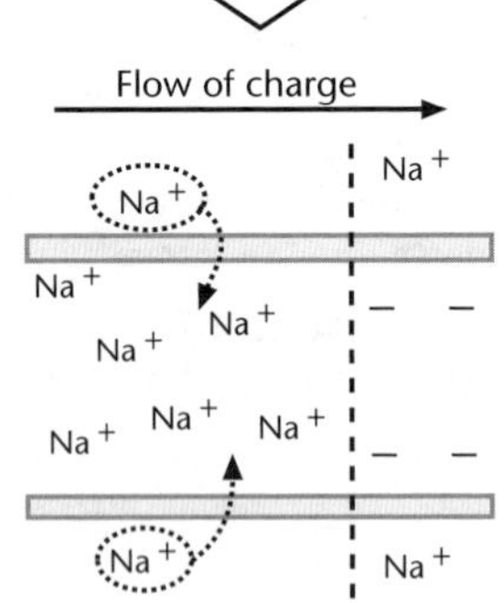

Neuron after stimulation

The nerve impulse works on an 'all or none' principle – ie the stimulus needs to be above a certain minimum (or threshold level) to generate an impulse, which then passes along the axon at a certain (high) speed. The size or speed of the impulse does not change with the strength of the stimulus; a stronger stimulus triggers more impulses (per second). On arrival at the end of the axon, the impulse travels across the synapse *chemically.* Chemicals called *neurotransmitters* (eg acetylcholine) are released and diffuse rapidly (take less than a millisecond) across the synapse to trigger a nerve impulse on arrival at the dendrites of the adjacent neuron. The neurotransmitter is then broken down.

Nerve network system of cnidaria (eg *Hydra*)

The nervous system of the cnidaria (eg marine jellyfish and the freshwater hydra) is *not* centralised – the nerves form a simple network in between the two cell layers of the body. The flow of information is *non-directional* – a nerve impulse can be transmitted in either direction across a synapse.

- Sensory receptors are *specialised cells* found in the outer cell layer (ectoderm) of the body – they detect touch (allows for detection of prey and predators) and chemicals (allows for detection of prey).
- Effector organs are the muscle-like fibres in the form of contractile 'tails' of cells of the ectoderm and endoderm (inner layer). Contraction of the ectoderm cell fibres shortens the body; contraction of the endoderm cell fibres lengthens the body. Stinging cells (*cnidoblasts*) are found throughout the ectoderm; they shoot out and paralyse small prey animals that touch them.
- Neurons between the ectoderm and endoderm form a network. The *axons* of the neurons are unusually short and do not occur in bundles. This shortness of the axons means synapses occur frequently. Passage across synapses is dependent on the number of impulses that arrive at the synapse. Enough impulses need to arrive so that their effect adds sufficiently to give transmission across the synapse. Passage of an impulse is comparatively slower than in animals with a centralised system. As the cnidaria are either slow moving (jellyfish) or largely sedentary (hydra), this system is adequate to meet their needs.
- There is no accumulation of nerves into ganglia, and therefore a simple 'brain' is not present.

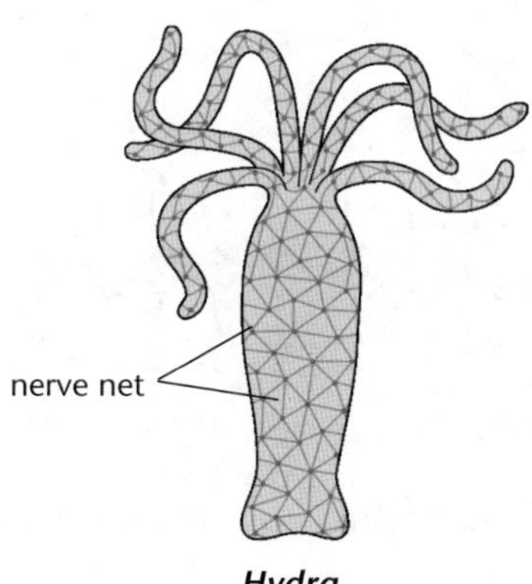

Hydra

Simple central nervous system (eg earthworm)

The central nervous system (CNS) of earthworms is relatively simple.

Sensory receptors are *specialised cells* that detect:

- Light – sensory receptors concentrated in the dorsal surface of the skin – worms need to avoid light and burrow in the soil to keep moist (in order to breathe) and avoid diurnal (day-active) predators (eg birds).
- Chemicals – sensory receptors concentrated in the anterior – worms need to avoid harmful chemicals (eg acidic soils).
- Vibrations/touch – sensory receptors concentrated in the anterior – worms need to avoid predators and locate other worms for reproduction.

Effector organs are muscles (eg the circular and longitudinal muscles that move the worm; protractor and retractor muscles that move the setae) and glands (eg mucus-secreting glands in the skin).

A central nerve cord of giant fibres runs from the head end (anterior) to the rear end (posterior) along the *ventral* surface of the body just above the skin – this *ventral nerve cord* (VNC) is typical of invertebrate animals (eg annelids, arthropods). It allows for rapid, co-ordinated movements, as the nerve impulse is transmitted *directionally* from the receptors to the effectors.

Concentrations of nerve cells (**ganglia**, singular ganglion) appear as swellings on the VNC in each segment. These receive sensory information for the segment and control segmental movement.

A simple brain of two large ganglia is found in the head region – this serves to co-ordinate activity. The development of a simple brain and the concentration of sensory receptors in the head region are the start of **cephalisation** – ie the development of a specific 'head', which in higher animals contains most of the sense organs and a brain. As the head in most animals (humans are an exception) is the part of the body that first enters a new environment, it is essential that the sensory receptors are here (to detect harmful conditions, locate food). The location of the brain close by means transmission of information from receptors is rapid, so an appropriate response (eg run away, pounce on prey) can also be rapidly made.

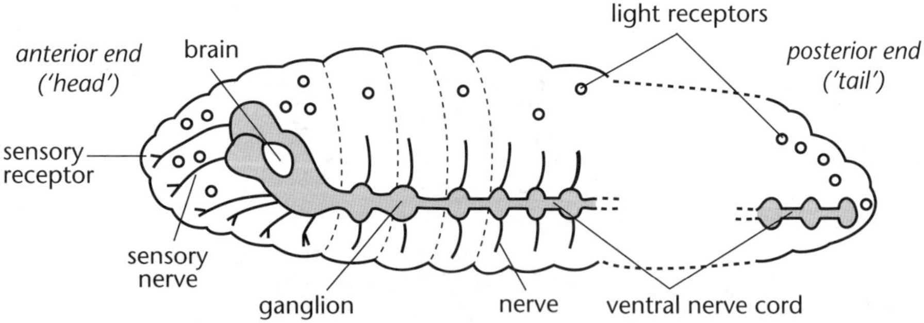

The relatively unspecialised sensory system of the earthworm.

Complex central nerve system (eg mammals)

The nerve system of mammals is complex, with three parts – *central nervous system* (CNS), *peripheral nervous system* (PNS), *autonomic nervous system* (ANS).

Sensory receptors

Sensory receptors are typically complex *sense organs* concentrated in the head.

- Eye – detects light – black and white only detected in all mammals, colour detected in some (eg primates).
- Ear – detects sound waves.
- Nose – detects chemicals (in the form of 'smell').
- Tongue – detects chemicals (in the form of 'taste').
- Skin – has specialised receptor cells in the dermis for temperature, pain, touch, pressure.

The sense organs allow for gathering of comprehensive information about the environment, facilitating an organism's ability to find favourable conditions (eg food, mates, water, suitable temperature and pH, etc) and avoid unfavourable conditions (eg predators, temperature extremes, dehydrating conditions, etc). The mammalian eye is an example of a sophisticated sense organ. For humans, 'seeing is believing', as vision provides the most important sensory information – unlike many mammals, for which the sense of smell and/or hearing are most important.

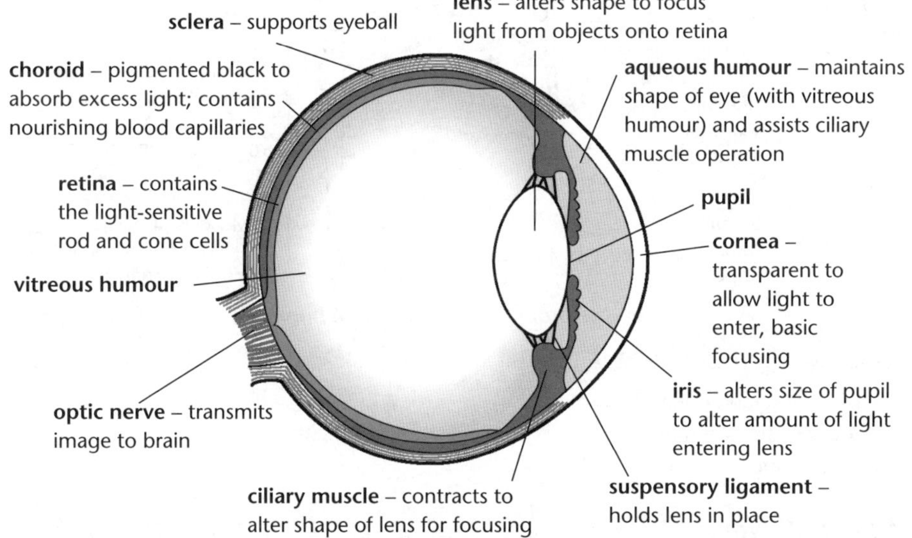

Cross-section of the human eye.

The shape of the *convex lens* is altered by the ciliary muscles to allow focusing on both near and far objects. Light from objects is focused on the light-sensitive cells of the *retina* – rod cells receive black and white, cone cells receive colour. An image is sent to the brain along the optic nerve – we 'see' the object. The ability to detect colour provides us with more information about the environment – eg whether fruit is ripe (edible) or not (inedible); objects of different colours stand out more from the background, so are easier to detect; cryptic colouration in the form of stripes (eg tigers), patches (eg leopards) is less effective if colour can be detected.

Effector organs

Effector organs are the muscles (voluntary, involuntary, cardiac), ducted glands (eg tear ducts, sweat gland in dermis) and ductless glands of the **endocrine** ('hormone') **system**.

Central Nerve System (CNS)

The CNS is made of a complex *brain* in the head and a *spinal* (dorsal) *nerve cord* enclosed in the vertebral column ('spine'). The brain is made of billions of nerve cells (*pyramidal neurons*), which co-ordinate all the body's activities, interpret stimuli from the sense organs, store information; brain also used for thought processes (in humans).

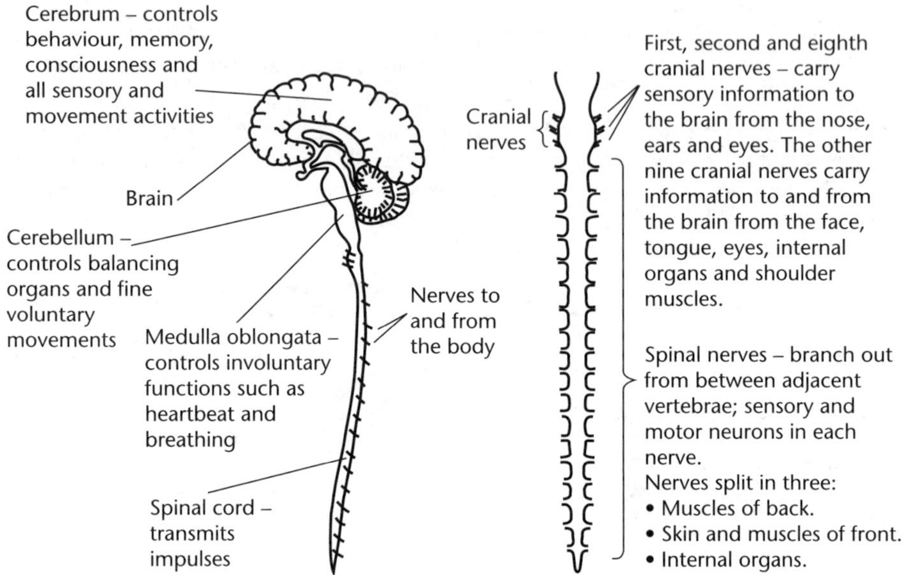

CNS – the brain and spinal cord in humans.

Relay neurons (also called *interneurons*) in the spinal cord link:

- Sensory and motor neurons.
- Sensory neurons and the brain.
- The brain and motor neurons.

Action pathways

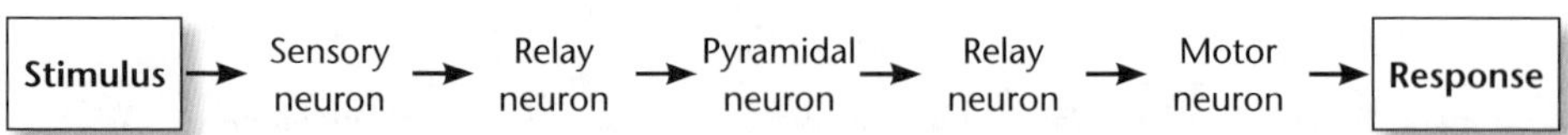

Actions may be *reflexive* – ie bypass the brain. As the brain is not initially involved in a reflex action, it is a very quick response to a stimulus.

Some reflexes never involve the brain (eg the knee jerk reflex), while others involve the brain only after the reflex action has occurred (eg when you stand on a thumb tack the reflex action lifts the foot immediately, but 'later', pain receptors will begin sending impulses to the brain to 'remind' you to remove the tack from your foot).

Reflex actions are *automatic, unlearned responses*. They increase the survival chances of organisms, by removing them immediately from danger.

Conditioned reflexes are modifications of reflex actions – they are learned responses. Dogs can be 'taught' or conditioned to salivate in anticipation of food when a bell is rung. The unlearned reflex response is salivating when food appears; the learned, or conditioned response is salivating at the sound of the bell.

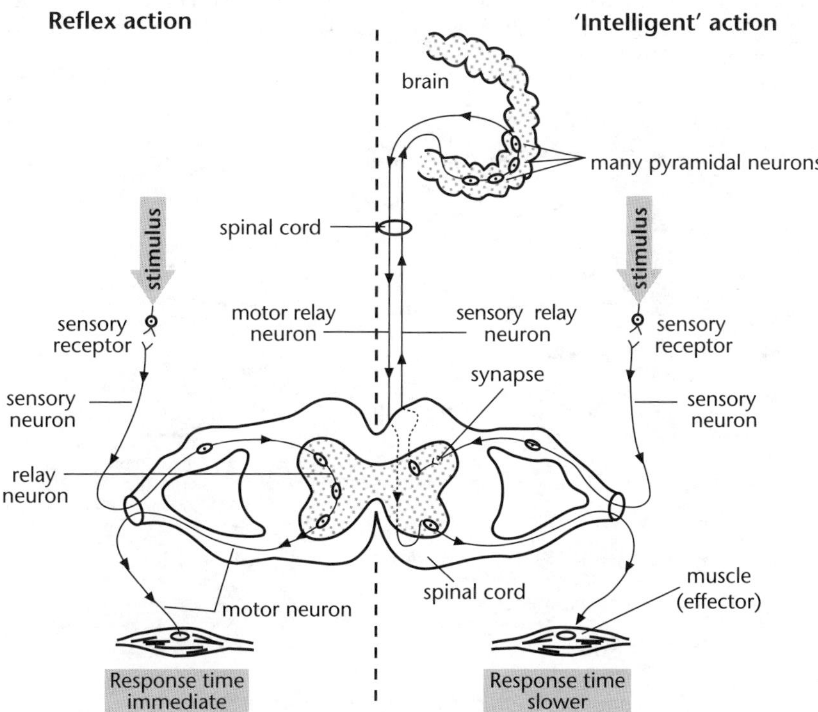

Peripheral Nerve System (PNS)

The PNS receives, and responds to, stimuli, through the nerves of the sensory receptors, skin, organs, and skeletal muscles. The PNS involves most of the body's nerves and allows for efficient collection from receptors and transmission to the brain.

Autonomic Nerve System (ANS)

The ANS is under the *subconscious* control of the brain and regulates the actions of glands, smooth muscles (eg contractions of the pupil of the eye) and heartbeat. The ANS also controls internal sensations such as hunger, thirst, and the desire to urinate.

The ANS consists of the *sympathetic* system and *parasympathetic* system, which work in opposition to each other. In general, the sympathetic system tends to speed functions up, whereas the parasympathetic system tends to slow things clown. Most organs are supplied with nerves from both these systems, to control organ function (eg the sympathetic system increases heartbeat rate, while the parasympathetic system decreases heartbeat rate).

Endocrine system

The endocrine glands produce and secrete hormones – chemical messengers that control or help to control some function elsewhere in the body. Hormonal control is usually slow acting, but its effect is long lasting.

Hormones control the internal environment, body processes and sexual development. They are important in maintaining stable conditions in the body (*homeostasis*). Homeostasis is described in Topic 3 of this Unit (p. 243).

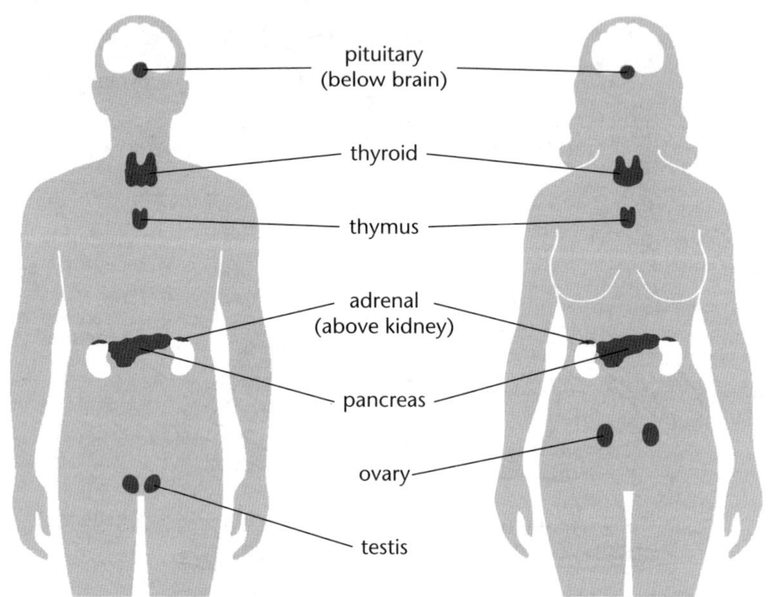

The glands associated with the human endocrine system

The **pituitary** is called the 'master gland', because hormones secreted by the pituitary regulate the activities of the other endocrine glands. The pituitary is controlled by an area of the brain called the hypothalamus. Hormones are released directly into the blood, which transports them to the target organs (this is much slower than nervous transmission). Receptors in the hypothalamus monitor levels of chemicals in the blood (eg water). This information is transmitted to the endocrine organs, which will stimulate or inhibit production of the appropriate hormone (eg ADH). Low water levels in the blood signal the pituitary to release ADH. The target organ is the kidney, which responds by reabsorbing more water from the filtrate.

Many hormones operate by feedback control – ie the production of a hormone is adjusted by its own output and/or the level of a chemical (eg glucose, water) in the blood.

Example

Feedback control involved in production of the hormone TSH

The production of thyroid stimulating hormone (TSH), by the pituitary, is an example of feedback control.

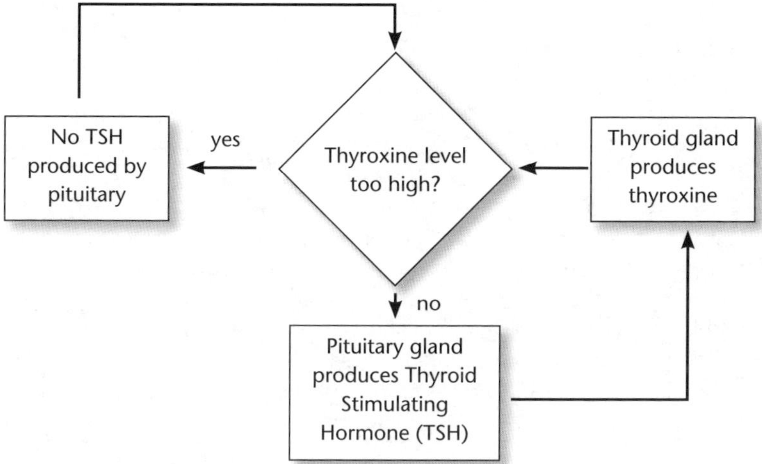

TSH from the pituitary gland stimulates the thyroid gland to produce the hormone *thyroxine*, which regulates the speed of most body processes (by controlling the rate of cellular respiration). When thyroxine levels in the blood become too high, the pituitary no longer produces TSH, preventing the production of thyroxine until the level of thyroxine drops again.

Other glands of the endocrine system

Gland	Hormone(s)	Function of hormone produced
Parathyroid	Parathyroid hormone	Regulates blood calcium levels.
Pancreatic islets	• Insulin • Glucagon	• Changes glucose to glycogen. • Changes glycogen to glucose.
Adrenals	Adrenalin	Prepares body for an emergency – 'flight or fight'.
Ovary	• Oestrogen • Progesterone	• Controls female sex characteristics. • Prepares uterus for egg.
Testes	Testosterone	Controls male sex characteristics.

The endocrine and nervous systems are linked to each other, either directly (eg the *pituitary* is stimulated by the *hypothalamus* of the brain), or indirectly (eg *adrenalin* and the *vagus nerve* both stimulate the heart).

Unit 11.5 Activity 2A: Sensitivity and coordination in animals

1. Describe the structure of neurons.
2. Explain how information is transmitted along nerve networks.
3. Describe the functions of:
   a. Sensory organs.
   b. Effector organs.
4. Explain the significance of cephalisation.
5. Explain the advantages of a simple centralised nervous system over a network system.
6. Explain the advantages of a complex centralised nervous system over a simple centralised system.
7. Discuss the role and operation of the endocrine system in humans.
8. Select three different taxonomic or functional groups, and, using named animals as examples, discuss the structure of their nervous systems and the reasons for their differences. In your answer:
   - Describe the system for each of the named animals.
   - Explain how each of these systems operates.
   - Explain the differences in the systems in relation to the different ways of life of the three animals.

Unit 11.5 Response to Stimuli

Topic 3: Maintaining stable internal conditions – homeostasis

Authors: Martin Hanson with Takis Solulu

To achieve the learning outcomes listed on p. 19 of the Syllabus, students are encouraged to develop an understanding of coordination and regulation in living organisms. This Topic looks at:

- The regulation of internal conditions in the body.
- Osmoregulation – maintenance of body water content.
- Thermoregulation – maintenance of body temperature.
- Regulation of blood glucose; diabetes.
- Regulation of salt.

Homeostasis

Cells are very delicate – they can only survive in a very narrow range of conditions. Human cells are quite incapable of surviving outside the body for more than a very short time (except in special culture solutions). Inside the body, cells are bathed in a liquid called **tissue fluid**, which acts as a kind of 'internal environment'. The properties of the tissue fluid and blood – its temperature, pH, and the concentrations of water, oxygen, CO_2, salt, and glucose – change little. This maintenance of a stable internal environment is called **homeostasis**.

There are three essential things needed for homeostasis:

- Receptors to detect a change from the normal value, or 'set point'.
- Effectors to bring about the appropriate corrective response.
- A communication system to carry the information from receptors to effectors. Communication is either by hormone or by nerves.

In some cases receptors are in an unusual place (eg the receptors for osmoregulation and thermoregulation are in the brain). Effectors also may be 'unorthodox' (eg the kidney acts as an effector in osmoregulation, and the liver does so in glucose homeostasis).

Regulating water content (osmoregulation)

Although water is essential to life, too much is as harmful as too little. Soon after drinking a large volume of water, urine output rises and we feel the need 'to go'; on the other hand, when short of water, we produce less urine. Regulating body water content is called **osmoregulation**.

The organs that get rid of surplus water are the *kidneys*, so osmoregulation is closely connected with excretion.

Most of the water filtered by the kidneys is reabsorbed from the first twisted tubule (*see* Structure of the kidney, p. 184). This occurs regardless of whether the body is over-hydrated or dehydrated.

- If the body is dehydrated, most of the remaining water is reabsorbed from the collecting ducts, resulting in a small volume of concentrated urine.
- If the body is over-hydrated, little or no more water is reabsorbed and a large amount of dilute urine is produced.

Role of the brain in osmoregulation

By itself, a kidney cannot control the water concentration of the blood – it can only respond to signals it receives from the brain. The brain communicates with the kidneys by **anti-diuretic hormone** (**ADH**). ADH is secreted by the **pituitary gland**, a pea-sized organ at the base of the brain. It stimulates the reabsorption of water from the collecting ducts.

Like all hormones, ADH is continuously being destroyed. At any given time, its concentration in the blood depends on the relative rates of its secretion and its breakdown.

If the output of ADH by the pituitary falls, so does its concentration in the blood.

- When the water concentration of the blood is too low, ADH output increases – more water is reabsorbed from the urine so the urine becomes more concentrated.
- If the blood becomes too dilute, ADH output falls – less water is reabsorbed, and the urine becomes more dilute.

It takes about half an hour for urine output to reach a maximum after drinking a large volume of water – it takes time for the gut to absorb the water, then the existing ADH in the blood has to be destroyed.

Although the pituitary gland secretes ADH, it does not actually sense changes in water concentration of the blood. This is the function of the **hypothalamus**, a part of the brain just above the pituitary gland.

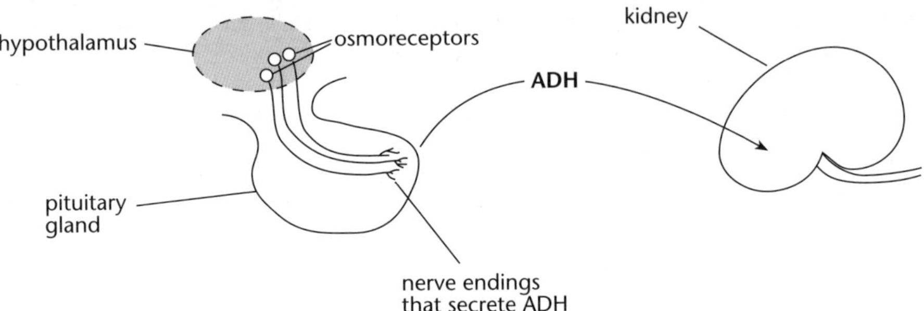

Control of the pituitary by the hypothalamus.

When the water concentration of the blood falls, the hypothalamus sends nerve impulses to the pituitary, causing it to secrete more ADH.

The release of ADH is inhibited by alcohol, which is why alcohol is a *diuretic*, stimulating urine production.

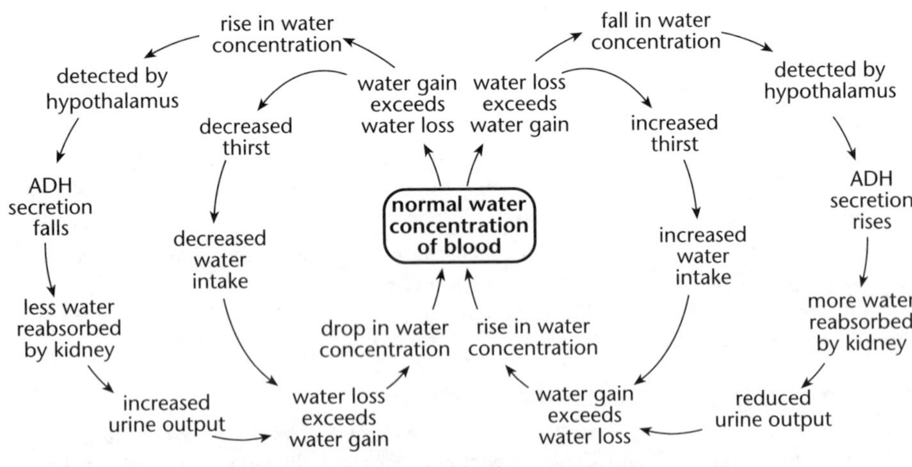

Feedback control of body water content.

Thirst

As important as regulating water *loss* is the control of water *intake*. When the body is dehydrated, the hypothalamus detects this, causing the feeling of thirst, which usually leads to drinking. The hypothalamus therefore controls body water concentration by regulating the balance between water loss by the kidney, and water uptake by drinking.

Unit 11.5 Activity 3A: Regulation of body water content – osmoregulation

1. For each of the phrases **1–6**, write the letter **A–F** of the term to which it applies.

Phrase	Applied term
1. Detects changes in water concentration of blood	A. Anti-diuretic hormone
2. Liquid that bathes cells	B. Homeostasis
3. Maintenance of near-constant internal conditions	C. Hypothalamus
4. Promotes the reabsorption of water from the collecting ducts	D. Osmoregulation
	E. Pituitary gland
5. Regulation of water concentration of blood	F. Tissue fluid
6. Secretes ADH	

2. After drinking a large volume of water, the following changes occur, but they are in the wrong order. Re-write the letters of the changes in the correct sequence.

A. pituitary reduces secretion of ADH

B. increase in water concentration of blood is detected by the hypothalamus

C. urine output increases

D. water concentration of blood falls

E. water concentration of blood rises

F. less water is reabsorbed from kidney tubules

3. It is recommended a person drinks 6 to 8 glasses of water per day to avoid dehydration. Discuss how the body of a dehydrated person tries to return to a normal fluid balance. You need to refer to specific body parts in your answer.

Regulation of body temperature (thermoregulation)

Metabolism is the collective name for all the chemical processes/reactions going on in the body's cells. The enzymes that catalyse these chemical reactions are very sensitive to temperature. As a rough rule, up to about 40°C, the rate of metabolism approximately doubles for every 10°C rise in temperature.

Example

You may have noticed how much more difficult it is to catch a butterfly or swat a fly on a hot summer's day than in the cool of the early morning. This is because an insect is about twice as active at 25°C as it is at 15°C, and four times as active at 25°C as at 5°C.

Like other mammals (and also birds), humans are **homeothermic** – they can regulate their body temperature at a near constant level. Normal body temperature in humans averages 37°C (it is slightly higher in the evening and a little lower in the early morning).

Though body temperature is regulated extremely well, the mechanisms cannot always cope with extreme conditions.

- If body temperature rises too high (**hyperthermia**), enzymes may begin to be inactivated (denatured), resulting in heat exhaustion, heat stroke and if severe, death. The remedy is to cool an affected person with cold, wet cloths.
- **Hypothermia** occurs when body temperature falls to dangerously low levels (eg after prolonged exposure to cold, wet weather with unsuitable clothing, or prolonged immersion in cold water). The remedy is to gently raise body temperature (eg by a rescuer providing body heat, or by immersion in lukewarm water).

Example

Advantages and disadvantages of homeothermy

The advantage of homeothermy (maintenance of a constant body temperatures – 'warm-bloodedness') is that enzymes can function at around their most favourable, or optimum, temperature. In poikilothermic ('cold-blooded') animals, the rate of metabolism depends on the temperature of the environment. Thus a bandicoot (a mammal) can remain active in cold weather and at night, when its prey, such as bush crickets, are too sluggish to escape.

The disadvantage of homeothermy is that the extra heat production costs a lot of energy, so mammals and birds have to eat more food than 'cold-blooded' animals like lizards.

Body temperature is regulated by balancing heat loss and heat gain.

Losing heat – the skin

The skin is the largest organ in the body, and plays an essential role in the regulation of heat loss. It consists of two distinct layers, an outer **epidermis**, and an inner **dermis**.

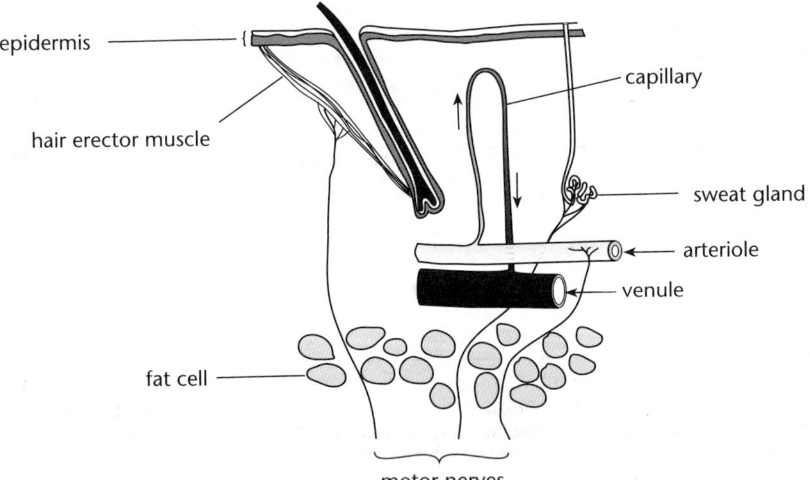

Section through the skin, showing structures concerned with thermoregulation.

The **dermis** contains a number of structures concerned with thermoregulation:

- *Blood vessels* – the wider they are, the more blood flows through them and the more heat is brought to the skin.

- *Sweat glands* – these deep intuckings of the epidermis open on the outer surface of the skin – they secrete sweat, which, when it evaporates, cools the skin. (We feel hotter in humid conditions because sweat does not evaporate so easily.)
- *Sensory nerve endings* – some are sensitive to changes in temperature.
- *Hair follicles* (intuckings of the epidermis from which hairs grow) – attached to each follicle is a small **erector muscle** which raises the hair; in most mammals this has the effect of thickening the coat and increasing insulation. (In humans these muscles have no function, though they still contract when the body is cold, producing 'goose pimples'.)

A fall in body temperature is detected by the hypothalamus. This sends out more impulses to the skin arterioles via sympathetic nerves, causing the arterioles to constrict, reducing the flow of blood (and therefore heat), to the skin. A rise in temperature causes the hypothalamus to send more impulses to the sweat glands, stimulating them to step up their sweat output.

Beneath the dermis is a layer of **adipose** (fat-storage) tissue. The cells are huge, each containing a single enormous fat droplet. Besides acting as an energy reserve, adipose tissue acts as a heat insulator.

Controlling heat production

Even when asleep, energy is needed for keeping warm and for other essential processes. This minimum rate of metabolism is called the **basal metabolic rate** (**BMR**).

The BMR is under the direct control of the hormone **thyroxine**, secreted by the **thyroid** gland in the neck. Thyroxine contains iodine, which is why this mineral is needed in the diet. In winter, thyroxine output and BMR are higher than in summer, so appetite tends to increase in winter.

The thyroid gland does not monitor body temperature – this is the job of the hypothalamus in the brain. The hypothalamus stimulates the **pituitary gland** just beneath it, via a hormone. In turn, the pituitary gland stimulates the thyroid by (another) hormone called **TSH** (thyroid stimulating hormone):

hypothalamus stimulates pituitary ⟶ pituitary stimulates the thyroid.

The thyroid is prevented from being over-stimulated by negative feedback – thyroxine *inhibits* the secretion of TSH, so a rise in thyroxine tends to inhibit its secretion, and vice versa.

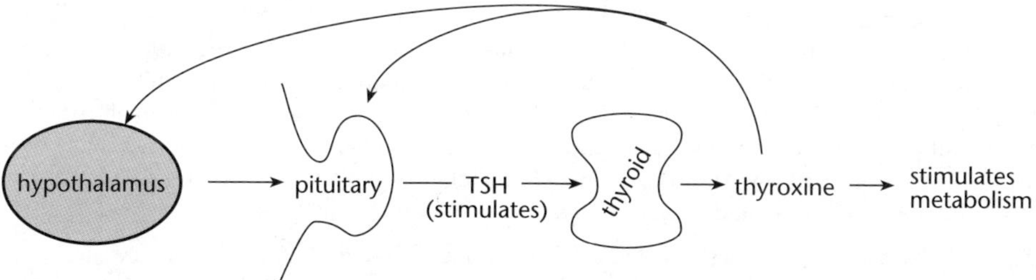

How the brain controls the thyroid gland.

The overall control of body temperature is summarised below.

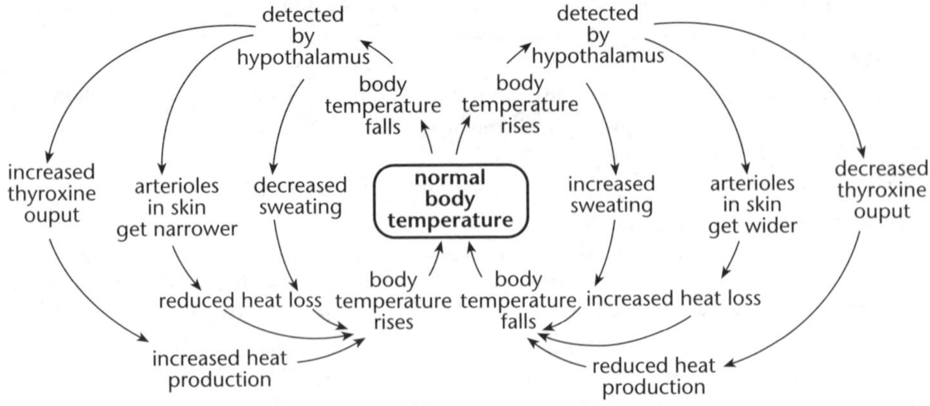

Mechanisms controlling body temperature.

Unit 11.5 Activity 3B: Temperature regulation

1. When a person exercises vigorously, their face can become red or 'flushed', and their skin becomes covered with sweat in order to cool the body.
   a. Describe the effect on the body of prolonged or excessive sweating.
   b. Explain why the face becomes 'flushed' when exercising.
   c. Discuss how the body helps to restore normal water balance after vigorous exercise. Include a flow diagram to support your answer.
2. Humans need to maintain a core body temperature of 37°C. Homeostatic mechanisms in the body are in action throughout the day to keep the temperature at this level.
   a. Define homeostasis.
   b. Explain why hands and feet may have a much lower temperature than 37°C.
   c. Explain two ways in which a cold body tries to return to normal body temperature.

Maintaining blood sugar levels

Although most tissues can use fat as an alternative source of energy, the main fuel is glucose (the brain can't use anything else). The need for glucose is continuous, but the supply from the intestine is intermittent. For a few hours after a meal, the blood leaving the small intestine is rich in glucose. During a short fast (eg at night while asleep), the gut supplies little or no glucose to the rest of the body. Despite this, in a healthy person the blood glucose concentration rarely rises above 0.44% or falls below 0.08% by weight. To maintain a stable glucose level the body has to store glucose after a meal, and release it back into the blood during a fast.

The body has two 'storehouses' for glucose – the liver and the skeletal muscles. After a meal, the liver cells remove glucose from the blood and convert it to **glycogen**, which is rather like starch. During a fast, the process is reversed.

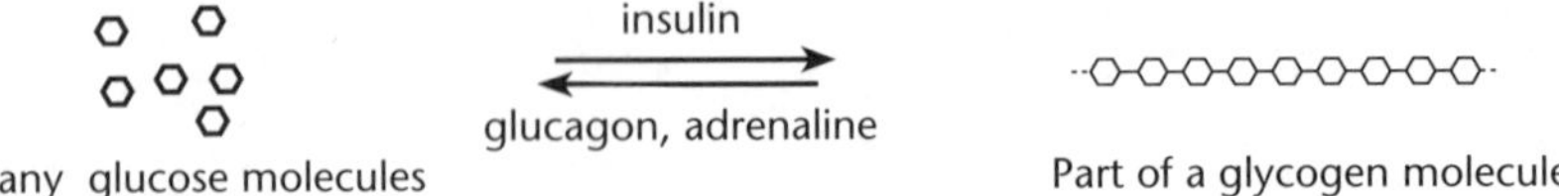

Inter-conversion of glucose and glycogen.

The glycogen stores in the muscles are normally only used during vigorous exercise. During recovery, the glycogen stores are replenished from blood glucose.

The storage and release of glucose is under the control of hormones secreted by the pancreas. The pancreas contains 1–2 million tiny clusters of **endocrine** cells called the **islet of Langerhans**. These secrete the hormone **insulin**, which promotes the conversion of glucose to glycogen and fat; under normal conditions, the reverse is stimulated by the hormone **glucagon**. When the liver glycogen store is full, surplus glucose is converted into fat and stored under the skin. In emergencies, **adrenaline**, secreted by the **adrenal glands**, powerfully stimulates the breakdown of glycogen to glucose.

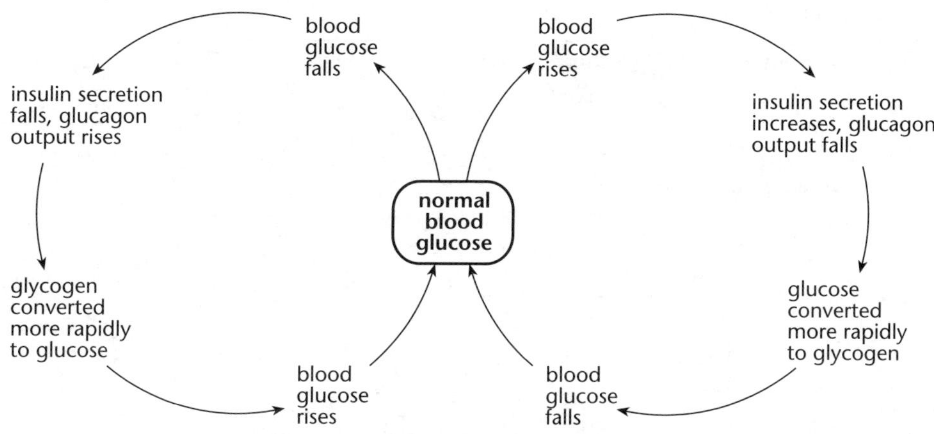

Feedback control of blood glucose level.

Glycogen stores in the liver are sufficient to provide the resting body with glucose for about 24 hours. During a longer fast (eg a '40-hour famine'), the fat stores under the skin are used.

- Fat is converted into glycerol and fatty acids. Mammals cannot convert fatty acids into carbohydrate, so they have to use the fatty acids directly in respiration.
- Glucose can be produced from protein (by deamination of amino acids) – it is this source of glucose that the brain depends on for energy during fasting. In cases of prolonged starvation, when all fat reserves have been used up, all blood glucose is obtained from body protein.

Glucose homeostasis also involves the control of the supply of glucose via the gut. Part of the hypothalamus – the 'hunger centre' – controls appetite and thus the entry of glucose into the body. When blood glucose level falls, the hunger centre stimulates the desire for food.

Diabetes

Diabetes, one of the most common diseases in wealthy countries, results in the inability of the body to control blood glucose. In a healthy person the capacity to convert surplus glucose into fat is so great that it is rare for the blood sugar level to rise above about 0.44%, even after a very large carbohydrate meal.

In diabetics, most of the cells of the body are unable to use glucose, and their blood concentration may rise to 10 times normal (**hyperglycaemia**). There may be so much glucose that the kidney cannot reabsorb it all from the filtrate, and some appears in the urine. The glucose in the urine makes it harder for the kidney to reabsorb water, so the urine output rises and the person becomes perpetually thirsty. Even though the blood glucose level may be

10 times normal, the cells cannot use it without insulin, so the body has to use fat. The use of fat in large quantities generates acidic waste products, and the resulting fall in blood pH may cause diabetic coma and, without treatment, death.

There are two kinds of diabetes.

- **Type I**, or **early onset**, diabetes usually develops in childhood, and is caused by the death of the insulin-secreting cells of the pancreas. Treatment involves regular injections of insulin, coupled with a carefully managed diet low in carbohydrate.
- **Type II**, or **late onset**, diabetes tends to develop in middle age. Insulin production is normal but the body cells are insensitive to it. Treatment involves a carefully managed diet to control carbohydrate intake.

Unit 11.5 Activity 3C: Maintaining blood sugar levels

1. A glucose drink is taken after hard exercise. Describe the effect the drink has on glucose levels in the blood.
2. A hormone called insulin helps control the blood-sugar level of the body. The diagram shows events that take place to control glucose levels in the blood. Use the information in the diagram to explain how insulin helps to maintain the glucose levels in the blood after finishing a drink high in glucose.

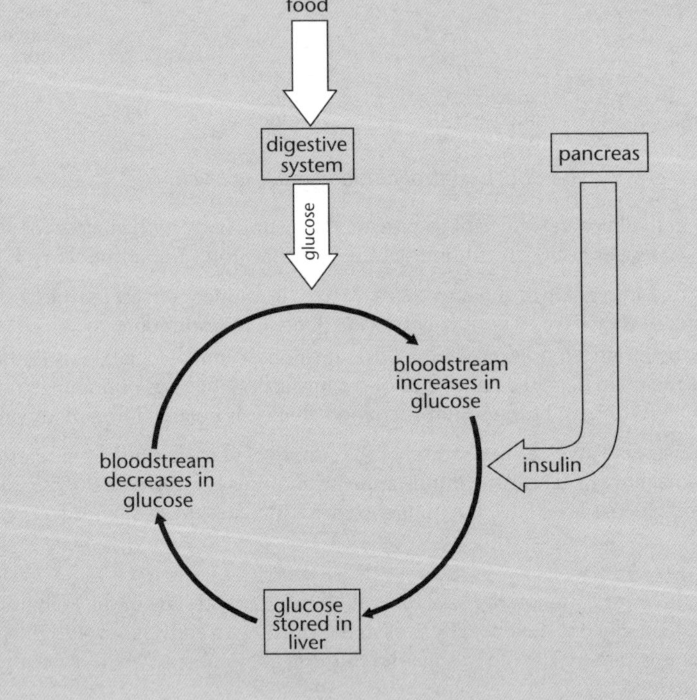

Regulation of salt

The levels of salt in the blood are regulated along with the levels of water in the blood.

The presence of too much salt in the blood is detected by the hypothalamus in the brain and the following responses occur:

- Anti-diuretic hormone (ADH) is released from the pituitary gland and causes water to be reabsorbed into the blood from the kidney tubules.
- Another hormone is also released that causes sodium reabsorption in the kidney tubules to decrease.

These two responses by the body result in the production of small volumes of dark, concentrated urine. At the same time, the person feels thirsty, and the fluid consumed helps restore the salt and water balance in the blood.

Too little salt in the blood has the opposite effects. ADH secretions are reduced, meaning less water is reabsorbed from the kidney tubules and instead, passes out of the kidneys and into the bladder. Sodium is reabsorbed from the kidney tubules, increasing the salt levels in the blood.

The endocrine system

The human body is made up of many organs and millions of cells that need to communicate with each other.

Example

A hot stove top is touched by the fingertip and the brain must detect this stimulus and coordinate the muscles concerned to remove the finger from the stove as quickly as possible.

In the *Example* above, the communication system is the *nervous system*, in this case a **reflex arc** that involves only three nerve cells that very rapidly communicate the message from the finger to the brain and back to the finger.

Example

Cells in the brain detect the presence of too much salt in the blood and this message must get to the kidneys, the organs responsible for regulating the blood salt concentration.

In the *Example* above, the communication system is the **endocrine system**, which produces slow-acting and long-lasting (relative to the nervous system) responses to stimuli. It does this by manufacturing substances called hormones in the **endocrine glands** in the body. Often, the site of production of the **hormone** is not where it acts (eg ADH is produced in the brain, but acts in the kidneys), and some hormones are produced to control the action of other hormones.

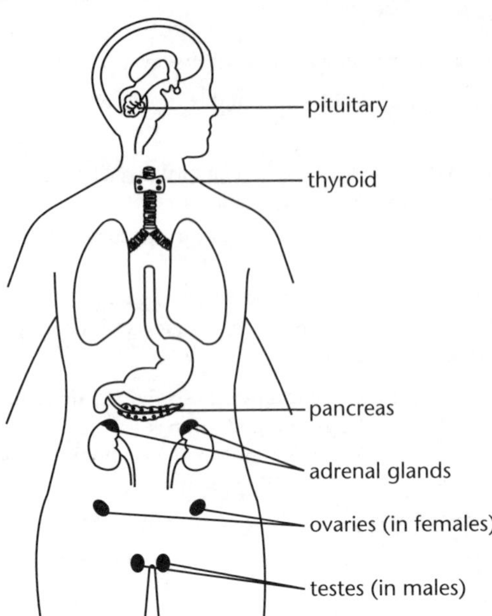

Human endocrine glands.

Hormones and the endocrine glands

There are about 50 hormones produced by the endocrine glands, but only the most important ones are summarised in the table below.

The pituitary gland in the brain is often called the 'master gland' as it produces hormones (indicated by an asterisk in the table) that control the production of hormones from three of the other glands. The three other glands are:

- The thyroid.
- The adrenal.
- The ovaries and testes.

Hormone	Production site	Action site	What the hormone does	Disorder in absence
Growth hormone	Pituitary	All over the body	Causes growth in height	A type of dwarfism
Thyroid stimulating hormone (TSH)	Pituitary	Thyroid gland	Stimulates thyroid gland to make thyroxine if levels are low	
Anti-diuretic hormone (ADH)	Pituitary	Kidneys	Regulates water and salt levels in the blood	Diabetes insipidus (large volumes of dilute urine)
Oxytocin	Pituitary	• Uterus • Breast tissue	• Causes contractions in birth • Causes milk flow	

Hormone	Production site	Action site	What the hormone does	Disorder in absence
Prolactin	Pituitary	Breast tissue	Milk production	
FSH and LH	Pituitary	• Ovaries • Testes	• Follicular growth and ovulation • Sperm and testosterone production	
ACTH	Pituitary	Adrenal gland	Stimulates adrenal gland to produce cortisol	
Thyroxine	Thyroid	Body cells	• Controls basal metabolic rate (BMR) • Promotes growth in children	Hypothyroidism
Insulin	Pancreas	• Liver • Body cells	Regulates blood sugar levels, converts glucose to glycogen for storage	Diabetes mellitus
Glucagon	Pancreas	Liver	Converts glycogen to glucose	
Adrenalin	Adrenal	• Muscles • Heart • Blood vessels	Readies the body for 'flight or fight' responses: • increases heart, breathing and metabolic rates • increases brain alertness • converts glycogen to glucose	
Cortisol	Adrenal	Muscle cells	Converts muscle protein to amino acids and then to glucose (in the event of starvation)	
Oestrogen	Ovaries	Sex organs	Secondary sexual characteristics	
Progesterone	Ovaries	Uterus	Prepares the uterus lining for implantation of zygote	
Testosterone	Testes	Sex organs	Secondary sexual characteristics	

Responses to change

Certain substances (eg alcohol, drugs and tobacco), and exercise can challenge the mechanisms involved in homeostasis. Responses made by the body can be both short term and long term.

Alcohol

While alcohol use may, in the short term, make a person feel happy, alcohol is classified as a depressant. Depressants are drugs that decrease the activity of the brain and nervous system, making the person feel drowsy and sleepy, and slowing the reaction rate. In addition, alcohol consumption results in increased aggression and can cause antisocial behaviour. Alcohol inhibits the production of ADH from the pituitary gland, causing the kidneys to work harder than normal to produce large volumes of dilute urine. In the long term, this can result in kidney damage, while the *hangover* is a symptom of the dehydration caused by excess alcohol.

The liver also must work harder to remove toxins produced from the breakdown of the alcohol from the body. Chronic consumption of alcohol can permanently damage the liver as the fats produced from the breakdown of alcohol destroy the liver cells (*cirrhosis*). Alcohol irritates the gut lining, causing peptic ulcers and may contribute to malnutrition as digestion and absorption cannot occur effectively.

Alcohol causes dilation of the blood vessels supplying the skin, making the person look 'flushed'. As the amount of alcohol consumed increases, the person may go through the stages of dizziness, nausea and vomiting, disorientation, sleepiness, unconsciousness and, in extreme cases, death by alcohol poisoning or asphyxiation (from inhaling vomit).

Drugs

There are five classes of drugs, all of which can have severe effects on the body.

- *Narcotics* – eg, opium. Narcotics generally initially give feelings of well-being, euphoria, and decreased tension, anxiety and aggression. However, excessive use of narcotics can result in sleepiness, dilation of the blood vessels and death through the slowing down of the area of the brain that controls breathing. Narcotics are highly *addictive*, and with long-term use, a person's *tolerance* to the drug increases, meaning that a greater amount must be taken for the same effects. Higher doses increase the chance of an *overdose* occurring, where death may be the result.
- *Depressants* – eg, alcohol. Depressants slow down the activity of the nervous system and the functioning of the brain, resulting in a sedative effect. They make a person feel sleepy and drowsy.
- *Stimulants* – eg, caffeine, nicotine, cocaine, methamphetamine ('P'). Stimulants have the opposite effects to depressants by increasing the activity of the nervous system. They result in feelings of euphoria, high levels of alertness and concentration (so much so that the person may not sleep for days), paranoia or hallucinations, and suppress the appetite (the person may not eat for days). Other physical effects include increased heart rate (often with palpitations or irregular heartbeats), increased sweating, vomiting and tremors. Stimulants are highly addictive and long-term use can result in death as the drug interferes with normal cardiac and temperature-regulation systems. There is also the risk of liver and kidney damage, weight loss and malnutrition (through decreased appetite and increased activity), and impotence.
- *Hallucinogens* – eg, LSD. These drugs cause either the distortion of sensory experiences (eg, foods may taste wildly different from reality), or they may cause the sensory experience of something that is not actually there (a hallucination), or they may intensify the mood of

the person taking the drug. While the drugs themselves may not be particularly dangerous, the behaviours they cause can be. For example, a person may think they are extremely strong and physically attempt to stop traffic. Or, if they are already feeling depressed, the hallucinogen may cause them to become suicidal.

- *Anabolic steroids*. These drugs are made from or closely resemble the male sex hormones – in particular, testosterone. Anabolic steroids are well known because of their use by some athletes who want to increase their performance in competitive sports. Anabolic steroids promote the formation of male features, particularly the growth of muscle tissue. Effects are reversed once the drug is no longer taken, but long-term use can result in death through damage to the liver, kidneys and heart.

Tobacco

Cigarettes contain many harmful substances, a significant number of which can, on their own, result in death. Short-term effects include:

- *Increases in heart rate, alertness and energy, and suppression of appetite* – as nicotine is a stimulant.
- *Increased heart rate* – the carbon monoxide inhaled diffuses easily into the blood, then binds to the haemoglobin molecules reducing the amount of oxygen able to be transported around the body. The heart rate must increase to supply sufficient oxygen to the body tissues.

Long-term effects include:

- *Addiction* – the nicotine in tobacco is highly addictive, resulting in various *withdrawal* effects if the body and brain are deprived of nicotine when a person attempts to give up smoking.
- *Various cancers*, particularly of the lungs, airways, mouth and throat – may develop as a result of the toxic chemicals present in the tobacco.
- *Staining* – of the teeth and fingers from the tar present.
- *Various disorders and infections of the airways* – including bad breath, 'smoker's cough', bronchitis, and increased chances of asthma attacks if the person has asthma. The tar irritates and paralyses the cilia lining the airways, leading to an over-production of mucus. This has two effects. Firstly, as the cilia no longer function effectively to remove foreign material, the lungs and airways become coated with substances such as dust that reduce the surface area of the alveoli, reducing the efficiency of the gas exchange system. Secondly, it is much easier for pathogens to enter the lungs and airways, becoming trapped in the sticky, warm, moist mucus and causing infections.
- *Increased risk of developing infections in general* – the chemicals in tobacco impair the immune system by affecting the levels of antibodies and white blood cells in the blood, and by affecting the functioning of some white blood cells so they can no longer destroy pathogens that enter the body.
- *Development of peptic ulcers in the gut.*

Exercise

Exercise is a challenge to the homeostasis systems of the body. Both short- and long-term effects result.

Short-term effects include:

- *Increase in heart rate* – as the muscle cells use up oxygen and glucose through respiration to provide energy for the body, more must be supplied. An increased heart rate increases the delivery of oxygen- and glucose-rich blood to the muscle cells.

- *Increase in breathing rate and depth of breathing* – these are linked to the increase in heart rate as more oxygen is required by, and more carbon dioxide is produced by, the muscle cells.
- *Red flush of the skin* – core body temperature increases as the rate of respiration increases (one of the by-products of this is heat energy). The blood vessels close to the surface of the skin dilate, with the result that they are closer to the surface of the skin and have increased blood flow. The skin becomes flushed red. As the blood passes closer to the surface of the skin, heat energy is lost to the outside and the body temperature is maintained at normal.
- *Increased sweating* – to keep body temperature normal. Sweat sitting on the skin surface uses heat energy from the body to evaporate, decreasing the amount of heat energy in the body. It is important to replace fluid lost from the body as a result of sweating, otherwise *dehydration* may occur. The loss of fluid results in lower than normal blood volume, which is detected by *osmoreceptors* in the hypothalamus. ADH is released by the pituitary gland, stimulating the kidneys to reabsorb water from the kidney tubules, restoring blood volume. The person also feels thirsty, drinking more fluids.

Long-term effects include:

- *Strengthening of the heart and muscles associated with breathing* – exercise causes these muscles to increase in size, which means that while the breathing and heart rates will still increase with exercise, each heart beat and breath will be more efficient at supplying working muscles with oxygen-rich blood. Resting heart and breathing rates will also decrease as a result.
- *Increase in skeletal muscle size in the limbs affected by the exercise* – eg, a cyclist will develop much larger leg muscles, particularly in the upper leg as the muscle cells in these muscles increase in size, becoming more efficient.
- *Weight loss (so long as energy intake is less than energy output)* – fat stores beneath the skin are broken down and used as an energy source for working muscle cells. In addition, the larger muscle cells require more energy when at rest, increasing the basal metabolic rate and contributing to further weight loss.
- *Decrease in blood pressure* – as the heart muscle becomes more efficient, it doesn't have to work as hard at pumping blood into the arteries. In addition, stronger skeletal muscles surrounding veins result in blood being returned to the heart more efficiently.

Unit 11.5 Activity 3D: Endocrine system and challenges to homeostasis

1. The pituitary gland is often called the 'master gland' of the endocrine system.
  a. Name two hormones produced by the pituitary gland.
  b. Explain how the pituitary gland works as a 'master gland'. Support your answer with an example.
  c. **i.** Name the gland that produces adrenalin.
    ii. Describe where the gland named in part **i.** is found in the body.
  d. **i.** Describe two effects the release of the hormone adrenalin has on the body.
    ii. Explain how one of these effects prepares the body for an emergency.

2. **a.** Describe one short-term effect of alcohol on the body.
  b. Discuss how long-term use of either alcohol or tobacco may influence the body's ability to maintain its normal functioning.

3. **a.** The diagram shows the location of the main endocrine glands. In the following table, describe a function **i.–v.** for each of the hormones.

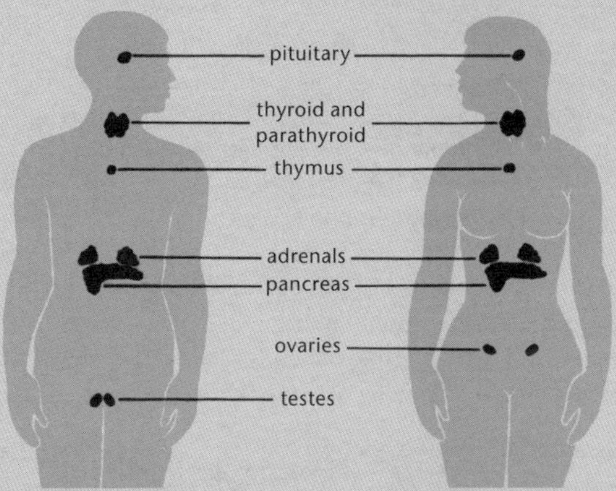

Gland	Hormone	Function
Pituitary	ADH (anti-diuretic hormone)	**i.**
Pituitary	GH (growth hormone)	**ii.**
Thyroid	Thyroxine	**iii.**
Pancreas	Insulin	**iv.**
Adrenals	Adrenalin	**v.**

b. The diagram shows how the thyroid gland releases thyroxine. This is an example of a feedback system.

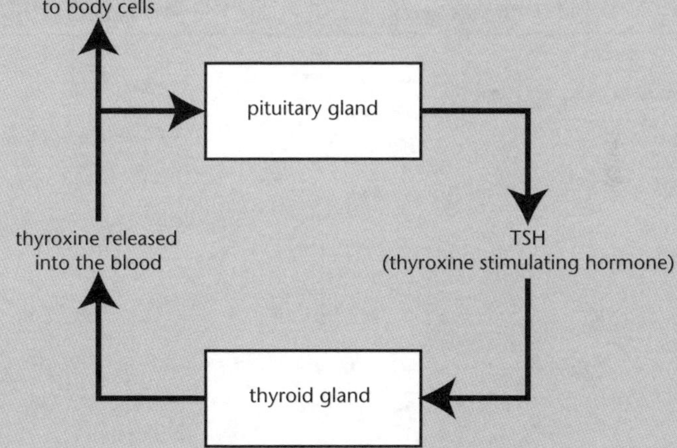

i. Explain how this feedback system maintains thyroxine levels.

ii. Thyroxine contains iodine. Discuss the effects a shortage of iodine in person's diet will have on this feedback system.

4. Energy drinks are typically advertised as:
   - Energising body and mind.
   - Combating stress and fatigue.
   - Perfect for getting you through the day, the night, the morning after the night before.

   The 'energy' in these drinks is sugar and caffeine. There is more caffeine in a can of energy drink than in an ordinary cup of plunger coffee. Each 250 mL serving of these drinks contains approximately 7 teaspoonfuls of sugar and provides close to 500 kJ of energy. Cans may have a warning label such as 'Usage: 2 cans max daily'.

   a. Describe how your body responds to the sudden increase of sugar after drinking an energy drink.

   b. Describe two short-term effects of caffeine on the body.

   c. Discuss whether a sportsperson in training should drink water or energy drinks during a training session. You need to compare both water and energy drinks in your answers.

5. The diagram shows a cross-section of a person's skin while he or she is exercising.

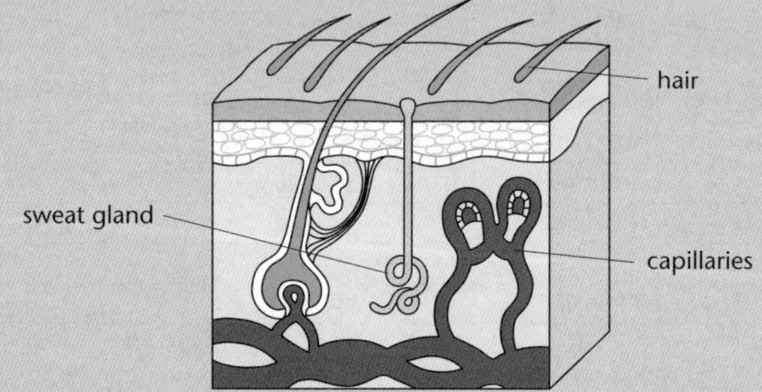

   a. Describe two changes that take place in the skin during exercise.

   b. Discuss how a cold body tries to return to a normal body temperature.

Unit 11.6 Reproduction

Topic 1: How animals reproduce

In line with the Biology Syllabus (p. 23), this Unit begins with the sexual and asexual reproduction in animals (we look at plants in Topic 3) and the reproductive structures in males and females. Topic 1 deals with:

- The reproductive cycles of animals.
- The structure and function of the parts of the reproductive systems of animals.
- Courtship behaviours, nutrition and care of embryo/young.
- Reasons for the differences in structure and function between different groups – these could relate to mobility and energy needs, size, habitat, life cycle, way of life.

Note that reproduction in micro-organisms is covered in the Supplementary Unit (p. 339).

Members of a species need to *reproduce* themselves to ensure the survival of their genes (and, incidentally, their species); however, the reproductive system is the one body system that an *individual* can exist without.

The simpler and quicker method of reproduction is **asexual**; the more complex and slower is **sexual**. All large and complex animals reproduce sexually (usually in response to adverse environmental conditions), though asexual reproduction occurs as well throughout many of these groups (eg in fish, reptiles, insects).

Asexual reproduction

In asexual reproduction, a single parent reproduces offspring (eg budding in hydra, parthenogenesis in aphids) by *mitotic* cell division. The offspring are all genetically identical to each other and the parent. The process is economical and typically produces large numbers of offspring rapidly, to take advantage of good environmental conditions facilitating spread and survival of the species (eg aphids). The big disadvantage of asexual reproduction is the lack of variation in the offspring; the only source of genetic variation being mutation in the DNA. This limits the species' ability to evolve to cope with environmental change.

Simple asexual reproduction – *Hydra*

Hydra is a freshwater cnidarian, with a simple, sac-like body made of two cell layers enclosing a middle, jelly-like layer. Cells in the outer cell layer (the ectoderm) are capable of undergoing mitosis when environmental conditions are favourable to form 'buds' of cells that develop into new *Hydra*. These eventually detach from the parent and lead an independent life by attaching to a firm substrate in a freshwater pond. As a result, large numbers of *Hydra* may be rapidly formed.

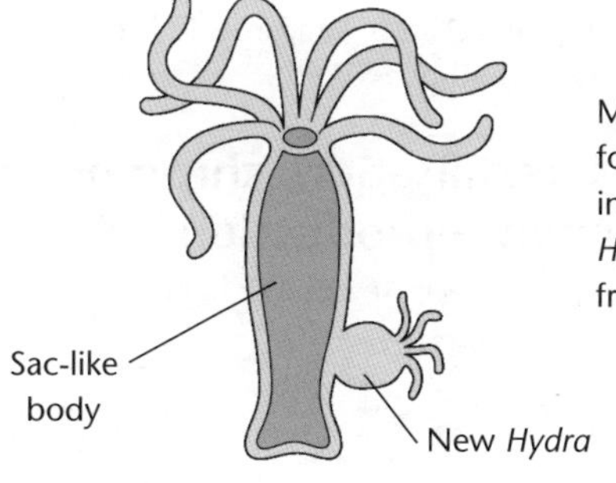

Mitosis in the cells of the ectoderm forms a ball of cells that differentiate into gut/mouth/tentacles of a typical *Hydra*; the new individual detaches from the parent to start life on its own.

Advanced asexual reproduction – aphids

Aphids are small, parasitic insects that feed from the sap of plants such as rose bushes.

Both male and female aphids occur, but in summer months when temperatures are warm and food is plentiful, *female aphids reproduce large numbers of female offspring asexually* (known as **parthenogenesis**). *Ova* ('eggs') are produced in the ovaries by *mitosis*, and these develop into small female aphids within the mother's body (known as **viviparous** development). These developing female offspring already have developing female ova inside them (a female can have up to 100 ova developing at any one time) – allowing huge numbers of aphids to be rapidly produced. Females may be winged or wingless; wingless forms tend to appear in crowded conditions – this allows more effective distribution of the aphids, reducing the chances of over-infestation of the plant(s) they parasitise.

When autumn approaches, day length and temperatures decrease, and food supplies become limited. This change in environmental conditions provokes *sexual* reproduction in aphids. Male ova are formed in the female aphid (by the loss of a sex chromosome – females are XX, males are XO). The hatched males produce two kinds of sperm (X sperm and O sperm) through *meiosis*, but the O sperm do not survive. Therefore, mating with females will produce only female (XX) offspring. Fertilised females will lay eggs (known as **oviparous** development), and these will overwinter, then, in spring when conditions are again favourable, hatch as parthenogenic females. This sexual reproduction allows for genetic variation, so aphids may survive changing environmental conditions (and, incidentally, the species evolves). Sexual reproduction in aphids shows incomplete metamorphosis.

Sexual reproduction

In sexual reproduction, two parents are needed (male and female, except for **hermaphrodite** species – hermaphrodites have *both* male and female sex organs). **Gametes** (*sex cells*) are produced in the sex organs by *meiotic* cell division, which halves the chromosome number (from 2n to n; *diploid* to *haploid*). Female gametes are the **ova** (singular **ovum**), made in the **ovaries**. Male gametes are the **sperm**, made in the **testes** (singular testis). Sperm are *motile* and swim to the ovum for **fertilisation**. The ovum is much larger than the sperm, as it has cytoplasm containing organelles. The fertilised ovum (the **zygote**) has a full set of chromosomes (2n, diploid). *Mitotic* divisions of the zygote form the **embryo**.

Offspring formed from sexual reproduction are *genetically different* from each other and their parents. It is these differences that evolution (natural selection) acts on, allowing the continuation of the species as it responds to changing environmental conditions through individuals surviving.

Different animals have different diploid numbers (eg gorilla 48, kangaroo 12, domestic hen 36), and most animals spend nearly all their life in the diploid phase.

Relationship between gametes, fertilisation, chromosome number and cell division in human reproduction

In human cells, the diploid (2n) number of chromosomes is 46 and the haploid (n) number is 23.

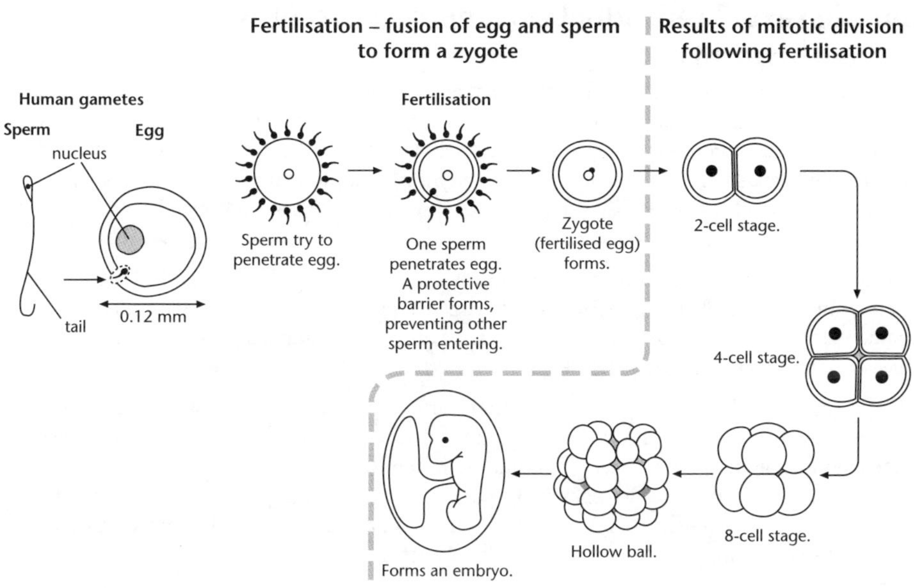

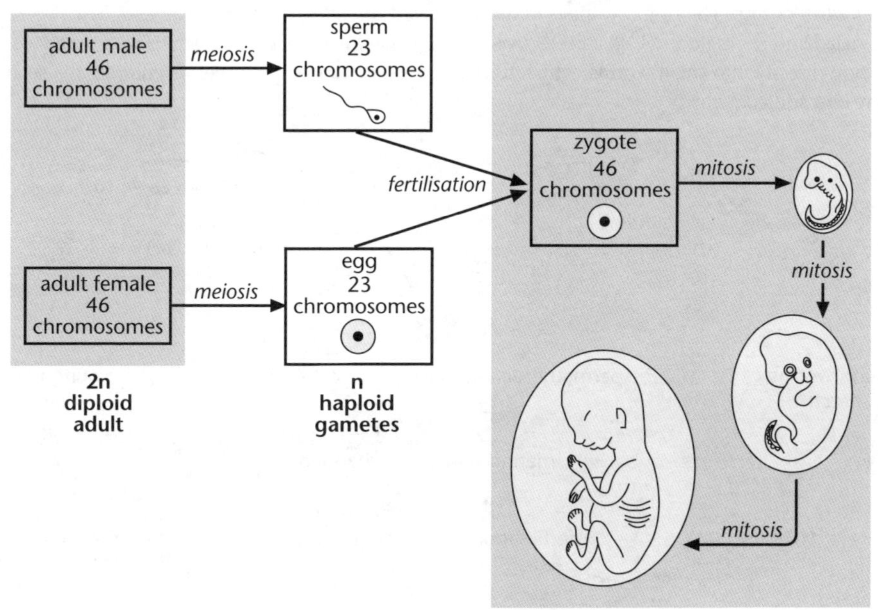

Diploidy and haploidy in humans

Simple sexual reproduction – hermaphrodites (eg earthworm)

Members of both the annelids (eg earthworms) and molluscs (eg garden snail) are *hermaphrodite* – individuals have *both* male and female sex organs, so any individual is capable of mating with any other sexually mature individual; this enhances the chances of finding a suitable mate in species which are small, slow moving, live solitary lives, and don't have complex sensory organs. When hermaphrodites mate, they exchange sperm which fertilise ova produced in the female system (ie cross-fertilisation still occurs).

Earthworms are segmented. The male sex organs are found in segments 10 and 11, producing sperm which pass back along sperm ducts to two openings on the ventral ('bottom') side of segment 15. The female sex organs producing ova are found in segment 13. The mature ova are released into eggs sacs in segment 14.

Mating occurs on the surface of the soil during warm, moist nights, when two worms lie side-by-side with their anterior ('head') ends pointing in opposite directions. Each worm secretes large amounts of *mucus*, which wraps around both from segment 9 to segment 37. The sperm released from the male pores in segment 15 pass forward in a slime tube formed between the mucus and the skin to segments 9 and 10, where sperm receptacles called *spermathecae* are found. The sperm are stored in these while the ova ripen.

When the ova are mature, the clitellum (a 'saddle' found between segments 32–37) makes a second mucus tube, which is slid forward towards the anterior end of the body. When it passes the female openings in segments 14, the mature ova pass out of the body and stick to the mucus ring. The ring keeps moving forward and receives sperm from the spermatheca when it passes segments 9 and 10. *Fertilisation* occurs. The ring continues to move forward until it slips off the head end. The two ends are sealed and drying occurs, so a protective case ('cocoon') is formed in the soil. The zygotes develop directly into small worms, which feed on the remaining egg case before entering the soil to grow into adult worms.

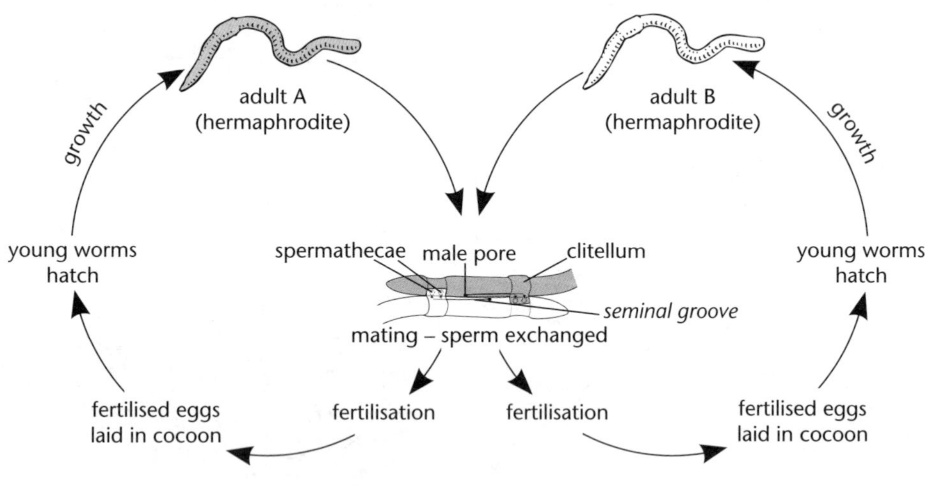

Life cycle of an earthworm

Simple sexual reproduction – metamorphosis (eg insects)

Insects have separate sexes and may exhibit simple *courtship behaviour*. In courtship behaviour, the male advertises for a female partner typically by some type of display/release of scent (pheromones)/ vocalisation (eg cicadas and crickets produce distinctive noises – 'stridulation'). Noise is typically produced by rubbing together parts of the body (eg legs, wings, wing covers, depending on the species). The purpose of courtship is to:

- Attract a partner of the *same species* so that mating occurs between members of the same gene pool and the species retains its *reproductive isolation*.
- Bring *members of the species together* (into the same area), *increasing chances of mating*.
- Provide a *choice of mates*, so *enhancing fitness* – the strongest, healthiest males 'display the best', so attracting the most females.

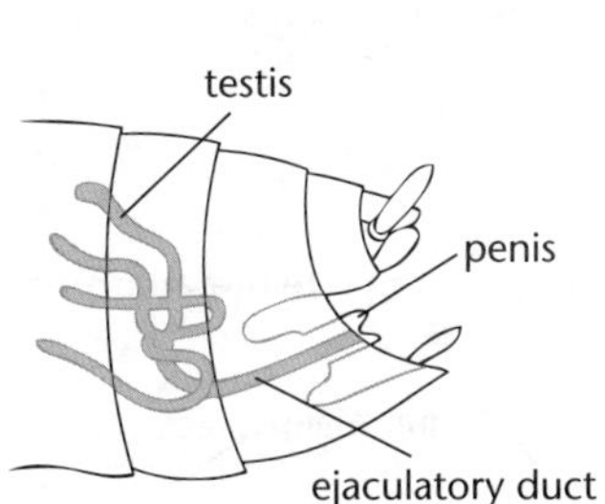

Male insect sex organs

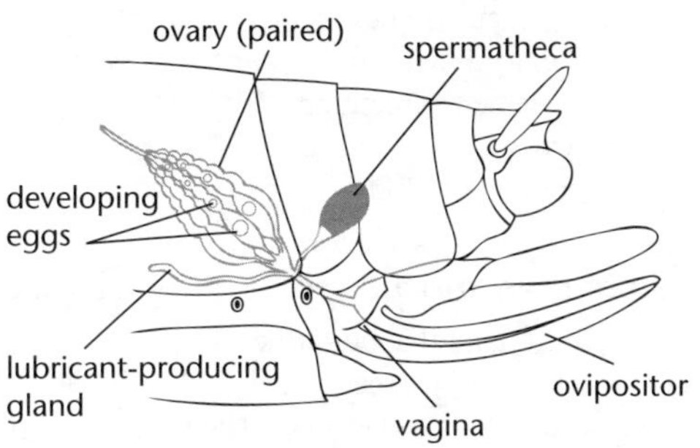

Female insect sex organs

Males have paired testes that produce sperm which are then stored in the seminal vesicles. Females have paired ovaries which lead into short oviducts. Males typically clasp the female in mating, and, using a short, blunt penis, insert a packet of sperm (*spermatophore*) into the oviduct. The sperm are stored in a spermatheca (as also happens in the case of earthworms) until the ova mature. Fertilisation is internal (an advantage over earthworms) – frees insects from dependency on moist conditions for successful reproduction. The fertilised eggs are laid (eg in the soil; attached to leaves; on, or inside dead bodies or faecal material) using an *ovipositor* (typically a tube from the end of the female's abdomen).

The eggs of many species overwinter (typically in the soil – eg cricket), hatching in spring; the adults typically die off before winter (eg crickets, cicada) giving a life cycle of a year. Other insects such as the tree weta may survive 2–3 years. The eggs of many species are able to survive extremes of climate (eg hot/cold temperatures, dryness), a factor in the success of insects as a group.

Incomplete metamorphosis

Eggs that hatch into immature adults (*nymphs*) grow and mature into adults by regularly moulting their exoskeleton (ecdysis). The new exoskeleton is soft for a few hours, allowing a growth spurt before hardening takes place. Each nymph stage is called an *instar*, and there may be 10 or more, depending on the species. In winged species, the wings develop in the later instars, starting as small buds. This type of life cycle is called **incomplete metamorphosis**. Apart from being smaller and sexually immature, nymphs are not different from the adults, occupying the same ecological niche, feeding on the same food (so intraspecific competition with the adults may occur) – eg crickets, cicadas, grasshoppers, aphids.

Insect life cycle – Incomplete metamorphosis

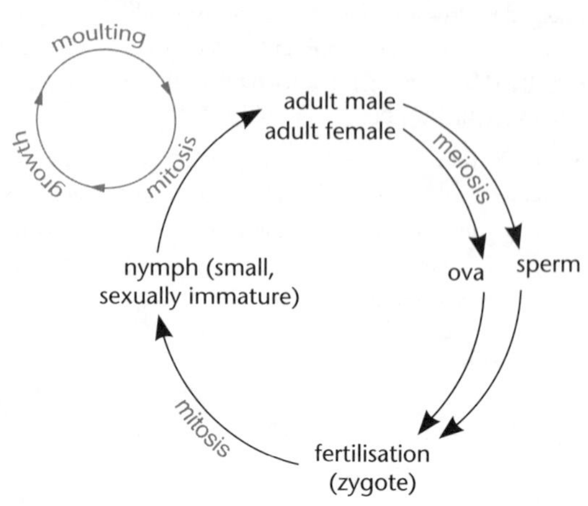

Complete metamorphosis

In other species (eg moths, butterflies, house flies, bees, beetles, mosquitoes), the eggs hatch into *larvae* ('caterpillars', 'maggots'). The larva (singular) feeds and grows rapidly, through a series of moults. The final moult turns the larva into an immobile, non-feeding stage, the *pupa* ('chrysalis', 'cocoon'). During this stage, the larva changes (*metamorphosis*) into the adult form, with the breakdown of old tissue (by enzymes) and the formation of new tissue. The pupal stage may also be an over-wintering stage. When the change is complete, the adult emerges as the 'skin' of the pupal case splits. The adult disperses the species, and reproduces the next generation.

Insect life cycle – Complete metamorphosis

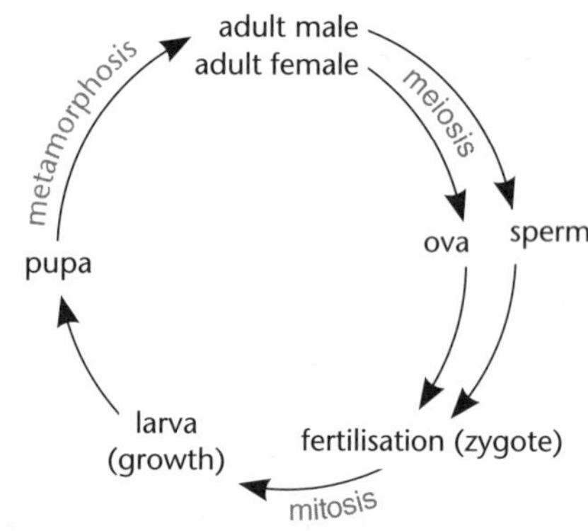

This type of life cycle is **complete metamorphosis**, as the adult and larval stages are completely different. Their ecological niches differ, including both habitat and food – intraspecific competition between adult and offspring is avoided.

Life cycle of the Cabbage White Butterfly (*Pieris rapae*)

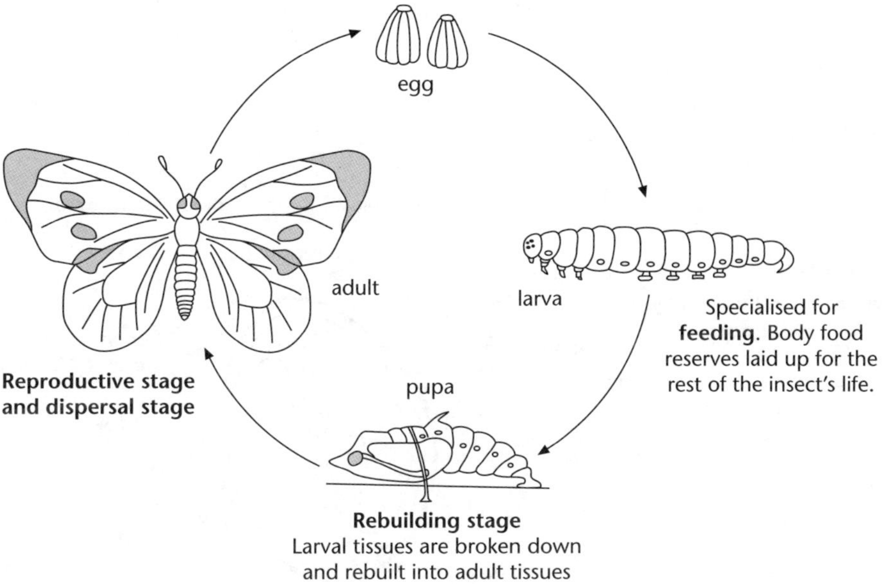

egg

larva

Specialised for **feeding**. Body food reserves laid up for the rest of the insect's life.

pupa

Rebuilding stage
Larval tissues are broken down and rebuilt into adult tissues

adult

Reproductive stage and dispersal stage

Complex sexual reproduction – vertebrates

Sexual reproduction increases in complexity within the vertebrate classes from fish to mammals. This reflects the evolutionary movement from an aquatic habitat to a terrestrial habitat. Trends include:

- An independence from water for breeding.
- Internal fertilisation.
- Parental care of offspring.

Fish

Fish generally release their gametes directly into the water surrounding them, so fertilisation takes place outside the body (*external fertilisation*). External fertilisation is very wasteful of gametes, so, to increase the chances of fertilisation and to reduce gamete loss, many fish release their gametes simultaneously, responding to environmental stimuli such as phases of the moon and tides, or to behavioural stimuli such as courtship.

Some fish, such as sharks and rays, reduce gamete wastage to a minimum by *internal fertilisation*. The male introduces sperm directly into the female reproductive tract by modified pelvic fins.

The number of eggs laid by a female depends upon the care given to the eggs after fertilisation. If no care is given to the eggs, juvenile mortality is extremely high – to compensate for this, large numbers of eggs (even millions) must be laid.

Freshwater fish generally lay fewer eggs, because their eggs stick to any substance they happen to collide with for protection. Some fish, such as carp, give their eggs even greater protection by deliberately attaching them to weeds. This habit has been developed still further in fish such as sticklebacks – sticklebacks make nests, and need to lay only a few eggs, since they are jealously guarded by the adults.

Amphibians

The productive systems of amphibians have no special modifications for land, as the eggs are soft and have no waterproof shell – so amphibians must return to water (ie need damp conditions) for breeding.

In the common green frog (*Hyla aurea*), males return to water in spring, attracting mates by calling. Eggs from the female are fertilised by a cloud of sperm released from the male frog who clings onto the back of the female.

The fertilised eggs hatch into tadpoles, which *metamorphose* into the adult form (the frog).

Reptiles

Reptiles have a *waterproof shelled egg*, allowing independence from water for reproduction.

Basic structure of a reptile (and bird) egg

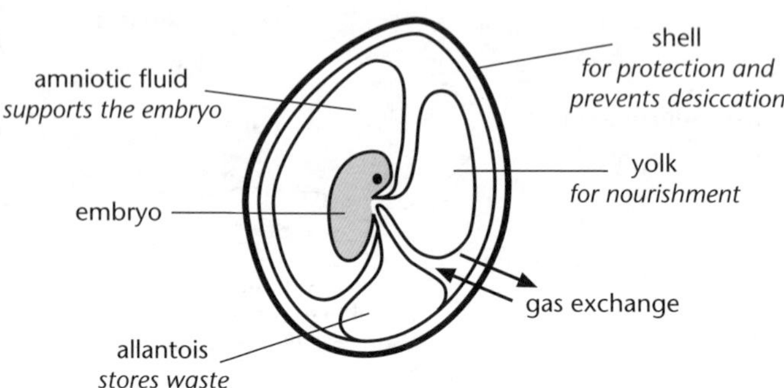

In some reptiles that produce live offspring by retaining eggs inside the female, the eggshell is reduced or absent.

Reptiles, like most vertebrates, are capable of inflicting serious injuries on each other. Detailed courtship behaviours have evolved in some species to ensure the 'good intentions' of prospective mates.

Courtship behaviour has other advantages as well:

- Ensures greater chances of fertilisation.
- Ensures mating occurs between members of the same species.
- Increases bonding links, so nest building and care of the young can be shared, which increases the survival chances of the offspring.

Birds

Birds have evolved from reptiles, and produce eggs similar in structure to those of reptiles.

Like reptiles, young birds possess an egg tooth to help them break out of the egg during hatching.

Reproductive organs are usually only developed during the breeding season. Their absence at other times of the year reduces body weight, making flying easier.

Birds show the most elaborate colours, displays, ornamentation and calls of all the vertebrates during courtship. Bird song acts to bring the sexes together and acts as a stimulus for **ovulation** (egg production) and copulation (the physical act of mating).

Male birds deposit sperm directly into the female reproductive tract when the cloacae of the male and female birds are pressed together during copulation.

Mammals

Because of the protection and parental care given to their offspring, mammals usually produce only a few offspring.

Egg-laying mammals, the monotremes (echidna and duck-billed platypus), produce eggs that are large and yolky, and the embryo retains the primitive egg tooth. These animals produce milk from specialised sweat glands.

Marsupials, such as wallabies and possums, produce young which are born at a very immature stage, within only a few weeks of conception. They are fed milk from a teat and are protected inside the mother's pouch during development.

In **placental** mammals, females possess a womb or **uterus**, specialised for protecting and nourishing the embryo. The placental mammal egg is very small, since it contains little yolk. Specialised mammary glands produce milk.

The development of the placenta represents the most sophisticated form of mammal reproductive evolution, because it allows an extended period of internal development and protection for the embryo.

Unit 11.6 Activity 1A: Reproduction in animals

1. Explain the advantages and disadvantages of:
   a. Asexual reproduction.
   b. Sexual reproduction.
2. Compare the processes of complete and incomplete metamorphosis in insects.
3. Explain why certain groups of animals are hermaphrodite.
4. Explain the significance of internal fertilisation.
5. Explain why animals such as fish and amphibians produce such large numbers of eggs in comparison with groups such as mammals.
6. Discuss the significance of courtship in reproduction.
7. Discuss the occurrence of metamorphosis in animal reproduction.
8. Select three different taxonomic or functional groups, and, using named animals as examples, discuss the structure of their reproductive systems and the reasons for their differences. In your answer:
   - Describe the system for each of the named animals.
   - Explain how each of these systems operates.
   - Explain the differences in the systems in relation to the different ways of life of the three animals.

Unit 11.6 Reproduction

Topic 2: Human reproduction

Author: Martin Hanson with Takis Solulu

In Topic 2 we describe sexual reproduction in humans and learn about male and female reproduction structures (as outlined in the Syllabus, pp. 23–24) and related structures to do with sex hormones, pregnancy and embryonic development by looking at:

- Anatomy of the male and female reproductive systems.
- Menstrual cycle.
- Hormonal control of reproduction.
- Fertilisation, implantation and pregnancy.
- Birth and lactation.

The beginning of a human life

All humans begin life as a single cell called a fertilised egg or **zygote**. This is produced by joining together of two **gametes**, special cells of very different size:

- The egg or **ovum**, is just large enough to be seen with the naked eye.
- The spermatozoon or **sperm**, is the smallest human cell.

The human male reproductive system

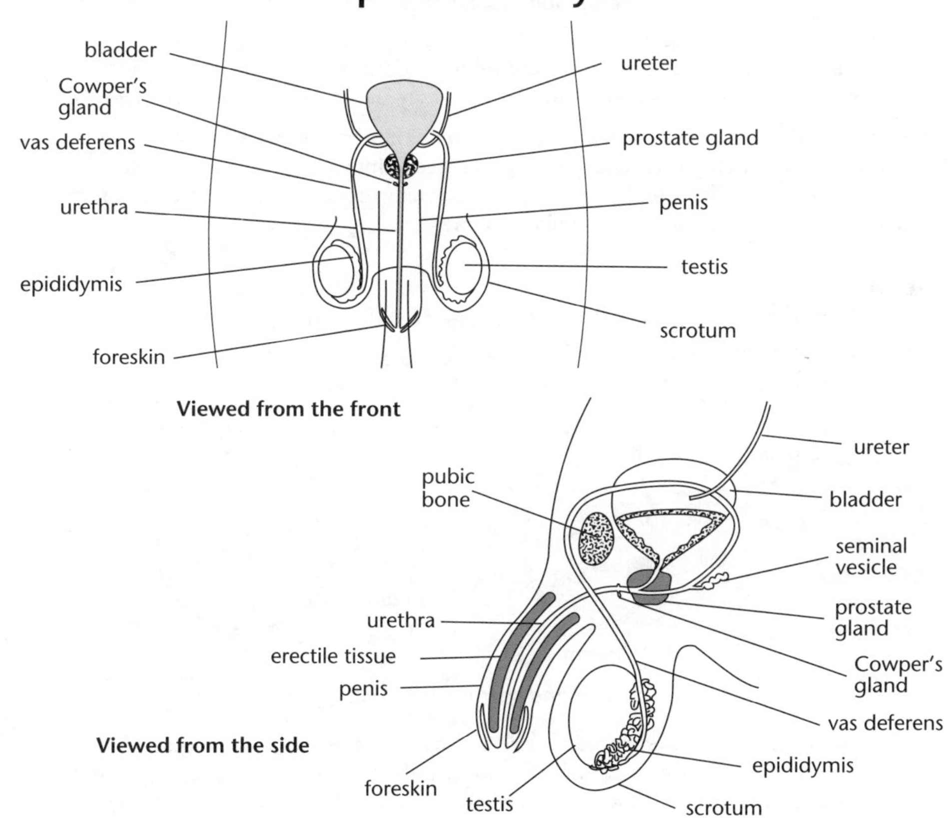

The male reproductive system.

The testes

The **testes** (singular testis) are about 5 cm long, and contain about 300 compartments, each with about three **seminiferous tubules**. The walls of these tubules produce sperm.

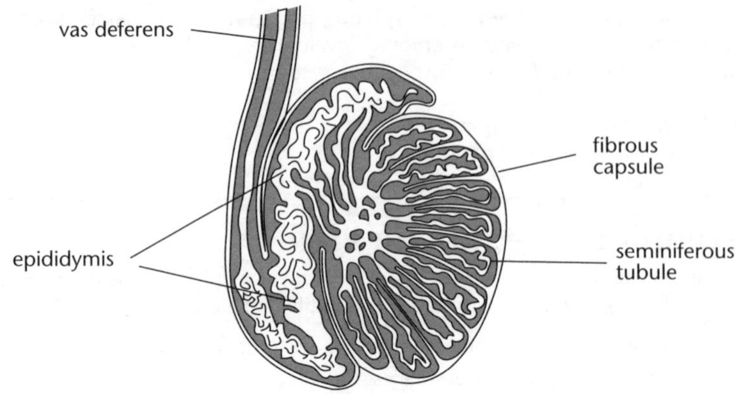

Testes are also called *testicles* in humans.

Longitudinal section of testis.

When first formed, a sperm is immature and cannot fertilise an egg. Their development is completed in a long coiled tube called the **epididymis**. Cells between the seminiferous tubules produce *male sex hormone* or **testosterone**. The testes are therefore also **endocrine glands.**

During the development of a boy in his mother's womb, his testes lie in the abdomen (as do a girl's ovaries), but shortly before birth the testes descend into a bag called the **scrotum**. This enables them to stay about 2°C cooler than body temperature – the lower temperature is necessary for sperm to develop (when puberty arrives).

Sperm

Each sperm looks like a tadpole, with a head containing the nucleus, a middle piece, and a long tail with which it uses to swim.

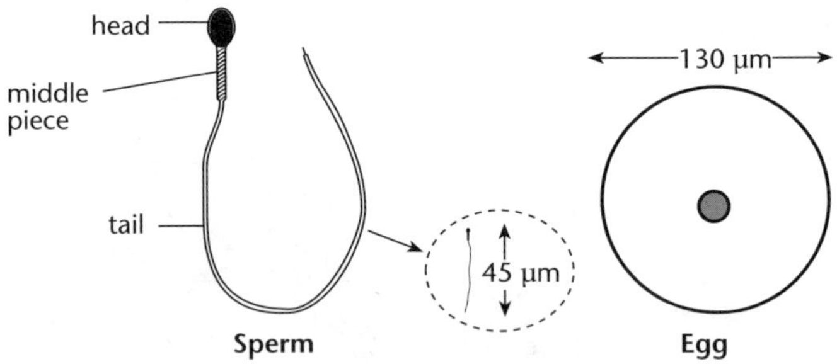

Egg and sperm compared.

At the tip of the head is an **acrosome**, a vesicle or bag containing enzymes needed to penetrate the egg. Sperm are produced continuously from puberty until old age. Sperm production (**spermatogenesis**) occurs in the walls of the seminiferous tubules. On average, about 200 million sperm are produced per day.

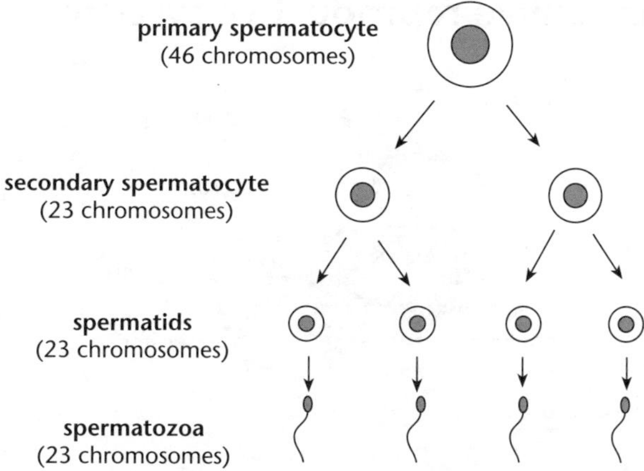

Sperm production (spermatogenesis).

The male tubes and glands

The **vas deferens** or sperm duct is a muscular tube carrying sperm from each testis to the base of the bladder. It loops over the bladder and ureter, because as each testis descends into the scrotum before birth, it drags the vas deferens with it.

Example

Vasectomy

One way a man can avoid having children is to have a *vasectomy*. This is a simple operation in which the vas deferens is cut. The operation does not prevent testosterone leaving the testes via the blood – the only effect is to prevent the release of sperm.

The **seminal vesicles**, **prostate** and **Cowper's gland** make different contributions to **seminal fluid** in which sperm swim. Seminal fluid contains fructose sugar, which provides the energy for the sperm to swim.

The **urethra** is a dual-purpose tube, carrying urine from the bladder and semen from the sperm ducts. The **penis** contains three spongy blood spaces called **erectile tissue**.

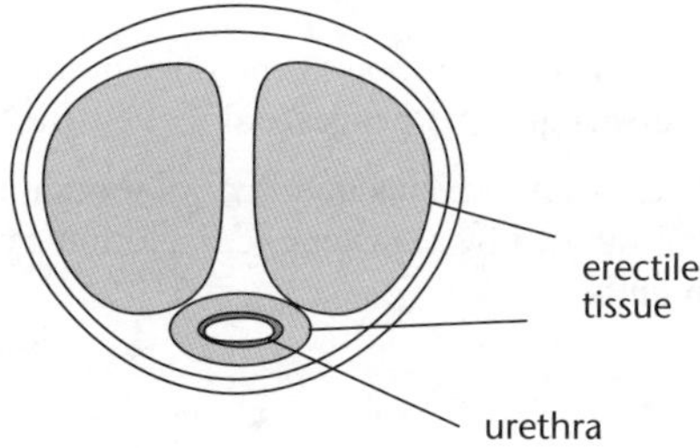

Cross-section of the penis.

During sexual excitement the spongy blood spaces become inflated with blood, causing the penis to become hard and erect, enabling it to enter the vagina during intercourse. The end of the penis (the *glans*) is covered by the *foreskin*. When this is too tight to allow easy urination, it may be removed by a simple operation called *circumcision*. Circumcision is also performed for religious reasons.

The human female reproductive system

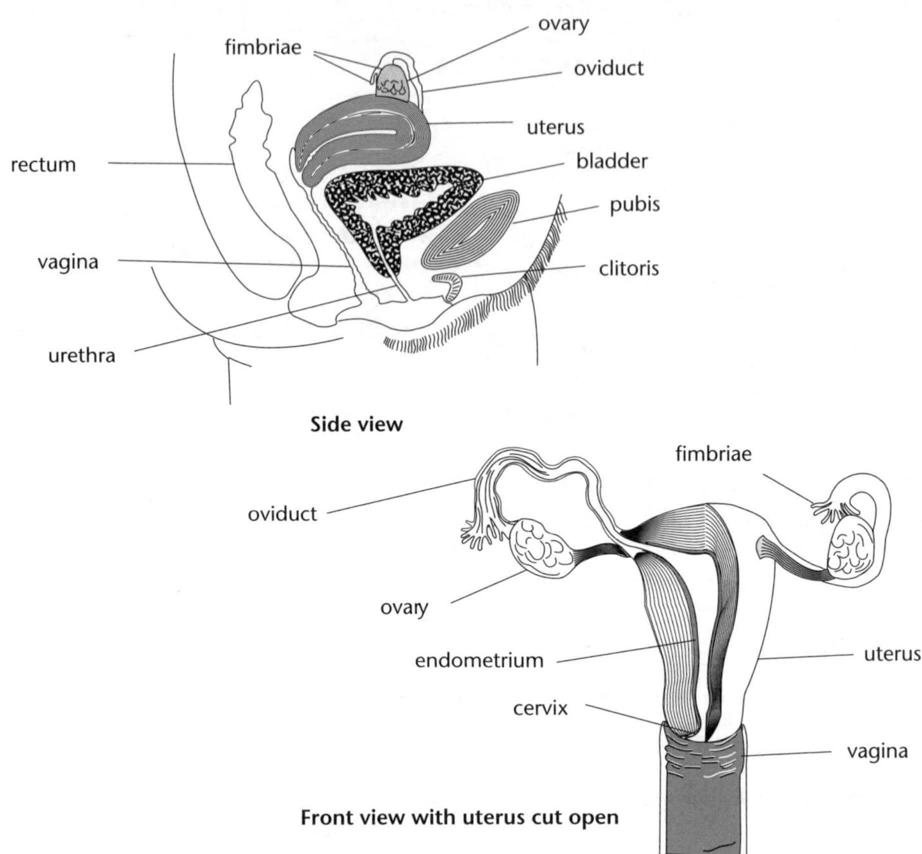

Human female reproductive system.

The **ovaries** are about 3 cm long and produce eggs. They also produce two *female sex hormones* – **oestrogen** and **progesterone**. Like the testes, the ovaries are therefore also *endocrine glands*.

Each ovary contains hundreds of thousands of immature eggs. Each developing egg is surrounded by a cluster of cells, forming a **follicle**. Eggs usually mature one at a time, once a month.

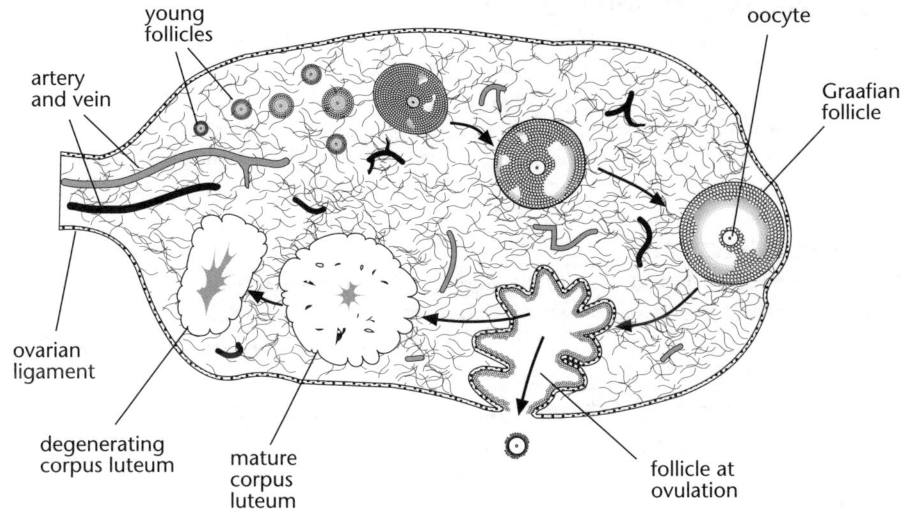

Section through a human ovary.

The follicle containing the egg begins to enlarge, eventually reaching the size of a pea. It is then called a *Graafian follicle*. The egg is released by rupture of the follicle at the surface of the ovary. This process is called **ovulation**, and occurs about every 28 days from puberty until about 50 years of age.

Although some of the follicle cells remain around the egg after ovulation, the rest stay in the ovary and develop into a **corpus luteum** ('yellow body'). This begins to secrete progesterone, which stimulates the lining of the womb to prepare for the arrival of a fertilised egg.

The female gametes (ova)

Ova are among the largest cells produced by humans (about 0.13 mm or 130 μm across) – just large enough to be seen with the naked eye. The process of egg production (**oogenesis**) differs in several ways from spermatogenesis:

- In females, the mitotic divisions that precede meiosis cease early in life, so that no more potential eggs are produced. (In males, the mitotic cell divisions that precede meiosis are continuous from from puberty to old age.) At birth, a human female has about 2 million cells that have the potential to produce eggs, but by puberty most have died; only about 400 000 are left.
- The cytoplasm is distributed very unevenly between the four products of meiosis, only one ovum being produced. The other three products are very small and functionless, and are called **polar bodies**.
- Egg production is intermittent, one egg normally being produced each month.

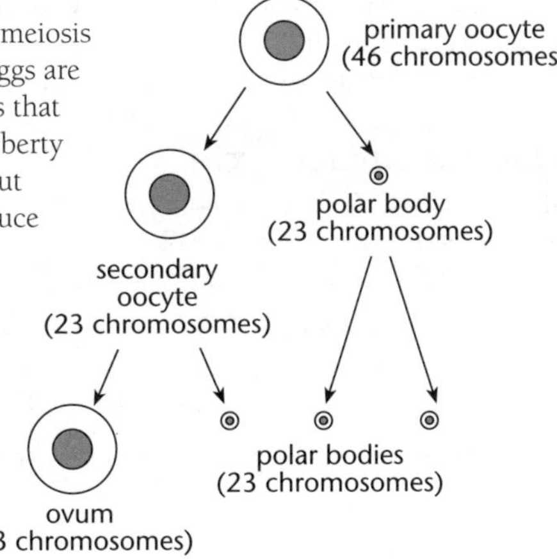

Egg production (oogenesis).

The oviducts (Fallopian tubes)

These transport eggs from the ovaries to the *uterus* or *womb*. One end of each oviduct has a funnel-shaped opening fringed by tentacles or **fimbriae**. These tentacles, and the lining of the oviduct, have millions of tiny hair-like **cilia**, which beat towards the uterus. At the time of ovulation, the tentacles slowly move over the surface of the ovary and the egg is slowly propelled down the oviduct by the cilia and also by weak peristaltic action of the muscle in the oviduct walls. It normally takes an egg about three days to reach the womb.

The uterus (womb)

This is where a baby develops. Most of the uterine wall is the **myometrium**, which consists of smooth muscle and functions only to expel the baby during birth. The inner lining, or **endometrium**, is glandular and is concerned with the anchorage and nutrition of the embryo. The neck or **cervix** of the uterus opens into the **vagina**.

The endometrium helps to form the **placenta**, which is important in feeding the baby while it is developing.

The vagina

This is the channel through which the penis enters the female during intercourse, and through which a baby is born. Close to the opening of the vagina is the **clitoris**, a small structure containing erectile tissue and a rich network of sensory nerve endings, which give pleasurable sensations when stimulated. The opening to the vagina is covered by folds of tissue forming the **vulva**.

Role of the pituitary and hypothalamus

The activity of the ovaries and testes are under control of two hormones secreted by the **anterior pituitary gland:**

- **Follicle-stimulating hormone** (FSH) – stimulates growth of follicles in the ovary and sperm production in the testes.
- **Luteinising hormone** (LH) – stimulates ovulation and the growth of the corpus luteum in the female, and testosterone secretion in the male.

The pituitary gland is controlled by the **hypothalamus**. This is located in the brain, immediately above the pituitary. Hormones secreted by the hypothalamus are carried directly to the anterior pituitary by a special system of blood vessels.

The menstrual cycle

About every 28 days, the smooth muscle in the walls of some of the arteries in the endometrium go into spasm, closing off their blood supply. As a result, most of the endometrium dies and comes away as blood and dead cells.

This process of **menstruation** ('a period') begins with pain in the abdomen, and during the next 4 or 5 days about half a cupful of blood is lost. The inner part of the endometrium does not die, and from this part, a new lining grows, until *it* dies and a new period begins, and so on.

Menstruation occurs regularly, so it is called the menstrual *cycle*.

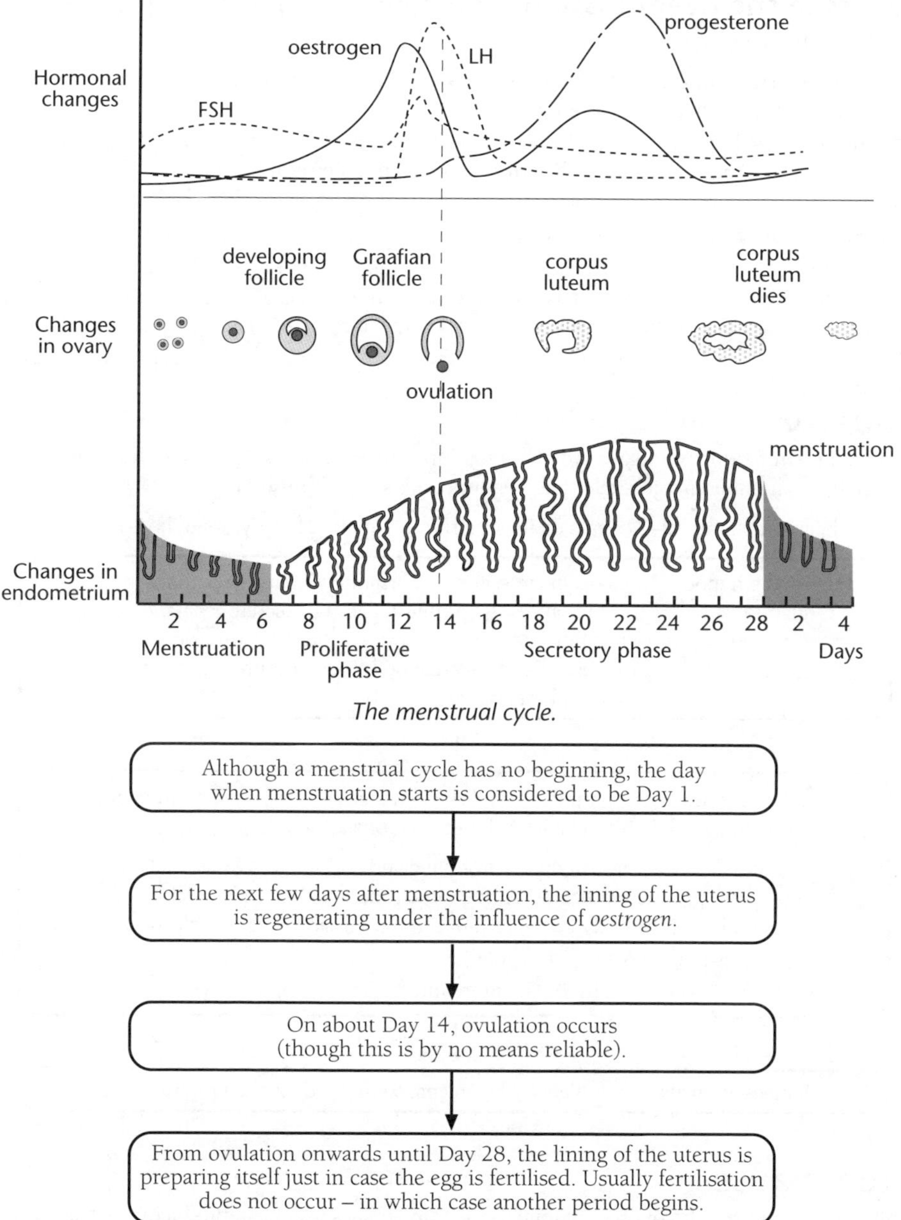

The menstrual cycle.

The variation in the duration of the cycle is in the first part (menstruation to ovulation); the second phase (ovulation to menstruation) is almost constant, varying by no more than a day.

Events in the ovary (assuming fertilisation has not occurred)

Paralleling the changes in the womb are changes in the ovary.

- LH is necessary for the growth and maintenance of the corpus luteum, which in turn is needed for maintenance of the endometrium.
- Towards Day 28 the pituitary produces less LH and the corpus luteum dies. Progesterone is no longer produced, causing the arteries in the endometrium to shut down and menstruation begins.

One of the effects of oestrogen is to inhibit the secretion of FSH by the pituitary. With the death of the corpus luteum, oestrogen output falls, so the pituitary is no longer inhibited. This causes one or sometimes more follicles to begin to grow. Later in the cycle, oestrogen begins to *stimulate* the output of FSH and LH by the pituitary, and the sudden surge in LH output brings about the next ovulation.

Puberty

Although the reproductive organs are present at birth, they are small and cannot function. The change from an immature, juvenile state to adulthood is called **puberty**.

Puberty begins at about 10–12 years of age in girls, and about 11–13 years in boys.

Firstly, the hypothalamus begins to secrete a hormone which passes to the pituitary gland just below it. This causes the pituitary gland to secrete FSH and LH.
As the testes begin to mature they start to secrete testosterone.
Similarly, the ovaries begin to secrete oestrogen and progesterone, and menstrual cycles begin.

↓

The sex hormones cause a number of marked changes in the body and in behaviour
In both sexes, hair thickens under the armpits and in the pubic region.

- In males, the penis, sperm ducts, prostate gland and seminal vesicles all enlarge, and sperm production begins. At the same time, hair begins to thicken on the face and body, the body becomes more muscular, and the voice deepens.
- In females the breasts begin to enlarge.
- Males and females begin to take more interest in the opposite sex.

↓

The changes are usually complete by late teens, so the body is then physically mature.

Menopause

Unlike males and females of other mammals, human females become reproductively inactive long before old age. Menstrual cycles cease and the output of oestrogen falls. One of the effects of this 'change of life' or **menopause** is partial de-mineralisation of the bones, so it is important that women have adequate calcium in their diet.

Sex hormones

The effects of oestrogen

- Enlargement of the uterus and vagina.
- Growth of the ducts of the mammary glands.
- Broadening of the hip girdle.
- Deposition of fat in the hips and breasts.
- Increased interest in the opposite sex.
- Inhibition of FSH secretion by the anterior pituitary.
- During pregnancy it inhibits milk secretion.

The effects of progesterone

- Stimulates growth of the glands and blood vessels of the endometrium.
- Stimulates growth of the secretory tissue of the mammary glands.
- Inhibits secretion of FSH by the pituitary.
- Inhibits contraction of uterine muscle in pregnancy.
- During pregnancy, inhibits milk secretion.

The effects of testosterone

- Growth of the penis and reproductive tract (eg epididymis, vasa deferentia, prostate, seminal vesicles).
- Growth of coarser and longer hair on the chest, armpits, pubic region, legs, and face.
- Enlargement of the larynx and deepening of the voice.
- Increased growth of bone and muscle.
- Increased interest in the opposite sex.

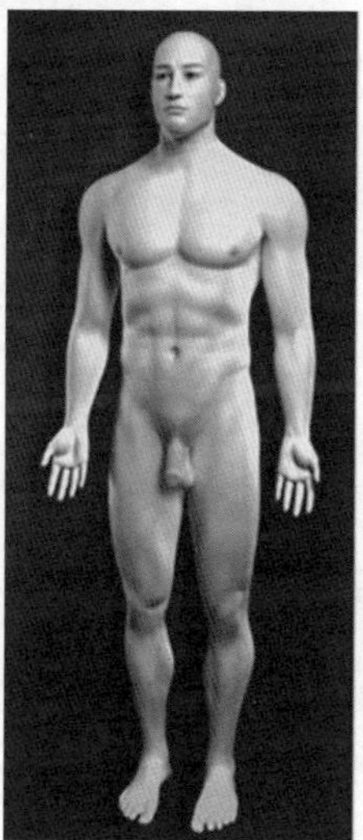

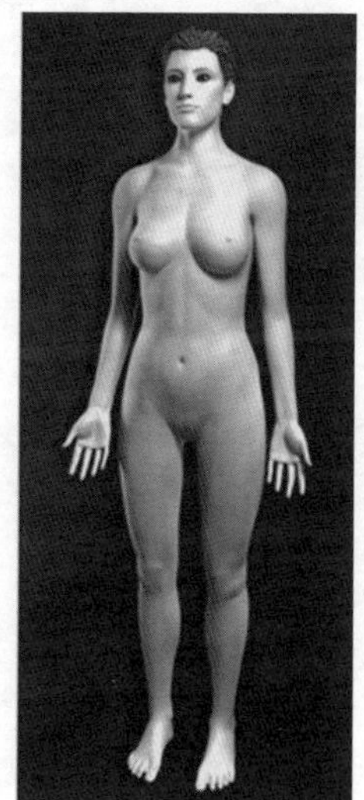

The differences that make the person on the left male and the person on the right female result from the sex hormones; testosterone in male and oestrogen in female.

Unit 11.6 Activity 2A: The reproductive systems

1. For each of the phrases **1–20**, write the letter **A–T** of the term to which it applies.

Phrase	Applied term
1. A female gamete	**A.** Epididymis
2. An egg or sperm	**B.** Endometrium
3. Conveys egg from ovary	**C.** Fallopian tube
4. End of reproductive life of a woman	**D.** FSH
5. Holds testes	**E.** Gamete
6. Lining of uterus	**F.** Hypothalamus
7. Makes eggs and female hormone	**G.** LH
8. Makes part of seminal fluid	**H.** Menopause
9. Makes sperm and male hormone	**I.** Oestrogen
10. Part of the brain that controls the pituitary	**J.** Oogenesis
11. Prepares uterus for arrival of fertilised egg	**K.** Ovary
12. Process of egg production	**L.** Ovum
13. Process of sperm production	**M.** Pituitary
14. Produce sperm	**N.** Progesterone
15. Produces FSH and LH	**O.** Prostate
16. Where sperm mature	**P.** Scrotum
17. Stimulates development of follicles	**Q.** Seminiferous tubules
18. Stimulates development of male characters	**R.** Spermatogenesis
19. Stimulates ovulation	**S.** Testis
20. Stimulates repair of endometrium	**T.** Testosterone

2. Oestrogen, progesterone, follicle stimulating hormone (FSH) and luteinising hormone (LH) have important roles in the functioning of the female reproductive system. Discuss how these hormones work together to produce the menstrual cycle. Your answer should consider the function(s) of each hormone and the interactions between the hormones.

3. The diagram shows the human male reproductive system.

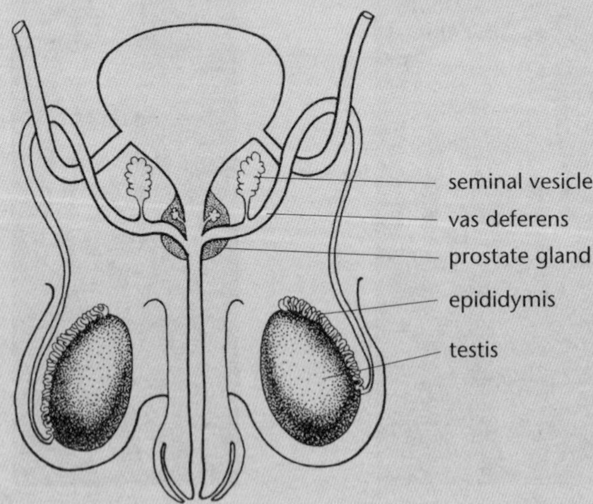

a. Describe the function of each of the following parts:

i. Testis.

ii. Epididymis.

iii. Prostate gland.
iv. Seminal vesicle.
v. Vas deferens.

b. Describe two physical changes that occur in the body of a 13-year-old boy beginning puberty.

c. Explain why the physical changes have started to occur in his body.

Fertilisation

Fertilisation, or the joining of the egg and a sperm, normally occurs in the oviduct. Before this can happen, intercourse must take place.

- During intercourse, the penis is inserted into the vagina.
- Movement of the penis causes **ejaculation**. In this process, sperm that have been maturing in the epididymis are propelled down the sperm ducts by contraction of muscles in walls of the sperm ducts. At the same time the prostate gland and seminal vesicles contract, adding more liquid to the sperm.
- In one ejaculation there are about 200–300 million sperm, in about 2–4 mL of seminal fluid.

The seminal fluid is deposited at the upper end of the vagina.

- The sperm still have to travel through the uterus and up one of the oviducts, a distance of about 10–15 cm. The vast majority of sperm die without reaching the egg.
- If an egg is not fertilised within 36 hours of ovulation, it dies. Sperm can survive in the female reproductive tract for about two days.
- Eventually, one sperm may join with the egg. The nuclei of the sperm and egg join, forming the single nucleus of the zygote.

Within seconds of entry of a sperm, the plasma membrane of the egg undergoes a change that prevents entry of any more sperm.

Implantation

In the oviduct, the zygote begins to divide, forming first two, then four, then eight cells, and so on, until a solid ball of cells or **morula** is produced. By the time the dividing ball of cells has reached the uterus it has developed a fluid-filled cavity and is called a **blastocyst**.

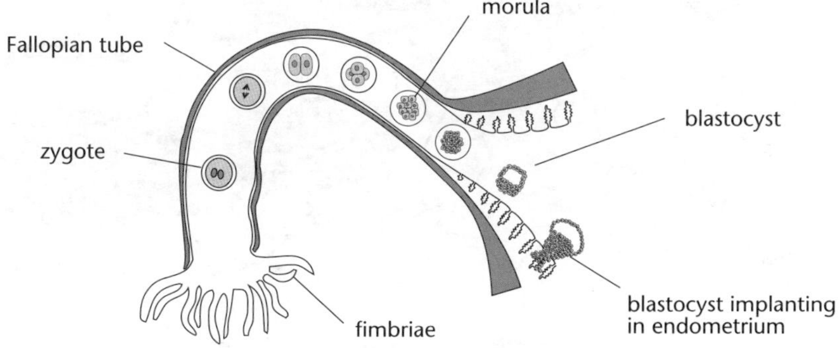

The first stages of human development.

The blastocyst is slowly propelled down the oviduct by cilia, and eventually reaches the uterus. It becomes buried in the lining of the uterus, a process called **implantation**. This marks the beginning of pregnancy, and the woman is said to have *conceived*.

Pregnancy

As the blastocyst becomes implanted, its outer layer (the **trophoblast**) develops branched tentacle-like **villi**. These increase the surface area for absorption of food. Inside the blastocyst there develops an *embryonic disc*, from which the **embryo** develops. As the embryo grows, so do the villi, and they eventually become concentrated in one area called the **placenta**.

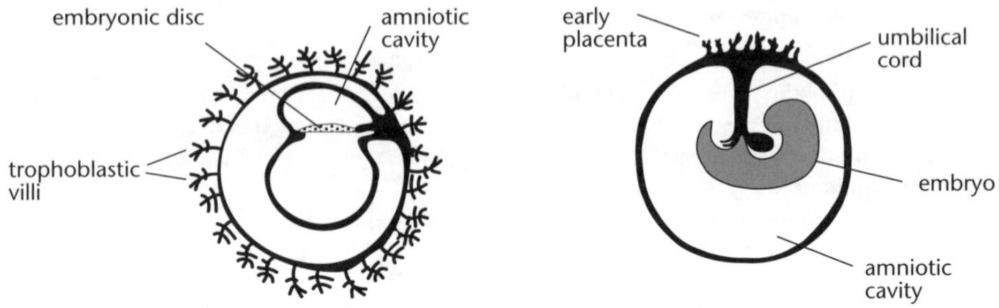

Early nutrition of the embryo.

The placenta

This is formed partly from the baby and partly from the mother, and when fully developed it is about 20 cm across and 4 cm thick.

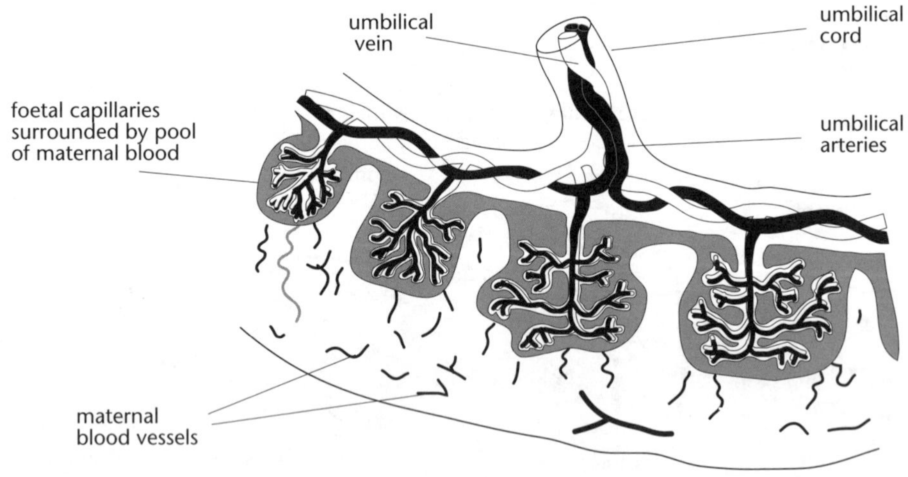

Diagrammatic view of part of the placenta.

The placenta has two functions:

- It promotes the exchange of substances between mother and baby.
- It produces hormones essential for the maintenance of pregnancy. In early pregnancy oestrogen and progesterone are produced by the ovary, but later the placenta itself produces the greater quantity.

In the placenta the blood of the mother and the baby come very close together, but *do not actually mix*. It is vital that they do not mix for two reasons:

- Disease-causing organisms would be able to pass from mother to baby (this can still happen in the case of rubella).
- The mother and the baby are often of different blood groups. On the mother's side of the placenta the walls of the blood vessels break down so that the villi are bathed in the mother's blood. Because there are so many villi and they are so highly branched, they have a *very large total surface area* (about 10 square metres). Oxygen and foods (eg glucose, amino acids, vitamins and minerals) pass from the mother's blood to the baby's blood. Wastes such as carbon dioxide and urea pass in the opposite direction – from the baby's blood to the mother's blood.

Simple substances with small molecules (eg oxygen, carbon dioxide and urea) cross between baby's and mother's blood by *diffusion*. Other substances such as amino acids move by *active transport*. The placenta is connected to the baby by the **umbilical cord**. This consists of two **umbilical arteries** carrying blood from the baby to the placenta, and an **umbilical vein** carrying blood back to the baby.

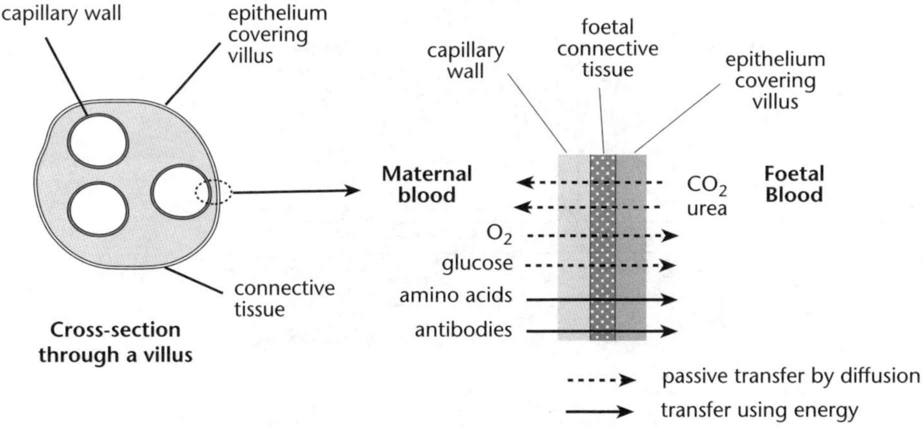

Transfer of substances across the placenta.

The transfer of oxygen is helped by the fact that foetal haemoglobin holds oxygen more strongly than the haemoglobin of the mother. As a result, the blood in the umbilical veins carries more oxygen than the blood in the uterine veins of the mother.

A human embryo is very delicate, and needs support and protection against any blows to the mother's abdomen. This is the function of the litre or so of **amniotic fluid** which surrounds the embryo. It is contained within a bag called the **amnion**.

Embryonic development

If sperm are present in the oviduct during ovulation, fertilisation may occur, forming a zygote. The zygote will begin dividing and travelling along the oviduct to the uterus, where it implants itself in the wall of the uterus.

In the ovary, the yellow body continues to mature, producing more oestrogen to prevent further egg releases and stimulates greater thickening and growth of the uterus.

No egg releases and retention of the uterus wall mean that the monthly period ceases during pregnancy.

The embryo in the uterus forms a **placenta** as small outgrowths from the embryo fuse with the wall of the uterus. Oxygen and food diffuse from the mother to the embryo across the placenta, and wastes and carbon dioxide are transferred from the embryo to the mother.

Capillaries in the placenta form into a large vein and artery (the *umbilical cord*) that takes materials to and from the embryo. The blood systems of mother and embryo never mix, despite the fact that materials are transferred between the two blood systems.

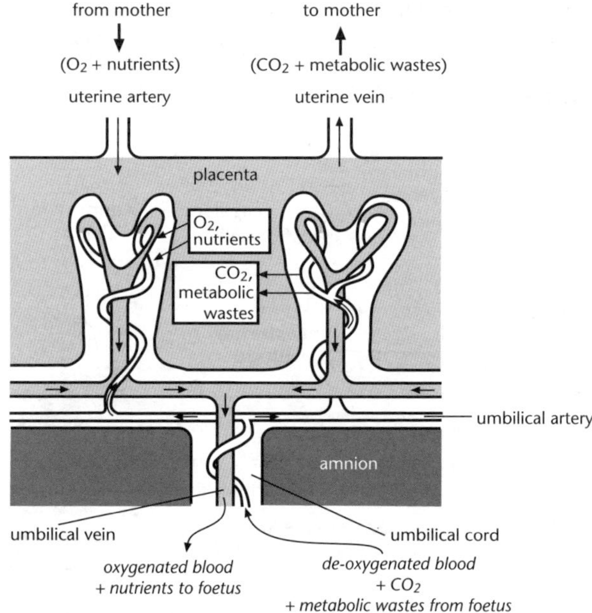

Section of a placenta; also shows the start of the umbilical cord

As the embryo grows, it becomes surrounded by the *amnion*, a protective fluid-filled sac.

Cell differentiation (development of cells into specific cell types, organs, etc) is complete by about 8 weeks. From this stage, the embryo is called a **foetus**.

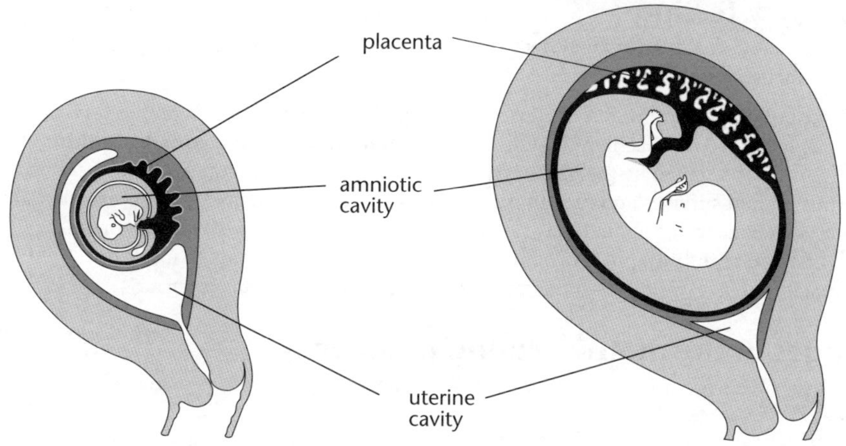

The foetal membranes.

The period of pregnancy is called **gestation**. In humans, it takes about 38 weeks (40 weeks after the beginning of the last menstrual period), or just over nine months. By the time it is a month old, the embryo is about 7 mm long and has the beginnings of a gut, kidneys, brain, and a heart which is already beating. By two months old the foetus is about 13 mm long, with a mass of about 1 gram and is like a miniature human, with all the main organ systems. It is now called a **foetus** and can move slightly.

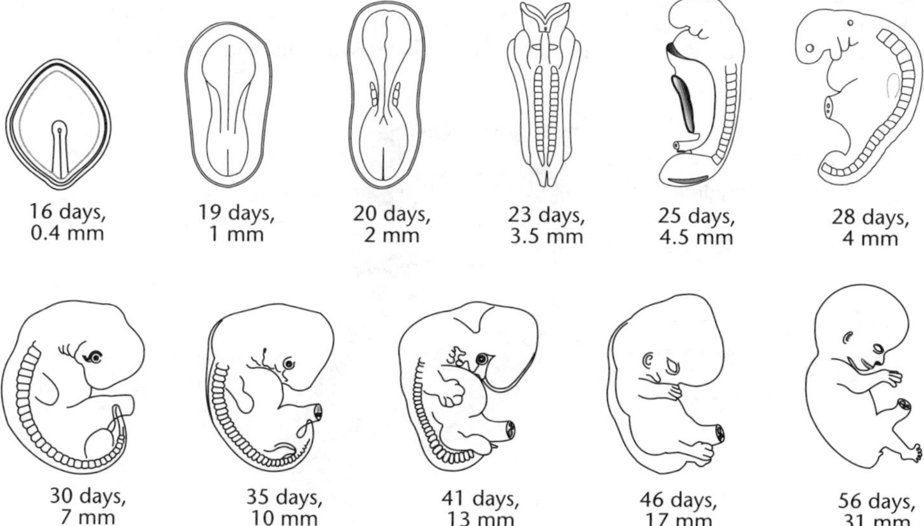

Changes in the appearance of the human embryo during development.

Pregnancy testing

At the end of a normal menstrual cycle, the level of LH declines and the corpus luteum dies, leading to menstruation.

In the event of pregnancy, the survival of the implanted blastocyst depends on the survival of the corpus luteum. This is achieved by the hormone **human chorionic gonadotrophin** (hCG). hCG begins to be secreted by the blastocyst a few days after implantation, with the result that the corpus luteum survives and the output of progesterone is maintained. hCG is excreted in the urine and can be detected about 8 days after implantation – this is the basis for a pregnancy test. For the first two months of pregnancy hCG is essential for pregnancy to be maintained, but after that the placenta itself produces enough progesterone for its own maintenance.

Changes in the mother during pregnancy

Important changes occur in the mother, under the influence of hormones secreted by the placenta.

- Oestrogen and progesterone stimulate the growth of the uterus to keep pace with that of the baby. They also stimulate the growth of the milk-secreting tissue of the breasts in preparation for feeding the baby.
- Relaxin, secreted in the last two months of pregnancy, causes the ligaments binding the pubic bones and the sacro-iliac joint to soften, allowing the birth canal to widen to allow the baby's head to pass through.

Amniocentesis

This is one of a number of tests available for detecting genetic and other abnormalities of the foetus. At 16–18 weeks of gestation, a needle is inserted through the abdominal wall into the amniotic cavity, and 10–20 mL of amniotic fluid removed. This fluid contains foetal cells that can be examined and tested for conditions such as Down's syndrome and haemophilia.

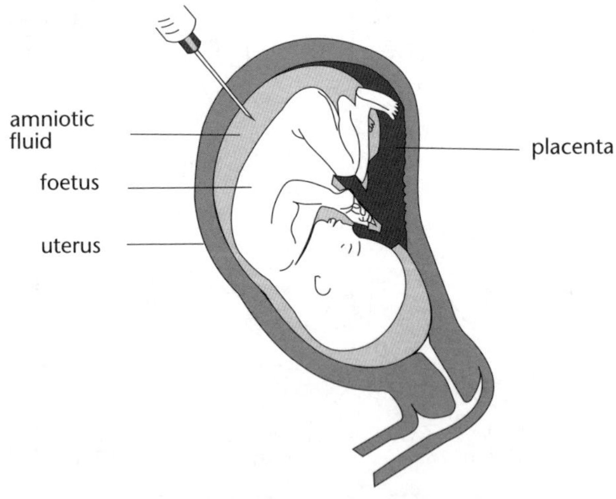

Amniocentesis.

Childbirth

A few weeks before birth, the foetus usually turns in the uterus to face downwards, with its head just above the cervix.

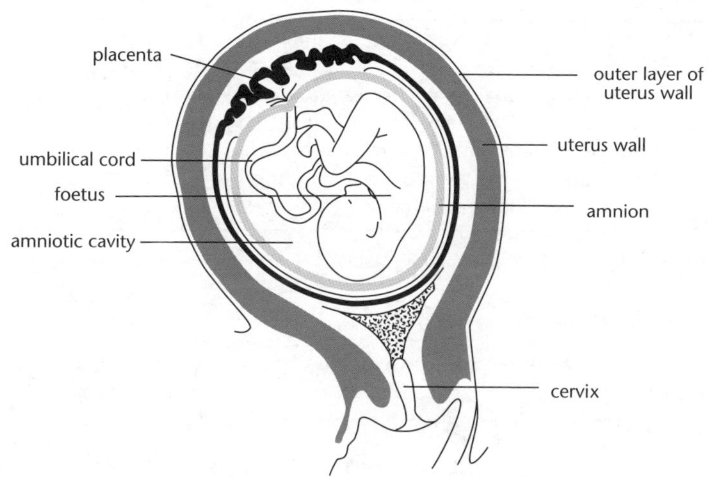

Diagram of a foetus in the womb shortly before birth

The uterus, which by now has become highly muscularised, begins to rhythmically contract as labour begins. (The exact cause of the onset of labour is not known, but it is known that progesterone levels drop and oestrogen levels rise.)

During labour, the amnion ruptures and the fluid contents are released. This is known as 'breaking of the waters'. As labour progresses, the cervix and vagina dilate (become larger) and *contractions* intensify until the baby is forced out of the uterus and through the vagina. The baby remains attached to the placenta by the umbilical cord. This cord is cut immediately after birth, freeing the baby from the placenta. Within an hour of birth, the placenta detaches from the uterus and is itself delivered as the *afterbirth*.

A baby is dependent on (the mother's) milk for nourishment for the first few months of its life, before it is weaned onto solid foods. A child is then dependent on its parents for about the next 15 years for learning the skills necessary to be successful in the complex society that humans have evolved. Human offspring have the longest period of dependency of all animals, which reflects the importance of *learnt behaviour* in humans compared with the importance of *innate behaviour* in most other animals – as well as the extent of *cultural evolution* that has occurred in our recent history, especially the last 10 000 years.

Unit 11.6 Activity 2B: Mother and foetus

1. For each of the phrases **1–7**, write the letter **A–G** of the term to which it applies.

Phrase	Applied term
1. Attachment of blastocyst to endometrium	A. Amniotic fluid
2. Fertilised egg	B. Blastocyst
3. Protects foetus against mechanical damage	C. Foetus
4. Stage at which implantation occurs	D. Gestation
5. Stage of development after organ formation	E. Implantation
6. The connection between foetus and mother	F. Placenta
7. Period of development within the mother	G. Zygote

2. The diagram shows a section through an ovary.

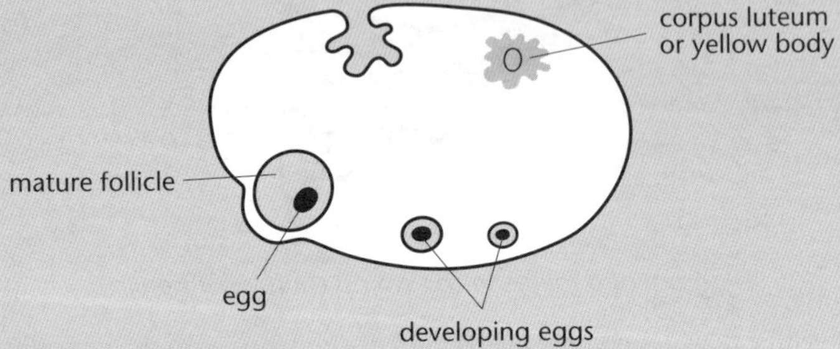

a. Name and describe the process that has occurred.

b. **i.** Name the hormone the corpus luteum produces.

ii. Describe the function of this hormone.

c. Describe where fertilisation occurs in a woman's body.

d. Describe the process of fertilisation.

e. **i.** Explain why the testes need to be outside the body to make healthy sperm.

ii. Give a function of the epididymis.

iii. Explain why a man must have a high sperm count if his partner is to become pregnant.

3. The diagram shows some changes that occur in the uterus during the last four months of gestation.

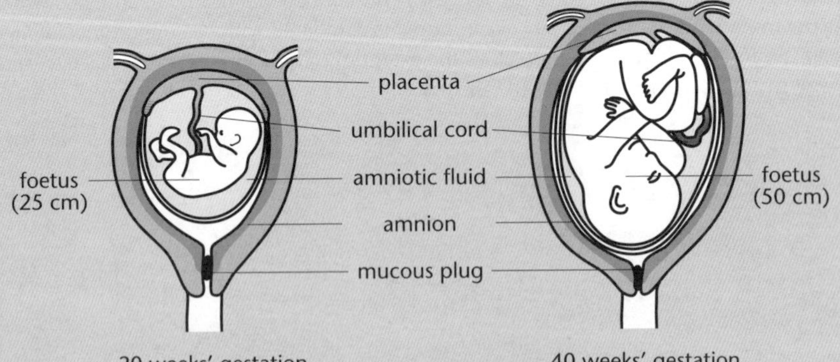

Not drawn to scale.

a. Complete the table below to describe the function of the two structures.

Structure	Function
Amnion	i.
Mucus plug	ii.

b. The lungs of a foetus are filled with amniotic fluid, so the foetus cannot breathe. Explain how the foetus gets its oxygen.

c. Discuss how a woman's reproductive system functions to allow a 20-week-old foetus to grow into a 40-week-old foetus during gestation. Your answer should consider the role of the placenta, the uterus and of oestrogen and progesterone during gestation.

4. The diagrams below outline the steps required to help a couple have a baby when the woman has discovered that both her oviducts (Fallopian tubes) are blocked.

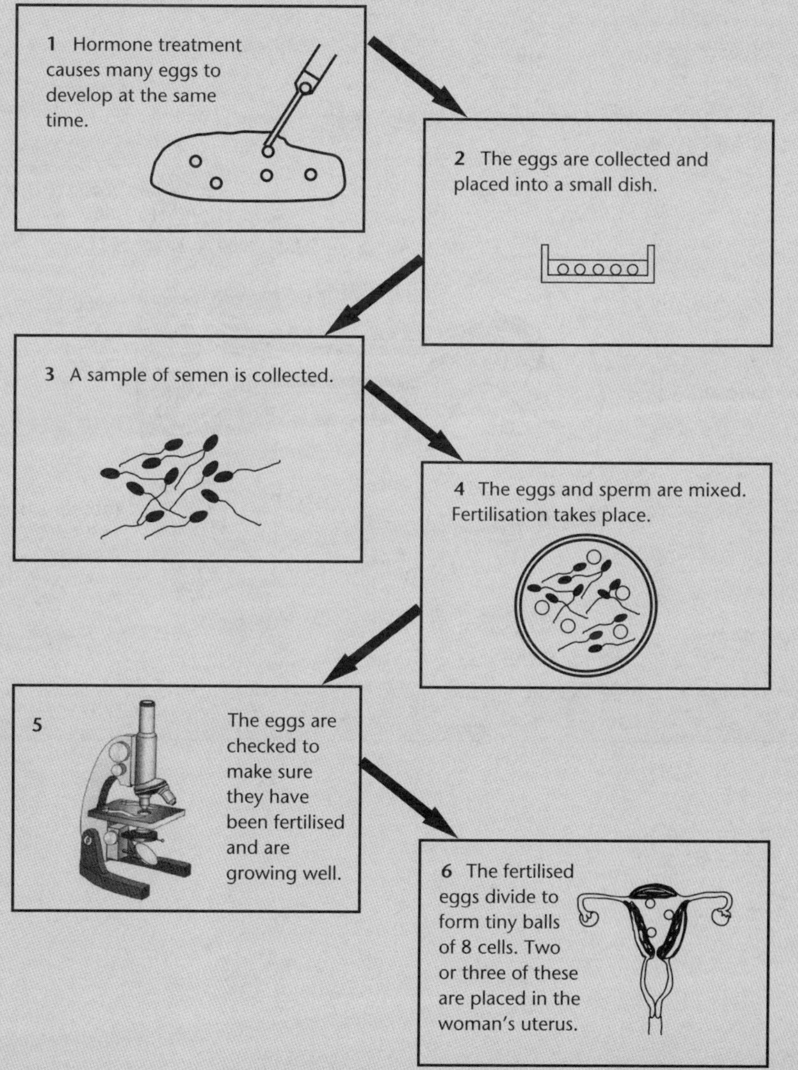

- **a.** Step 1 illustrates the development of many eggs. State where this occurs in a woman's body.
- **b.** What is semen made up of?
- **c.** Explain what happens when fertilisation takes place.
- **d.** Explain why two or three embryos (the tiny balls of 8 cells) are placed into the uterus as shown in Step 6 of the diagram.

5. Use the information in the following statements to help you answer the questions below:

- The ovaries and the uterus lining (endometrium) change during a menstrual cycle.
- The female hormones oestrogen and progesterone have different roles in the menstrual cycle.
- One of the first signs of pregnancy is that menstruation stops.

- **a.** Describe how pregnancy affects progesterone levels.
- **b.** Discuss the effects of pregnancy on the mother's body during the first six weeks of pregnancy.

6. Use the information in the diagram to answer the following questions.

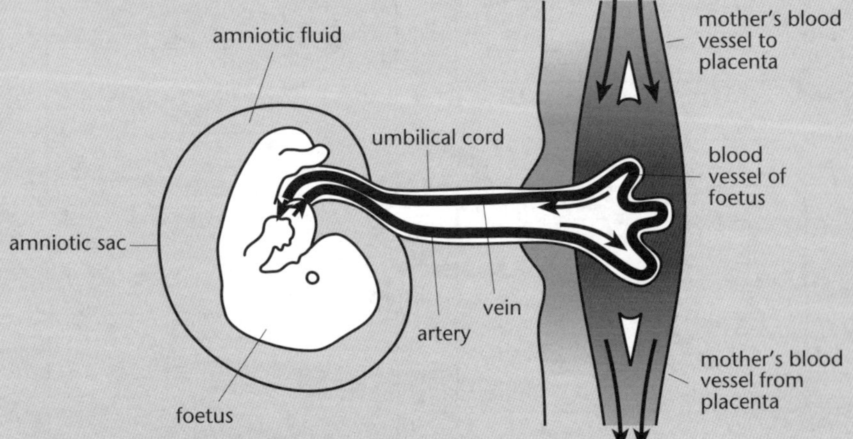

- **a.** Describe the function of the amniotic fluid.
- **b.** Describe the function of the vein in the umbilical cord.
- **c.** The foetus is linked to the mother by the umbilical cord and the placenta. Discuss how the developing foetus is nourished during the woman's pregnancy.

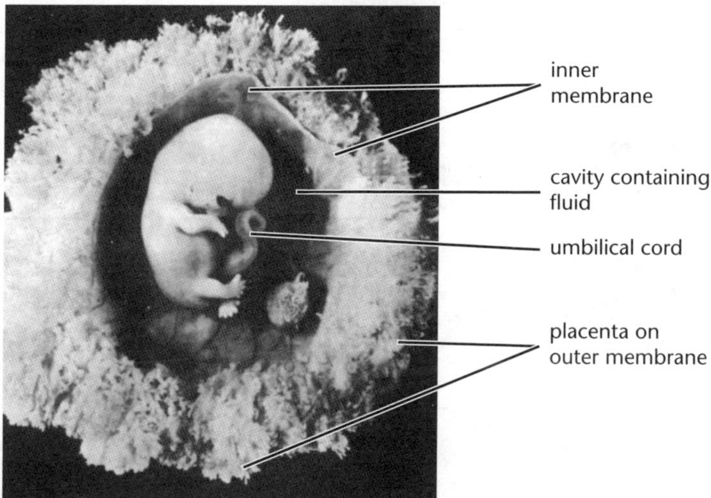

Photograph of a baby at 8 weeks' gestation with its membranes in the uterus. The baby is in the transition period from embryo to foetus. Seven months of growth lie ahead before its birthday.

Birth

During the last few weeks of pregnancy, the baby has usually come to lie with its head down against the **cervix**, or neck of the womb.

When the baby is ready to be born, a chemical signal is sent across the placenta to the mother. As a result, the muscles of her womb begin to contract strongly about once every 15 minutes. These contractions are stimulated by a hormone called **oxytocin**, produced by the pituitary gland.

↓

The contractions become stronger and more frequent, pressing the baby's head against the cervix. Early in labour the amnion breaks and the amniotic fluid pours out – this is called 'breaking the waters'. During the next few hours, the cervix is slowly stretched from its normal 5 mm or so to about 10 cm, causing 'labour pains'. This is the *first stage of labour*, and by far the longest.

↓

When the cervix is fully widened or dilated, the *second stage of labour* begins. The mother begins to push with her abdominal muscles as well, forcing the baby out. The umbilical cord is cut and normally the baby takes its first breath very soon thereafter.

↓

About half an hour after the baby has been born, the placenta is passed out – this is the *third stage of labour*.

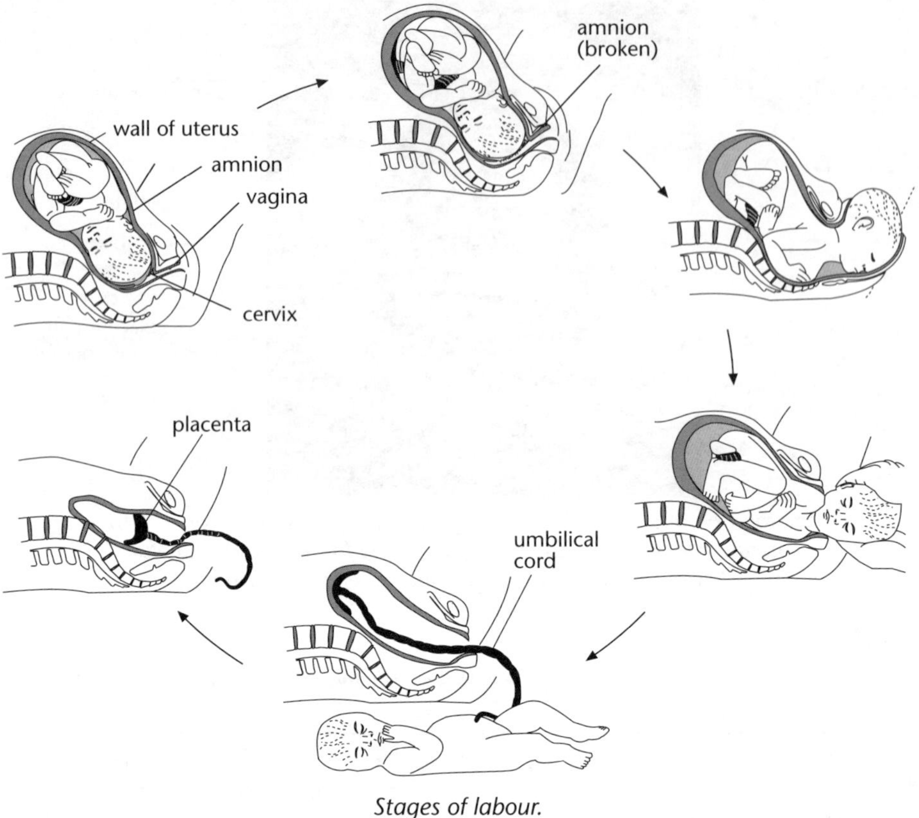

Stages of labour.

Extension: Changes in the baby's circulation at birth

One of the most amazing things about birth is the change that must occur in the baby's circulation – the blood system has to undergo a fundamental change within seconds of the baby taking its first breath.

After birth the two sides of the heart are quite separate, the right side pumping deoxygenated blood through the lungs and the left side pumping an equal volume of oxygenated blood through the rest of the body.

In the uterus, oxygenated blood reaches the *right* side of the heart, yet the baby's brain, which has the greatest need for oxygen, is supplied by the *left* side. The switch of blood from one system to the other (ie from the right side to the left side of the heart) is accomplished by two shortcuts, which are closed off after the baby takes its first breath.

- The *foramen ovale*, which connects the right and left atria, is guarded by a flap valve which allows blood to flow from the right side to left, but not in the reverse direction.
- The *ductus arteriosus,* carries blood from the pulmonary artery to the aorta.

Blood enters the foetal right atrium in two streams:

- A deoxygenated stream from the head and arms.
- A partially oxygenated flow from the placenta, abdomen and legs.

Mixing of these two streams is only partial. The most oxygenated blood passes through the foramen ovale into the left side of the heart, which pumps this blood to the brain and arms. The less oxygenated flow passes into the right ventricle and the pulmonary artery. Only a small proportion of this stream goes to the lungs, because the lungs, in their unexpanded state, offer a high resistance to flow. Most of the blood crosses to the aorta via the ductus arteriosus, and then to the legs, abdomen and placenta. Because of these two 'shunts', the left and right sides of the heart work in parallel (rather than in series as they do *after* birth), and so do not have to pump equal volumes of blood.

After birth, the shunts close and the left and right sides of the heart become separated.

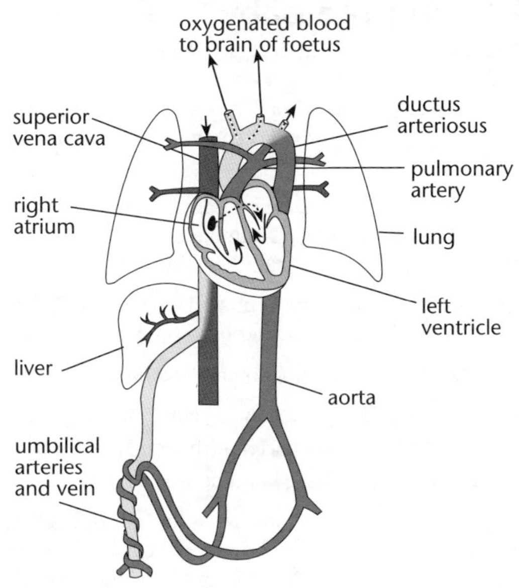

Foetal circulation

Firstly, the smooth muscle in the umbilical arteries contracts, closing them. The CO_2 level in the baby's blood therefore begins to rise, stimulating the breathing centre in the baby's brain. As it takes its first breath, the expansion of the lungs lowers the resistance of the pulmonary vessels, increasing the flow of blood through them and raising the pressure of blood in the left atrium. This reverses the pressure gradient across the foramen ovale, pushing the flap valve closed. Over the next few months, closure becomes permanent.

In some babies closure is incomplete, leading to reduced separation of oxygenated and deoxygenated blood.

The ductus arteriosus also closes soon after birth, though for a different reason.

Before birth, the placenta takes 40% of the cardiac output. The loss of placental circulation increases the resistance to blood flow in the systemic circuit, so the pressure of blood in the aorta rises. This, coupled with the fall in pressure in the pulmonary artery, reverses the blood flow through the ductus arteriosus. Oxygenated blood from the aorta now flows through the ductus arteriosus, and this rise in oxygenation stimulates the smooth muscle in the ductus arteriosus to contract, closing it off.

Birth also triggers a switch from the production of foetal haemoglobin to adult haemoglobin, though it takes 3–4 months for all the foetal red cells to be replaced. Since adult haemoglobin binds oxygen less strongly, it releases it more readily in the tissues, allowing them to function at higher oxygen concentrations.

Feeding the baby

After the baby has been born, the milk-producing tissue of the mother's breasts begins to secrete milk under the influence of the hormone **prolactin** released from the pituitary gland. This process is called **lactation**. The first milk to be produced is called **colostrum**, which is a clear yellow fluid, rich in protein. The normal, white milk is not produced until several days after birth.

When the baby suckles, nerve impulses travel to the mother's brain, causing her pituitary gland to produce **oxytocin**. This stimulates the muscle in the walls of the milk ducts to contract, helping to squeeze milk into the baby's mouth.

Suckling also helps to prevent ovulation – a mother is therefore less likely to conceive (become pregnant) if she is feeding her baby on her own milk.

Breastfeeding is better than bottle feeding for several reasons:

- Human milk contains the essential nutrients in the right proportions.
- The milk is free of bacteria (though HIV can be transmitted via breast milk).
- It promotes a strong bond between baby and mother.

Milk contains all that a baby needs for four or five months, and is especially rich in calcium to help its bones grow. Soon, however, the baby needs solid food, because milk contains no fibre and practically no iron (a new-born baby has enough iron stored in its liver to last it only several months). Milk is also rather poor in vitamin C. (See also the last paragraph on p. 285.)

Twins

Although producing one baby at a time is the normal situation in humans, two or even more can be produced in one pregnancy.

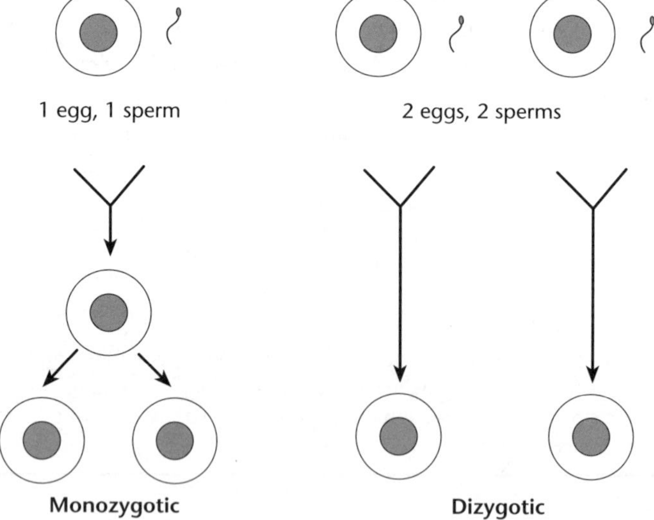

Formation of monozygotic and dizygotic twins.

Identical twins

These are produced when the zygote splits into two cells that fail to stick together, and each separated cell grows into a baby. Because they develop from a *single* fertilised egg, the individuals that form are called **monozygotic** twins, and are genetically identical.

Identical twins are never *quite* identical because no two people have identical *environments* (living conditions). Siamese twins are monozygotic twins in which separation has been incomplete.

Non-identical twins

These are produced when more than one egg is released at ovulation (probably one from each ovary). Since they develop from *two* fertilised eggs, they are called **dizygotic** ('di' = two) or **fraternal** twins, and are no more alike than other brothers and sisters.

Unit 11.6 Activity 2C: Human reproduction

1. The diagram shows a baby just before birth.

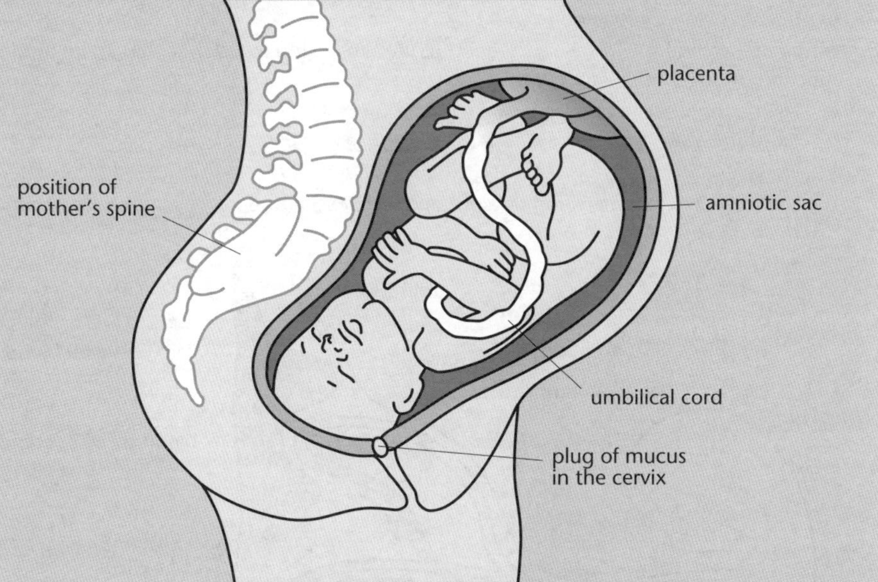

a. At the start of labour the waters break. Explain what happens inside the mother's body for the waters to break.

b. Describe two changes that occur in the mother's reproductive system during delivery of the baby. In your answer, you will need to refer to specific body parts, and do not refer to 'the waters break'.

2. The diagram below shows the human female reproductive system.

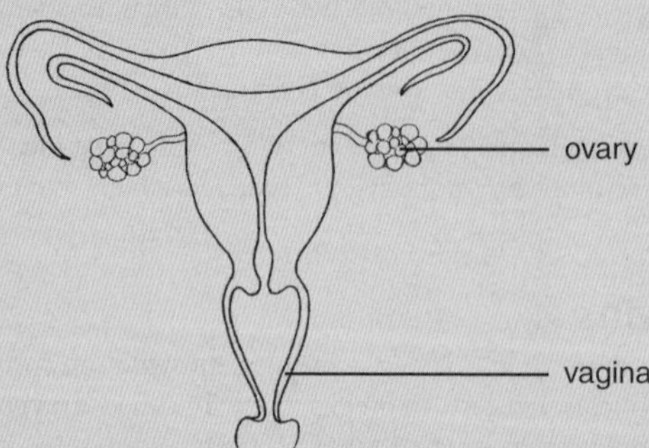

a. In the correct order, name the structures that sperm must pass through in a woman to reach and fertilise an egg.
b. A woman is pregnant with twins (a boy and a girl). Explain how the twins were formed in terms of ovulation and fertilisation.

3. Gestation takes about 38 weeks from fertilisation to birth. Progesterone has an important function throughout gestation.
a. Discuss how progesterone is produced during gestation and the function progesterone has throughout gestation.
b. Name two waste substances that the developing embryo needs to get rid of.
c. Use the diagram to explain how substances are exchanged between the embryo and the mother.

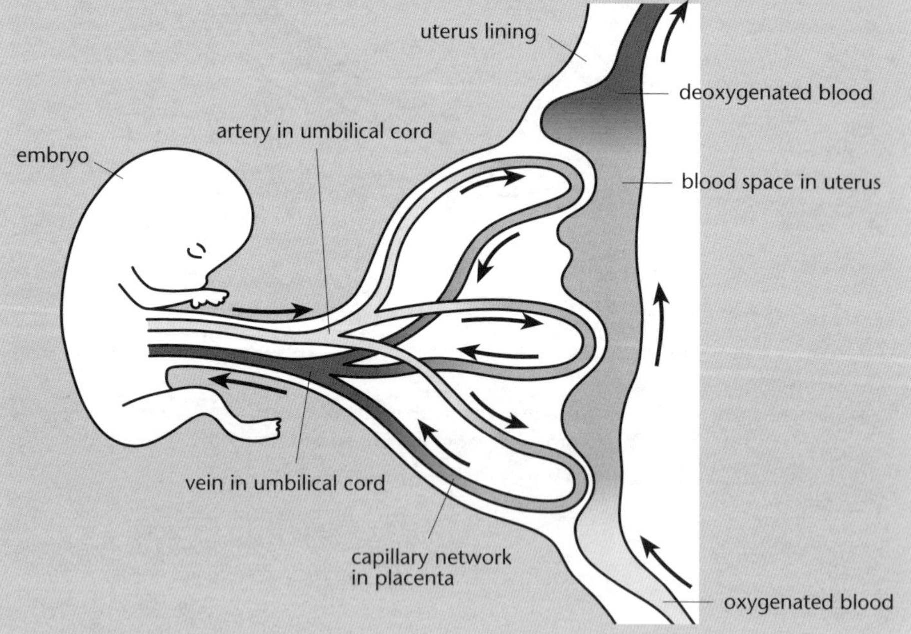

d. After gestation, a baby is born. Hormones, such as oxytocin in the blood, control the birth of a baby. Discuss the birth process and the role of oxytocin during labour.

Position of baby immediately before birth

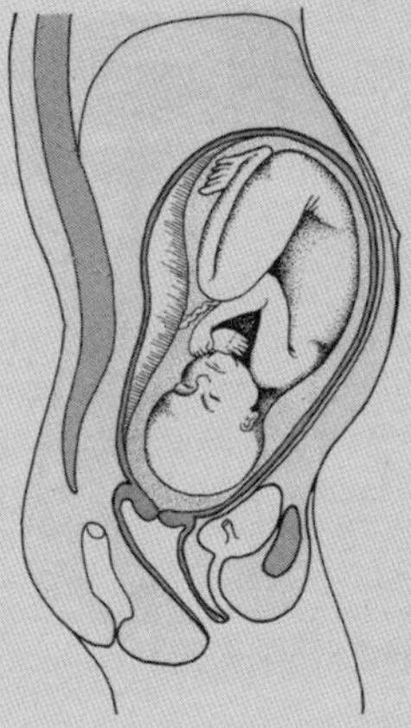

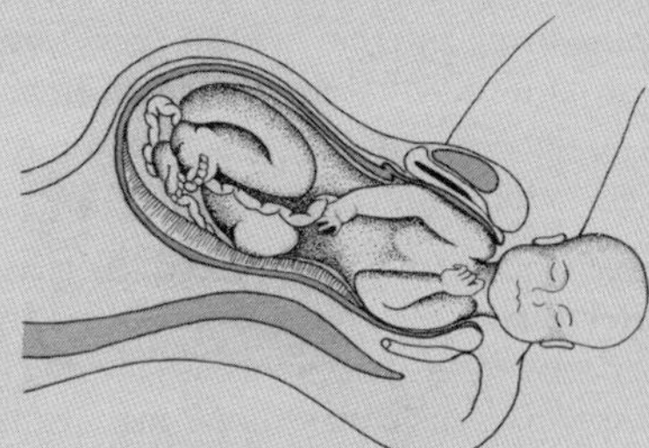

Baby is born head first

Family planning methods

Birth control in humans involves a variety of techniques and strategies that affect various stages of reproduction. These include:

- *Behavioural methods.*
- *Physical barriers* that are used to block the sperm from reaching the egg.
- *Chemical treatments* that prevent egg release or success of the sperm. Treatments can also be used to reduce the success of embryological development.

The table on the opposite page lists the various birth control methods available and their levels of effectiveness for family planning purposes.

Birth control methods not only prevent pregnancy from occurring but also involve techniques to treat **infertility**, which is the failure of a couple to achieve pregnancy after a year of regular unprotected sex. Male infertility is usually due to a low sperm count and/or a large number of abnormal sperm. Female infertility is mainly influenced by body weight. Sometimes causes of infertility can be corrected by medical intervention, or females can be given fertility drugs so that couples can have children.

Some couples will use assisted reproductive technologies (ART) to increase the chances of pregnancy where sperm and/or eggs are obtained from testes and ovaries, and fertilisation takes place in a clinical or laboratory setting. Two common ART are:

- **Artificial insemination** *by* **donor** (AID) – harvested sperm are placed in the vagina by a medical doctor or trained health worker.
- **In-vitro fertilisation** (IVF) – conception occurs outside the body, usually in a laboratory, where immature eggs are grown to maturity in a petri dish and sperm is added for fertilisation. Embryos are then transferred to the woman in the secretory phase of the menstrual cycle. In IVF, excess embryos may be frozen for later use.

Birth control methods

This table summarises common types of birth control methods. (Modified from Mader S.S. 2010 & Hoefnagels M. 2009 – see details, below.)

Birth Control Method	Procedure	Likelihood of Effectiveness (%)*
Behavioural		
Abstinence	Refrain from sexual intercourse; *Advantage - avoids contracting STDs*	100%
Rhythm method	No intercourse during fertile times	79%–87%
Withdrawal (= *coitus interruptus*)	Penis withdrawal before ejaculation	75%–91%
Barriers & Spermicides[#]		
Condom and spermicide	Worn over penis or inserted into vagina, keeps sperm out of vagina, and kills sperm that escape	*95%–98%*
• *Male condom*	*Latex sheath fitted over erect penis*	*About 85%*
• *Female condom*	*Polyurethane linear fitted inside vagina*	*About 85%*
Diaphragm and spermicide	Kills sperm and blocks cervix	*83%-97%*
• *Diaphragm*	*Latex cap inserted into vagina to cover cervix before intercourse*	*With jelly about 90%*

Birth Control Method	**Procedure**	**Likelihood of Effectiveness (%)***
Barriers & Spermicides#		
Cervical cap and spermicide • *Cervical cap* Spermicidal foam or jelly • *Jellies, creams, foams* • *Vaginal sponge*	Kills sperm and blocks cervix *Latex cap held by suction over cervix* Kills sperm and blocks cervix *Spermicidal products inserted before intercourse* *Sponge permeated with spermicide is inserted into vagina*	*80%–95%* *About 85%* *78%–95%* *About 75%* *About 90%*
# Spermicides are compounds formulated with specific chemicals to kill sperm. Physical barriers coated/covered with spermicides before sexual intercourse improves their effectiveness – for instance, refer to condoms with and without spermicide above.		
Hormonal		
Combination birth control pill	Prevents ovulation and implantation, thickens cervical mucus	90%–100%
Depo-Provera	Prevents ovulation, alters uterine lining	99%
Norplant	Prevents ovulation, thickens cervical mucus	99.8%
Surgical/Sterilisation		
Vasectomy	Cuts *vas deferens* and tied, so sperm never reach urethra	Almost 100%
Tubal ligation	Cuts oviduct and tied, so oocytes never reach uterus	Almost 100%
Other		
Intrauterine device (IUD)	Prevents implantation of pre-embryo	95%–99%
Douche	Vagina cleaned after intercourse	Less than 70%
Natural family planning	Day of ovulation determined by record keeping; various methods used.	About 70%
** Based on percentage of sexually active woman per year who will not get pregnant using this method.*		

Table modified from:

Hoefnagels M. *Biology – Concepts and Investigations* McGraw-Hill, New York, 2009.

Mader S.S. *Essentials of Biology* (2nd edn) McGraw-Hill, New York, 2010.

Sexually transmitted infections

Sexually transmitted infections (STIs) are due mainly to the micro-organisms: *bacteria*, *viruses* and *fungi*. Sexually transmitted diseases (STDs) are the *diseases* or *symptoms* resulting from infection by these micro-organisms.

A sexually active person who usually has unprotected sex is more likely to get an STD than a person who engages in safe/protective sex. Abstinence is the best protection against STDs while a latex condom offers some protection. Bacterial infections can be treated with antibiotics but while viral STDs can be treated with drugs, they cannot be cured. The table on the opposite page lists the common STDs in Papua New Guinea.

Common STIs in Papua New Guinea

This table lists common STIs in Papua New Guinea caused by bacteria, viruses and fungi, and their respective symptoms, treatment and medical problems. (Modified from Miller & Solien 2007 – see details below.)

Infection (STDs)	**Symptoms**		**Treatment**	**Medical problems if STI not treated**
• **Bacterial STDs**				
Gonorrhoea (bacteria)	***In women***	***In men***	✓ Cured with antibiotics	• Infertility or sterility • Blindness in babies if woman giving birth has gonorrhea • Painful swelling of joints • Damage to heart and liver
	Pain when urinating	Heaviness, pain and inflammation of the testicles. Heavy **pus**-like discharge and pain urinating.		
	Many people have gonorrhea but have no symptoms at all. A person without any symptoms can still pass the infection on and may develop complications from the infections.			
Chlamydia (bacteria)	***In women***	***In men***	✓ Cured with antibiotics	• Infertility or sterility • Eye damage to babies if woman giving birth has Chlamydia
	Usually no symptoms – increased vaginal discharge or irritation, irregular bleeding.	Usually no symptoms – sometimes pain during urination and discharge from penis. Can cause heaviness and inflammation of the testicles and a small, hard area of painful swelling at the base of the testicles.		

Infection (STDs)	Symptoms	Treatment	Medical problems if STI not treated
• **Bacterial STDs**			
Donovanosis (bacteria)	Small red bumps on the penis or vagina and around the anus which bleed easily. The sores might be painful swelling at the base of the testicles.	✓ Cured with antibiotics	• Ulcers will become larger and parts of the genitals will be destroyed • Infection may spread to other parts of the body
Syphilis (bacteria)	**3 Stages:** **1. Primary syphilis:** A small pimple appears where the bacteria entered the body, usually on the penis or inside of the vagina. A colourless, infectious liquid oozes from the pimple. The sore disappears by itself without medication, but the bacteria will spread to other parts of the body. **2. Secondary syphilis:** A few weeks or months after the sore disappears the following symptoms may appear: • Fever • Swelling in the groin, armpits and neck • Sores appear in the most parts of the body (mouth, genitals and armpits) • Skin develops a dry scaly rash These symptoms disappear within a couple of weeks if left untreated, but the bacteria are still in the body. **3. Tertiary syphilis:** The infection continues to attack the body and may affect organs such as the heart, brain and bones.	✓ Cured with antibiotics	If left untreated, syphilis may result in blindness, birth defects or stillbirth, heart trouble, poor mental health and death

Infection (STDs)	Symptoms		Treatment	Medical problems if STI not treated
• Viral STDs				
HIV (**h**uman **i**mmunodeficiency **v**irus)	Infected people show no symptoms for many years (may have flu-like symptoms shortly after infection). Lifelong damage to immune system and AIDS conditions begin between 1 and 20 years after infection.		**X** *No vaccine or cure.* ✓ Antirettroviral therapy keeps people healthier for longer	• AIDS-related illnesses such as TB, pneumonia and diarrhea
Genital herpes (virus)	Discomfort or itching with small blisters appearing in infected areas of the skin, usually the genital area. Fever can occur. After a few days, blisters form a thin yellowish crust which disappears in 10–12 days. Blisters can recur.		Symptoms can be treated with drugs, but the virus remains in the body and the infection cannot be cured.	• Urinary problems, possible meningitis in the most severe cases • Increased risk of cancer
Genital warts (**h**uman **p**apilloma **v**irus, or HPV)	Tiny painless lumps (cauliflower-like) around vagina, penis or anus. Sometimes no symptom.		✓ Treated with freezing or special paint ✓ Virus remains in the body and can reappear later	• Linked to cervical cancer
• **Fungal STDs**				
	In women	***In men***		
Thrush (*Candidia* - yeast)	Creamy thick discharge, smelly, itchy and inflamed vagina. Can also be caused by stress or use of antibiotics.	Itchy rash on penis or anus. Can be found in mouth and throat.	✓ Antifungal creams and other natural options.	–

Table modified from:

Miller and Solien *HIV/AIDS and STIs in Papua New Guinea* Oxford University Press, Melbourne, 2007.

Unit 11.6 Reproduction

Topic 3: How plants reproduce

In Unit 11.5 Topic 1 (p. 223) we looked at how plants respond to stimuli. In this Topic we consider how plants reproduce by looking at:

- Vegetative propagation in plants.
- Parts of a flower.
- Pollination, fertilisation, formation of seeds and fruits and their dispersal.
- Structure and germination of seeds.
- Structures involved in plant reproduction – may be at cell level (eg pollen, sperm, ovules, spores) or at organ level (eg cones, flowers).
- Processes involved in plant reproduction (pollination, fertilisation).
- Reasons for the differences in structure and function between different groups – these could be related to habitat, size, life cycle, way of life.

Asexual reproduction in plants

Asexual reproduction is reproduction *without* meiosis and fertilisation. The only kind of division involved is mitosis, so there is no reshuffling of genes into new combinations. The offspring are *genetically* identical to the parents. (Genetically identical plants growing under natural conditions will not be phenotypically identical, because they will not have identical environments.)

Vegetative propagation

Vegetative propagation in flowering plants occurs when an axillary bud grows out into a lateral shoot which develops its own roots and becomes an independent plant. While it is connected with the parent plant, it continues to obtain nourishment from it, but when the connection with the parent decays, the young plant becomes independent.

Runners

A runner is a long lateral shoot that grows over the surface of the ground, producing a new plant at the tip, eg strawberry, creeping buttercup.

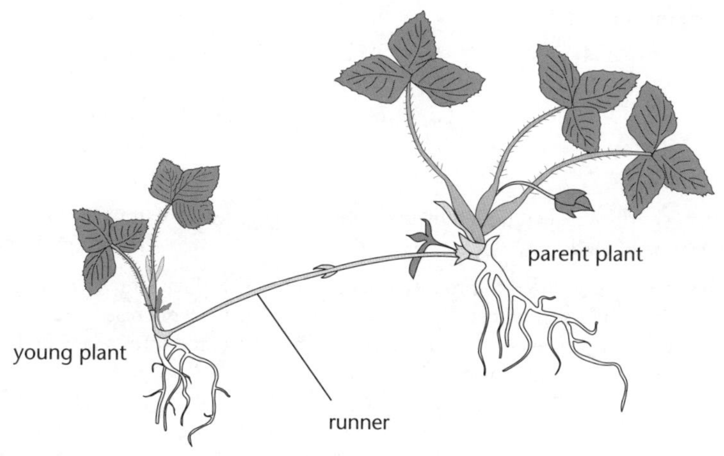

A strawberry runner.

Rhizomes

Like a runner, a rhizome is a horizontally-growing stem, but it grows *below* ground, eg couch grass (a common weed) and iris.

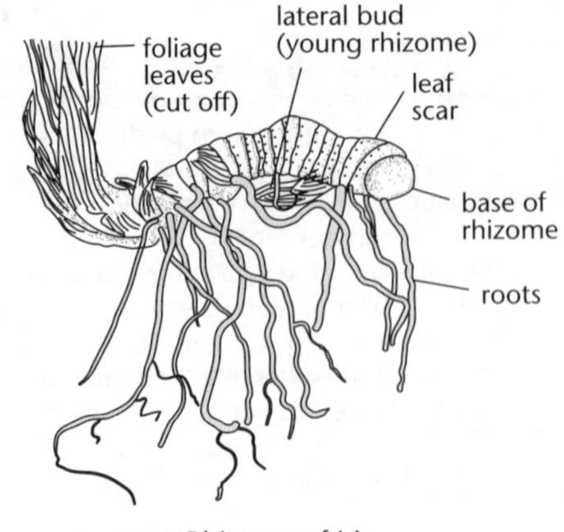

Rhizome of iris.

Stem tubers

A potato is a stem tuber.

The tiny buds and very tiny scale-like leaves show that it is a stem rather than a root. When a potato is planted, the buds grow into aerial shoots that photosynthesise. Rather than growing upwards, some of the lower lateral branches grow down into the soil. During the summer, the tips of these shoots become swollen with starch made from sugar produced in the leaves. If these new tubers are not harvested, each would grow into a new plant the following spring. Each potato plant produces many tubers.

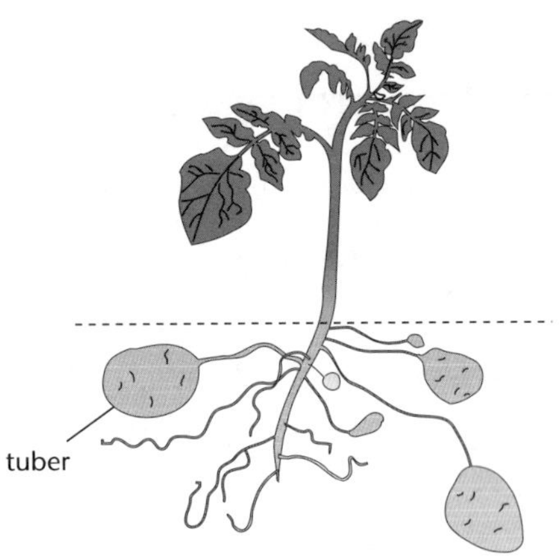

Tubers of potato.

Bulbs

A bulb is a very short stem surrounded by the bases of foliage leaves that have become swollen with energy reserves ('food').

In spring, these reserves are used to produce the new shoots. As the energy reserves are used up the scales become thin and wither. During the summer, new energy reserves are transported down to the bases of the leaves of a lateral bud, which swell out and form a new bulb. Each of the two or three lateral buds can form a new bulb.

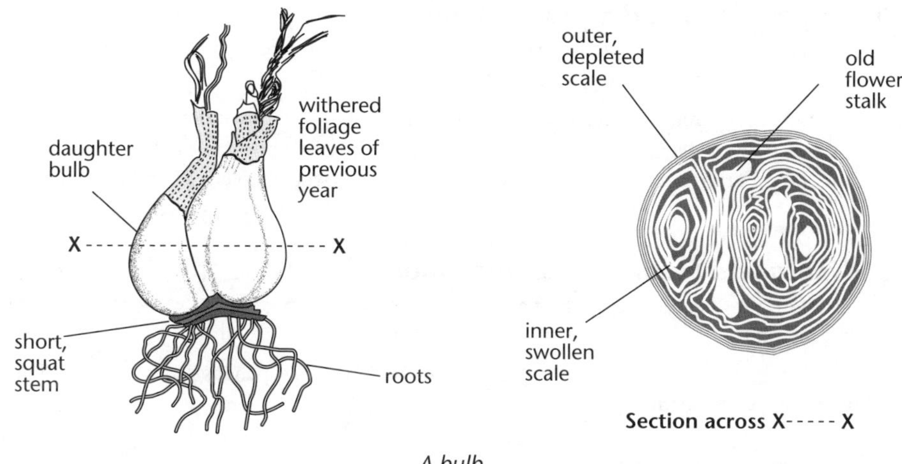

A bulb.

Corms

A corm looks like a bulb, but it is really quite different, because the energy reserves are stored in a stem rather than the bases of leaves.

A corm is covered by the withered bases of the previous year's foliage leaves, and has several lateral buds. In the spring, each bud produces a shoot, and during the summer the energy reserves are passed down the stem, forming a new corm.

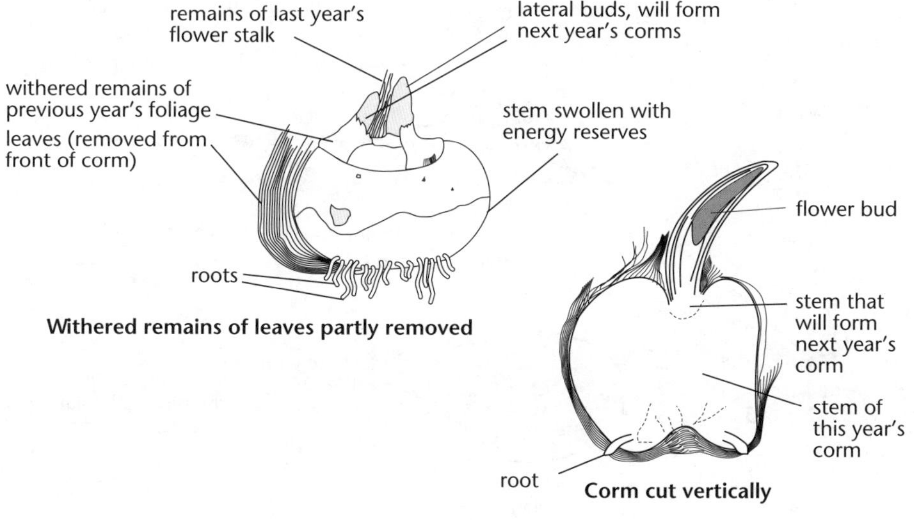

A crocus corm.

Advantages of vegetative propagation

Since there is only one parent, plants that undergo vegetative propagation do not have to rely on animals or wind for pollination or for dispersal.

- The young plants remain attached to the parent until they are fully independent.
- New plants are genetically identical to the parent, so if conditions do not change much from one generation to the next, the offspring will all be well adapted. This is particularly useful in horticulture, in which distinct varieties can be multiplied and maintained.

Disadvantages of vegetative propagation

New plants are produced close to the parents and may therefore compete with them for light, water and minerals.

Because offspring are genetically identical to the parents, if there is a change in conditions that is bad, conditions will be bad for *all* of them.

Artificial vegetative propagation

Once a new plant variety has been produced, horticulturalists often want to mass-produce genetically identical copies of the new plant variety for sale. Typically, small fragments will be taken from the desired plant, and grown into new plants. The easiest method is to cut off small shoots (cuttings) and to add an artificial plant hormone that stimulates root production. Many amateur gardeners use this method.

A more sophisticated way is to remove tiny fragments such as discs cut from leaves, and to use these to produce new plants. Leaf discs cannot by themselves grow into new plants – they must first be treated with the appropriate plant hormones in a sterile growth medium. These hormones stimulate the disc to produce a mass of new cells called a **callus**, from which a new plant can grow.

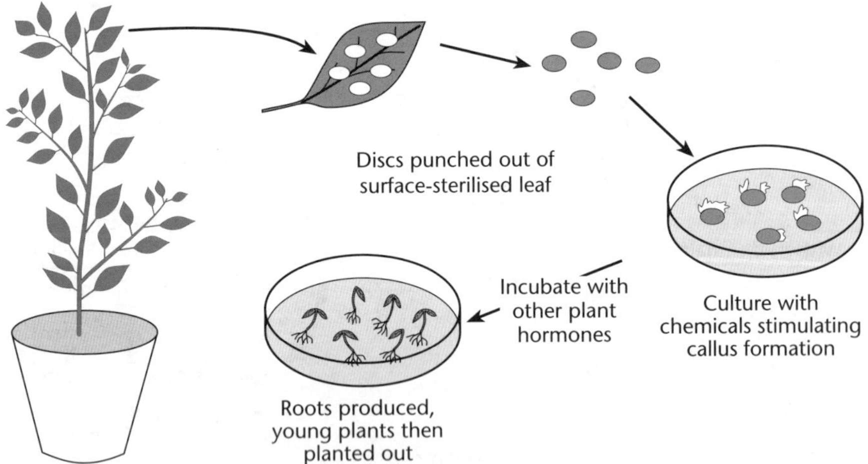

A method of artificial vegetative propagation.

Perennation

'Perennation' in plants means 'surviving from year to year'.

The energy reserves of bulbs, corms, tubers and rhizomes enable them to survive the winter and from year to year. Plants which survive from year to year are called **perennials**.

- In herbaceous perennials, all the aerial parts die back at the end of the year and new growth is resumed from buds at or below ground level.
- In woody perennials, growth is resumed from buds well above ground, so the plant continues to increase in height each year.

Biennials live for two years, the first being devoted to storage of energy reserves, and the second to flowering.

Domesticated biennials, such as the carrot, are harvested at the end of the first year, before they get the chance to use up their energy stores on flowering.

Annuals complete their entire life cycle in a year or less.

Sexual reproduction in flowering plants

In plants, sexual reproduction always involves two key cell processes:

- **Meiosis** – the division of a diploid cell (with two sets of chromosomes) into four haploid cells (each with one set of chromosomes).
- **Fertilisation** – the joining together of two haploid **gametes** to form a **zygote**. In most organisms, the gametes are of two kinds – large female gametes or **eggs**, and small male gametes or **sperm**.

The advantage of sexual reproduction is that the offspring are slightly different from each other. Without this *variation*, evolution could not occur. In flowering plants, the part concerned with sexual reproduction is a special reproductive shoot, the **flower**.

In flowering plants the male gametes are produced inside cells called **pollen grains**. Before fertilisation can occur, pollen must be transferred from male to female organs, a process called **pollination**. Pollen transport is achieved by animals (insects or birds) or by wind.

Structure of a hibiscus flower

Flowers show immense variation, so there is no such thing as a 'typical flower'.

A hibiscus, however, shows the basic features of an insect-pollinated flower.

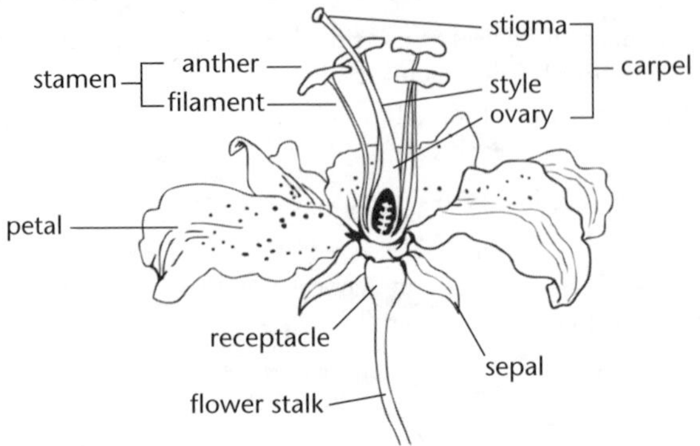

Vertical half of a hibiscus flower.

The flower parts of a hibiscus are arranged in four layers, all anchored to the top of the flower stalk, the **receptacle**.

The outer layer is called the **calyx** which protects the flower in the bud stage. It is usually made up of five **sepals**. The next layer in is the **corolla**, whose function is to make the flower conspicuous to insects. It usually has five **petals** which can come in many different colours, including red, orange, yellow, blue, purple, pink, white and sometimes more than one colour. Each petal has a **nectary** at its base. The nectary is a patch of cells which secrete (pour out) a sugar solution called **nectar**, which acts as food for insects. Without this 'reward', insects would not visit the flower, so pollination would not occur. (In many flowers the petals are joined together to form a tube, so that only insects with long tongues can reach the nectar.)

Inside the petals is the **androecium**, consisting of 30–40 stamens. Each stamen consists of an **anther**, held up by a long stalk or *filament*. The anther consists of four **pollen sacs**, where pollen grains are produced.

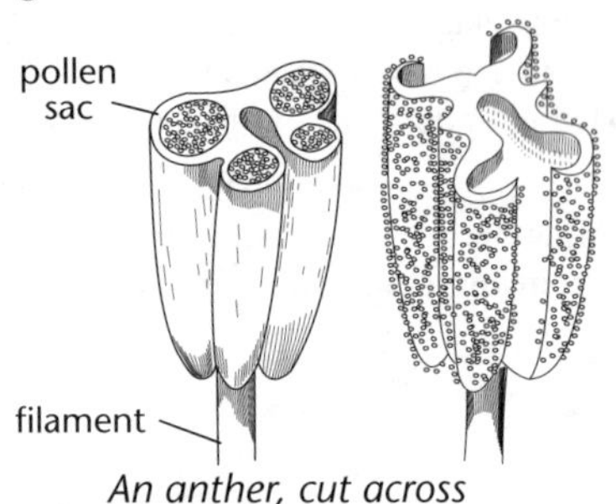

An anther, cut across to show pollen sacs.

Pollen grains are produced by meiosis. When the anther is ripe it splits down the middle to release the pollen. The **gynoecium** consists of 30–40 **carpels**. A carpel is the female equivalent of a stamen and consists of three parts:

- At the top of the **style** is the **stigma**. This is the part which receives pollen from the body of a visiting insect or bird. It secretes sugar, so pollen sticks to it. The sugar is also a signal for the pollen grains to start growing.
- Above the ovary is a short style, which supports the stigma.

- The **ovary** contains and protects a single **ovule**, which later develops into a **seed**. The ovule has two coats or **integuments**, with a small hole in it called the **micropyle**. Inside the ovule, one cell divides by meiosis, and after further divisions a female gamete or **egg** is produced. Another nucleus in the ovule is the **endosperm nucleus** which, after fertilisation, gives rise to a food reserve (**endosperm**) for the embryo. In some seeds (eg pea, bean) the endosperm is used up before the seed is ripe.

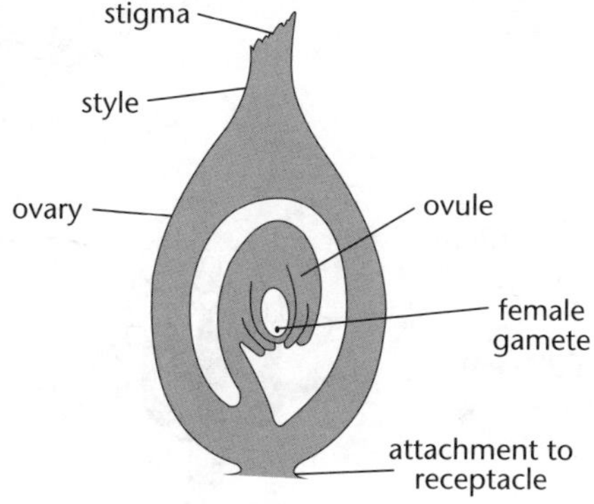

Longitudinal section of a carpel.

Some flowers have only one carpel, but in most there are several. These may be separate or joined.

Pollination

Pollination is the transfer of pollen from stamens to stigma.

- In **cross-pollination** the pollen is transferred from a stamen to a stigma of a *different* plant.
- In **self-pollination**, pollen is transferred from a stamen to a stigma on the *same* plant. It makes no difference whether or not the pollen is transferred to a stigma on the same flower, or onto a stigma on a different flower on the same plant.

Cross-pollination is more common than self-pollination, and usually gives more *variable* offspring.

Although most flowers are hermaphrodite (have both male and female organs), most are adapted to prevent self-pollination:

- Male and female parts ripen at different times; most often the stamens release pollen before the gynoecium is ready to accept pollen.
- Pollen cannot grow on the stigma of the same plant, eg apple, clover.

Some plants are not hermaphrodite, ie the male and female flowers are on different plants.

Though flower colour may be 'beautiful' or scent 'lovely', to an insect, colour and scent are simply advertisements. It is the nectar that is the reward. When searching for nectar, insects brush against the anthers, picking up pollen on their hairy bodies. When visiting another flower, some of the pollen comes off the insect and sticks to the sugary stigma.

Self-pollination involves the transfer of pollen within the same flower (eg peas and cereals) or between flowers on the *same* plant (eg oak).

Self-pollination

Cross-pollination involves the transfer of pollen between flowers on *different* plants of the same species (eg clematis), and produces *greater genetic diversity than self-pollination.*

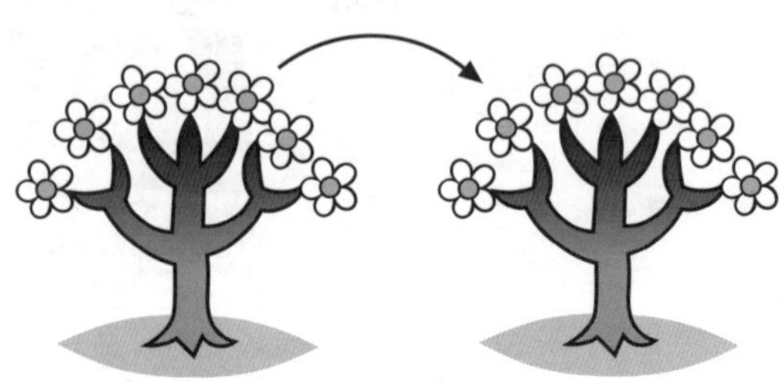

Cross-pollination

The two main agents of pollination are animals and the wind, but some plants use water for pollination.

Insect pollination

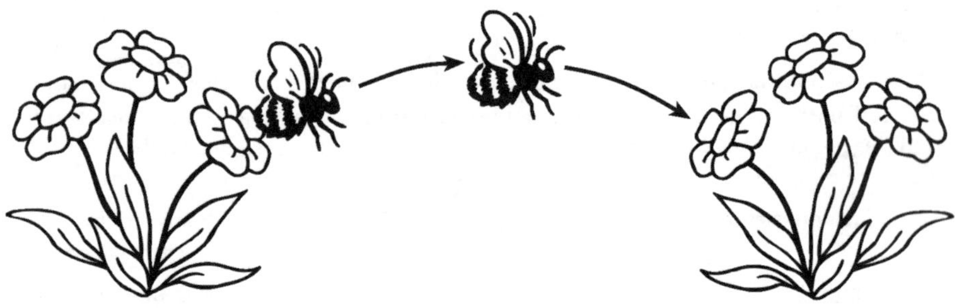

Flowers have large colourful petals; large pollen sticks to insects

Flowers that have *animal pollinators* typically have brightly coloured petals and nectaries full of sweet fluid ('nectar'); scent may be produced from scent glands to attract pollinators. Pollen grains are large and sticky; small amounts are produced (compared with wind-pollinated flowers). Pollinators are mainly insects – bees are common daytime pollinators, moths are common night-time pollinators. However, birds can act as pollinators.

Example

Some birds and bats can pollinate flowers when they feed from the flower's nectaries. When the beak or the head is dipped into the nectary, some pollen sticks to it. This is transferred to the stigma of the next flower that they feed at.

When insects crawl into a flower to get nectar, the stigma and stamen are usually enclosed within the petals, with the stigma above the anthers – this allows an insect to brush pollen onto the stigma as it enters the flower, then it collects that flower's pollen from the anthers on its way to the nectary; this assists with cross-pollination (to prevent self-pollination, the anthers and stigma of some plants ripen at different times). Animal-pollinated flowers tend to produce less, but heavier, pollen, than wind-pollinated flowers, and the pollen may be rough or sticky to attach to the pollinator. Some plants have evolved very complex mechanisms to effect insect pollination.

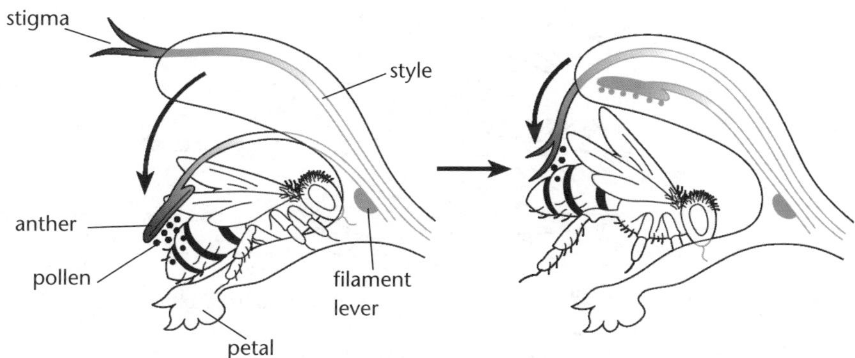

Collapsing stigma of the plant ensures pollination when a bee triggers the filament lever as it searches for nectar

Wind pollination

Plants such as grasses and many trees are pollinated by wind.

The stigmas of wind-pollinated flowers are usually feathery, giving a large surface area so that pollen is more likely to land on it. Wind-pollinated flowers have non-sticky pollen grains. This ensures that the pollen does not stick to leaves, branches or other obstacles.

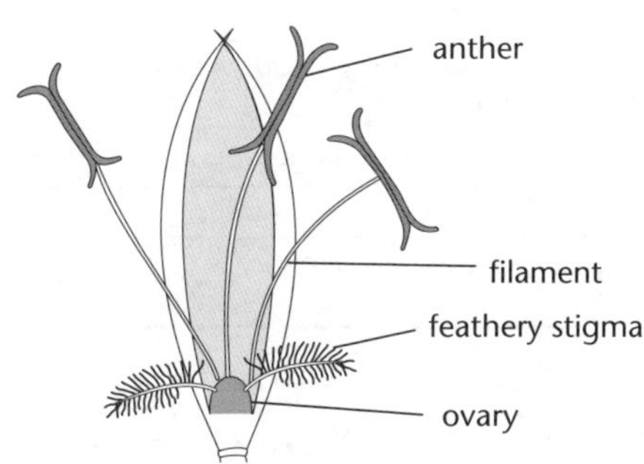

Rye grass flower, a wind-pollinated flower.

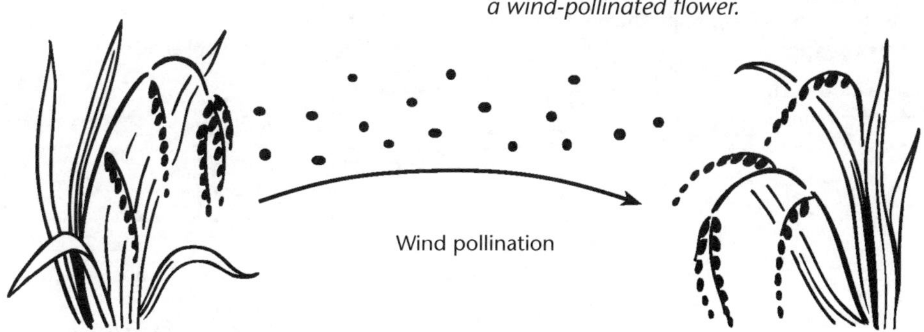

Flowers are small and inconspicuous; small, light pollen blows to other flowers.

Flowers that have wind pollination typically do not have colourful petals, and they may be small and inconspicuous (eg grasses). Nectaries are absent, and no scent is produced. The anthers hang outside the petals, so are exposed to the wind; the stigma are typically feathery and protrude from the petals, to catch pollen. Large amounts of light pollen is produced to increase the chances of the pollen being blown to another flower of the same species.

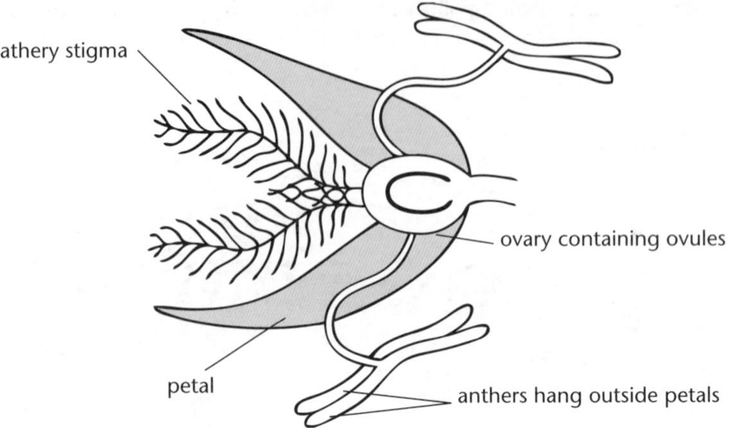

Structure of a wind-pollinated flower.

Insect-pollinated flowers	Wind-pollinated flowers
Usually produce nectar	Never produce nectar
Often produce scent	Never produce scent
Flowers are conspicuous with large, often colourful petals	Flowers are small, green and inconspicuous
Pollen grains are large, rough and sticky	Pollen grains are very small and smooth
Anthers do not hang loose	Anthers hang loose in the wind
Stigmas are not feathery	Stigmas often feathery and project from the flower

Differences between insect-pollinated plants and wind-pollinated plants.

Fertilisation

About the time of pollination, the single nucleus in the pollen grain divides into two. One of these nuclei then divides again, producing two male gametes or **sperm**.

Fertilisation occurs inside an ovule, several days after pollination.

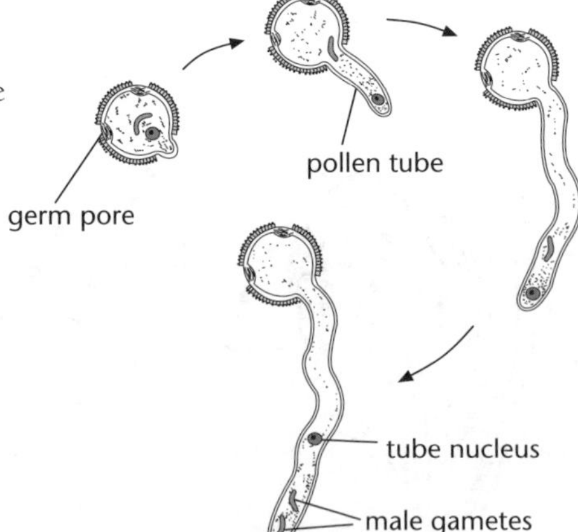

Germination of a pollen grain.

The pollen grains are stimulated by the sugar on the stigma to germinate.

↓

The thick outer layer round each pollen grain has several pores or holes. A pollen tube grows out through one of the pores. The tube grows down into the stigma and style, and secretes enzymes which digest the complex foods in the style. The resulting simple foods such as sugars and amino acids are absorbed and used by the pollen tube to grow.

↓

When the pollen tube reaches the ovary, it grows towards an ovule. It is attracted by chemicals produced by the egg. When the pollen tube reaches the egg, its tip breaks down. One of the sperms then joins with the egg to form a zygote.

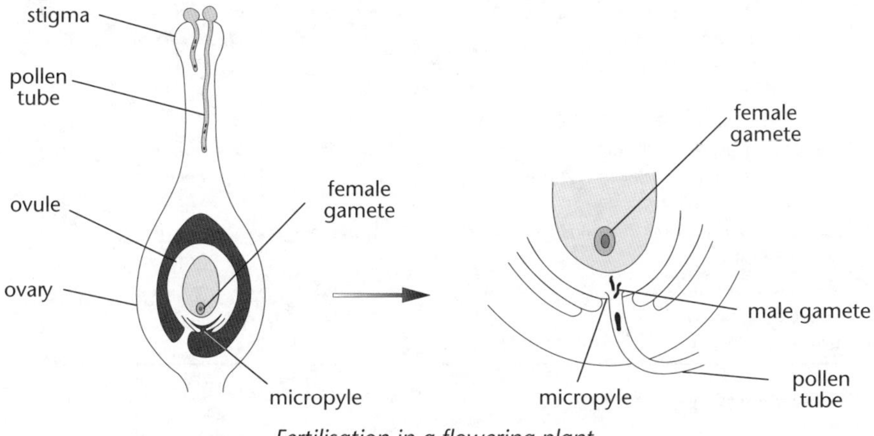

Fertilisation in a flowering plant.

In most plants, the ovary contains many ovules, so many pollen grains will be needed if all the ovules are to develop into seeds.

Bypassing fertilisation

Some plants regularly produce seeds without fertilisation, eg dandelion, blackberry. This is called **parthenogenesis**, and results in offspring genetically identical to the parent.

Unit 11.6 Activity 3A: Flowers, pollination and fertilisation

1. For each of the phrases **1–26**, write the letter **A–Z** of the term to which it applies.

Phrase	Applied term
1. After fertilisation, develops into a seed	**A.** Androecium
2. Coat round ovule	**B.** Anther
3. Collective name for the petals	**C.** Calyx
4. Collective name for the sepals	**D.** Carpel
5. Collective name for the stamens	**E.** Corolla
6. Connects stigma to ovary	**F.** Egg
7. Consists of anther and filament	**G.** Endosperm
8. Development of an egg without fertilisation	**H.** Fertilisation
9. Egg or sperm	**I.** Filament
10. Female equivalent of a stamen	**J.** Gamete
11. Female gamete	**K.** Gynoecium
12. Female part of the flower	**L.** Integument
13. Fertilised egg	**M.** Micropyle
14. Food reserve present in some seeds	**N.** Nectar
15. Hole in integument	**O.** Ovary
16. Makes insect-pollinated flowers conspicuous	**P.** Ovule
17. Male gamete	**Q.** Parthenogenesis
18. One of several parts that protects the flower in bud	**R.** Petal
19. Pollen-producing part of a stamen	**S.** Pollination
20. Process that produces a zygote	**T.** Receptacle
21. Protects ovules	**U.** Sepal
22. Secretes sugar as stimulus for pollen germination	**V.** Sperm
23. Stalk of stamen	**W.** Stamen
24. Sugar solution acting as 'reward' for pollinating insects	**X.** Stigma
25. Tip of flower stalk, to which all flower parts are attached	**Y.** Style
26. Transfer of pollen from anther to stigma	**Z.** Zygote

2. The diagram below shows two types of flower.

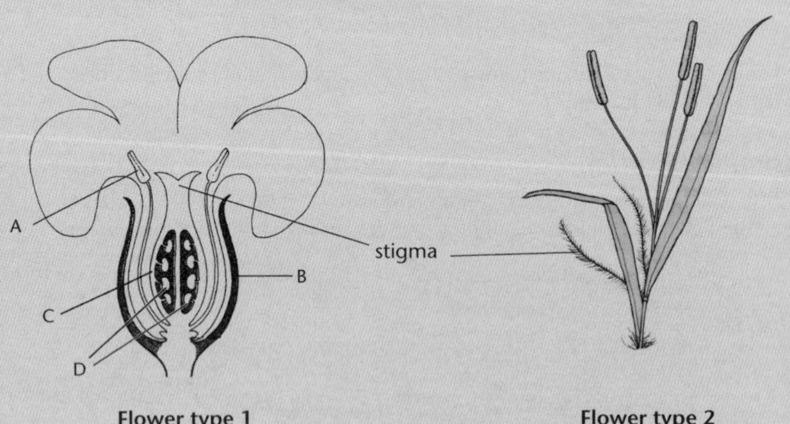

a. In the table, name and describe the function of any three of the flower parts labelled A to D in the diagram for Flower type 1.

Letter	Name of flower part	Function

b. Explain why the parts labelled stigma in Flower type 1 and Flower type 2 are different.

c. The diagrams below show the stages in plant reproduction that occur after pollination. Explain how the growth of a pollen tube leads to fertilisation.

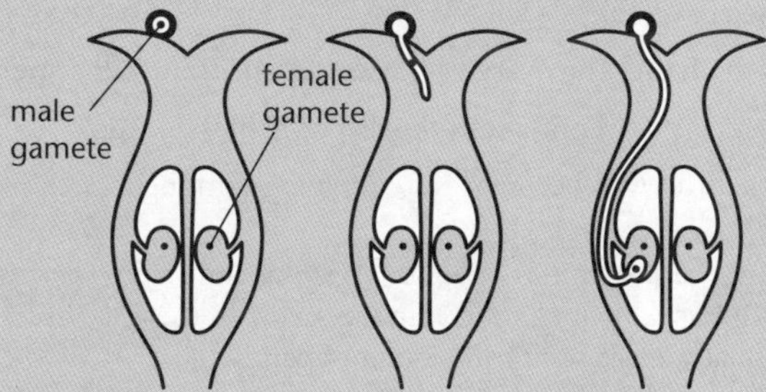

3. Give three characteristics of flowers that would attract insects for pollination.

4. Pollination is the transfer of pollen from male flower parts to female flower parts. Fertilisation is necessary for the production of seeds, and follows soon after successful pollination. Explain the role of the pollen tube in the process of fertilisation in a flower.

5. The diagram shows reproduction in a sweet potato. The kaukau is capable of reproducing using two different methods – seeds and tubers. Discuss the advantages and/or disadvantages to the kaukau plant of each type of reproduction.

Development of the fruit and seed

As soon as fertilisation has occurred:

- The ovule develops into a **seed**. The coverings of the ovule grow into the **testa** or seed coat.
- Inside the ovule, the zygote begins to divide and grows into the young plant or **embryo**.

Eventually, the embryo develops three parts:

- A young root or **radicle**.
- A young shoot or **plumule**.
- One or two *seed leaves* or **cotyledons**.

In most flowers, the sepals and petals fall off.

The ovary grows into the **fruit**. The ovary wall becomes the fruit wall or **pericarp**.

In many plants, the pericarp is important in the dispersal of the seeds:

- In some plants, the fruit wall becomes soft and juicy, forming a *succulent fruit* which is eaten by animals.
- In *dry fruits*, the pericarp becomes quite tough and is not eaten. Dry fruits that contain more than one seed usually open to release the seeds. In these *dehiscent* fruits, the seeds are dispersed individually. *Indehiscent* fruits do not open.

Dispersal of fruits and seeds

When seeds are fully developed, they are released from the parent plant. In most cases they are spread around or *dispersed* some distance from the parent. This has two advantages:

- The young plants do not have to compete with one another or with the parent for light, water and mineral salts.
- The seeds *may* reach a better habitat (place to live) than the parent.

How fruits and seeds get dispersed

Fruits and seeds can be dispersed in a number of different ways.

- By wind – many fruits and seeds have feathery or wing-like outgrowths which increase their surface area and make them fall more slowly.

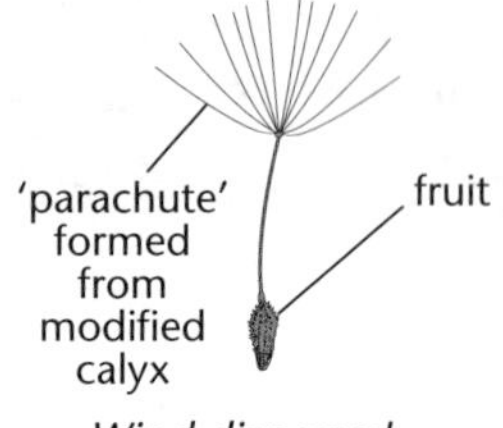

Wind dispersed.

- By animals – either by being eaten and passing through the gut (eg tomato), or by becoming attached to the animal.

Animal dispersed.

- By water – eg mangrove. In the mangrove, the seeds begin to germinate before leaving the parent.

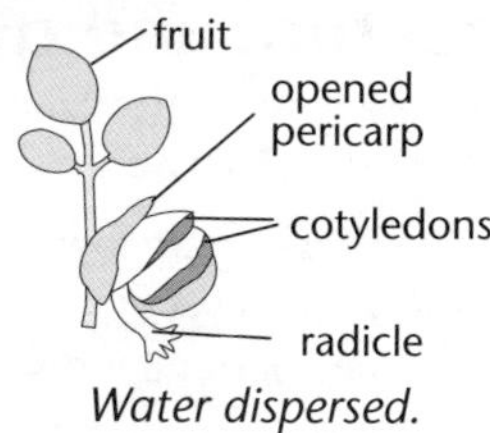

Water dispersed.

The **fruit** surrounding the seed may be large and fleshy (eg pears, apples), or dry (eg the pods of peas, lupins). The fruit aids *dispersal* of the seed, eg:

- Large, fleshy *succulent* fruits attract animals to eat them. The seeds pass through the intestine and emerge with the faeces, which assist germination by acting as a fertiliser.

- Pods dry out then split open violently, to eject the seeds up to several metres from the parent.

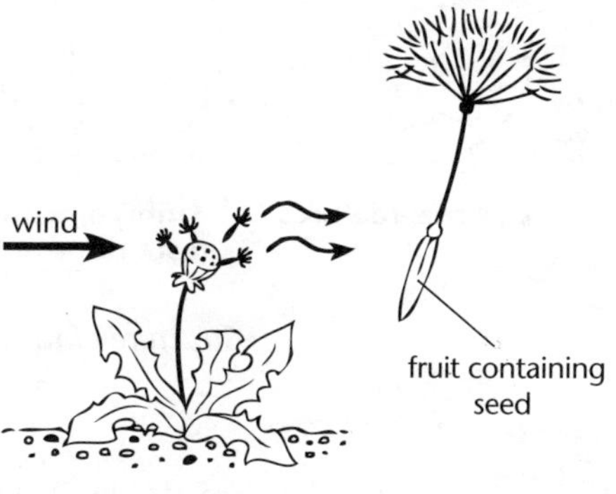

- Some become 'hairy', to form a parachute-like structure that catches the wind.

- Some form a thin, wing-like structure, to catch the wind.

Structure and germination of seeds

A seed is a very young plant (embryo) protected by a seed coat. When they leave the parent plant, most seeds have the following characteristics:

- They have a very low water content. Actively growing plant tissue contains about 70–90% water, but most seeds contain about 10%.
- The rate of respiration is so low as to be undetectable.
- All seeds have a store of energy (in the form or starch or fat) and raw materials (in the form of protein).

Pea seed

The seed is protected by the **testa** or seed coat. This has a scar left by its attachment to the fruit. Just next to the scar is a tiny hole in the testa, the **micropyle**.

After soaking in water (eg for 24 hours), the testa becomes soft and is easily removed. Underneath is the **embryo**. This consists of two very large **cotyledons** or seed leaves, the **radicle** or young root, and the **plumule** or young shoot, tucked in between the cotyledons. The cotyledons do not look like leaves – they are very thick because they are swollen with stored starch and protein.

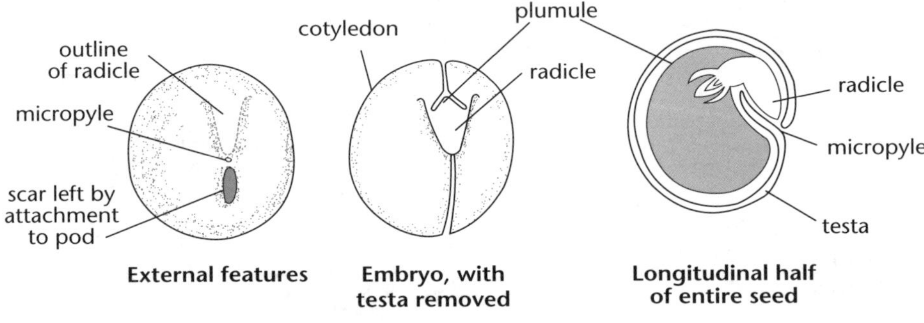

Structure of the pea seed.

Corn grain

Like all cereals, corn differs from peas in some important ways:

- Cereals do not store food in the embryo, but in a tissue called the endosperm. This is packed with starch, and its outer layer stores protein. (It is from the endosperm of wheat that flour is made.)
- There is only *one* cotyledon – cereals and other grasses are *mono*cotyledons. (The pea belongs to the *di*cotyledons – it has two cotyledons.)
- A cereal grain is actually a one-seeded fruit (the outer layer is the pericarp and testa fused together).

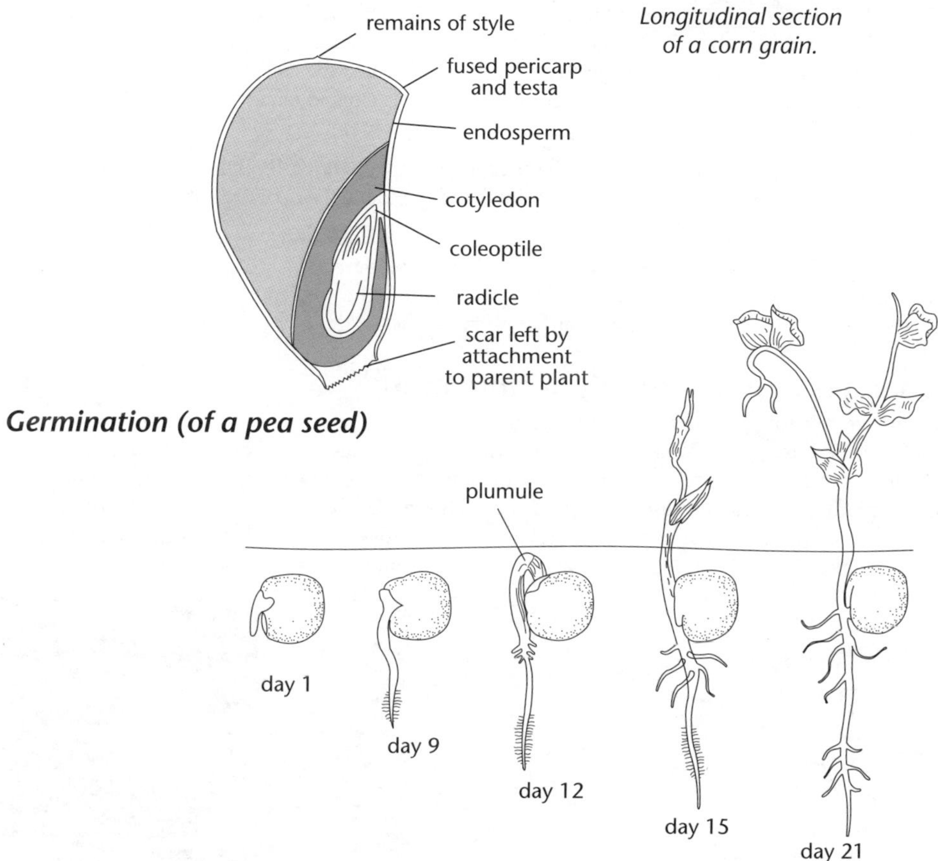

Longitudinal section of a corn grain.

Germination (of a pea seed)

Germination of a pea seed.

For the first 24 hours after being planted in moist soil, a pea seed absorbs water and swells, and the testa softens.

Next, the starch and protein in the cotyledons are digested by enzymes. Starch is converted into sugar and proteins are converted into amino acids. (These simpler, soluble substances can be transported from the food store to the radicle and plumule, where cells are dividing. In these growing areas, amino acids are used as building materials to make new proteins, and sugar is used in respiration to supply energy.)

The first visible signs of germination occur after a few days.

- First, the radicle grows through the testa. No matter which way the seed falls, the radicle always grows downwards, towards gravity. As it reaches deeper soil where the water supply is more permanent, lateral branches begin to develop.
- A day or two after the radicle begins to grow, the plumule grows out from between the cotyledons. It always grows *away* from gravity, and therefore towards the soil surface. Until it reaches the surface it is pale yellow with very small leaves. It is also hook-shaped, which protects the delicate leaves as it grows through the soil.

The embryo uses the energy stored in the cotyledons. The plumule must therefore reach the light so that photosynthesis can begin before all the energy is used up. This is why small seeds (which store less energy) can only germinate successfully near the soil surface. As soon as the plumule reaches the light it straightens out, the leaves expand, and it becomes green as chlorophyll develops.

Until the plumule reaches the surface, the seedling is dependent on food it obtained from its parent. Once photosynthesis begins, the plant is independent. The cotyledons eventually wither and die as all the energy reserves are removed.

Some seeds (eg the runner bean) germinate in a slightly different way. The region between the radicle and cotyledons (the hypocotyl) elongates, pushing the cotyledons *above* ground. The cotyledons then become green and begin to photosynthesise.

Changes in mass during germination

During germination, the mass of living tissue in the seed – the *live mass* (or fresh mass) – increases because of the uptake of water. However, the *dry mass* – the mass of the seed minus water – decreases, because energy reserves are being used up in respiration. The dry mass of the plumule plus radicle increase, but there is an even greater decrease in dry mass of the cotyledons. When the plumule reaches the surface, photosynthesis begins and the plant is no longer dependent on food received from the parent. From then on, dry mass increases.

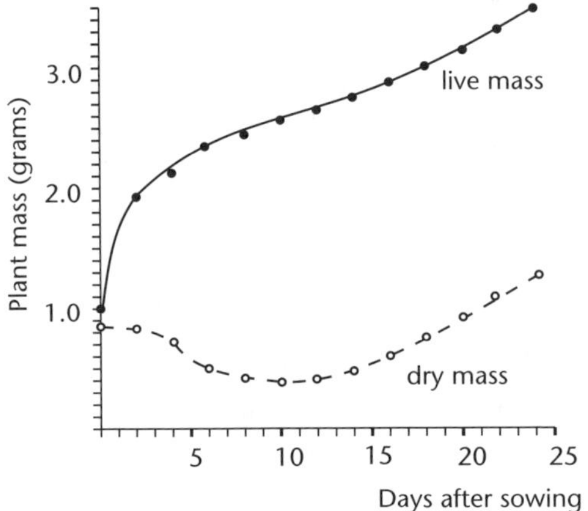

Changes in mass of a bean during germination.

Planting seeds prior to germination – maximum dry mass of seed cotyledons.

Unit 11.6 Activity 3B: Germination and seed dispersal

1. For each of the phrases **1–15**, write the letter **A–O** of the term to which it applies.

Phrase	Applied term
1. Develops from ovule after fertilisation	A. Auxin
2. Food-storage tissue in cereals and some other seeds	B. Coleoptile
3. Fruit wall	C. Cotyledon
4. Growth response to gravity	D. Dehiscent
5. Growth substance that stimulates elongation in shoots	E. Embryo
6. Growth towards light	F. Endosperm
7. Leaf of embryo plant	G. Fruit
8. Minute hole in testa	H. Gravitropism
9. Opens to release seeds	I. Micropyle
10. Ripened ovary	J. Pericarp
11. Root of embryo plant	K. Plumule
12. Seed coat	L. Positive phototropism
13. Shoot of embryo plant	M. Radicle
14. Tubular sheath that protects foliage leaves of grass seedling	N. Seed
15. Young plant protected by seed coat	O. Testa

2. During a class discussion, the teacher said 'pumpkins and lemons are both fruits'. Use the structure and development of fruit to explain why the pumpkin is biologically a fruit.
3. The diagram below compares a typical monocotyledon and a dicotyledon seed.

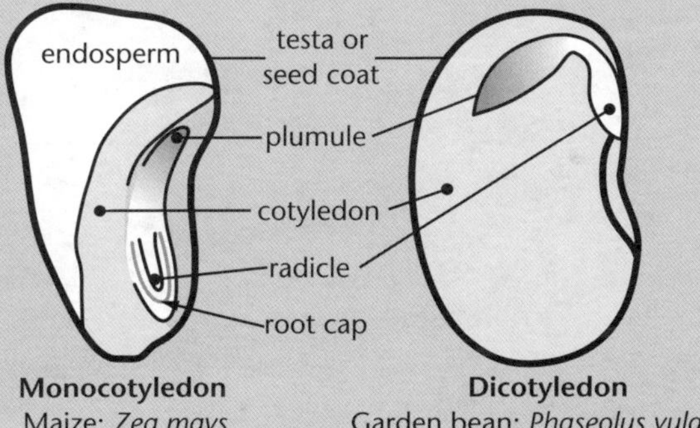

   a. Describe the function of the endosperm of a monocotyledon seed.
   b. A seed only germinates in suitable environmental conditions. Describe the two main environmental conditions necessary before a seed can germinate.
   c. Explain how one structure of a seed enables it to survive for long periods of time, before it finally germinates.

4. a. The pea pod is the fruit of the pea plant. When mature, the pod contains 8–10 seeds.

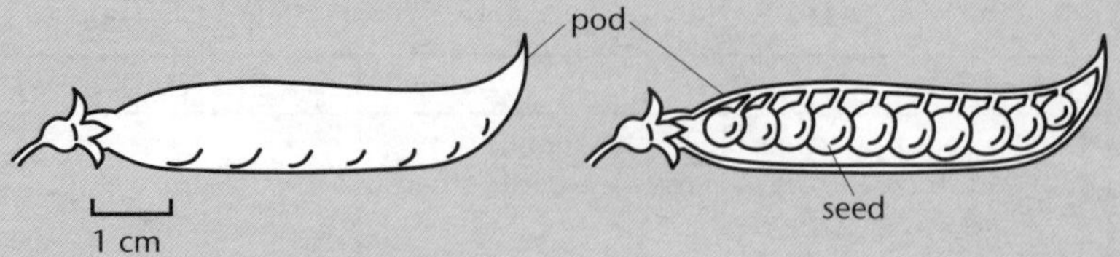

Name and describe how the parts of a pea flower develop into:

i. The pea pod.

ii. The pea seeds.

b. Explain how the production of fruit provides an advantage to the survival of flowering plants.

5. The following series of diagrams shows a typical example of germination and early seedling growth.

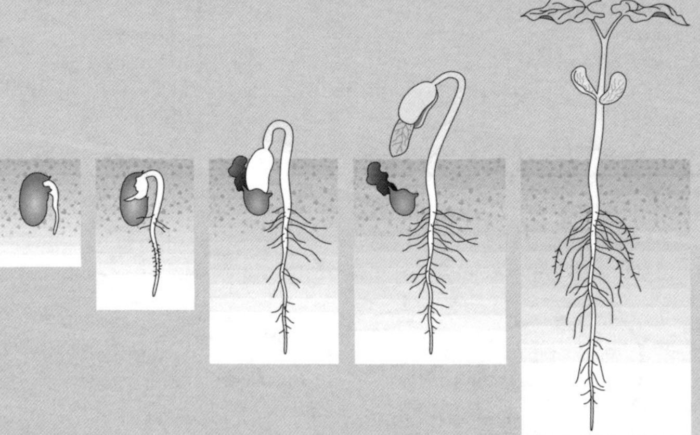

a. Describe how early seedling growth is affected by:

i. Gravity.

ii. Light.

b. Explain why the seed has a store of carbohydrate.

c. Discuss how gravity and light are important for successful early seedling growth.

6. The class arranged some fruits into three groups based on the method of seed dispersal.

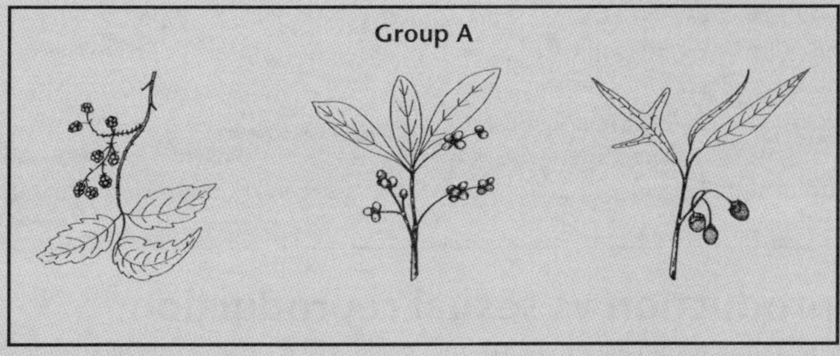

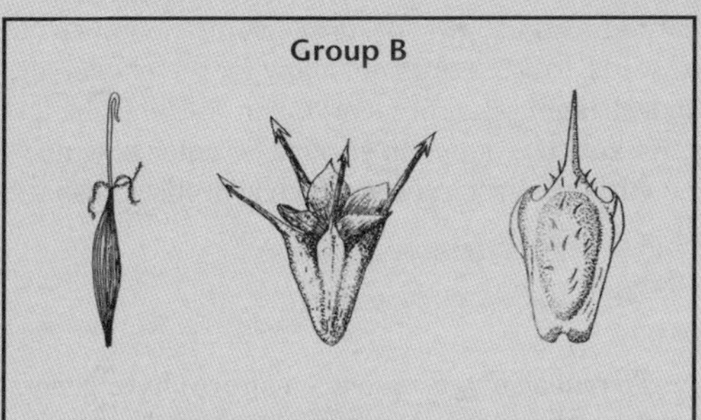

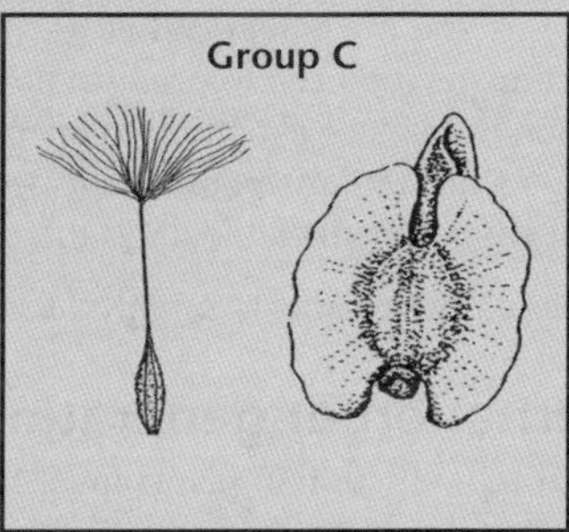

In the table provided, describe each method used to disperse the seeds.

Group	Method of seed dispersal
A	a.
B	b.
C	c.

7. The diagram shows a pea seed cut in half to show the internal structure of a typical dicotyledon seed. Letters A to D indicate the important parts of the seed.

testa
A
D
B
C

a. Describe the internal structure of the seed. You should include the names of the parts, indicated by the letters A to D, in your description.

b. Describe the function of one of the parts labelled A, B, C in the germination of a seed.

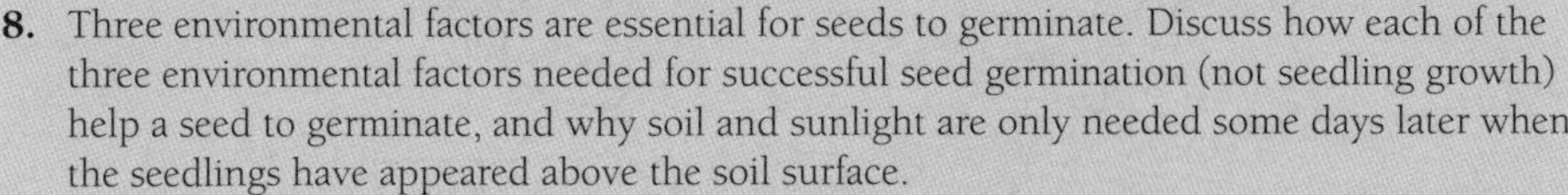

8. Three environmental factors are essential for seeds to germinate. Discuss how each of the three environmental factors needed for successful seed germination (not seedling growth) help a seed to germinate, and why soil and sunlight are only needed some days later when the seedlings have appeared above the soil surface.

9. a. Pollination and fertilisation are both important processes in the sexual reproduction of flowering plants.

i. Draw a labelled diagram to describe the process of cross-pollination.

ii. Describe the role of fertilisation in the sexual reproduction of flowering plants.

iii. Wind-pollinated flowers produce large amounts of smooth, light pollen. Explain why the pollen of wind-pollinated flowers is different from the pollen of insect-pollinated flowers.

b. After pollination, seeds form in either a dry or fleshy fruit. Different types of fruit use different methods to disperse the seeds away from the parent plant. Discuss how the structure and features of two different fruits help to disperse the seeds they contain. You may use diagrams in your answer.

Asexual reproduction vs sexual reproduction

Offspring from asexual reproduction are produced from *mitotic cell division*, so are all identical to each other and the parent. While asexual reproduction allows for *rapid* colonisation of favourable environments (no need to produce flowers, seeds, fruit), it does not allow for the phenotypic variation needed for evolution to operate in a changing environment. Environmental changes may therefore eliminate the population/species. Asexual reproduction also makes colonising new or distant environments difficult, as there is no effective dispersal stage such as seeds and spores.

Therefore, most plants also reproduce *sexually*.

Alternation of generations

Sexual reproduction in plants involves the *alternation of generations* – a **sporophyte** generation and a **gametophyte** generation. ('Phyte' = plant, therefore 'sporophyte' = plant that produces spores; 'gametophyte' = plant that produces gametes.)

The **gametophyte** generation is **haploid** (**n**) and produces the **gametes** (ova and sperm) from *mitotic* cell division (this is in contrast to animals, in which the gametes are produced from meiosis). At fertilisation, the fusion of the ovum and sperm produces a *zygote* that grows into the sporophyte generation.

The sporophyte generation is **diploid** (**2n**) as a result of fertilisation, and produces the **spores** from *meiosis*. The spores disperse and germinate to grow into the gametophyte generation.

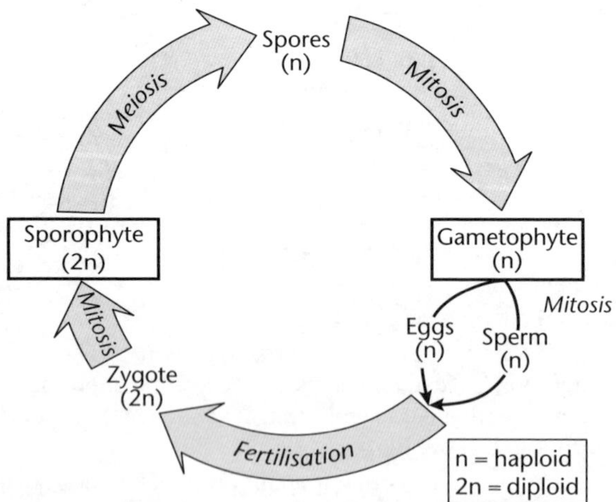

The plant life cycle – alternation of haploid and diploid generations.

The diploid sporophyte generation has twice the genetic information of the gametophyte generation, because there are twice the number of chromosomes (as homologous pairs). This is advantageous because:

- If there is a disadvantageous mutation in an allele on one chromosome, then there is a 'back-up' allele on the other member of the chromosome pair (which can act as a template for correcting the faulty allele since it has the 'correct' information).
- The spores produced from meiosis are genetically unique, and having paired alleles for every characteristic increases the amount of variation in the offspring. (Variation in the offspring is the raw material for the evolutionary process.)

The trend in plant evolution has been for the sporophyte generation to become dominant at the expense of the gametophyte generation. Along with this has been a trend away from dependence from water for reproduction as plants have evolved to a full terrestrial existence.

Gymnosperms and angiosperms

The gametophyte generation in ferns, while small and short lived, is nevertheless an independent plant. However, in the **gymnosperms** and **angiosperms**, the gametophyte is reduced to a few cells which are contained *within the sporophyte* – the sporophyte generation is now completely dominant. As the sporophyte is diploid (2n), it contains twice the genetic information of the haploid gametophyte, allowing for much more variation in the phenotype – both gymnosperms and angiosperms show much more diversity than ferns or mosses.

The sporophyte no longer produces small unicellular **spores**, but large, multicellular **seeds**, which contain a food supply – this allows seeds to germinate *within the soil*, where water is more likely to be available. Evolution of the *pollen tube* has meant *fertilisation is completely internal*, so gymnosperms and angiosperms are *now independent of water* for reproduction and can live a completely terrestrial life. Both gymnosperms and angiosperms can grow into large, long-lived trees. Conifers ('pine trees') are the most common representatives of the gymnosperms.

Gymnosperms – conifers

The large tree is the diploid sporophyte and its reproductive organs are **cones** ('pine cones' or 'fir cones'). Male cones are smaller than female cones; both are borne on the same tree.

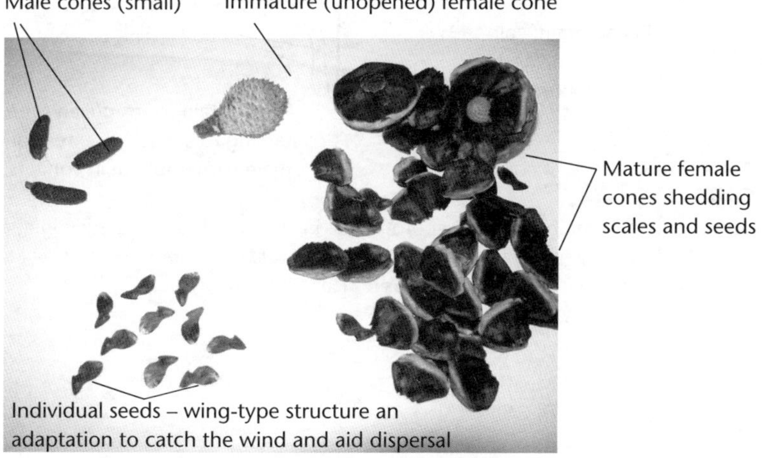

Cones and seeds

Cones

Cones are made of woody scales – the **sporophylls** or reproductive leaves.

Male cones

Male cones have pollen sacs enclosed in the sporophylls, and these produce large numbers of light, haploid, **pollen grains** from meiosis. The nucleus of the pollen grain divides by mitosis to form four nuclei, two of which become the *male gametes/sperm* and the remaining two are the *tube nuclei*.

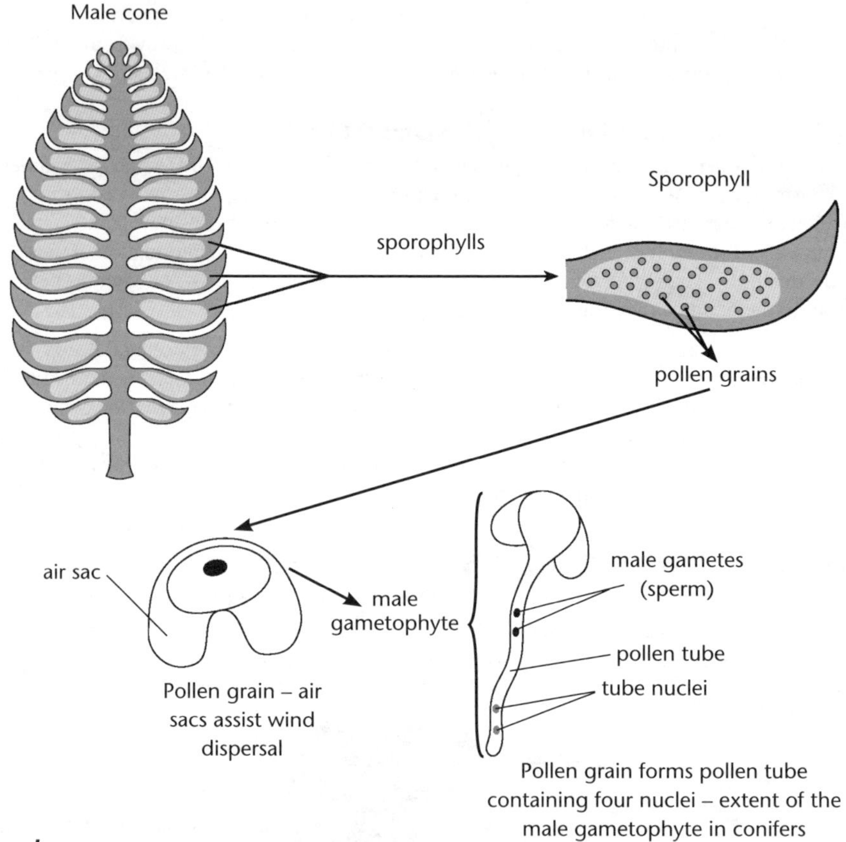

Pollen grain – air sacs assist wind dispersal

Pollen grain forms pollen tube containing four nuclei – extent of the male gametophyte in conifers

Female cones

Female cones have ovules enclosed in the sporophylls. Each ovule has two archegonia, each containing a haploid ovum/egg produced in meiosis. *The ovule and its ova are the extent of the female gametophyte*. A tiny opening (the **micropyle**) leads into the ovule.

Female cones and sporophylls are much larger than male cones. Sporophylls spread to receive pollen and release seeds (eg the large 'fir cones' often burnt on fires).

Ovule forms at base of each sporophyll on the upper side – extent of the female gametophyte in conifers.

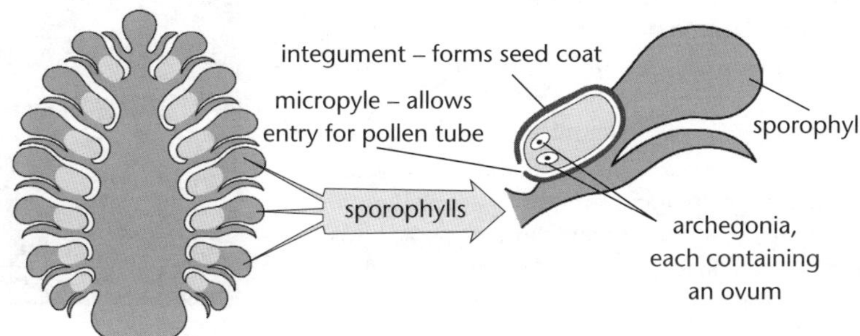

Pollination

Wind blows the pollen grains around, some of which will meet a female cone to land on an ovule. The tube nuclei control the development of a pollen tube, which grows towards, and into, the ovule, through the micropyle. *The pollen tube with its four nuclei is the extent of the male gametophyte.*

Fertilisation

When a pollen tube enters an ovule, each male gamete fuses with an ovum to form the diploid zygote. The zygote divides repeatedly (mitosis) to form the embryo, contained in a **seed** (develops from the *integument* that surrounds the ovule). The seed has a thin, wing-like section that assists in its dispersal by the wind when released from the female cone. When conditions are suitable, the seed will germinate and grow into the large sporophyte that will develop cones when mature, so completing the life cycle.

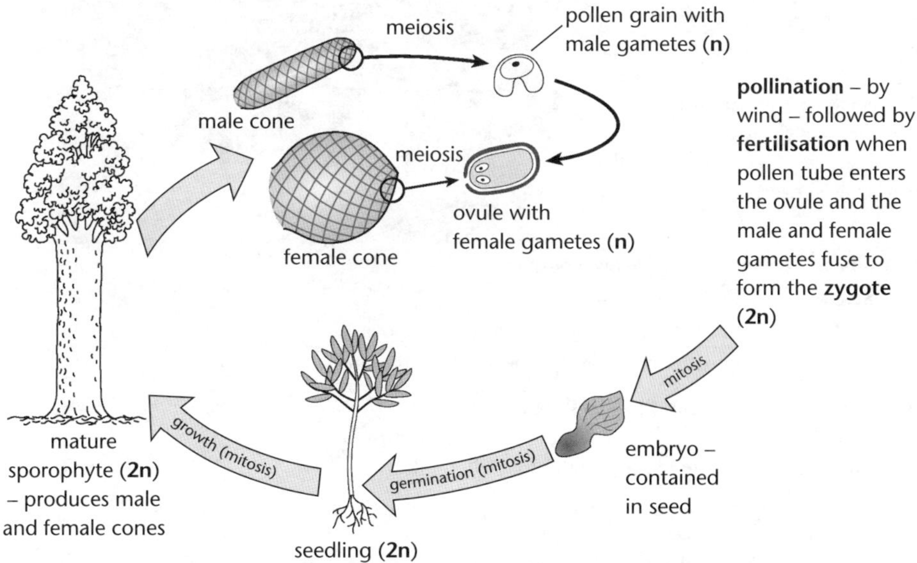

Angiosperms

Angiosperms are considered the most successful and most highly evolved group of plants, because:

- They have been able to *colonise all habitats* in the terrestrial environment.
- They show 'reverse evolution', in that some types (known as *hydrophytes)* have re-colonised aquatic habitats (eg *Elodea*, water lilies).
- The reproductive organs, the *flowers*, show great diversity of colour and structure – adaptations for *pollination by animals* (mainly insects, some birds). Animal pollination is more reliable than wind pollination.
- The structure of seeds allows them to remain dormant for a long time (about two years in some species) awaiting suitable conditions for germination.

The monocotyledons (eg grasses, lilies, orchids) are considered to be more highly evolved than the dicotyledons (eg all shrubs).

Life cycle

The large conspicuous plant is the diploid sporophyte; the haploid gametophyte generation is reduced to a few microscopic cells *within the flower*.

Flowers

Most species have both male and female sex organs in each flower, but some species may have separate male and female flowers on the same plant (*monoecious*, eg beech trees, nikau palms), while other species have male and female flowers on different plants (*dioecious*, eg pawpaw, *Coprosma*, *Spinifex*) – in dioecious species, it is only the female plant that will bear fruit/berries.

Carpels are the female sex organs, and have:

- **Stigma** – sticky, to trap pollen.
- **Style** – holds the stigma aloft and connects to the ovary.
- **Ovary** – contains the ovules and is the site of fertilisation.

Stamen are the male sex organs, and have:

- **Anther** – produces the pollen.
- **Filament** – holds the anthers aloft for pollen release.

Petals enclose the sex organs, and are colourful in plants that are animal pollinated.

Sepals enclose the petals, etc, as a *bud* as the flower develops; typically green.

Nectary may be present, adjacent to the ovary, in animal-pollinated species. Scent glands may also be present.

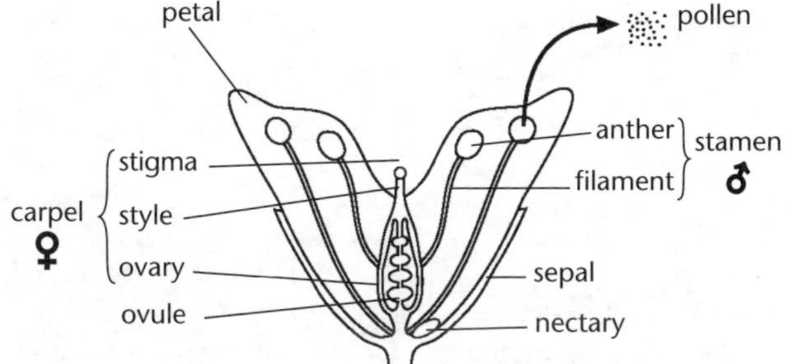

Structure of a typical insect-pollinated flower.

There are very few 'typical' flowers and huge variations in flower structure are seen. It is these differences that are used to classify the different angiosperm groups.

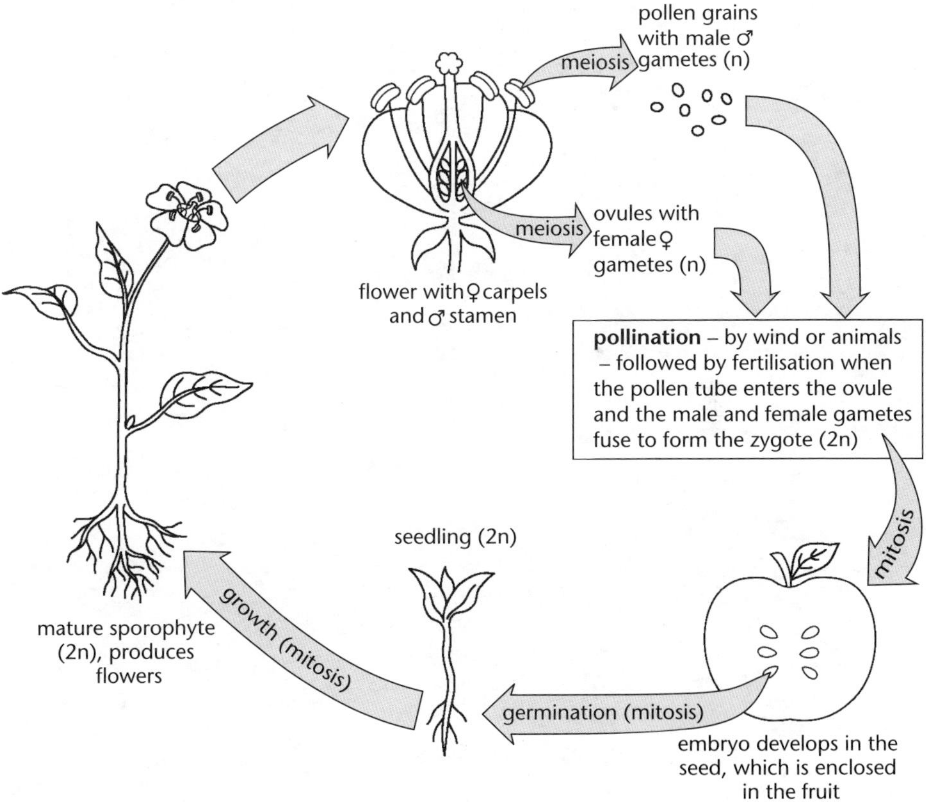

Unit 11.6 Activity 3C: Plant reproduction

1. Distinguish between:
   a. Asexual and sexual reproduction.
   b. Gametophyte and sporophyte.
   c. Meiosis and mitosis.
   d. Meiosis and fertilisation.
   e. Pollination and fertilisation.
   f. Spores and seeds.
   g. Haploid and diploid.
   h. Seed and fruit.
2. Explain why mosses and ferns are restricted to moist habitats.
3. Explain why the trend in plant evolution is for the sporophyte generation to become dominant over the gametophyte.
4. Describe the changes that occur in the gametophyte from mosses to ferns to angiosperms.
5. Explain the need for cross-pollination, and describe how it is achieved.

Supplementary Unit: Micro-organisms

Topic 1: The structure and function of micro-organisms

Authors: Martin Hanson with Takis Solulu, Beatrix Marjen-Waiin and Diaiti Zure

This Supplementary Unit looks at micro-organisms in relation to nutrition, transport systems, respiration and gas exchange, responses to stimuli, and reproduction. In other words, this Unit draws together the information already presented in Units 1–6 in this book and, by focusing on micro-organisms, demonstrates how inter-related these aspects of Biology are to the existence of living things. Much of the content of this Unit is not spelt out in the Syllabus but an understanding of micro-organisms is important in terms of relating the study of Biology to real life, real issues and the local environment.

Topic 1 deals with:

- Bacteria – their structure, nutrition, respiration and reproduction.
- Fungi – structure, nutrition and reproduction of a common mould; harmful and useful fungi.
- Viruses – their structure and reproduction.

Organisms too small to see with the naked eye are called **micro-organisms**, or **microbes**. They include bacteria, viruses and many fungi. To measure such tiny organisms, scientists have to use smaller units than millimetres. Microbes are measured in **micrometres** (μm) or in the case of viruses, **nanometres** (nm).

$$1 \text{ mm} = 1\,000 \text{ μm} = 1\,000\,000 \text{ nm}$$
$$1 \text{ μm} = 10^{-3} \text{ mm} = 1\,000 \text{ nm}$$
$$1 \text{ nm} = 10^{-3} \text{ μm} = 10^{-6} \text{ mm}$$

The smallest thing that can be seen with the naked eye is about 0.1 mm, or 100 μm. An average bacterium is about a hundred times smaller than this, and most viruses are about a thousand times smaller than bacteria!

Bacteria

Though very small, bacteria affect our lives in a number of important ways:

- Bacteria cause decay by feeding on dead matter. Though it is a nuisance when food goes 'off', decomposition of dead bodies and sewage is beneficial because it leads to the release of CO_2 and minerals, which are the raw materials for plant growth. Without decay, plants would run out of nutrients.
- Bacteria are important in many industries, eg making cheese and yoghurt and, in recent years, in the chemical industry.
- Some microbes live with another organism in a mutually beneficial partnership that may play a vital role in agriculture. For instance, nitrogen-fixing bacteria living inside the roots of leguminous plants (such as clover) provide plants with nitrogen in a form the plant can use. Even more important are the cellulose-digesting microbes living inside the gut of sheep and cattle, for without them the animals cannot survive.
- Some bacteria cause disease.

Shapes of bacteria

Bacteria used to be classified according to their shapes. Most bacteria conform to four basic shapes.

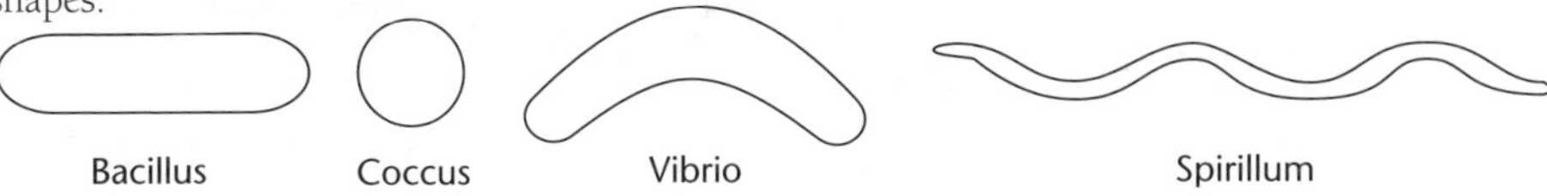

The most common bacterial shapes.

- Spherical bacteria, called **cocci** (singular, coccus), eg *Staphylococcus* (causes boils and pimples), *Streptococcus* (causes sore throats).
- Rod-shaped, called **bacilli**, eg *Salmonella typhimurium* (the cause of typhoid).
- Bent rods, called **vibrios**, eg *Vibrio cholerae* (causes cholera).
- Spiral-shaped, called **spirilla**, eg *Treponema pallidum* (causes syphilis).

Bacteria are also divided according to whether or not they stain pink with Gram stain. Those that do are said to be *Gram positive*, while those that do not are *Gram negative*.

Structure of bacteria

Bacteria are single-celled organisms whose cells differ from those of other living things in several key ways:

- There is no clearly defined nucleus – the chromosome is not enclosed in a nuclear envelope. Instead it occupies a less well-defined area called the *nucleoid*. This is why bacteria are called **prokaryotes**, which means 'before the nucleus'. Other living things, which have distinct nuclei in their cells, are called **eukaryotes**.
- The single main chromosome forms a closed loop – instead of being open-ended as in other organisms.
- They are very small (0.5–2 µm) – eukaryotic cells are usually larger than 5 µm.

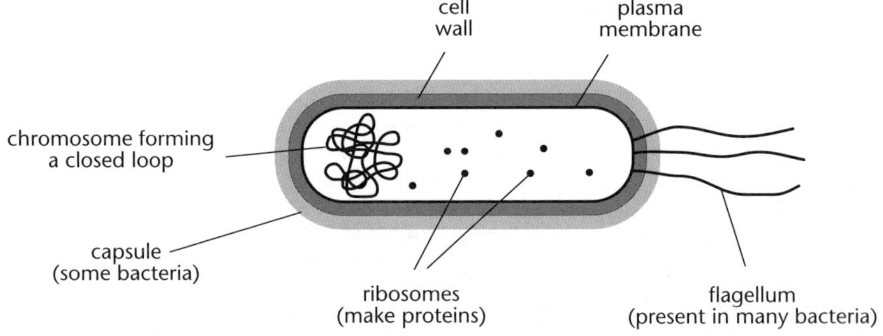

The basic structure of a bacterial cell.

In some bacteria, an outer **slime capsule** gives protection against white corpuscles, eg *Bacillus anthracis* (causes anthrax). The cell wall gives rigidity and protects the cell from swelling and bursting in dilute solutions. The **plasma membrane** is partially permeable and regulates the movement of substances into and out of the cell.

Not all bacteria can move. Those that can – the *motile* ones – are generally propelled by **flagella** (singular *flagellum*). Flagella are thread-like and project from the cell wall. They are fundamentally different from the cilia of eukaryotes.

How bacteria feed

Bacteria fall into two main groups – autotrophs and heterotrophs. The distinction depends on whether or not they can make their own energy-rich materials.

Autotrophs

These build up their own organic matter ('food') from carbon dioxide and water. Some of these autotrophic bacteria are like plants in that they make carbohydrate by photosynthesis. Others use the energy released when chemicals in their environment are oxidised. Some of these *chemosynthetic* bacteria are important in the nitrogen cycle.

Heterotrophs

These use organic matter produced by other organisms. Bacteria feeding on dead organic matter are **saprophytes**, and cause decay.

Saprophytic bacteria secrete digestive enzymes into the surrounding food. In digestion, starch is converted to glucose, proteins are broken down into amino acids, and fats into fatty acids and glycerol. Because digestion occurs *outside* the cells, it is said to be **extra-cellular**.

The simple products of digestion cannot diffuse through the partially permeable plasma membrane. Instead, food molecules are pumped across the membrane by *active transport*. Once inside, the food is used in either of two ways:

- Some is used in growth by being built up into complex compounds, such as proteins.
- Some is used for energy by being broken down into simpler substances (these simpler substances can be used as raw materials for plant growth – in this way, microbes play an important role in recycling of nutrients).

Decay causes the spoilage of food – resulting in its unpalatability or food poisoning (caused by, for example, *Staphylococcus aureus* and *Clostridium botulinum*). *Clostridium botulinum* produces one of the most lethal toxins known, causing a frequently fatal paralysis called **botulism**.

Most bacteria feeding on living matter are **parasites**, doing at least some harm to the host organism. Some bacteria are **mutualists**, living in a partnership with another organism in which both benefit, eg those in the gut of herbivores. Some do neither harm nor good to the host organism and are called **commensals** (eg the many bacteria that live on the human skin and in the human gut).

Bacteria and oxygen

Many bacteria can live without oxygen, and are said to be **anaerobic**, eg those that live in mud at the bottom of ponds and the bacteria in our large intestines. Bacteria that use oxygen are said to be **aerobic**.

Most anaerobic bacteria obtain their energy by **fermentation**, eg the bacteria that cause milk to go sour ferment glucose to lactic acid:

glucose ⟶ pyruvic acid ⟶ lactic acid + energy

Some bacteria obtain energy by **respiration**, in which the food is oxidised:

pyruvic acid + oxygen ⟶ carbon dioxide + water + energy

Some bacteria carry out **anaerobic respiration**. After converting glucose to pyruvic acid, the pyruvic acid is oxidised by a chemical other than oxygen, eg the *denitrifying* bacteria important in the nitrogen cycle oxidise pyruvic acid using nitrate:

pyruvic acid + nitrate ⟶ carbon dioxide + water + nitrogen gas + energy

Anaerobic bacteria can be important to human health. If heat sterilisation of food during canning fails, anaerobic bacteria can grow inside the sealed tin can. *Clostridium botulinum*, which normally lives in the soil, can cause a potentially lethal form of food poisoning.

Reproduction

Bacteria reproduce by **binary fission**, or dividing into two.

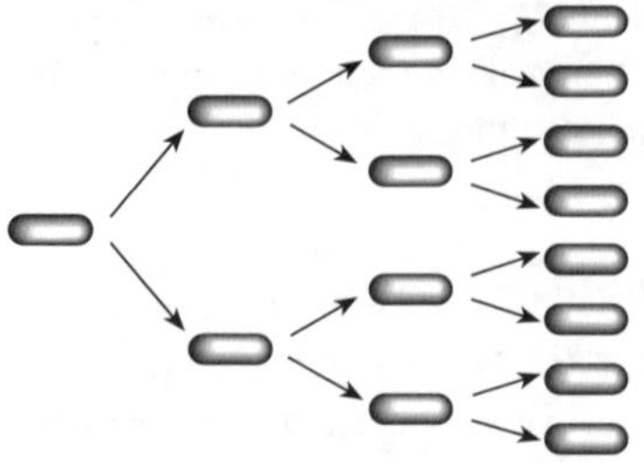

Binary fission in bacteria.

Under ideal conditions, bacteria can reproduce every 20 minutes (a few species can divide every 10 minutes!) – the population doubles three times (an 8-fold increase) every hour. This **exponential growth** cannot continue indefinitely, because:

- Food is used up.
- Waste products accumulate.
- In natural environments, bacteria are eaten by other organisms.

Extension: A closer look at bacterial growth

Population of bacteria growth (in table format).

Growth phase	Time (hrs)	Numbers	Increase	% increase
Period of exponential growth	0	1		
	1	2	1	100
	2	4	2	100
	3	8	4	100
	4	16	8	100
	5	32	16	100
Relative growth rate slowing	6	64	32	100
	7	120	56	87
	8	200	80	67
	9	300	100	50
	10	420	120	40
Absolute growth rate decreasing	11	500	80	19
	12	550	50	10
	13	580	30	5
	14	590	10	1.7
	15	595	5	0.85
Growth rate zero	16	595	0	0

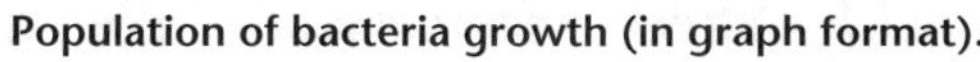

Population of bacteria growth (in graph format).

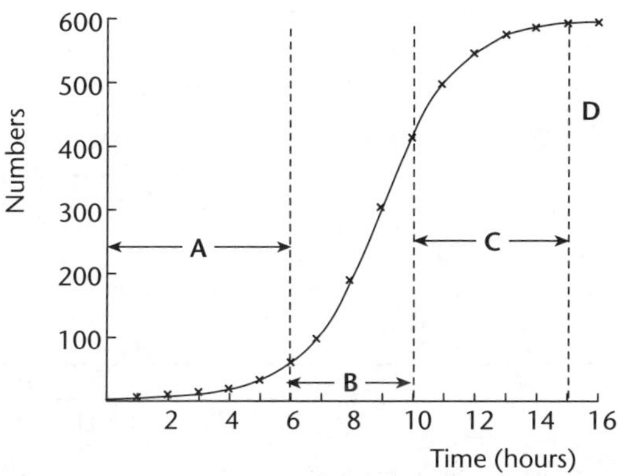

Growth such as that shown over a 16-hour period, starting from a single cell, might occur in a bottle of milk; the growth assumes each bacterium can divide once every hour. (The actual numbers would, of course, be in billions.)

Growth of an imaginary population of bacteria.

At first the bacteria in above Figure divide every hour, so the population doubles every hour. Growth in numbers can be divided into four phases.

Phase 1

Period of *exponential growth* (A on the graph above) – each individual bacterium is reproducing at a constant rate so the *percentage* increase per unit time (the *relative* growth rate) is constant. In this case, there is a 100% increase every hour.

Phase 2

Each individual bacterium is taking longer to grow and divide (B on the graph above) – increasing competition for food, and accumulation of waste products. The relative growth rate (percentage increase per unit time) is slowing. But, the *absolute* growth rate (increase in numbers per unit time) continues to increase (graph still getting steeper). This is because the slower reproductive rate of each bacterium is more than offset by the increasing number of 'parents'.

Phase 3

The absolute growth rate is decreasing (C on the graph above).

Phase 4

Growth rate is zero (D on the graph above) – division has ceased altogether.

Fridges and your health – effect of temperature on bacterial growth

Keeping food fresh is important – not just because it tastes better, but because it reduces the danger of contamination from disease-causing bacteria. Bacterial growth is extremely sensitive to temperature, so the easiest way of keeping food fresh is by refrigeration.

Bacterial growth rate roughly doubles for every 10°C rise. Above a certain *optimum* (most favourable) temperature, further increase slows growth. At higher temperatures, bacteria are killed. For most bacteria, death occurs above 55–60°C.

Most domestic refrigerators operate at about 5°C. At 15°C, bacteria grow twice as fast as at 5°C, and at 25°C (a typical summer's day) they grow *four* times as fast. At 37°C (body temperature) they grow over *eight* times faster than at 5°C.

Suppose two bacteria land on two pieces of meat, one bacterium on each piece of meat. One piece of meat is transferred to the fridge, the other is kept in a sunny area at 25°C.

Suppose that at 25°C each bacterium can divide every hour:

- After 16 hours at 25°C, there would be 65 536 bacteria.
- At 5°C, the bacteria would be dividing four times slower – every 4 hours rather than every hour. After 16 hours there would be only 16 bacteria.

By growing four times faster, after only 16 hours at 25°C there would be 4 096 times as many bacteria as those on the meat in the fridge! After 32 hours there would be 4 096 × 4 096 or over 16 million times as many.

If the bacterium had been *Salmonella*, keeping the meat in the fridge would have made a huge difference to stopping food poisoning.

Time (hrs)	Numbers	
	5°C	25°C
0	1	1
1		2
2		4
3		8
4	2	16
5		32
6		64
7		128
8	4	256
9		512
10		1 024
11		2 048
12	8	4 096
13		8 192
14		16 384
15		32 768
16	16	65 536

Growth of an imaginary population of bacteria at two different temperatures.

Enzymes are only slowed down by low temperatures and will speed up again if warmed, so as soon as food is taken out of a fridge, the bacteria will 'get going' again.

Conjugation

Besides the main chromosome, most bacteria also carry smaller circular 'mini-chromosomes' called **plasmids**. These can carry genes that produce resistance to antibiotics. A plasmid (and a considerable part of the donor chromosome) are transferred to other bacteria by *conjugation*, a sexual process in which two bacteria become connected.

Endospores

Most bacteria are quickly killed by high temperatures, but some (eg anthrax bacteria) are able to form resistant **endospores**. These have a very low water content and can survive drying. They can also survive boiling for several hours at 100°C. To kill endospores, an **autoclave** is used.

Fungi

All fungi are *heterotrophic*. Some are *parasitic*, and many cause serious diseases of crop plants. A few cause disease in humans, such as athlete's foot and thrush (both skin infections). Other fungi are *saprophytic*.

Most fungi have a body consisting of a network of fine threads or **hyphae**. The whole network is called the **mycelium**. The walls of the hyphae usually consist of **chitin** (which also forms the skeleton of insects).

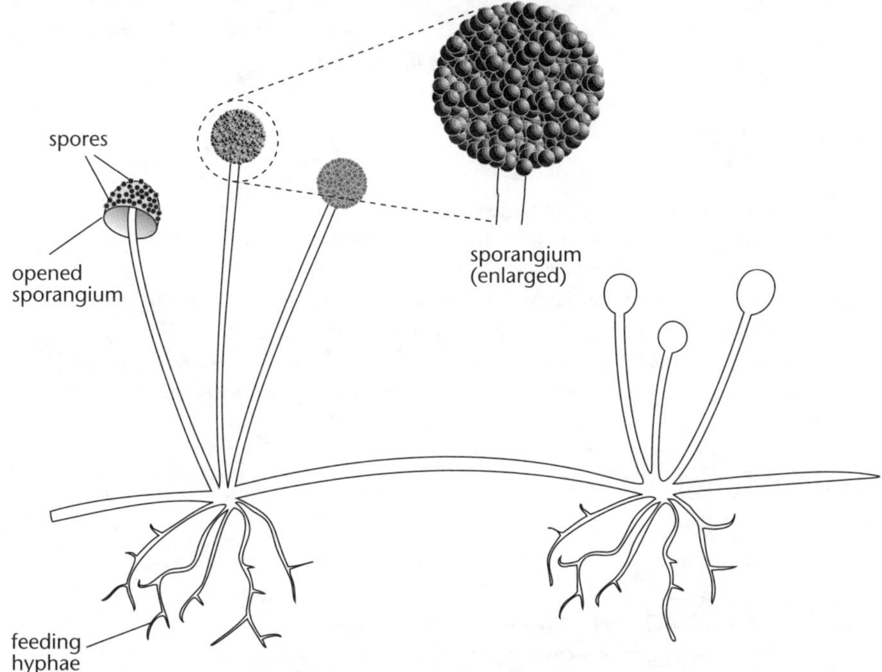

Rhizopus is often found growing on old bread or rotting fruit. *Mucor* is another, similar type of fungus.

Rhizopus, *a common mould.*

Unlike many other fungi, the hyphae of *Rhizopus* are not divided into cells by cross walls.

The cytoplasm contains many nuclei and there is a large central vacuole.

Hyphae grow and branch at their tips, enabling the fungus to rapidly invade food.

Fine feeding hyphae grow down into the food and secrete enzymes into it. These convert complex foods into simpler substances which are absorbed by active transport and used either as raw materials for growth or used in respiration for energy. Because digestion occurs outside the body, it is said to be **extra-cellular**.

Before its food is used up, a fungus must reproduce and disperse offspring to other food sources. Reproduction is mainly *asexual*; the offspring are produced by *mitosis*. Certain hyphae grow upwards and produce swellings called **sporangia** (singular *sporangium*) at their tips. Inside each sporangium large numbers of **spores** are produced, genetically identical to the parent. The spores have a waterproof wall so they can resist drying. In most terrestrial (land-living) fungi, spores are dispersed by wind. Their tiny size means they have a very large surface with which to catch the air compared with their tiny mass, so they are easily carried on air currents. If a spore lands on a suitable food supply, it germinates to form a new mycelium.

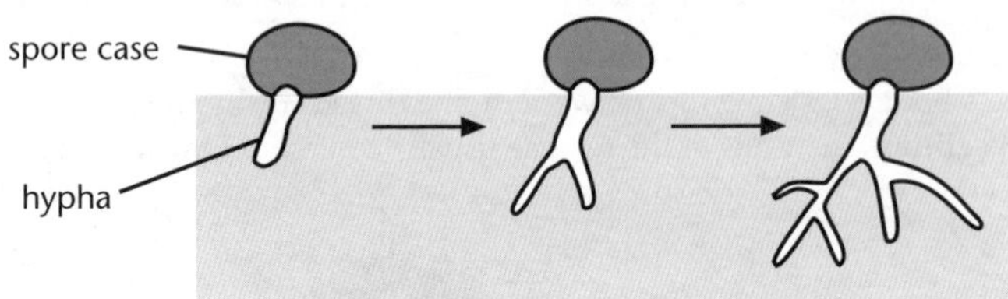

Germination of a fungal spore.

Harmful fungi

Some fungi are harmful.

- Some serious diseases of crops and other plants are caused by fungi, eg Dutch elm disease. The great Irish potato famine of 1845–1849, in which a million people perished, was caused by the fungus-like *Phytophthora infestans*.
- Some human diseases are caused by fungi, eg athlete's foot, thrush, and ringworm.
- Fungi cause tremendous damage to timber unless it is chemically treated.
- Fungi can spoil food and render it dangerous to eat (some species of *Aspergillus*, when growing on peanuts, produce aflatoxin, a powerful carcinogen (cancer-producing substance)).

Beneficial fungi

Some fungi are beneficial.

- In causing decay, fungi promote the recycling of nutrients.
- Yeast, a single-celled fungus, is used in bread-making, brewing and wine-making. The yeast cells ferment glucose, forming ethanol and carbon dioxide:

  glucose ⟶ ethanol + carbon dioxide

  The carbon dioxide is used in brewing and bread-making, and the ethanol is the alcohol of brewing and wine-making.
- Many antibiotics are produced by fungi, eg penicillin.
- Some fungi are edible, and some are greatly sought-after delicacies, eg truffles.

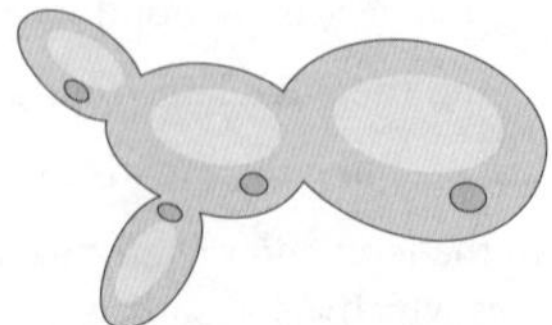

Yeast cells budding.

Culturing bacteria and fungi

Bacteria are usually cultured on a jelly-like medium called **agar**, a complex carbohydrate extracted from seaweed. Most bacteria cannot digest agar, so nutrients such as glucose and minerals are added. The agar is poured as a hot liquid onto a sterile **Petri dish** and allowed to set after the lid has been placed back on again. A Petri dish containing agar is called an *agar plate*. Once the agar has set, Petri dishes are stored upside down.

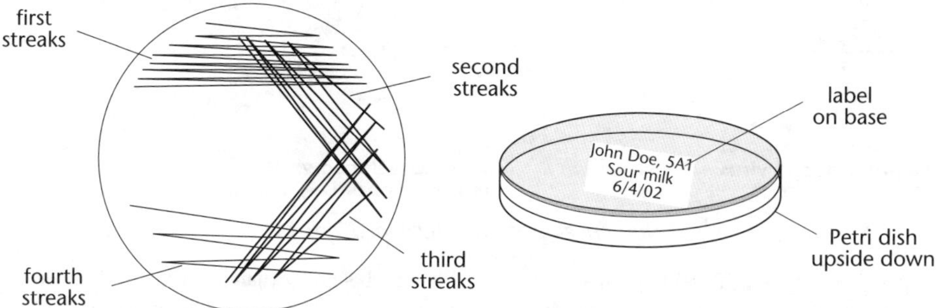

Inoculating an agar plate.

Bacteria are introduced on to the agar using an inoculating loop that has been heat-sterilised in a Bunsen flame. Each bacterium gives rise to a population of descendants called a *colony* which, after 24 hours or so in an incubator at about 30°C, is visible to the naked eye. Thus, by counting the colonies, the number of bacteria inoculated can be determined. (This is only feasible if the bacteria were very thinly dispersed in the first place. This can be ensured either by previous dilution of the bacterial culture, or by repeated streaking with the inoculating loop. Liquid water on the agar allow motile bacteria to spread, so no distinct colonies would be formed. This is why Petri dishes are stored upside down – condensed water might drop onto the agar.)

Different kinds of colony can be distinguished by their colour, texture and size (at a given temperature, differences in size indicate different growth rates). Most bacterial colonies have a glistening appearance.

The hyphae of fungi form 'woolly-looking' colonies.

Viruses

Viruses are the simplest things capable of reproduction – this is the only characteristic of life they all possess.

Almost all forms of life are susceptible to virus infection, including humans and other animals, and plants. Even bacteria are attacked by viruses (called *bacteriophages*). Viruses cause enormous annual losses of crops and domestic animals, and have been responsible for countless diseases since ancient times.

Viruses are not cells, but consist of a protein coat or **capsid**, surrounding a core of **nucleic acid**. In some viruses the nucleic acid is DNA (eg hepatitis B virus), in others it is RNA (eg cold virus and Human Immunodeficiency Virus or HIV).

In some viruses, the capsid is surrounded by an envelope of lipid and proteins, derived from the plasma membrane of the previous host cell.

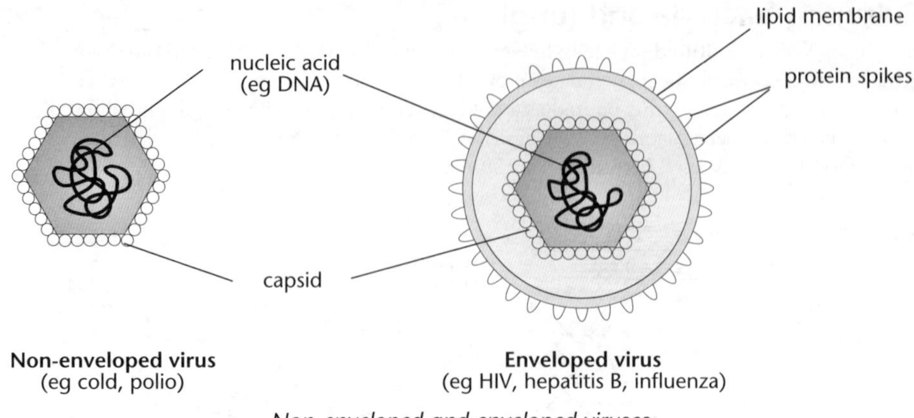

Non-enveloped and enveloped viruses.

Viruses are typically 20–300 nm in size, and are thus much smaller than bacteria. All but the very largest can only be seen with the electron microscope. Each virus particle or **virion** has a characteristic size and shape.

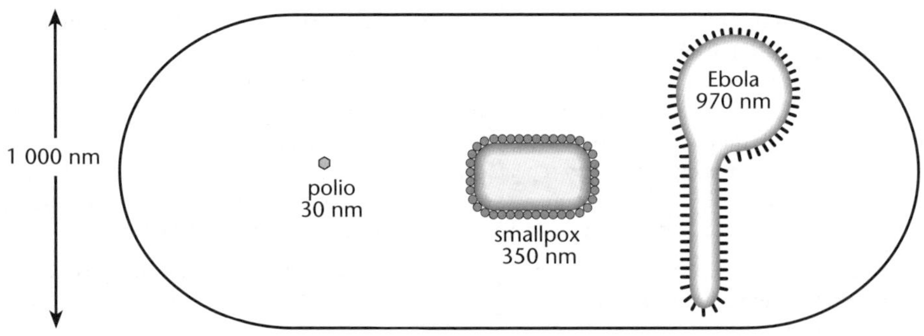

The viruses are drawn inside the outline of an average-sized bacterium to give an idea of how small they are.

Viral shapes and sizes.

The outer layer

This is the capsid or, in the case of enveloped viruses, the proteins in the envelope. Its function is to 'recognise' and gain entry to a host cell by combining with chemicals in the surface membrane of the host cell. Viral proteins act like a key that must fit molecules on the host cell before the virus can get in. Viruses are host-specific – most will only attack a particular species, and many will only attack certain kinds of cell in the host (eg yellow-fever virus attacks human liver cells, HIV attacks one kind of cell in the human immune system).

The proteins in the outer coat (ie the capsid or the envelope) form the part of the virus that is recognised by the immune system. Some viruses (eg influenza) mutate frequently, resulting in different strains of the virus, with slightly differing protein coats. This is why you can have viral diseases such as flu and colds more than once – a second attack is by a strain you haven't had before.

Reproduction

Viruses can only reproduce inside a host cell – they have no chemical processes of their own. Raw materials, energy and enzymes to produce new virus particles are supplied by the host cell.

Only the 'blueprint' for making new virus particles is supplied by the virus – in the form of its genetic information. (This is why viruses cannot be grown on nutrient agar like bacteria and fungi, but must be grown on a medium consisting of living cells.)

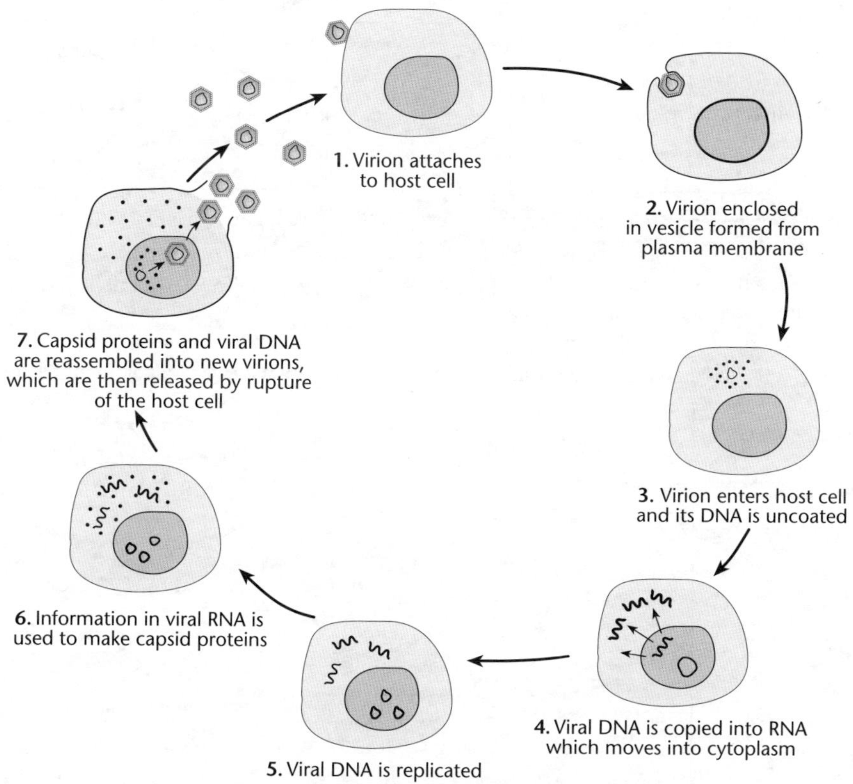

Life cycle of the virus that causes warts.

The reproductive cycle of an animal virus (such as the one that causes warts) can be divided into stages:

- *Attachment* – virus particle attaches to specific proteins in the plasma membrane.
- *Penetration* – plasma membrane forms an intucking, enclosing the virus particle in a vesicle.
- *Uncoating* – the capsid is digested away, exposing the viral nucleic acid.
- *Biosynthesis* – the host cell nucleic acid is used to make more viral nucleic acid, followed by capsid proteins.
- *Maturation* – viral nucleic acid and proteins are assembled into new virus particles.
- *Release* – virus particles are transported to the plasma membrane and then released by the bursting open of the host cell. In some viruses, the new virus particles become coated by plasma membrane, forming the envelope, before being released. Only the intact virus particle is capable of infecting another cell.

The life cycle of a virus that attacks bacteria differs in some details – eg it injects its nucleic acid *through* the cell wall rather than being taken in by infolding of the plasma membrane.

Supplementary Unit Activity 1A: Micro-organisms

1. a. Complete the following table with simple drawings to describe the shape of these bacterial groups. One column has been completed for you.

Type	Bacillus	Vibrio	Spirillum	Coccus
Shape				

b. Many bacteria are pathogenic; they cause disease. Give an example of a pathogenic bacteria and the disease it causes.

c. Some bacteria are classified as saprophytes, and others as parasites. Explain the difference between these groups.

d. Bacterial colonies often grow very quickly.

i. Sketch a graph below to show the population growth curve of a colony of bacteria.

ii. Discuss how the reproduction of bacteria leads to rapid growth, and why such growth cannot continue indefinitely.

Supplementary Unit: Micro-organisms

Topic 2: Human responses to micro-organisms

Authors: Martin Hanson with Takis Solulu, Beatrix Marjen-Waiin and Diaiti Zure

Much of the content of this Topic is not in the Syllabus but it is important in terms of relating the study of Biology to real life, real issues and the local environment. This Topic deals with:

- How disease-causing organisms are prevented from entering the body.
- Natural and acquired immunity.
- The human immune system and its components.
- Defences against disease (vaccine, passsive immunity).
- Symptoms – inflammation.

Parasites and pathogens

Most illnesses are *infectious* – ie they can be passed on from one person to another. Infectious diseases are caused by **parasites**. The parasites we hear most about are the 'less successful ones' – those that harm their hosts enough to cause *disease*. These parasites are called **pathogens**.

The most important pathogens are bacteria, viruses, single-celled eukaryotes, fungi and various kinds of parasitic worms.

Just as parasites are adapted to live in and on their hosts, so too are hosts adapted to resist their parasites – there is a constant evolutionary 'arms race' between them. The most successful parasites are those that have evolved ways of exploiting the host by doing least harm to the host (ie their food supply). The most successful hosts are those best at resisting parasites.

Humans have two basic ways to protect themselves from pathogens:

- Preventing them getting into the body in the first place.
- Killing those that do manage to get in.

How pathogens are discouraged from entering the body

To reach its food supply, a pathogen has to penetrate the various outer defences of the body.

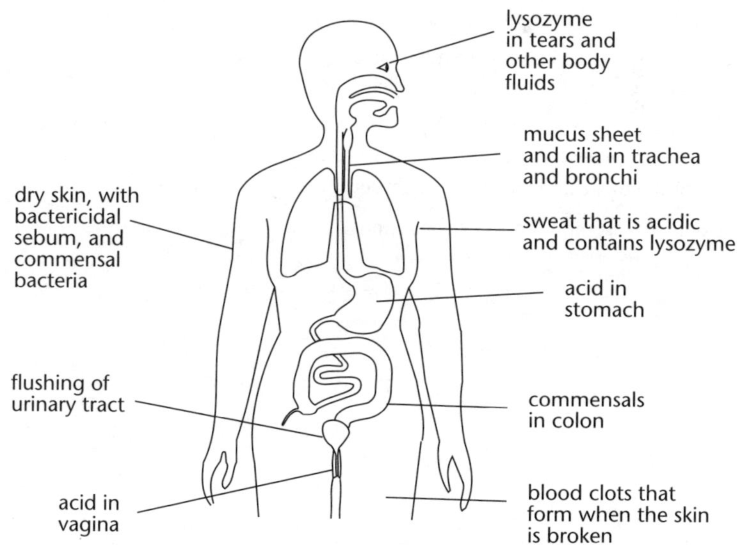

The outer defences of the body.

The inner body tissues are separated from the outside world by protective layers called **epithelia**. An epithelium is a layer of cells that forms a covering of an outer surface such as the skin, or the lining of an internal cavity such as the linings of the gut and lungs. All epithelia have a protective function.

- The outer layer of skin (epidermis) contains a tough, water-resistant protein called **keratin**. The greasy **sebum**, secreted by the sebaceous glands, has an anti-microbial effect, as does the wax secreted by the skin lining the outer ear cavity.
- The stomach secretes an extremely acidic **gastric juice** (pH 1–2) which kills most bacteria (though not agents of food-borne diseases). The rather less acidic conditions in the vagina also inhibit the growth of many micro-organisms.
- The lining of the nose, trachea and bronchi secretes a layer of **mucus** that traps bacteria and dust. Tiny **cilia** beat to push the trapped pathogens up to the nose or throat, from where they are blown out of the nose or swallowed.
- Tears and some other body fluids contain an enzyme called **lysozyme** which digests the cell walls of many bacteria.
- **Interferon** is a protein produced by cells infected by viruses – it acts to prevent reproduction of viruses after entry to a host cell.

Fever results from the release by white blood cells of **pyrogens**. Pyrogens alter the setting of the 'thermostat' in the brain, and the resulting raised body temperature is believed to inhibit the growth and reproduction of some micro-organisms (eg syphilis bacteria).

'Friendly' bacteria

We gain some protection from certain harmless bacteria that live on our skin or in our large intestine, which are indirectly beneficial because they prevent pathogens getting established. If these bacteria are harmed, parasitic fungi 'may get a hold'.

'Thrush', a fungal infection, often strikes after a course of antibiotics has killed bacteria, including those bacteria that normally keep the fungus 'in check'.

There are also large numbers of bacteria in the large intestine, some of which successfully compete with potential pathogens.

The gut and the lungs seem ideal entry routes for pathogens – both have large areas with a very thin, moist lining, but they are actually well defended.

- The acid in the gastric juice secreted by the stomach kills most bacteria – though not all, because some diseases are spread by drinking water and food.
- The lining of the breathing passages is guarded by a layer of mucus.

Despite these defences, some pathogens sometimes get in.

Route of entry	Kind of pathogen	Disease
Breathing passages	bacteria	diphtheria, tuberculosis
	viruses	influenza, measles, chickenpox
Gut	bacteria	cholera, typhoid
	viruses	hepatitis A, mumps, polio
	*unicellular eukaryotes	giardia
Reproductive tract	bacteria	gonorrhoea, syphilis
Skin wounds	bacteria	tetanus, plague
	viruses	rabies, yellow fever
	*unicellular eukaryotes	malaria

* 'unicellular eukaryotes' represent all other unicellular organisms other than bacteria

Various ways in which pathogens enter the body.

Immunity

Immunity includes a variety of ways in which the body can resist infection by pathogens. It is quite distinct from *non-susceptibility*, which results from micro-organisms being unable to harm the body because some condition essential for their growth and reproduction is *absent* (eg humans are not susceptible to rabbit calicivirus (RCD) because human cells lack the specific proteins in their surface membranes that the virus needs to gain entry).

It would be incorrect to say that humans are *immune to RCD*. Whereas non-susceptibility is due to the *absence* of something, immunity is due to the *presence* of something.

The immune system fights pathogens in two different ways:

- 'Hand-to-hand combat' – white cells attack bacteria and kill them at close quarters.
- 'Chemical warfare' – uses special defence proteins called *antibodies*. These are carried by the blood throughout the body and thus can act a long way from where they are produced.

There are many mechanisms involved in immunity, but they fall into two general categories – *natural* (or innate) immunity, and *acquired* immunity.

Natural immunity

The various mechanisms of natural immunity share two distinguishing features:

- They are *non-specific* – each mechanism acts against a variety of pathogens.
- They are *present from birth* – they do not develop as a result of exposure to a pathogen.

Many of the mechanisms of natural immunity are concerned with preventing the entry of pathogens into the body, and thus represent the first line of defence.

Acquired immunity

Acquired immunity has three important features:

- It only develops *after* contact with a pathogen.
- It is *specific* to particular kinds of micro-organism (eg immunity to measles does not protect against chickenpox).
- It has *memory* – a second exposure to a pathogen produces a more powerful response than the first one.

Parts of the immune system

The immune system extends throughout the entire body and consists of cells in the **bone marrow**, **lymph nodes**, **thymus**, and **spleen**.

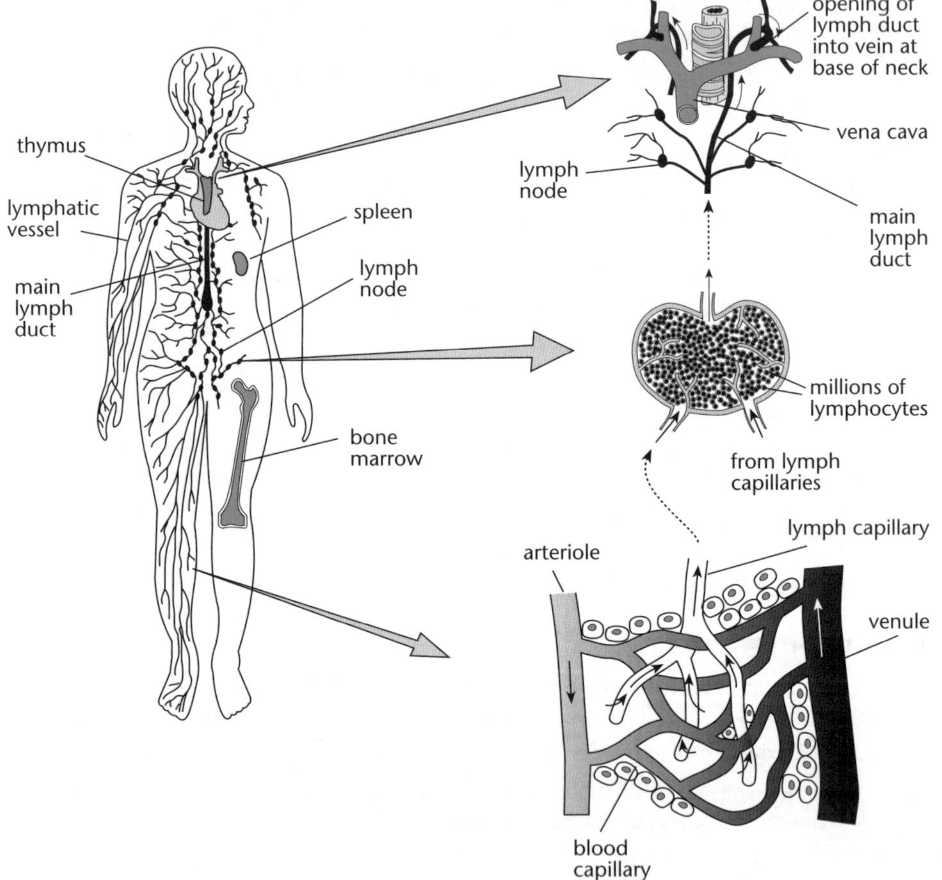

The 'twigs' and 'trunk' of the lymphatic tree are shown enlarged.

The immune system.

Lymphatic system

The lymphatic system acts as a kind of drainage system for tissue fluid. The lymphatic vessels all converge to two main vessels, which open into the veins near the base of the neck, returning the tissue fluid back to the blood. Lymph vessels have valves, which ensure one-way flow during body movement.

At various points the lymph passes through **lymph nodes**. These nodes are packed with lymphocytes and phagocytes and act as filters, since all the lymph has to pass through them on its way back into the blood. During infection, the number of white cells greatly increases, causing the lymph nodes to swell and become painful ('swollen glands').

Bone marrow

The bone marrow is where the immune cells are produced. These consist of various kinds of white blood cell or **leucocytes**:

- *Phagocytes* – eat bacteria, enclosing them in small spaces called vacuoles, in which they are digested. Phagocytes are transported to sites of infection by the blood.

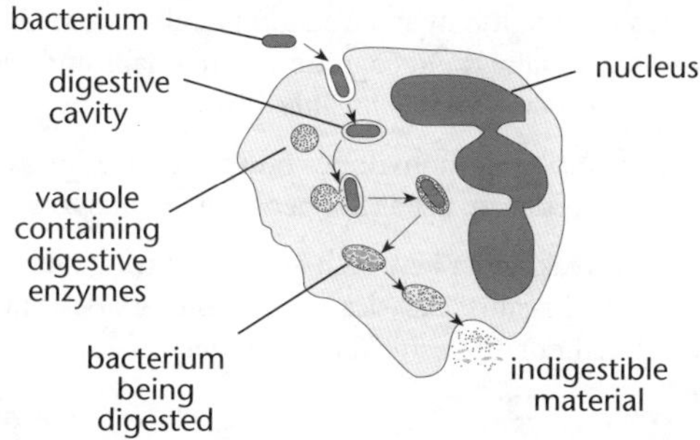

A phagocyte digesting a bacterium.

- *Lymphocytes* – these small, insignificant-looking cells are present in huge numbers in the lymph nodes and spleen. If they come into contact with an appropriate antigen (foreign substance) they become activated. Lymphocytes are 'born' in the bone marrow, but before they can be of use they have to go through a maturation process. They are of two main types – B and T lymphocytes.

B lymphocytes

These mature in the bone marrow (think of 'B' for **b**one marrow) before travelling in the blood to the lymph nodes and spleen. On activation, some B lymphocytes develop into **plasma cells** which secrete antibodies.

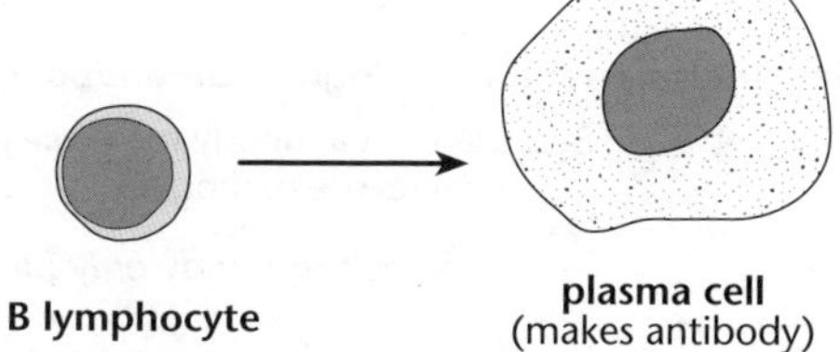

Lymphocyte and plasma cell.

Other activated B lymphocytes become *memory cells*, ready 'to spring into action' if there is a second attack. This is the reason why we don't normally get the same infectious disease twice.

T lymphocytes

These mature in the **thymus gland** above the heart ('T' for **t**hymus), and then travel in the blood to the lymph nodes and spleen.

- On activation, some develop into cells that kill virus-infected cells or cancer cells.
- Others develop into cells that help other immune cells to function.
- A third group remain 'on watch' as memory cells. (One type of T cell is attacked by HIV, so people affected by HIV lose their ability to fight disease.)

Antibodies and antigens

Many body chemicals are the same in all of us (eg haemoglobin), but some are unique to each individual. It is these unique chemicals that enable the immune system to distinguish the body's own cells ('self') from those of an invader ('non-self').

The surface of every bacterium or virus contains substances that the body can recognise as foreign. These stimulate T lymphocytes to develop into plasma cells which produce special defence proteins called **antibodies**.

Any substance that stimulates production of an antibody is called an **antigen**. The vast majority of antigens are foreign, but occasionally the recognition system fails and the body produces antibodies to its *own* chemicals, causing an *autoimmune* disease.

Most antigens are proteins or complex carbohydrates on the surfaces of bacteria and viruses. The toxic waste products of some bacteria can act as antigens (eg tetanus toxin, diphtheria toxin).

Antibody and antigen are like a lock and a key, each antibody fitting only one antigen. This explains why antibodies produced against measles virus cannot work against chickenpox – an antibody to the measles virus will not 'fit' a chickenpox antigen.

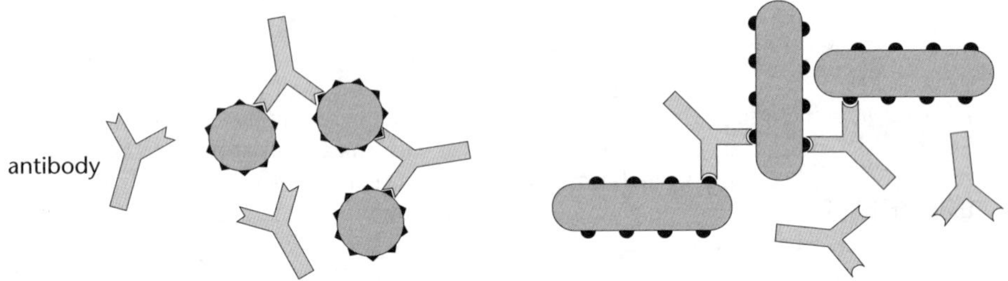

Meningitis bacteria + meningococcus antibody **Typhoid bacteria + typhoid antibody**

Because each antibody molecule has more than one binding site, it can cause pathogens to clump together (agglutinate).

Each antibody only binds with one antigen.

By binding with an antigen, antibodies can have five kinds of effect:

- Bacteria covered with antibody are more 'palatable' to phagocytes, which are stimulated to 'eat' them.
- Bacterial toxins are neutralised by antibodies. Antibodies that do this are called **antitoxins**.
- Antibodies may bind to bacterial flagella, immobilising the bacteria and making it harder for the bacteria to invade new tissues.

- Antibodies may cause bacteria to clump together (agglutinate).
- By binding to the protein coats of viruses, antibodies prevent the virus entering host cells.

Mutating viruses

Some diseases (eg influenza) can be caught more than once. This is because the pathogen exists in several different types or *strains*. The influenza virus mutates from time to time, producing a new strain to which there is no previous immunity. It takes time for the body to make antibodies against each strain of virus, by which time a new strain may have arisen.

Supplementary Unit Activity 2A: Antibodies, antigens, lymphocytes and lysozymes

1. For each of the phrases **1–15**, write the letter **A–O** of the term to which it applies.

Phrase	Applied term
1. A substance, normally foreign to the body, that stimulates production of antibody	A. Antibody
2. An antibody that neutralises a poisonous substance produced by a pathogen	B. Antigen
3. Anti-microbial secretion produced by sebaceous glands	C. Antitoxin
4. In the absence of micro-organisms	D. Bone marrow
5. Defence protein produced by plasma cell, combines with a specific foreign substance (antigen)	E. Epithelium
6. Enzyme that digests cell walls of bacteria	F. Interferon
7. Organ where T lymphocytes mature	G. Leucocyte
8. Parasite that causes disease	H. Lymph node
9. Protective slime secreted by lining of breathing passages	I. Lymphocyte
10. Secretes antibody	J. Lysozyme
11. Small swellings on lymphatic channels, packed with phagocytes and other immune cells	K. Mucus
12. Small white blood cell that becomes activated after contact with antigen	L. Pathogen
13. Tissue that acts as a protective barrier to infection	M. Plasma cell
14. Where immune cells are produced	N. Sebum
15. White blood cell	O. Thymus

2. Explain why a person never becomes immune to the common cold.
3. a. Young children are vaccinated against polio. Explain how the human body achieves active immunity from this vaccination.
   b. A child has recovered from measles. She has immunity from measles and is unlikely to suffer measles again. Explain how the child's immune system is now able to provide her with immunity from measles.
   c. Discuss how a mother provides her newborn baby with immunity to some diseases during the first few months of its life.

4. The diagram shows some of the places where the body's natural defences work against bacterial infection.
Discuss how the body's natural defences prevent bacteria from entering the bloodstream. In your answer, you should refer to at least three of the places shown in the diagram.

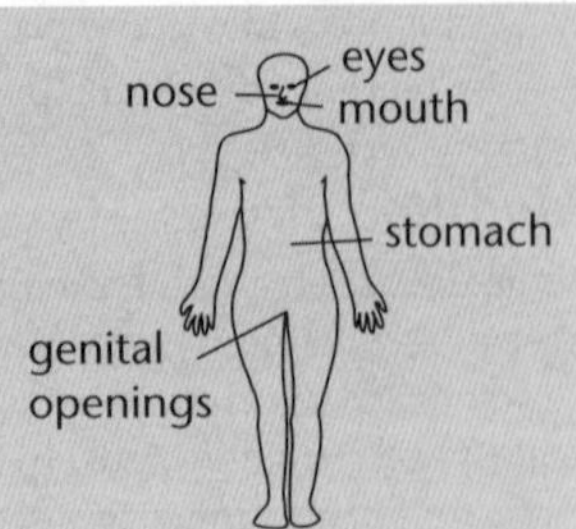

Defence against disease

Vaccines

The ability of the immune system to 'remember' previous infections is the basis for **vaccination**.

Vaccination usually involves the introduction into the body of either a weakened (*attenuated*) pathogen or one that has been killed without losing its ability to stimulate antibody production.

Another kind of vaccine consists of a **toxoid** – a toxin that has been chemically treated to render it harmless, but which retains its antigenic properties – eg tetanus toxoid.

Passive immunity

Antibodies are obtained from another organism. Since all proteins in the blood (including antibodies) are normally continually broken down and re-synthesised, passive immunity only lasts a few weeks.

Antibodies can be introduced into the body in two ways:

- From the mother via the placenta. A newborn baby is thus temporarily immune to all the diseases to which its mother is immune. In some mammals, antibodies are present in the first milk to be secreted (the **colostrum**).
- By injection of antibodies made by an animal (eg anti-tetanus antibodies made by a horse). This is done after *possible* infection, as a *precaution*. Repeated doses of antibody made by another animal can be dangerous, because these antibodies are themselves 'foreign', so they provoke an immune response. This tends to become more powerful with successive doses and can cause a kind of shock. It is therefore advisable to obtain active immunity to tetanus by vaccination rather than repeatedly relying upon passive immunity.

There are two types of protective injection:

- A preventative one given before infection and leading to active immunity.
- Given after possible infection, leading to passive immunity.

Boosters

When a pathogen is first introduced into the body, there is a delay of several days before any antibody can be detected in the blood.

Peak antibody levels are reached after 2–3 weeks and then decline to zero. After a second infection there is a shorter delay and antibody level rises more steeply, reaching a higher level and lasting longer. This is the basis for the 'booster' dose in vaccination.

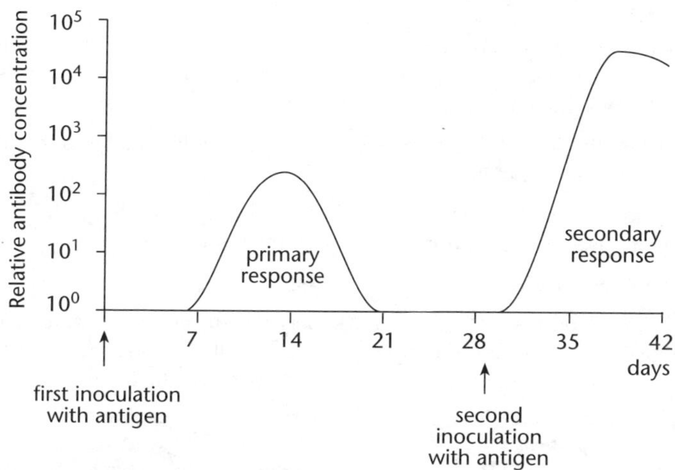

Note logarithmic vertical scale.

Primary and secondary responses to vaccination.

Summary of types of immunity

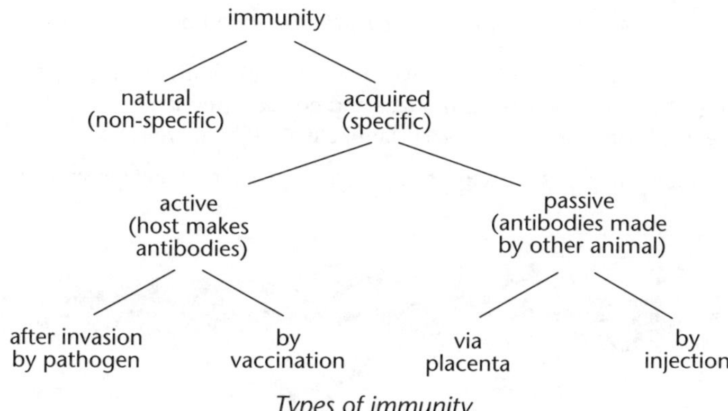

Types of immunity.

The importance of vaccination

Despite the proven benefits, some parents still do not have their children vaccinated against whooping cough, measles and diphtheria.

- Some people believe that vaccination does not protect against disease.
- Some are fearful of the rare, deleterious side-effects of some vaccines, such as measles. (Harmful effects of measles itself are more common and more severe.)
- Some people 'cannot be bothered'.

Provided that a high proportion of the population is immunised, epidemics are unlikely because the chances of a pathogen finding a susceptible host are greatly reduced.

Inflammation

Inflammation is a defensive response to tissue damage resulting from mechanical injury, harmful chemicals, or microbial attack. Even if damage is purely mechanical, it can render tissues more vulnerable to pathogens. Inflammation prepares the body to deal with such localised dangers.

Inflammation enables the cells of the immune system to be rapidly brought to the affected area by the blood and concentrated there.

Inflammation has four symptoms – redness, swelling, heat, and pain. These effects are due to the release of chemicals such as **histamine**, by cells at the injured site.

Histamine has a number of effects:

- Widening of the blood vessels (vasodilation), especially the arterioles, which increases the blood flow to the area and thus the supply of defence cells. The increased pressure in the capillaries increases the formation of tissue fluid, causing swelling.

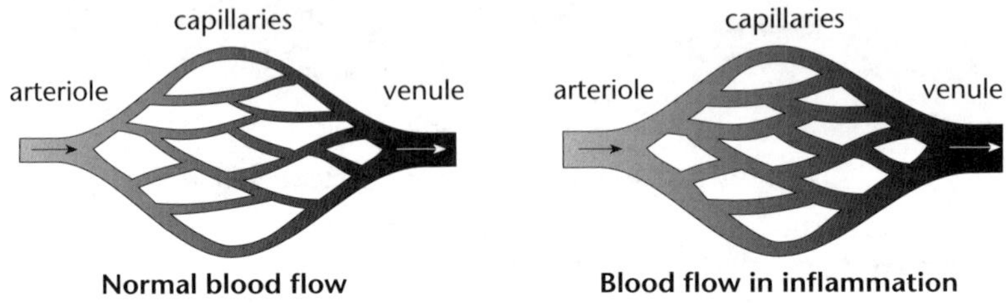

Increased blood flow occurs during inflammation.

- It makes capillaries more 'leaky' than usual, enabling antibodies to escape into the surrounding tissues. Antibodies are proteins and normally too large to get through the tiny pores between the capillary wall cells. Histamine makes these pores larger.

Other chemicals stimulate phagocytes to leave the blood by squeezing between the cells lining the venules.

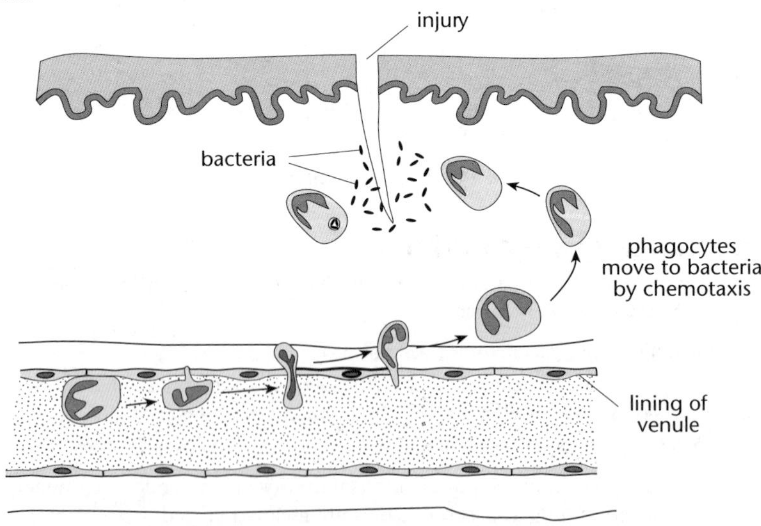

Phagocytes leaving venule in an inflamed area.

Phagocytes migrate towards chemicals released by cells at the source of infection (chemotaxis). The arrival of so many phagocytes may eventually be visible as an area of whitish **pus**. Another effect is the increased sensitivity of nerve endings, causing pain, which helps to protect the affected part from further injury.

Supplementary Unit Activity 2B: Defence against disease

1. A scratch on the hand from a cat can cause an inflammatory response.

- **a.** Give two symptoms of this inflammatory response.
- **b.** Discuss what happens during an inflammatory response to help stop the spread of pathogens through the body.

2. Hepatitis B is a disease caused by the hepatitis B virus (HBV). Infection with this virus can cause liver damage. A vaccine can be given to people to protect them from this virus. The hepatitis B vaccine is given as *a series of 3 injections over 6 months.*

- **a.** Describe why more than one injection is needed.
- **b.** Discuss how the hepatitis B vaccination works to prevent a person from being infected with the virus.

3. **a.** **i.** Describe what is meant by the term 'pathogen'.

- **ii.** Describe three common symptoms that indicate pathogens have entered the body.

b. John cuts his hand on a rusty nail in the backyard. Pathogens have entered the bloodstream.

- **i.** Explain how the body responds to pathogens in the blood.
- **ii.** John's hand becomes infected. Explain why lymph nodes under his armpit have become swollen.
- **iii.** Explain why John is advised to have a tetanus vaccination.

4. **a.** Hoani needs a new tetanus vaccine. This is an example of passive immunity. Describe passive immunity.

- **b.** Explain why a person needs to be vaccinated against tetanus every twenty years.
- **c.** A child has gained active immunity after recovering from chickenpox. Explain how activity immunity is gained.
- **d.** The diagram on page 352 shows the immune response that occurred when the chickenpox virus first entered the child's body.

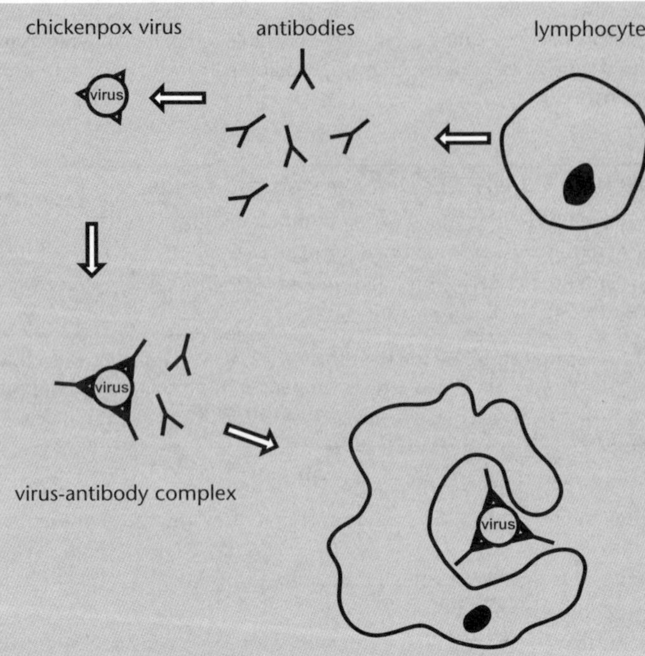

i. Describe the role of the lymphocyte.

ii. Discuss how antibodies help the body to gain immunity to chickenpox.

5. a. The human body uses defence mechanisms to stop pathogens from entering. Explain how defence mechanisms in the mouth or throat prevent pathogens from entering the body.

b. A common cause of food poisoning is *Salmonella* bacteria. A person suffering from *Salmonella* poisoning may have diarrhoea, stomach cramps, vomiting, nausea and fever.

i. Describe one method to prevent the spread of *Salmonella* bacteria.

ii. Give two possible symptoms of food poisoning.

iii. Discuss why the body could not stop the *Salmonella* from causing food poisoning.

Supplementary Unit: Micro-organisms

Topic 3: Respiration and gas exchange in micro-organisms

The content of this Topic is an extension of Unit 11.4 Respiration and Gas Exchange.

A microbial cell makes a whole organism, in contrast to other living organisms that are made up of multiple cell types (multicellular). Therefore, movement of gases, particularly oxygen and carbon dioxide, across the cell membranes can take place by passive diffusion or, in some cases, facilitated diffusion. Gaseous exchange is vital for the metabolic reactions to take place within the microbial cell.

Micro-organisms are tiny single-celled organisms that do not have any specialised breathing organs. Oxygen is carried across the plasma membranes and cell walls for those that use it.

Microbes differ in their needs for oxygen. Obligate or strict aerobes require oxygen. In contrast, obligate or strict anaerobes are poisoned by oxygen. Facultative anaerobes are more adaptable than the aerobic and anaerobic microbes. They can use oxygen when available in aerobic respiration processes. They can use anaerobic respiration or fermentation as an alternative means if oxygen is not available.

Aero-tolerant microbes do not use oxygen, but are not poisoned by the presence of oxygen. Micro-aerophilic microbes use oxygen in very small quantities.

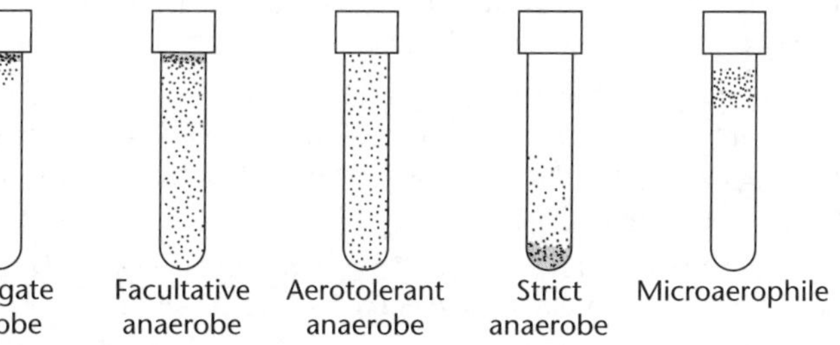

Test for oxygen needs of microbes in test tubes.

Aerobic respiration requires oxygen, while anaerobic respiration does not need oxygen, but needs other inorganic molecules.

Anaerobic respiration in yeast and bacteria

In yeast and bacteria, anaerobic respiration is a way of life. In yeast, the pyruvate is broken down into *ethanol*, C_2H_5OH, and *carbon dioxide*. This process is also known as **fermentation**. The small energy yield of 2ATP molecules is sufficient to meet the lifestyle needs of these organisms.

Word equation

glucose ⟶ ethanol + carbon dioxide + ATP + heat

Formula equation

$$C_6H_{12}O_6 \longrightarrow 2C_2H_5OH + 2CO_2 + 2ATP + heat$$

Fatty acids and *glycerol* (from the digestion of fats), and *amino acids* (from the digestion of proteins), can also be used in respiration (entering the Krebs cycle after prior processing) to generate ATP.

Some species like single-celled green algae and photosynthetic bacteria behave like plants and carry out photosynthesis with their very own photosynthetic pigments. These photoautotrophic micro-organisms are responsible for production of over half of the earth's atmospheric oxygen. The remainder is produced by green plants.

Many other photosynthetic eubacteria that do not have chlorophyll a or b use other types called bacteriophylls. These eubacteria belong to the Purple and Green groups of bacteria that do not use water molecules as well in their photosynthesis, and therefore cannot produce molecular oxygen as a by-product. Photosynthesis for the Green and Purple bacteria is strictly anaerobic.

Many protista and fungi are aerobic microscopic organisms that do not have specialised organelles for gas exchange, making the general cell membrane serve as the gas exchange surface.

In the bacteria, cellular respiration takes place on the inner part of the membrane called the mesosome. In protists and fungi, the same respiration takes place in the cytoplasm and mitochondrial matrix.

Most other species of micro-organisms that do not use oxygen use alternative sources to complete their metabolic processes. These microbes utilise anaerobic respiration or fermentation. They are largely called chemoheterotrophs.

Anaerobic respiration and fermentation is used by micro-organisms in waterlogged or very deep soil depths, ocean bottoms, stagnant ponds or animal intestines. Inorganic substances like nitrate or sulphate are used by the microbe as a substitute of molecular oxygen as the terminal acceptor in the electron transport chain.

The fermentation process does not need the electron transport chain step. Certain bacteria and some fungi (yeasts) utilise this system to yield enough energy to sustain their metabolic needs.

Certain types of bacteria (*Rhizobium, Azotobacter, Cyanobacteria*) have the ability to fix atmospheric nitrogen via structures called heterocyst in anaerobic conditions. Often times they cause 'blooms' when they grow out of control, making the water appear green and giving it a foul odour. Archaebacteria species that live in hot springs, volcanic mouths, deep sea vents, extreme dry conditions, extreme salty conditions, bogs, marshes, swamps and underground all use some form of anaerobic respiration or fermentation process in their cellular metabolism.

Answers for many questions include a 'Marking Guide':

- **A** ('Achievement', meaning 'satisfactory achievement'). **A** answers are for *descriptions*.
- **M** ('Merit', meaning 'high achievement'). **M** answers are for *explanations* (but it is possible to be given **A** if an answer includes a description without an explanation).
- **E** ('Excellence', meaning 'very high achievement'). **E** answers are for *discussions* (but it is possible to be given **M** if the answer only explains or an **A** if the answer only describes). **E** answers require a *discussion* – bullet points are not appropriate.
- Note: '/' in an answer (eg feature/trait/protein) refers to acceptable alternative answers.

We hope this Marking Guide will be a help to students who are striving for the best possible results.

Unit 11.1 Activity 2A: Structure and function of cells (Page 23)

1. Plant cells have a (cellulose) *cell wall* surrounding the cell membrane. They may also have chloroplasts and large vacuoles in their cytoplasm. (***A***)

2. **a.** Both cilia and flagella are microtubules that extend from the cytoplasm (of unicellular organisms) and are capable of movement; however, cilia are much shorter and more numerous than flagella. (***A***)

b. Mitochondria are cell organelles associated with respiration, chloroplasts are cell organelles associated with photosynthesis. (***A***)

c. Rough ER has ribosomes attached to its membranes, smooth ER does not have attached ribosomes. (***A***)

d. Ribosomes are the site of protein synthesis, lysosomes are vacuoles that contain enzymes. (***A***)

e. Vacuoles are storage sacs in the cytoplasm of cells, contractile vacuoles are found only in unicellular organisms and are the organelles responsible for osmoregulation/collect excess water and expel it to the outside. (***A***)

3. Cytoplasm is made of a fluid called cystol which is mainly water containing many dissolved substances (eg sugars, mineral ions). (***A***)

4. **a.** **i.** Golgi body – responsible for the modifying, packaging, and excretion of proteins made by the rough ER (has bound ribosomes). (***A*** – both name and function)

ii. Chloroplast – site of photosynthesis (in plant cells). (***A*** – both name and function)

b. Ribosome – site of protein synthesis. (***A*** – both name and function)

5. **a.** Cell membrane of each cell has been greatly increased/extended /elongated/SA increased. (***A***)

b. The extended cell membrane *increases its SA* for absorption (***A***); which allows both cells to *increase the (rate of) diffusion into the cell* of needed substances (eg mineral ions for root hair, nutrients for intestinal cell). (***M***)

6. **a.** Cilia and flagella. (***A*** – need both)

- **b.** Movement needs *energy* from the *mitochondria* (***A***). Energy is released from (aerobic) *respiration* which occurs in the mitochondria. The greater the need for energy, the more mitochondria a cell will have. (***M***)

7. **a.** Rough ER differs from smooth ER because it has ribosomes attached to its membranes (and smooth ER doesn't). (***A***)

- **b.** Ribosomes make proteins and the cells of the pancreas make (large amounts of) proteins, but the gamete-producing cells do not. (***A***) The cells of the pancreas are actively *making proteins called enzymes*, which are secreted into the small intestine for the digestion of foodstuffs – therefore they need a large number of ribosomes/rough ER. Could also mention the production of hormone insulin (which is a protein) by pancreatic cells. Gamete-producing cells are not protein secretory cells and do not produce large amounts of protein, therefore do not need rough ER. (***M***)

8. **a.** Ribosomes are located on the ER. (***A***)

- **b.** The cells make a large number of proteins and secrete them. (***A***) Large numbers of ribosomes are needed by cells that make large amounts of proteins, and the Golgi body modifies and packages the proteins into vesicles for secretion. Therefore, the cells of the small intestine are *actively making and secreting proteins* (ie the enzymes for digestion in the small intestine). (***M***)
- **c.** The microvilli increase the *SA* of the membrane (***A***); this allows the *rapid secretion* of (large amounts) of the enzymes needed in digestion. (***M***)
- **d.** The cell is likely to have large numbers of mitochondria as it needs *large amounts of energy* for these processes/making proteins/secreting proteins. (***A***) The cell makes and secretes large amounts of protein/enzymes for digestion. Synthesis and secretion of proteins needs energy; therefore, large numbers of mitochondria are needed in the cell to supply the energy demands. (***M***)

9. The cell walls are for *support* and the vacuoles are for *storage*. (**A** – need both) The large central vacuole acts as a *fluid skeleton* when *the cell is full of water/turgid* and, together with the rigid cell wall, is the support material/system for non-woody plants. (***M***)

10. The cell membrane is very thin and made of a bilayer of phospholipid molecules (made of a hydrophilic 'head' of glycerol-phosphate and a hydrophobic 'tail' of fatty acids). Within the bilayer are large transport proteins which may extend partially or all the way through the bilayer. On the outside of the bilayer are special receptor proteins. The function of the membrane is the transport of substances into and out of the cell and to maintain the internal environment of the cell. (***A*** – need a description of both the structure and the function) The membrane is semipermeable, allowing only certain substances to pass through it. Small molecules (eg O_2) diffuse through the membrane in passive transport. Large molecules need to be *actively* taken in using the transport carrier proteins in the membrane *or* by the infolding of the membrane in phagocytosis. Special transport proteins may facilitate the diffusion of molecules (eg glucose). Receptor proteins on the surface recognise self from foreign and provide receptors for molecules such as antibodies and neurotransmitters. (***M*** – relates structure to transport of materials and semipermeable nature) The bilayer is fluid and allows the membrane to move (eg phagocytosis and pinocytosis); break and rejoin (eg vesicle secretion); reassemble itself (eg cell division), and repair damage. (***E*** – comprehensive answer details structure and function, *linked*)

Unit 11.1 Activity 3A: Movement of materials (Page 34)

1. **a.** Passive transport is the movement of substances across the membrane from high to low concentration and does not require energy (eg diffusion, osmosis); active transport is the movement of substances across the membrane from low to high concentration and/or large substances which needs energy (eg exocytosis and endocytosis). (***A***)

- **b.** Diffusion is the (passive) movement of substances from high to low concentration; osmosis is the (passive) movement (diffusion) of water from high to low concentration across a SPM. (***A***)
- **c.** Hypertonic refers to solutions with large amounts of dissolved solute; hypotonic refers to solutions with small amounts of dissolved solute. (***A***)
- **d.** Dilute solutions have small amounts of dissolved solute; concentrated solutions have large amounts of dissolved solute. (***A***)
- **e.** Exocytosis is the process of removing (bulk) substances out of the cell by active use of the cell membrane; endocytosis is the process of taking (bulk) substances into the cell by active use of the cell membrane. (***A***)
- **f.** Phagocytosis is the active process of engulfing large substances by enfolding them with the cell membrane; pinocytosis is the active process of 'drinking' liquid substances, using invaginations of the cell membrane. (***A***)

2. Facilitated diffusion *speeds up the diffusion* of substances across the cell membrane by using the *transport/carrier proteins* in the membrane. These proteins are specific in their action, and the process does not require energy (is passive). (***A***)

3. **a.** Both the ions are moving by diffusion from high to low concentration. (***A***) This is because the concentration of K^+ is 152 inside the cell and 9 outside the cell, so K^+ moves from the higher concentration inside to the lower concentration outside. The concentration of Na^+ is 146 outside and 18 inside, so Na^+ diffuses in from the higher concentration to the lower concentration. (**M** – figures support answer)

- **b.** The water moves into the red blood cells from osmosis, causing them to burst. (***A***) This is because the cytoplasm and plasma were isotonic so there was no net diffusion of water; but the distilled water is hypotonic, causing the net inflow of water in osmosis (hence the bursting of the cells). (**M**)

4. Small cells have a high SA:V ratio, and large cells have a low SA:V ratio – so smaller cells have *more efficient diffusion* than larger cells. (***A***) This is because as the cell grows, its volume (the cytoplasm) grows faster than its SA (ie the cell membrane). The amount of diffusion is limited by the amount of membrane, so the size of the cell is limited by the ability of needed substances to diffuse into the cytoplasm through the membrane. (**M**)

5. Water moves by osmosis across the SPM from the right-hand side to the left-hand side, while solute diffuses across the SPM from the left-hand side to the right-hand side. (**A** – need both water and solute) This is because water is more concentrated on the right-hand side and the solute is more concentrated on the left-hand side, so the direction of *net* movement is to the lower concentration for each (solute particles are small enough to pass through the cell membrane). (**M**)

6. Bag X will shrivel up as water leaves, bag Y will stay the same, bag Z will expand as water enters. (***A***) Bag X shrivels up, as water moves by osmosis out from higher concentration inside the bag than in the beaker (solution in beaker is hypertonic); bag Y will stay the same, as water concentrations are the same in the bag and beaker (solutions are isotonic); bag Z will expand, as water moves by osmosis into the bag from higher concentration (solution in the bag is hypertonic) in the beaker. (**M**)

7. Water moves by osmosis across SPM from the right-hand side to the left-hand side, potassium chloride diffuses across SPM from left-hand side to the right-hand side, sucrose doesn't change/move. (***A***) Water moves from the right-hand side to the left-hand side as it is more concentrated on right-hand side; potassium chloride moves from the left-hand side to the right-hand side as it is more concentrated on the left-hand side; sucrose doesn't move as it is too large to diffuse across the SPM. (***M***) Both water and potassium chloride can move across the SPM, as they are small substances – unlike sucrose. Both substances move in *both* directions across the SPM, but the *net movement* is from high to low concentration. The net movement of water into arm X will cause *the level of the solution in that arm to rise* (and that in arm Y to fall). Net movement of the substances will stop when *concentrations on either side of the membrane are equal* (isotonic). (***E*** – in-depth explanations which consider the net movement of particles and the result of these)

8. **a.** Diagram needs to show plasmolysed cell, with cell membrane and contents pulled away from the cell wall. Cell wall, membrane, cytoplasm need labelling. (***A***)

b. Potato cubes J, K, L increased in weight, as their cells have gained water by osmosis; potato cube M has not lost or gained weight; potato cubes N and O have lost weight, as their cells have lost water in osmosis. (**A** – refers to all potato cube samples)

The solute concentration of the cytoplasm of potato cells in J, K, L is higher than in the salt solutions that the potatoes are in, therefore water moves from the salt solution into the cells and the potato cubes gain weight. In M, the concentrations of the cytoplasm of the potato cells and the salt solution are equal, so there is no net movement of water into or out of the cells, so there is no change in weight. In N and O, the solute concentration in the salt solution is now higher than that of the cytoplasm, so water moves out of the cells into the beaker and the potato cubes lose weight. (**M**)

At M, there is no net loss/gain of water, therefore the cytoplasm and salt solution are isotonic at 1.5% concentration of solute. Salt concentrations above 1.5% are hypertonic, so the water moves out of the cells; salt concentrations below 1.5% are hypotonic, so water moves into the cells. (***E*** – discussion is in depth, with reference to the *net* movement of water into/out of the cells and relates the loss/gain of water *to the data given on the graph* (% salt concentrations and weight loss/gain in grams))

9. **a.** High level of iodine in the cells is maintained by *active transport* from sea water (**A**); with transport proteins in the cell membrane using energy to take up the iodine *against a concentration gradient* from the sea water. (**M**)

b. Active transport of iodine into the cell needs energy from respiration – without respiration, this won't occur. (**A**) Without the process of active transport, iodine will *diffuse across the cell membrane in response to its concentration gradient*. If iodine is in higher concentration in the cytoplasm, it will diffuse out, and the level of iodine in the cytoplasm will decrease (and vice versa). (**M**)

10. Movement of glucose from the tubule into the cells is by *active transport*, which needs *energy* from mitochondria. (**A**) Uptake of glucose is against a concentration gradient (as *all* the glucose needs to be reabsorbed), and involves large amounts of glucose. This needs large amounts of energy – hence the large numbers of mitochondria in the tubule cells. (**M**)

Unit 11.1 Activity 4A: Enzyme activity (Page 40)

1. Enzymes speed up the chemical reactions in cells. (**A**) Enzymes are called 'biological catalysts' because they are found in the cells of *living things*/organisms and function to *speed up and control* the chemical/metabolic reactions of the cell. (**M**)

2. The substrate molecules A and B fit into the enzyme C the same way that a lock fits into a key, so get joined together. (**A**)

The shapes of substrate molecules A and B *correspond to that of the active site* of the enzyme C, in the same way that the shape of a key corresponds to that of the lock. An 'enzyme-substrate complex' is formed, and A and B bond together to form a new compound, D. (**M**)

3. Enzymes are essential to the survival of organisms because they *catalyse all the important reactions* that take place in the cells of the organism (eg respiration, photosynthesis). (***A***) Without the catalytic effect of enzymes, these essential chemical reactions would take place *too slowly* for life to exist. (**M**)

4. Co-factors are molecules or ions needed to *assist an enzyme in catalysing its reaction*. (***A***) Co-factors may bond to the active site of the enzyme along with the substrate, locking the enzyme and substrate more closely and promoting the formation of the product. (**M**)

5. Most reactions catalysed by enzymes occur within cells, so work best at a neutral pH of 7. (***A***)

Enzymes that catalyse reactions outside cells (such as in the digestive tract) work best at the pH of the region where they perform catalysis – eg enzymes in the stomach (eg pepsin) work best at (acid) pH 2, while those in the duodenum (eg lipase) work best at (basic) pH 10. (**M**)

6. Increasing the temperature *increases the rate* of the reaction up to an optimum temperature, after which the rate of the reaction decreases rapidly and then ceases. (***A***)

Above the optimum temperature, the enzyme is *denatured* and can no longer act as a catalyst. (**M**)

Above the optimum, the higher temperatures break the hydrogen bonds that hold the shape of the enzyme. The enzyme loses its shape/*active site*, so becomes denatured and can no longer catalyse its reaction, as the *active site no longer corresponds to the shape of the substrate*. The optimum temperature is body temperature in birds and mammals, which are homeotherms. The optimum temperature may vary greatly in poikilotherms, depending on their habitat – it can be low in cold-adapted organisms, and high in organisms adapted to hot environments. (**E**)

7. **a.** The shape of an enzyme is important because it has to fit the shape of the substrate (the way that a lock fits a key), so that it can act as a catalyst for the reaction. (***A***)

The active site of the enzyme must correspond to the shape of the substrate that it catalyses, so that the substrate and enzyme can form the enzyme-substrate complex which is essential for the reaction to occur. (**M**)

b. Enzymes do not work/act as catalysts at high temperatures such as 45°C. (***A***) Because they get denatured/lose shape/lose active site. (**M**)

c. Temperature of 26°C needed for the first hour, as this is the optimum temperature for the yeast to reproduce; and then at 35°C for the next two hours, as this is the optimum temperature for yeast to carry out the process of fermentation. (***A***)

At the start of the bread making, it is necessary to get large numbers of yeast to maximise the process (occurs at 26°C) – these yeast will have the highest fermentation rate if the temperature is increased to 35°C. (**M**)

If temperatures are too low (less than 20°C), the yeasts do not reproduce as quickly, as the low temperatures slow down the enzyme-controlled cell processes (eg respiration); and, if too high (greater than 40°C), yeasts do not reproduce as quickly as enzymes start to become denatured. Above 60°C, all enzymes will be denatured – so cell processes, and therefore reproduction, stop. Fermentation is anaerobic respiration, and its chemical pathways are enzyme-controlled too. Temperatures below zero (eg –20°C) will stop the chemical reactions of fermentation; if the temperature is increased above the optimum (35°C), then the enzymes will start to denature and fermentation will begin to stop. (**E** – in-depth explanation links temperatures of cell processes to enzyme control and the effect of temperature on this, especially denaturing)

d. An enzyme inhibitor can take the place of the substrate molecule, so stopping the normal reaction being catalysed. (**A**)

The inhibitor has the shape to lock onto the active site of the enzyme, and, having done so, stays there, which effectively deactivates the enzyme. If enough enzymes are deactivated like this, the cell process that the enzyme catalyses will no longer be able to be carried out. (**M**)

Unit 11.1 Activity 5A: Classification or Dichotomous keys (Page 48)

1. Red beech = *Nothofagus fusca*
Mountain beech = *N. solandri var cliffortioides*
Silver beech = *N. menziesii*
Hard beech = *N. truncata*
Black beech = *N. solandri var solandri*

2. Yorkshire fog = *Holcus lanatus*
Couch = *Agropyron repens*
Paspalum = *Paspalum dilatatum*
Browntop = *Agrostis tenuis*
Perennial grass = *Lolium perenne*
Kikuyu = *Pennisetum clandestinum*

Unit 11.1 Activity 5B: Classification (Page 50)

1. This is the system of naming organisms using two Latin words, the first word being the genus and the second word being the species that the organism belongs to.

2. The scientific name is written in *italics* or underlined. The first word begins with an upper case/capital letter, while the second word begins with a lower case/small letter.

3. Any group that an organism is *classified* into (eg phylum, order, genus), based on the organism's relationship to other organisms.

4. Prokaryotes are organisms which have cells in which there is no true nucleus, as the genetic material is not enclosed within a membrane. Eukaryotes are organisms which have cells that contain a true nucleus, as the genetic material/chromosomes is (are) contained within a membrane/nuclear envelope. Their cytoplasm contains membrane-bound organelles.

5. a. Prokaryote groups are the archaea (the archebacteria) and the eubacteria (bacteria and cyanobacteria). Also still acceptable to say prokaryotes are the monera (bacteria and cyanobacteria).

b. Eukaryote groups are the protista, fungi, plants, animals.

6. Fungi are heterotrophs not autotrophs – they do not make their food in photosynthesis, but feed on living and dead material using extra-cellular digestion. Their cell walls are made of chitin, not cellulose.

7. Autotrophs are organisms that are 'self-feeders', making their own food – either by photosynthesis (plants), or chemosynthesis (some bacteria). Heterotrophs are organisms that need a source of organic material for food, therefore they feed off other organisms or their remains (eg fungi and animals).

8. a. A group of organisms that look alike and are able to interbreed to produce fertile organisms. They belong to the same gene pool.
   b. Organisms produced from the breeding of members of two different species. Typically infertile, so cannot produce more hybrids.
   c. Not being able to reproduce/produce offspring.
9. MRS GREN refers to the seven characteristics of life (Movement, Reproduction, Sensitivity, Growth, Respiration, Excretion, Nutrition). All organisms are able to carry out these processes.
10. Viruses are not classified as living organisms as the only life process they carry out is reproduction, and they cannot do this independently, but are reliant on living cells as hosts.

Unit 11.1 Activity 8A: Unicellular organisms (Page 67)

1. a. The flagellum moves in a circular motion (like sculling an oar over the stern of a rowboat to move it along) to push the organism through the water. (***A***)
   b. The amount of shadow provides *Euglena* with information on the light intensity of the area it is in, so it can move into an area of high light intensity. (***A***) *Euglena* needs light for its chloroplasts to photosynthesise and so make glucose. Information from the eyespot will allow *Euglena* to move to an area of optimum light to allow a high rate of photosynthesis. (***M***)
   c. It is difficult to classify *Euglena* as either a plant or animal, as the cytoplasm contains chloroplasts which are used for photosynthesis (like plant cells), but it also has a flagellum and can move around which is an animal feature (***A*** – describes one animal and one plant feature of *Euglena*)

      Euglena photosynthesises and does not eat/make food vacuoles – typical of a plant. However, it is enclosed in a pellicle not a cell wall – all plant cells have a (cellulose) cell wall; all animal cells are enclosed in a membrane only. The presence of eyespots and contractile vacuoles are also indicative of animal cells. (***M*** – explains a distinguishing factor)

      (***E*** – discussion accounts for all the plant features of *Euglena* and all the animal features *linked* to it being impossible to classify *Euglena* as either a plant or an animal, as it has features of both groups)
2. a. The contractile vacuole is used for *osmoregulation*/control of the water content in *Paramecium*. (***A***)
   b. Water enters *Paramecium* more rapidly from the 1% salt solution, so the contractile vacuole has to work harder to empty out the water. (***A***)

      Because the 1% salt solution has a higher concentration of water than the 4% salt solution, the contractile vacuole therefore fills and empties the water more rapidly to maintain the water balance in the *Paramecium* / stops *Paramecium* bursting / to osmoregulate. (***M***)
   c. The contractile vacuole needs energy from respiration to contract and empty water out of the *Paramecium*. (***A***) Energy is provided in (aerobic) respiration, which needs O_2. If O_2 levels are low, respiration will be slowed, so there is less energy for the contractile vacuole; hence, it slows down. (***M***)
   d. Diffusion is the movement of particles from an area of high concentration to an area of low concentration. Osmosis is the movement of water across a SPM from an area of high concentration to an area of low concentration. (***A***) The difference between the processes is that diffusion is the (random) movement of all liquid and gas particles,

while osmosis is the movement of water only and its movement across a SPM. (**M** – specifies the *difference* between the two processes)

3. As *Paramecium* lives in fresh water, water enters by osmosis and must be removed by the contractile vacuole to stop *Paramecium* from bursting. (**A**)

   The concentration of water is higher outside *Paramecium*, so constantly enters by osmosis. This water is collected into the contractile vacuole, which fills then empties the water to the outside. This maintains the water balance in *Paramecium*. (**M**)

   The entry of water is determined by the concentration of solutes in the surrounding water; if solute concentrations are low, then water enters rapidly in osmosis (and vice versa). This means that the contractile vacuole will work faster to maintain water balance/osmoregulate. Osmosis is a passive process, with water entry dependent on its concentration; but the filling and emptying of the contractile vacuole is an active process, needing energy from respiration. (**E**)

4. Unicells have a *high SA:V ratio*, as they are small/microscopic and are dependent on the process of *diffusion* for the entrance and exit of substances (eg O_2, CO_2). (**A**)

   A small SA:V ratio is needed for *diffusion to be efficient* and allow substances to penetrate all the cytoplasm and its organelles, and to do so rapidly. A large cell/unicell would have a SA:V ratio too low for diffusion to be efficient, and the central area/cytoplasm/organelles of the cell would not receive the needed substances. (**M**)

   The size of the unicell is limited when its growth reaches the extent when the efficiency of diffusion is reduced, growth stops, and the organism divides to form two new small organisms (binary division), which now have a large SA:V ratio. The shape of unicells is often *elongated* (eg *Paramecium*, *Euglena*), as this increases the SA:V ratio and diffusion of substances to/from the central cytoplasm is rapid as the diffusion distance is short (eg as these organisms move rapidly, their mitochondria need large amounts of O_2 for respiration – this is assisted by their shape reducing the diffusion distance for O_2 to enter and CO_2 to leave the mitochondria throughout the cytoplasm). (**E**)

5. **a.** 1 = nucleolus, which produces the RNA needed for the ribosomes.

   2 = ribosome, which is the site of protein synthesis.

   3 = cell wall, which is rigid and provides shape and support for *Chlamydomonas*.

   4 = chloroplast, which is the site of photosynthesis.

   5 = mitochondrion, which is the site of (aerobic) respiration.

   (**A** – name and function for four out of five correct)

   b. *Chlamydomonas* is an autotroph because it makes its own food. It is not a heterotroph because it does not consume organic foodstuffs (**A**); this is because *Chlamydomonas* has a chloroplast and is able to make glucose in photosynthesis. (**M**)

   c. It osmoregulates using the contractile vacuole. (**A**) As *Chlamydomonas* lives in fresh water, water is more concentrated outside, so enters by osmosis. This water is collected up by the contractile vacuole and emptied to the outside, so maintaining the water balance. (**M**)

   d. It moves using the two flagella (which beat in the manner of a person swimming using breaststroke), pulling the *Chlamydomonas* through the water. (**A**)

   e. The eyespot detects light, so allows *Chlamydomonas* to move into areas of high light intensity for photosynthesis. (**A**)

   High light intensities are needed to maximise the rate of photosynthesis and so produce supplies of glucose, which is the energy source for the organism to live. (**M**)

   Chlamydomonas needs the energy from glucose (released in respiration) to fuel all its life processes (eg movement, growth, reproduction). If it were not able to detect and move to areas of high light intensity, then it could not photosynthesise or the rate of photosynthesis would be low/too low, and so *Chlamydomonas* would not be able to carry out these essential process and would not survive – hence the importance of the eyespot. (***E***)

6. *Paramecium* would have large numbers of mitochondria, as it moves rapidly and this requires large amounts of energy. (***A***) Aerobic respiration occurs in the mitochondria, breaking down glucose and providing ATP to fuel the cell's reactions. Cells such as *Paramecium* with high energy demands will have large numbers of mitochondria to provide the required supplies of ATP. (***M***)

Unit 11.2 Activity 1A: Leaves and photosynthesis (Page 78)

1. 1N; 2E; 3M; 4L; 5G; 6J; 7D; 8K; 9C; 10A; 11F; 12I; 13H; 14O; 15B;
2. **a.** Oxygen and water. (***A***)

   b. Answer for either cell layer 1 *or* cell layer 2.

   Cell Layer 1 – The palisade mesophyll cells are long and thin, with large numbers of chloroplasts close to the edge of the cells. (***A***) This means that the CO_2 does not have to penetrate deep into the cell to reach the chloroplasts where it is used in photosynthesis; thus the efficiency of photosynthesis is increased. (***M***)

   OR:

   Cell Layer 2 – The spongy mesophyll cells have large air spaces between them. (***A***) CO_2 diffuses into the leaf and because of the large air spaces in this layer, and the short distance the CO_2 has to travel (the leaf is very thin), diffusion is efficient – meaning more efficient photosynthesis. (***M***)
3. **a.** Guard cells are kidney-shaped cells which have a thickened inner wall. The small space between the inner walls of a pair of guard cells is known as the stoma. (***A***)

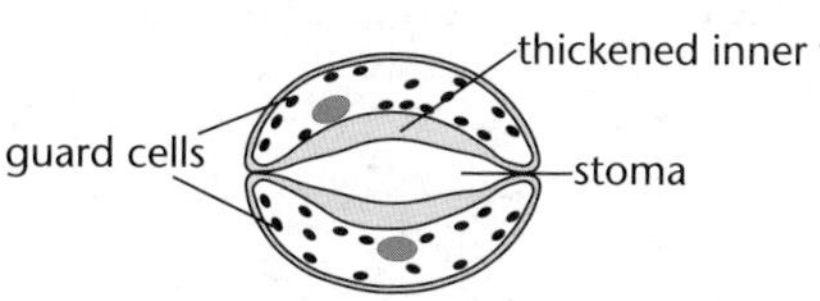

   b. When a plant has a plentiful water supply, the guard cells absorb water. Because the inner wall is thickened, when guard cells absorb water they bend, increasing the size of the stoma. The increased pore size allows more CO_2 to enter the leaf. CO_2 is required for photosynthesis; therefore, when the stomata are open, the rate of photosynthesis increases because there is more CO_2 available. When the plant lacks water, water is lost from the guard cells, reducing the size of the pore. This in turn reduces the amount of CO_2 entering the leaf and therefore reduces the rate of photosynthesis. (***A*** – description of how CO_2 availability alters rate of photosynthesis; ***M*** – explanation includes how water availability affects guard cells)
4. The results of the experiment show that both light and chlorophyll are necessary for starch production in a leaf. (***A***)

   Chlorophyll is required to trap the light energy required for photosynthesis. Light is required to provide the energy that is necessary to produce starch by means of photosynthesis. (***M***)

   The white part of the leaf has no chlorophyll. The green part of the leaf has chlorophyll. The green part of the leaf turned black when tested for starch, whereas the white part did not. This proves that chlorophyll is required for photosynthesis, which produces the starch.

The part of the leaf that had been covered in cardboard, stopping light reaching it, did not produce starch. The green part of the leaf that had been exposed to light produced starch. This proves that light and chlorophyll are required together to produce starch. (***E***)

5. **a.** Palisade mesophyll cells are the long, rectangular-shaped cells positioned just beneath upper epidermis (top layer of small rectangular cells). Label cells on diagram to reinforce answer. Palisade mesophyll cells are tall and thin with large central vacuole. Chloroplasts are arranged around outer edges of the cells (shown as clear circles in diagram). Nucleus can also be seen in diagram as a small black circle, also found near cell wall. (***A***)

b. CO_2 enters leaf via the stoma on the underside of the leaf and diffuses into the palisade cells/chloroplasts. (***A***) This diffusion occurs because the CO_2 concentration is greater outside the cells than inside/long rectangular shape of palisade cells means there is a large surface area to volume ratio for diffusion/the chloroplasts are located close to the inside of the cell wall so diffusion distances are short. (***M***)

c. *Shape*, *structure* and *position* of a leaf can all affect rate of photosynthesis. Ability of the leaf to capture sunlight is important. This can be affected by all three factors.

If the leaf *shape* provides a large surface area, this increases amount of sunlight the leaf is exposed to. The need for large surface area for trapping sunlight has to be balanced with the ability of the plant to maintain adequate water, not losing too much via transpiration. Plants exposed to high winds and/or high levels of sunlight are prone to high rates of transpiration, meaning loss of water. If leaf shape is kept small and narrow (such as pine needle or long narrow leaf shown in the diagram), the rate of transpiration can be reduced, while, due to the high levels of sunlight exposure, adequate light is given for good rate of photosynthesis. A thin leaf (from top to bottom) also increases rate of photosynthesis as sunlight does not have a great distance to travel to the mesophyll cells. Leaves that do not get a high level of exposure to sunlight (eg those found in lower layers of a forest) tend to be larger to maximise exposure to available light.

The *structure* of the leaf is designed so there is a transparent cuticle over the epidermis to allow sunlight to travel straight through this layer to the mesophyll cells which contain chloroplasts. Chloroplasts are contained within both palisade and spongy mesophyll cells which make up the layers under the epidermis. These cells all contain a large central vacuole with the chloroplasts occupying the space closest to edge of the cells. This structural adaptation enables efficient diffusion of sunlight into chloroplasts. Majority of stomata found on underside of the leaf. This adaptation enables plant to have high level of exposure to sunlight on top side of leaf and not lose excess water via transpiration as a result of position of stomata. Stomata are also essential for diffusion of gases required for photosynthesis into and out of the leaf.

The *position* of leaves can also be maximised for sunlight exposure. Leaves are arranged so they are spread out from each other, not covering each other, as in the first set of leaves in the diagram. This enables all leaves on plant to get good supply of sunlight for photosynthesis. (***A*** – descriptions related to each of the three features, may be linked to photosynthesis; ***M*** – explanations related to two of the three features; ***E*** – discussion includes linked descriptions and explanations for two of the three features related to *maximising* photosynthesis)

Unit 11.2 Activity 1B: How plants feed (Page 81)

1. Light energy is required for photosynthesis. Light is absorbed by chloroplasts, and the energy from the light used to drive photosynthesis. Both the colour and the amount of light will influence the rate of photosynthesis.

When white light (which is made of all the colours) strikes a chloroplast, the green light is transmitted and all red and blue light are absorbed. This is what makes the leaf green. If green light were to strike the leaf, all the light would be transmitted and none would be absorbed. This would produce a very low rate of photosynthesis due to lack of energy. Red and blue light produce high rates of photosynthesis as all the light is absorbed by the chloroplasts. Increasing the intensity (or amount) of light also increases the rate of photosynthesis. When there is more light, more energy is absorbed by the chloroplasts, providing more energy to drive photosynthesis. However, this only happens up to a point as other factors (such as CO_2) become limiting (ie in short supply), after which the photosynthetic rate will not increase. (**A** – descriptions related to colour of light *and* intensity of light; **M** – explanation of *either* colour *or* intensity; **E** – discussion of how *both* colour *and* intensity affect photosynthetic rate)

2. a. When the light intensity is low, photosynthesis is slow. As the light intensity increases, the rate of photosynthesis increases until it reaches a plateau (flattens out); at this point, the rate of photosynthesis is no longer increasing, despite the fact that light intensity is increasing. The flat section of the graph is described as a plateau. At this point, there must be other factors required for photosynthesis that are limiting the rate of reaction. (**A** – describes what is happening in sloping and flat sections of the graph)

b. In the flat section of the graph, the rate of photosynthesis is not increasing any further due to limiting factors. (**A**) These could be:

- Availability of carbon dioxide – there is inadequate CO_2, meaning that the rate of photosynthesis cannot increase any further.
- Number of chloroplasts available for the reaction. Photosynthesis takes place in the chloroplasts. Once all the chloroplasts are operating as fast as possible, increasing light intensity will not increase the rate of photosynthesis.

(**M** – identifies one limiting factor and explains why this factor would limit the rate of the photosynthesis reaction)

3. a. The inner wall of the guard cell is thickened with cellulose. When the guard cell is turgid, the inner wall does not expand, making a 'kidney bean shape' that creates the pore that is the stoma when the cells are turgid. (**A**)

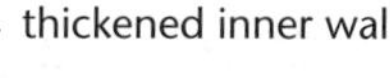

b. Stomata allow carbon dioxide, CO_2, and oxygen gas particles to pass into and out of the leaf. (**A**) This is important for the process of photosynthesis – carbon dioxide that diffuses into the leaf is used as a raw material in the process of photosynthesis.

(**M** – link stomata allowing CO_2 into the leaf with the fact that CO_2 is required for photosynthesis)

c. Photosynthesis can be summarised by the equation:

$$\text{carbon dioxide + water} \xrightarrow{\text{sunlight and chlorophyll}} \text{glucose + oxygen}$$

The rate of photosynthesis is dependent on availability of water, carbon dioxide, chlorophyll and sunlight. In addition, as a chemical reaction, the rate of photosynthesis will increase, to a point, with increased temperatures.

A combination of hot, dry, windy weather conditions increases rate of transpiration in the plant as water is evaporated more quickly from stomata on the surface of leaves in these conditions. As a result of this, the stomata may close, reducing the supply of CO_2.

The hot, dry, windy conditions should make you immediately think about increased evaporation and therefore increased transpiration and loss of water.

(**A** – describes that rate is dependent on some or all factors listed; **M** – explains effects of either temperature, wind or lack of water on the rate; **E** – must comment on combination of weather conditions)

Unit 11.2 Activity 2A: Plant nutrition (Page 92)

In Achievement Standard 90463 (Biology 2.7), it is expected that all answers will be *full and lengthy*, whether descriptions ('Achievement' – **A**), explanations ('Merit' – **M**), or discussions ('Excellence' – **E**). Students are also expected to *provide evidence from all three plant groups* studied. The answers given below for *aspects* of plant nutrition systems are not indicative of the *comprehensive* answers needed in the AS 90463 exam. Candidates should seek their teacher's guidance with this.

1. Photosynthesis needs H_2O and CO_2 as raw materials. CO_2 *diffuses* into the leaf from the air, in response to its concentration gradient. CO_2 diffuses through the stomata and air spaces around the mesophyll cells, dissolving in the moisture around the cells, then passing through the membrane and into the underlying chloroplasts. H_2O enters the root hairs from the soil in *osmosis*, in response to its concentration gradient. H_2O moves from cell to cell inside the root to reach the xylem. Water is carried up the xylem to the leaves, mainly by transpiration pull. In the leaves, water exits the xylem to pass through the membranes of the mesophyll cells and into the underlying chloroplasts. (**A** – describes entry of *both* CO_2 and H_2O into chloroplast; minimum of two factors described for each)
2. Leaves are adapted to photosynthesis by, eg, being thin, having a large SA:Vol ratio, being positioned around the stem so that overlap is minimised. (**A** – three factors described)
3. Cell layers are adapted by having a waxy cuticle to reduce water loss; epidermis (upper and lower) being one layer of transparent cells, to allow light to pass through; palisade mesophyll under the upper epidermis has large numbers of chloroplasts, to maximise photosynthesis; spongy mesophyll cells are loosely packed, providing air spaces to increase rate of diffusion of CO_2 and O_2; guard cells are present in the epidermis (mainly lower), to form the stomata to allow entry of CO_2/exit of O_2 into/from the leaf; veins of xylem and phloem pass through the cell layers to bring H_2O (and minerals) to the leaf and remove glucose. (**A** – minimum of four adaptations described; **M** – minimum of four explained)
4. Chloroplasts are found in the cytoplasm under the membrane of the mesophyll cells. They contain stacks of (thylakoid) membranes called *grana* and a fluid matrix called the *stroma*. It is on/in the grana and stroma that photosynthesis occurs. (**A**) The light-absorbing pigments (eg chlorophyll) are embedded in the membranes of the grana, which provide a large SA for the light-dependent reactions of photosynthesis to occur, generating ATP and splitting H_2O. The fluid stroma provides the matrix for the light-independent reactions/Calvin cycle, which use(s) the ATP to fix CO_2 and to produce glucose. (**M**)

5. Carnivorous plants are autotrophs, because they make their own food ('self feeders') in photosynthesis and heterotrophs because they eat insects. (***A***) They are autotrophs because their leaves have chloroplasts in their cells which carry out photosynthesis to produce glucose. However, the leaves also have adaptations to catch insects (eg pitchers, sticky hairs, hinge mechanisms), and enzymes to digest them. Insects are needed to provide essential nutrients (eg N), which are lacking in the environment where the plants live. In catching and digesting insects, the plants are acting like heterotrophs. (***M***)

6. **a.** Parasitic plants produce large numbers of seeds to find a new host. (***A***) Host plants are distributed throughout the community, so the more seeds produced by a plant, the greater the chances are that a seed will land on/by another host, germinate, and parasitise the host. In this way, members of the species spread and no host should become over-infested and killed. (***M***)

   b. Haustoria are modified roots that digest into the host's stem and form a connection with the host's phloem and xylem. (***A***) Haustoria are needed to obtain H_2O and minerals from the host's xylem and glucose from the host's phloem. If it did not obtain the water, minerals and glucose, the parasite would not survive, as it is no longer able to photosynthesise. (***M***)

   c. The vascular tissue, leaves and roots are greatly reduced or absent as the parasite no longer needs them, because it gets its food and water from the host. (***A***) Leaves are not needed as the parasite does not need to photosynthesise, as it gets its glucose from the host. The roots are no longer needed as the parasite is anchored to the host via haustoria, and gets its water from the host's xylem. The haustoria tap directly into the host's xylem and phloem, reducing/removing the need for transport tissue. (***M***)

7. *Comparisons* of the leaves could include:
   - Presence and location of the typical cell layers/structures are the same, eg epidermis, mesophyll cell layers, air spaces, cuticle, stomata.
   - Chloroplasts present in mesophyll cells, have same internal structure to carry out photosynthesis.
   - Chemical reactions in photosynthesis the same (light-dependent followed by light-independent).
   - Vascular tissue/veins/xylem and phloem present in leaves to transport substances.

   Contrasts of the leaves could include:
   - Leaves of sun-adapted plants typically smaller than shade-adapted plants.
   - Leaves of sun-adapted plants may be vertical, while those of shade-adapted plants horizontal.
   - Leaves of sun-adapted plants thicker than those of shade-adapted plants.
   - Thicker waxy cuticle in sun-adapted plants.
   - Leaves may be hairy in sun-adapted plants.
   - Chloroplasts smaller in sun-adapted plants, with less thylakoid membranes in the grana. Chloroplasts positioned more on vertical than horizontal walls of the cells.
   - Presence of CAM photosynthesis in some sun-adapted plants.
   - Shade-adapted plants may change the orientation of their leaves as light intensity changes.
   - Shade-adapted plants have only one palisade cell layer; spongy cells have more chloroplasts than the palisade cells. Sun-adapted plants may have two or more palisade layers, and these have more chloroplasts than the spongy cells.

(***A*** – descriptions with little attempt at comparisons/contrasts; ***M*** – more extensive explanations with weak comparisons/contrasts; ***E*** – comparisons and contrasts extended, with detailed descriptions and explanations, including both structures and functions)

Answers could be enhanced by clear, labelled, annotated diagrams – annotations could be descriptions (***A***), or explanations (***M***).

Correct terms should be used in all answers (eg stomata not 'holes', xylem not 'water tubes', phloem not 'food tubes').

8. Three clearly different taxonomic or functional groups (eg any three of sun-adapted plants, shade-adapted plants, carnivorous plants, parasitic plants) need to be selected.

For *each* group, the *structures* associated with the *nutrition* need to be *described fully and carefully using correct terms*, eg:

- Size, shape, thickness, position (vertical, horizontal) of leaves and location on stem.
- Location and internal structure of chloroplasts.
- Internal cell layers of leaves.
- Leaf specialisation (eg spines of cacti, pitchers of carnivorous plants).
- Vascular tissue (xylem and phloem).
- Root hairs.
- Haustoria.

For *each* group, *how* these structures perform their particular function needs to be *explained*, eg:

- How leaves of carnivorous plants lure, trap and digest insects.
- Transport of raw materials and products of photosynthesis (osmosis, diffusion, transpiration pull, translocation).
- Reducing water loss.
- How the chloroplasts perform photosynthesis, photosynthetic pigments.
- CAM photosynthesis.
- How parasites find a new host and tap into its vascular tissue.

For *each* group, reasons for *differences* between the groups needs to be *explained*, eg:

- Habitat – eg shade plants adapted to permanent, low-light intensities within forest; carnivorous plants to living in low-nutrient marshlands; cacti to living in deserts.
- Size – eg canopy trees have leaves high up the tree exposed to high light intensities, so these leaves are smaller than those lower down the tree that experience lower light intensities.
- Life cycle – eg changes in size, shape of leaves from juvenile plants living in lower light intensities (ground floor of forest) to adult plants living in higher light intensities (canopy of forest); need for parasites to find a new host, so large amounts of seeds produced.

(***A*** – descriptions of the structures for the three groups; ***M*** – explanations of how they function; ***E*** – discussions with comparisons/contrasts and reasons for differences)

Unit 11.2 Activity 4A: The digestive system (Page 106)

1. 1E; 2B; 3R; 4W; 5C; 6A; 7L; 8P; 9H; 10G; 11I; 12Q; 13T; 14F; 15J; 16M; 17X; 18S; 19Y; 20U; 21K; 22D; 23N; 24V; 25Z; 26O

2. a.

Name of tooth	Function
Molar	Crushing or grinding
Canine	Tearing or holding
Incisor	Cutting or biting

(***A*** – five correct)

b. The mouth contains an enzyme which digests starch. The mouth does not have the other enzymes to digest protein, fat, etc. (***A*** – enzymes involved in chemical digestion, not all present in the mouth)

The amylase enzyme is the only enzyme in the mouth, and it only digests starch. Other enzymes digest other types of food. (***M*** – identifies amylase as digesting starch, and that other enzymes digest other foods)

c. The stomach has a pH of 1–2 and this is where protein is digested OR the duodenum has an alkaline pH and this is where fats are digested. (***A***)

Each digestive enzyme works best in a different pH, eg protease enzymes that digest protein function best in acid pH. (***M*** – linking different pH levels to best conditions for enzyme activity, named example)

When the food goes from the stomach to the duodenum, the pH is changed from acid to alkaline by the bile that gets mixed into the food. This stops protease enzymes, from the stomach, digesting any more protein. The change to alkaline pH is also because the lipase enzymes that digest fats in the duodenum function best in an alkaline pH. (***E*** – comparison of effects of different pH conditions for two named enzymes and their substrates in two different parts of the digestive system)

3. a. The purpose of absorption is to get the digested food into the bloodstream. (***A***)

b. Villi are small folds in the lining of the small intestine that increase the surface area where absorption can take place. (***A*** – description of folds of villi or increase of surface area)

Also, the membrane of villi cells is folded into microvilli. Because there are many folds, much more surface area is in contact with the food, which means much more surface for food molecules to travel through and into the blood. (***M*** – how villi and microvilli significantly increase surface area for absorption of food)

c. Bile is a mix of chemicals produced by the liver that is alkaline/causes the emulsification of fats/breaks down fats. Pancreatic juice contains enzymes. (***A*** – description of what each substance does or is)

Emulsification causes large insoluble drops of fat to be broken down into smaller droplets. The lipase enzymes in pancreatic juice digest fat. (***M*** – explanation of what emulsification is and of function of enzymes in pancreatic juice)

Having many smaller droplets of fat means lipase enzymes in the pancreatic juice have much more surface area where they can attack the fats and break them down quickly so they are small enough to be absorbed into the blood. (***E*** – linking of emulsification to a large surface area for effective action of lipase)

4. a. Bowel cancer can cause blood in the faeces / diarrhoea or constipation / lower abdominal pain or swelling / always tired / person is pale and weak. (***A*** – description of two of the symptoms)

b. A person with diarrhoea can die through losing too much water from their body (dehydration) (***A*** – description of effect of diarrhoea); so the blood is thicker and harder to pump, putting strain on the heart / less sweat produced so less body heat lost or core body temperature can increase. (***M*** – links effect of diarrhoea to death of person)

c. An ulcer causes damage to the lining of the stomach / causes stomach pain / causes less efficient digestion. (***A*** – description of effect of an ulcer on the stomach)

Damage to the mucus lining of the stomach means stomach wall can be digested by stomach enzymes or acid / damage to the stomach wall means the muscle is not able to work as efficiently to churn food/peristalsis so digestion of food is reduced. (***M*** – reason for effect of ulcer on stomach function)

The mucus lining prevents stomach enzymes digesting the stomach itself (stomach wall is muscle, which contains protein). An ulcer is damage to this lining and the stomach wall caused by the enzymes breaking them down. When the muscle wall is damaged, it does not churn the food as efficiently / peristalsis is reduced, so digestion of food in the stomach not as efficient. (***E*** – discussion links effect of damage by an ulcer to stomach lining by the enzymes to function of the stomach)

5. a. i. Mixes with food to moisten, soften and lubricate / adds salivary amylase to begin digestion of starch and glycogen / kills bacteria that enter the mouth. (***A***)

ii. Contains acid that assists digestion, kills micro-organisms and provides an acid environment for pepsin / contains enzyme pepsin that breaks proteins into polypeptides. (***A***)

iii. Contains bile salts that aid digestion and absorption of fats by emulsification (ie breaking droplets of fats and oils into smaller particles) to allow greater surface area. (***A***)

b. Food contains different types of nutrients (eg protein, starch). (***A***) Enzymes are specific and only digest one type of nutrient (eg lipase digests only fats and oils); different enzymes needed to digest range of different nutrients in food. (***M***)

c. Pancreatic juice contains a wide range of enzymes, including amylase, nucleases, lipase. Enzymes from the small intestine function at optimum levels when in alkaline conditions. (***A*** – description of activity of enzymes in normal small intestine)

pH of chyme entering small intestine very acidic due to addition of hydrochloric acid when food is in stomach / bile and pancreatic juice both alkaline, so they neutralise acid pH of chyme from the stomach. Each enzyme has specific shape important to its function; each enzyme has certain environmental conditions (eg pH and temperature) required for optimum levels of activity. Enzymes denatured by altered pH levels. (***M*** – explanation compares pH of the two compartments)

With reduced bile, pH of acidic chyme will not become alkaline, therefore enzyme activity in small intestine reduced – digestion of carbohydrates, proteins and fats will be reduced. (***E*** – link made between effect of gallstones and activity of enzymes)

6. a. Absorbs water from the undigested food and forms the undigested food into faeces. (***A***)

b. As food becomes drier in colon, contractions of colon are able to push colon wall against the food which allows remaining water to be absorbed (***A***); without need for villi to increase surface area. (***M***)

7. a. Difficulty in passing faeces / passing very hard, dry faeces. (***A***)

b. Soluble fibre gives the faeces a softer, gel-like texture so that they are easier to pass *or* soluble fibre holds moisture which keeps the faeces softer *or* fibre gives the intestine something to push on (***A***) so that food can move through the intestine at a steady rate and not dry out too much. (***M***)

8. **a.** Breaks down food into small molecules that can be absorbed into the blood. (***A***)

b. *Any two of:*

- Salivary gland – produces saliva containing mucus, water and enzymes.
- Liver – produces bile which changes pH of the chyme in the duodenum / produces bile which breaks the drops of fats and oils into smaller droplets.
- Pancreas – produces enzymes.
- Large intestine – absorption of nutrients and water. (***A***)

c.

Organ	**(*A* – descriptions of structure and function)**	**(*M* – explanations relating to structure or function)**	**(*E* – discussion relating structure and function to digestive system)**
Stomach	• Sac shape to hold food. • Sphincters to control release of digested food into duodenum. • Smooth muscles in walls to churn food. • Cells in lining to produce gastric acid/ enzyme/ pepsin. • Holds food for several hours. • Churns/mixes food. • Digests proteins.	• Sphincters ensure food digested efficiently before moving to next stage. • Sphincters contain pepsin/ acid within the stomach – so it doesn't move into next stage where it may not work efficiently/cause damage. • Food churned for thorough mixing with pepsin/acid/ break large food molecules down into smaller ones to ensure efficient digestion. • Gastric acid kills bacteria in food/ensures correct pH for action of pepsin. • Pepsin digests large proteins into small amino acids for absorption into bloodstream.	The stomach is the starting point for the digestion of proteins and carries out further mechanical digestion. Large food molecules are broken down into small, soluble ones for efficient absorption. The smooth muscles lining the stomach wall contract rhythmically to churn and mix the food and pepsin enzymes ensuring thorough mixing and efficient digestion. The stomach has sphincter muscles at both ends that contain the pepsin and the gastric acid in the stomach. This keeps conditions in the stomach acidic enough for the pepsin, and prevents leakage of corrosive acid into the oesophagus or the duodenum.
Small intestine	• Very long thin folded tube. • Cells in the lining produce enzymes. • Supplied by ducts from the liver and pancreas. • Walls folded and have villi. • Most absorption occurs in ileum.	• Folds increase the length of tube that can fit inside the gut, so increasing the time food remains in the intestine. • Ducts supply digestive enzymes to the intestine only when needed to avoid wastage of enzymes. • Villi greatly increase the surface area for efficient absorption into the blood. • Alkaline conditions of the small intestine suit the enzymes that carry out digestion here, so absorption occurs efficiently as food digested.	The majority of the digestion and absorption of digested food occurs in the small intestine. It is a very long thin tube with many external and internal folds. The external folds ensure food stays in the intestine for as long as possible, while the internal folds and the villi ensure that maximum surface area is in contact with the digested food for absorption. The alkaline conditions suit the enzymes found here, again so that maximum enzyme activity and digestion occur.

9. **a.** Catalysts that break down large food molecules into small molecules. (***A***)

b. Enzymes have an optimum temperature at which they work. (***A***) When temperature was 50°C, the enzymes were denatured; a temperature of 22°C is below the enzymes' optimum temperature so enzyme activity was slow. (***M***)

c. Different enzymes work best in different pH. (**A**) Pepsin in the stomach works best in acidic conditions, while enzymes in the small intestine work best in alkaline conditions/are denatured in acidic conditions. (**M**)

Unit 11.2 Activity 5A: Nutrition in animals (Page 119)

1. a. Ingestion is the taking in of food into the (first part) of the digestive system. (**A**)
   b. Digestion is the breaking down of food by mechanical and chemical processes into its small soluble constituent molecules (eg glucose, amino acids). (**A**)
   c. Absorption is the process in which the products of digestion (eg glucose, amino acids) leave the gut and enter the transport system/body cells. (**A**)
   d. Egestion is the process of removing foods that cannot be digested (eg cellulose products) from the body (as faeces). (**A**)
2. Mechanical digestion is the *physical* breakdown of food (eg by teeth, gizzard); chemical digestion is the *chemical* breakdown of food (eg by enzymes, acids). (**A**)
3. Ectoparasites have piercing and sucking mouthparts to ingest their food. (**A**) The piercing part penetrates the skin and blood capillary/phloem tube, which allows the sucking part to suck up the liquid food (blood/sap) from the host. (**M**)
4. The differences in the teeth of mammals *reflects their diet*, with herbivores having large incisors and molars, carnivores having sharp incisors and molars and pointed canines, omnivores having all three teeth types very similar in shape and size. (**A**) Herbivores have large incisors to bite off (often hard) vegetation and large ridged molars to grind up the vegetation; canines are absent (as they are the piercing, stabbing teeth needed to kill and tear meat). Carnivores have canine teeth to kill prey and tear meat; their incisors and molars are very sharp and/or pointed to eat meat and crunch up/gnaw hard bones. Omnivores eat both vegetation and meat, so have all three types of teeth, but these teeth are more uniform in size and shape, which reflects the unspecialised diet. (**M**)
5. a. The teeth (eg mammals) and gizzard (eg insects and birds) are hard structures/ have hard surfaces to break down food. (**A**) They break down food mechanically, so *increasing its surface area and facilitating chemical digestion*. (**M**)
   b. The crop (eg insects and birds) and stomach (eg mammals) are both sac-like structures in the digestive system. (**A**) They are sac-like to allow them to *store food* while it awaits digestion/is digested. (**M**)
6. Filter feeders ingest their feed from water using a filtering device (eg baleen in whales, feathery limbs in barnacles, gills in shellfish). (**A**) The water flows over the filter (eg shellfish), or is forced through (eg baleen in whales), and the food particles (eg krill for whales, organic particles for shellfish) caught on the filter are then taken into the digestive system. (**M**)
7. Mechanical digestion (eg teeth, gizzard) is important in breaking the food into smaller pieces, *increasing its surface area* for the process of chemical digestion by enzymes and acids. (**A**) The larger the surface area of the food, the faster enzymes are able to act/catalyse its breakdown. Therefore, the greater the surface area of the food, the faster the process of digestion. (**M**)
8. Enzymes are the agents of *chemical digestion* of food, as they are *catalysts/catalyse the breakdown* of the food. (**A**) As catalysts, enzymes *speed up* the process of digestion, allowing it to *occur fast enough* to meet an animal's nutritional needs. (**M**) Many *different enzymes* are needed, as digestion is a step-by step-breakdown of food, with chemical digestion occurring in the mouth, stomach, small intestine (in mammals), and each enzyme is *specific* in its action (ie only catalyses one particular reaction – eg amylase digests starch in the

mouth, pepsin digests proteins in the stomach, lipase digests lipids in the small intestine). This is because the *active site* of the enzyme fits the shape of the specific food chemical (eg protein), so only that chemical can be catalysed. The end point of chemical digestion is the production of the small soluble constituent molecules from the large chemicals that are ingested as food. (***E***)

9. *Comparisons* of the two systems could include:

- Both groups have digestive systems that essentially are modified tubes that run from mouth to anus.
- The processes of ingestion, digestion, absorption, egestion occur.
- Digestion is both mechanical and chemical.
- Chemical digestion begins in the mouth with the action of enzymes; cellulose from the plant diet is digested by bacteria and other micro-organisms that live in the digestive tract.
- Digestive systems include sac-like storage and digestive structures (crop and stomach), digestive and absorptive tubes (mid-gut in insects, small intestine in mammals), egestive tubes where water is reabsorbed from wastes (hind gut in insects, large intestine/rectum in mammals), parts that contain cellulose-digesting bacteria and micro-organisms (caecae in insects, caecum in rodents, stomach in ruminants).
- Transport system (blood) used to transport the absorbed nutrients around the body.

Contrasts of the two systems could include:

- Insects ingest vegetation using biting mouthparts – palps to taste and handle the food (scissor-like), mandibles/jaws with hard, jagged edges of chitin to bite off bits of food. Mammals ingest vegetation using teeth and tongue – large sharp incisors of dentine bite off bits of food, tongue manipulates the food, large ridged molars chew/grind up the vegetation, which is mixed with saliva for swallowing.
- Mechanical digestion in insects uses the mandibles and the gizzard (lined with hard 'teeth' of chitin); in mammals, mechanical digestion involves the teeth and the churning action of the stomach.
- Presence of long, tube-like caecae attached to the mid-gut in the insect increase the surface area for digestion and absorption of digested food. The length of the small intestine in mammals gives a large surface area for digestion; the villi lining its epithelium increase the surface area for absorption of digested food.

(***A*** – descriptions with little attempt at comparisons/contrasts; ***M*** – more extensive explanations with weak comparisons/contrasts; ***E*** – comparisons and contrasts extended, with detailed descriptions and explanations, including both structures and functions)

Answers could be enhanced by clear, labelled, annotated diagrams (annotations could be descriptions (***A***), or explanations (***M***).

10. Three clearly different taxonomic (eg cnidaria, insects, mammals) or functional groups (eg parasitic insects, herbivorous insects, carnivorous insects *or* herbivorous mammals, carnivorous mammals, omnivorous mammals) need to be selected.

For *each* group, the *structures* associated with the processes ingestion, digestion, absorption, egestion, need to be *described fully and carefully using correct terms*, eg:

- The mouthparts (eg teeth, mandibles, piercing stylets, sucking tubes) present and their structure.
- The parts (eg stomach, crop, gizzard, small intestine, colon) of the digestive system and their structures.

For *each* group, *how* these structures perform their particular function needs to be *explained*, eg:

- How the mouthparts ingest food.
- How the parts of the digestive system mechanically and chemically digest food.
- The occurrence and function of enzymes.
- The location of absorption, and how it occurs.
- Mechanisms to increase SA for digestion and/or absorption.
- How digested nutrients are transported.
- Removal of water and packaging of wastes for egestion.

For *each* group, reasons for *differences* between the groups needs to be *explained*, eg:

- Differences in diet (eg types of mouthpart present, presence/absence of cellulose-digesting bacteria, length of intestines, types of enzyme present, presence of anti-coagulents in saliva or crop).
- Differences in size of the animals (eg need for a transport system for digested nutrients).
- Habitat – eg aquatic or terrestrial (filter feeding only in aquatic medium, terrestrial animals need to conserve water in wastes, fluid feeders need less complex digestive system).
- Life cycle (eg different diets at different stages of life cycle – larvae/adults in insects with complete metamorphosis such as butterflies, amphibians such as tadpoles/frogs). These differences also reflect differences in the ecological niche of an animal at different stages in its life cycle.

  (**A** – descriptions of the structures for the three groups; **M** – explanations of how they function; **E** – discussions with comparisons/contrasts and reasons for differences)

Unit 11.3 Activity 1A: Transport of materials in plants (Page 135)

1. **a.** Water and minerals are transported in the *xylem*. Mature xylem dies and becomes hollow, with a large lumen and lignified (bands or spirals) cell walls for strength. In *vessels*, the end walls break down, so a continuous tube is formed. In the more simple *tracheids*, the end walls remain, but are tapered and become pitted for passage of water. (***A***)

b. Sugars are transported in the phloem (sieve tubes). Mature phloem is living and consists of large *sieve tubes* and smaller, elongated *companion cells*. Sieve tubes have cytoplasm (but no nucleus) and the end walls (sieve plates) are perforated to allow passage of sugars. Companion cells are alongside the sieve tubes and have cytoplasm, a nucleus and organelles, and are believed to control activity of the tubes. (***A***)

2. **a.** Transport of water is via *osmosis*, from the soil into the root hairs then from cell to cell (via the cell walls rather than the cytoplasm) to the xylem. Transport up the xylem to the leaves results from the cohesive and adhesive properties of water, root pressure, and (mainly) *transpiration pull*. Water leaves the xylem (leaf veins) by osmosis and passes into the chloroplasts of the mesophyll cells. The direction of movement is in response to a *concentration gradient* (high to low). (***A***)

b. Transport of minerals is via *diffusion* (with the *concentration gradient*) or *active transport* (against the concentration gradient), from the soil into the root hairs then from cell to cell to the xylem. Minerals are transported *in solution in the water* in the xylem to where needed in the plant. (***A***)

c. Sugars are transported from sugar sources to sugar sinks by *translocation* in the sieve tubes of the phloem. The method is not clearly understood, but it is *active transport* and likely due to *pressure differences* between sources (high pressure) and sinks (low pressure). (***A***)

Gases (eg O_2/CO_2) move by diffusion in response to their concentration gradient from high to low. Diffusion is through stomata into/out of leaves and through lenticels into/out of stems from the air. (***A***)

3. Angiosperm leaves have *veins* of *xylem* and *phloem* that pass through the mesophyll cell layers. Water and minerals enter the leaf in the xylem and pass into individual cells for photosynthesis and other cell processes; sugars made in photosynthesis enter the phloem for transport out of the leaf. *Guard cells* are present in the epidermis (mainly the lower), and open and close *stomata*, which are the pores in the leaf that let gases in and out – CO_2, O_2, H_2O vapour. On the inside of the stomata are large *air spaces*, which are continuous with air spaces surrounding the mesophyll cells. These allow for rapid diffusion of the gases to and from the cells. (***A***)
4. Mosses are non-vascular plants because they do not have vascular tissue (ie no xylem and phloem). (***A***) This is because they are small and live in damp areas, so can survive on the (slow) passage of substances by diffusion from cell to cell or by capillary action (water) up the side of the stem. (***M***)
5. Transpiration is the loss of water from the leaves of the plant. As water is lost from the leaf, it draws more water up the xylem from the roots to replace it. (***A***) Water vapour lost from the leaf is replaced by water from the xylem. As there are *strong attractive forces* (which provide cohesion) between water molecules, water forms a *continuous column* up the xylem. As a water molecule is lost from the xylem in the leaf, another water molecule is pulled up to take its place. The loss of water in transpiration therefore generates a *transpiration pull* to draw water up from the roots. (***M***)
6. CO_2 is needed for photosynthesis, and O_2 is needed for respiration. Both enter the plant from the air in the process of diffusion – through stomata in the leaves, lenticels in the stem. Inside the plant, CO_2 diffuses into the cells and then into the chloroplasts, while O_2 diffuses into the cells and then into the mitochondria. (***A*** – processes of both gases, location of processes, process of diffusion described)

   The pathway that the gases take is dependent on their *concentration gradient*. When CO_2 is used in photosynthesis, it will always have a lower gradient inside the leaf, the cells and chloroplasts, so will move in that direction. This also applies to O_2 used in respiration. (***M*** – movement explained in terms of concentration gradients)

   However, photosynthesis only occurs during daylight hours, while respiration occurs all the time. Therefore, during daylight, the O_2 produced in photosynthesis will also diffuse into the mitochondria for respiration (in response to its concentration gradient), and the CO_2 released in respiration will also diffuse into the chloroplasts for photosynthesis (in response to its concentration gradient). At night time, without photosynthesis, the diffusion of the two gases is determined solely by the occurrence of respiration (O_2 diffuses in, and CO_2 out). (***E*** – explanations extended to include consideration of effects of both respiration and photosynthesis on gas transport)
7. H_2O is needed for photosynthesis, to keep cell membranes moist for gas exchange, and to cool the plant. It enters the plant from the soil through the root hairs in the process of osmosis. It is transported up the plant to the leaves in the xylem, mainly as a result of the pull of transpiration. (***A*** – processes water is needed for, its entry site, xylem, and processes of osmosis and transpiration pull described)

   Water is one of the two raw materials for photosynthesis, and is needed for the production of glucose. It is required to keep cell membranes moist, so that the gases CO_2 and O_2 can dissolve and can then diffuse across membranes in the processes or respiration and photosynthesis. Water is needed to cool the leaf so that essential, enzyme-controlled reactions can occur/enzymes are not denatured by excess heat. Water enters the root hairs and across the cortex to the xylem in response to its concentration gradient. Transport up the xylem is mainly by transpiration pull acting on the column of water (water forms a column in the xylem from its cohesive and adhesive properties), though root pressure is a contributory factor. As water molecules are lost in transpiration, other water molecules are drawn up the xylem to replace them. (***M*** – explanations of processes water needed for and its method of transport; ***E*** – comprehensive explanations of both the processes water is needed for and processes involved in its transport)

8. Three clearly different taxonomic (eg mosses, ferns, angiosperms) or functional groups need to be selected.

For *each* group, the *structures* associated with the *transport* need to be *described fully and carefully using correct terms*, eg:

- Location of the vascular tissue (roots, stem leaves), structure of xylem (tracheids, vessels) and phloem (sieve tube cells, companion cells).
- Internal cell layers of leaves (eg air spaces in mesophyll, veins, stomata and guard cells, waxy cuticle).
- Location and structure of root hairs.

For *each* group, *how* these structures perform their particular function needs to be *explained*, eg:

- Processes associated with the transport of CO_2 and O_2 for both respiration and photosynthesis (eg diffusion, concentration gradients, time of day).
- Processes associated with the transport of water from roots to leaves (eg osmosis, transpiration pull, cohesion and adhesion, root pressure, concentration gradient).
- Processes involved with the transport of minerals from roots to rest of plant (eg diffusion, active transport, concentration gradients).
- Processes involved in the transport of sugars made in photosynthesis to rest of the plant, storage of sugars (eg translocation, pressure differentials, active transport).
- Opening and closing of stomata (eg turgor pressure, time of day).

For *each* group, reasons for *differences* between the groups needs to be *explained*, eg:

- Habitat – eg need for damp environment for mosses, adaptations of angiosperms to a full terrestrial existence, xerophyte adaptations to conserve water.
- Size – eg mosses are small plants related to reliance on diffusion of materials in transport, very large sizes of some angiosperms related to development of transport and support systems.
- Life cycle – eg need of mosses and ferns for water for fertilisation places habitat restrictions on them.

(***A*** – descriptions of the structures for the three groups; ***M*** – explanations of how they function; ***E*** – discussions with comparisons/contrasts and reasons for differences)

9. E

10. B

11. D

12. **a.** false. **b.** false. **c.** true. **d.** true. **e.** false. **f.** false.

Unit 11.3 Activity 1B: Short answer questions (Page 133)

1. To increase the surface area for water absorption.
2. The guard cells lose water by osmosis and so the cells become flaccid, straightening and closing the pore/stomata.
3. Water molecules are attracted to, or adhere to, the walls of the xylem vessels. This adhesion helps water move through a plant.
4. Lignin.
5. Xylem vessels allow water and minerals transport upwards to the leaves and phloem vessels transport sugar and other organic materials up and down the plant.
6. Flowering plants need organic and inorganic materials for growth and repair.
7. Dicot leaves are flat and broad with a central and lateral network of veins. Monocot leaves are narrow and speared-shaped with parallel veins.

Unit 11.3 Activity 1C: Crossword puzzle (Page 134)

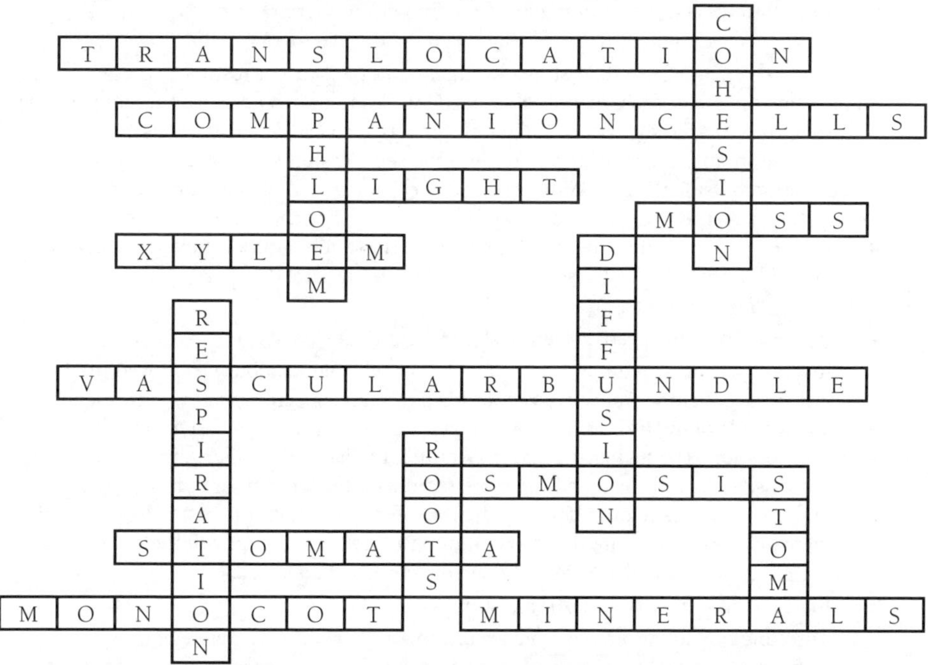

Unit 11.3 Activity 1D: True or false? (Page 135)

1. True
2. False
3. True
4. False
5. True
6. False

Unit 11.3 Activity 1E: Multiple choice (Page 135)

1. B
2. D
3. E
4. E
5. C
6. D
7. B

Unit 11.3 Activity 2A: Transpiration (Page 142)

1. Transpiration is the loss of water from the leaves of a plant. (***A***) Water that leaves the xylem enters the air spaces of the leaf and surrounds the cell membranes. Water molecules *vaporise* in the air (rate increases with increasing temperature) and then *diffuse* out of the leaf, as they are more *highly concentrated* in the air in the leaf than the air outside the leaf. (***M***)
2. Plants need water to keep cool, to keep cell membranes moist, for photosynthesis. (***A*** – minimum of two reasons) Water is an essential raw material for the production of glucose in photosynthesis. Cell membranes need to be kept moist to allow gas exchange – CO_2 and O_2 need to dissolve in order to diffuse across the membrane. Plants need to keep cool as excess heat can denature enzymes needed for essential cell processes. (***M*** – minimum of two reasons explained)
3. Transpiration helps the plant to keep cool and to draw water up the xylem from the roots to the leaves (transpiration pull) for photosynthesis. (***A***) The plant is kept cool because the water takes heat from the cells as it vaporises. As water vaporises and leaves the leaf, it exerts a pull (force) on the column of water in the xylem that draws the water up the xylem (against gravity). Therefore, water constantly enters the leaves. (***M***)
4. Guard cells open and close their stoma by changes in their turgor (osmotic) pressure. When water enters the guard cells, their thicker inside walls are drawn apart, so the stoma opens (and vice versa). (***A***) The guard cells have chloroplasts, so can photosynthesise. The sugar produced in photosynthesis draws water into the cell in osmosis, as water is in higher concentration outside the cell. Water entering the cell increases a cell's turgor pressure, so water entering the two guard cells swells them and they bend, opening the stoma. When photosynthesis stops, the sugar is converted to starch. Water is now more highly concentrated in the guard cells, so moves out. Turgor pressure drops, and the guard cells collapse, closing the stoma. (***M***)
5. **a.** Increases in wind speed, air temperature, light intensity. (***A***)

   Increasing wind speed increases the rate at which water vapour is removed from the leaf surface. Removal of water vapour keeps a high concentration between the inside and outside of the leaf, so the faster the vapour is removed, the faster the rate of transpiration. The higher the air temperature, the faster the vaporisation of water in the air spaces of the leaf, and the more rapid its diffusion through the air spaces and out of the leaf; therefore, the faster the rate of transpiration. Increasing light intensity increases air temperature (light, like heat, is a form of energy), with the same effects as have been just described. (***M*** – two factors correctly explained)

   b. Increases in humidity and soil moisture/water. (***A***)

   The more humid the air, the more water vapour it contains. Therefore, increasing the humidity reduces the concentration gradient for water vapour inside and outside the leaf. Therefore, increasing humidity, rate of transpiration falls. The more water in the soil, the lower the concentration gradient between the soil and the root hairs, therefore less water enters the plant. Therefore, less water enters the leaves, and so less is available for transpiration and the rate of transpiration falls. (***M*** – any one explained)
6. CAM photosynthesis occurs in some xerophyte plants when the stomata close during the day and open at night. Therefore, the CO_2 needed for photosynthesis can only enter the leaves at night, and must be stored until daytime when it can be used to make glucose. (***A***) This is an adaptation to conserve water in xerophytes which live in dry, hot climates, as it allows the stomata to be closed during the higher temperatures of daytime

(when transpiration is highest) but open in the cooler nights (when transpiration will be lowest). (***M***)

7. *Comparisons* could include:
   - Leaves are the photosynthetic organs in both groups (except cacti), and basic internal structure of the leaves essentially is the same (eg waxy cuticle, epidermis, mesophyll cells, guard cells and stomata).
   - Root structure and function essentially the same.
   - Stem present to hold up leaves, flowers.
   - Stomata function in same manner and can close to conserve water when necessary; most stomata found on lower epidermis.

   Contrasts could include:
   - Waxy cuticle is much thicker in xerophytes.
   - Leaves often reduced in size in xerophytes (eg spines in cacti).
   - Leaves often very hairy in xerophytes.
   - Stomata often sunken in pits in xerophytes.
   - Fewer stomata in xerophytes, and often confined to the lower epidermis.
   - Leaves may be rolled (lower epidermis on inside) in xerophytes.
   - Roots may grow much longer/deeper into the soil in xerophytes (eg marram grass).
   - Water storage tissue may be present in the stems of xerophytes (eg cacti).

   (***A*** – descriptions with little attempt at comparisons/contrasts; ***M*** – more extensive explanations with weak comparisons/contrasts; ***E*** – comparisons and contrasts extended, with detailed descriptions and explanations, including both structures and functions)

   Answers could be enhanced by clear, labelled, annotated diagrams – annotations could be descriptions (***A***), or explanations (***M***).

   Correct terms should be used in all answers (eg stomata not 'holes', xylem not 'water tubes', phloem not 'food tubes').

8. *Comparisons* could include:
   - Leaves are the photosynthetic organs in both groups, and basic internal structure of the leaves essentially the same (eg epidermis, mesophyll cells).
   - Stem present to hold up leaves and flowers (flowers above water in hydrophytes).
   - Roots serve to anchor the plant.
   - Gases enter and leave by diffusion, whether in water or in air.

   Contrasts could include:
   - Vascular and support tissues tend to be reduced in hydrophytes.
   - Roots tend to be reduced in hydrophytes.
   - Leaves and stems of hydrophytes tend to have large air spaces.
   - Cuticle and stomata absent in leaves of submerged hydrophytes, present on upper epidermis of floating leaves (eg water lilies).
   - Hydrophytes obtain water (osmosis), minerals (diffusion), from surrounding water.

   (***A*** – descriptions with little attempt at comparisons/contrasts; ***M*** – more extensive explanations with weak comparisons/contrasts; ***E*** – comparisons and contrasts extended, with detailed descriptions and explanations, including both structures and functions)

   Answers could be enhanced by clear, labelled, annotated diagrams – annotations could be descriptions (***A***), or explanations (***M***).

Correct terms should be used in all answers (eg stomata not 'holes', xylem not 'water tubes', phloem not 'food tubes').

9. Three clearly different taxonomic or functional groups (eg hydrophytes, mesophytes, xerophytes) need to be selected.

For *each* group, the *structures* associated with the *transpiration* need to be *described fully and carefully using correct terms*, eg:

- Size, shape, location of leaves.
- Location of stomata.
- Presence and location of hairs.
- Presence and structure of internal cell layers of leaves (eg cuticle, air spaces).
- Stem specialisations (eg water-storage tissue).
- Vascular tissue (xylem and phloem).
- Root system and root hairs.

For *each* group, *how* these structures perform their particular function needs to be *explained*, eg:

- Need for water (gas exchange, photosynthesis, cooling).
- Transport of water (osmosis, diffusion, transpiration pull).
- Factors affecting transpiration and how water loss is reduced.
- How and when stomata open and close.
- CAM photosynthesis.

For *each* group, reasons for *differences* between the groups need to be *explained*, eg:

- Habitat, eg adaptations to aquatic environment, arid environment, temperate environment.
- Size, eg need for support and transport systems in large terrestrial plants, little need for these in plants living in a buoyant aquatic environment; aquatic plants restricted in size, as dependent on diffusion of substances from water.
- Life cycle, eg need for aquatic plants to get their flowers above the water for pollination, rapid flowering and seeding of desert plants to take advantages of short periods of heavy rain, shedding of leaves in autumn by deciduous plants.

(***A*** – descriptions of the structures for the three groups; ***M*** – explanations of how they function; ***E*** – discussions with comparisons/contrasts and reasons for differences)

10. a. Friday

b. $30\ cm^2 \times 35$ open stomata = 1, 050 stomata

c. $30\ cm^2 \times 10$ open stomata $\times$ 20 leaves = 6,000 stomata

d. $30\ cm^2 \times 25$ open stomata $\times$ 20 leaves = 15,000 stomata

e. hot and dry or windy weather would have contributed to less stomata opening on Wednesday to prevent transpiration and reduce loss water loss by the plant.

Unit 11.3 Activity 3A: Transport systems in animals (Page 148)

1. A

2. C

3. B

4. B

5. E

6. D

Unit 11.3 Activity 4A: Blood composition (Page 152)

1. 1G; 2J; 3H; 4A; 5E; 6I; 7F; 8D; 9B; 10C.

2. White blood cells – fight disease / make antibodies / engulf and destroy pathogens.
Platelets – required to clot the blood.
Plasma – transports substances such as glucose and carbon dioxide.
Red blood cells – transport oxygen. (**A** – three correct)

Unit 11.3 Activity 4B: Human heart (Page 158)

1. 1K; 2H; 3B; 4L; 5C; 6A; 7F; 8I; 9J; 10E; 11G; 12D.

2. a. i.

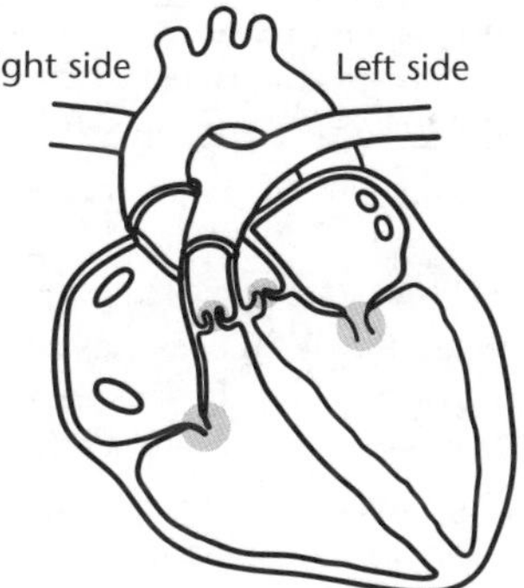

(**A** – position of all four valves must be correct)

ii. The valves prevent the backflow of blood / maintain blood flow in a forward direction. (**A**)

b.

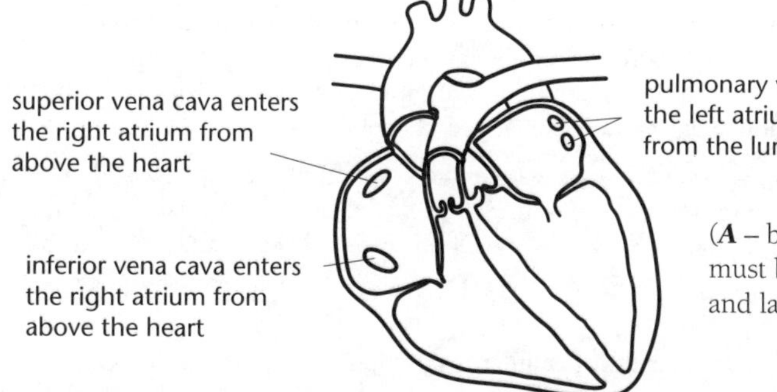

(**A** – both vessels must be drawn and labelled)

Unit 11.3 Activity 4C: Human circulation (Page 165)

1. **A** brain; **B** lungs; **C** intestine; **D** liver; **E** kidneys; **F** muscle; **G** bone marrow.

2.

a. (*A* – description of *two* differences)	b. (*M* – explanation of *one* difference)
One of:	*One of:*
Aorta walls are thick *and* vena cava walls are thin.	Because the aorta carries blood under high pressure *while* the vena cava carries blood under low pressure.
Aorta walls are elastic *and* vena cava walls are inelastic.	Because the elastic walls can smooth out the surge of blood that moves into the aorta with each heart beat *while* this surge is not significant by the time blood reaches the vena cava.
Aorta doesn't have valves *and* vena cava does have valves.	Because blood flow in the vena cava is under low pressure so valves prevent backflow of blood, *but* blood in aorta is under high pressure so it flows forwards.
Aorta diameter is small *and* vena cava diameter is large.	Small diameter maintains high pressure of blood in aorta *while* larger diameter in vena cava means little resistance to low-pressure blood returning to the heart.

c. Pulmonary vein carries oxygenated blood while pulmonary artery carries deoxygenated blood. (**A**)

d. Answer should contain the following points in paragraphs, not bullet points.

- Capillaries are narrow, thin-walled vessels that lie in beds between the arteries and veins in the circulatory system. The extensive capillary beds throughout the body ensure all tissues are in contact with a blood supply via the capillaries.
- Function of the capillaries is to exchange materials between the blood and the cells via diffusion across the single-celled walls of the capillaries.
- Materials that are exchanged from the blood to the cells include oxygen, nutrients (glucose, amino acids, fatty acids, glycerol, vitamins and minerals), water and hormones.
- Materials that are exchanged from the cells to the blood include carbon dioxide, urea, mineral salts and water.
- Capillaries sit in close contact with the cells and tissues in order to reduce the distance required for diffusion of substances exchanged between the blood and cells. (**A** – description of a function of capillaries; **M** – explanation of function in terms of how materials are exchanged in capillaries)

Use an example (such as the exchange of glucose from the gut to the blood and from the blood to the cells (eg in muscles), or the exchange of oxygen from the alveoli to the capillaries and from the blood to body cells) to explain in detail how the capillaries allow the substances to reach their desired destination. One such example follows.

Glucose is required by cells for cellular respiration, providing the energy required to run the cell. Food containing glucose is taken into the body and broken down to individual glucose molecules in the digestive tract. Large capillary beds close to the small intestine allow glucose to enter the blood by active transport. From the small intestine, the glucose-rich blood can continue on its journey through the circulatory system, enabling the subsequent delivery of glucose to cells throughout the body as the blood passes through capillary beds close to these cells. Glucose moves by active transport into the cells, where it is used to provide energy for respiration, which in turn provides energy for cellular processes. The structure of the capillary walls (one cell thick) allows effective diffusion into the cells. Extensive capillary beds in all parts of the body ensures all cells are in close proximity to capillaries and can therefore receive the glucose required for respiration.

Capillaries are also responsible for heat loss from the body. This is achieved because there are large capillary beds close to the skin. As the blood travels through these beds close to the skin, heat is lost from the blood to the cooler skin surface. When greater heat loss is required, the capillaries dilate; when reduced heat loss is required, the capillaries constrict. This mechanism allows effective control of heat loss from the body.

(***E*** – answer should include a description of the structure and purpose of the capillaries, an explanation of how capillaries function, discussion showing understanding of why capillaries are required within the circulatory system and how their structure enables them to carry out this function – use of specific examples strongly recommended)

3. The question requires that you explain the role of the *heart*, *plasma* and *red blood cells* and refer to at least four substances that are being transported. The four substances used in this answer are *oxygen*, *carbon dioxide*, *glucose* and *hormones*.

Substances are transported around the body via the circulatory system. This consists of a *pump* (the heart), and a series of closed vessels that transport the substances in the blood via the *red blood cells* or the *plasma*. The *heart* is central to transportation of substances around the body. Blood is pumped from the heart at high pressure and enters the arteries which have thick walls and a narrow lumen, enabling the high pressure to be maintained as the blood travels to the tissues. From the arteries, the blood enters the capillary beds, where, due to the thin walls of the capillaries, substances can be exchanged between the blood and the tissues. From the capillaries, blood travels at low pressure via the veins back to the heart.

Oxygen is transported via the *red blood cells*. The blood entering the lungs has a low concentration of oxygen, so oxygen diffuses from the alveoli into the capillary beds in the lungs. Once in the blood, the oxygen binds to the haemoglobin and is transported in this way to all cells in the body, as oxygen is required by the cells for cellular respiration. Once the blood reaches the capillary beds in the tissues, the oxygen will diffuse into the cells where there is a lower oxygen concentration than in the blood.

Substances such as *nutrients* (*glucose*, *amino acids*, *vitamins*, and *mineral salts*), *carbon dioxide*, *urea*, *water*, *hormones*, *antibodies* and *platelets* are transported in the plasma. As in the case of oxygen, many of these substances cross from the cell to the blood and vice versa due to concentration gradients. For example, *carbon dioxide*, a waste product of cellular respiration, enters the blood from the cells at the tissue level. It is dissolved in the plasma and is transported via the circulatory system to the lungs, where it diffuses out of the blood and into the alveoli where there is a lower concentration of carbon dioxide. *Glucose* is required in the cells for energy. It moves by active transport into the capillaries, where it is transported to the cells and again moves by active transport into the cells, where it is used in cellular respiration to provide energy. Some substances that are transported in the plasma are made in the body. *Hormones* are an example of chemicals that are made in the endocrine glands and transported in the plasma to sites in the body where they are required. An example is adrenalin – a hormone made in the adrenal glands in response to a fright. When it is transported in the blood, it causes an increase in heart rate, depth of breathing and alertness, so that a person can react effectively to the fright.

(***A*** – describes role of two of *heart*, *plasma* or *red blood cells*, and refers to *two* substances; ***M*** – explains how or why two of *heart*, *plasma* or *red blood cells* transport substances, with reference to at least *two* substances; ***E*** – a discussion of the role of two of *heart*, *plasma* or *red blood cells* which refers to at least *four* substances being transported)

4. **Arteries**:
   - *Function* – To carry oxygenated blood under high pressure away from heart to body (except for pulmonary artery, which carries deoxygenated blood from heart to lungs).
   - *Structures associated with function* – thick elastic walls (to withstand high pressure and to smooth out the surge of blood that moves into the artery with each heart beat), small lumen (to maintain high pressure).

   Veins:
   - *Function*: To carry deoxygenated blood under low pressure towards heart from body (except for pulmonary vein, which carries oxygenated blood from lungs to heart).
   - *Structures associated with function* – thin inelastic walls (as blood under low pressure), have valves (to ensure no backflow of blood).

   (***A*** – description of two features for either arteries or veins; **M** – explanation of each feature related to function; **E** – comparison of both arteries and veins relating structures to function)

Unit 11.3 Activity 4D: Internal transport in animals (Page 169)

1. In open transport systems, the blood is not carried in vessels but circulates in the body cavity, the coelom. In closed systems, the blood circulates in blood vessels (arteries and veins). (***A***)
2. Cnidaria have a simple, sac-like body structure and an inactive lifestyle. Simple diffusion is able to meet the nutrient needs of their cells. (***A***) Insects have a much more complex body structure and active lifestyles. Diffusion is inadequate in meeting the nutrient needs of insects' cells, thus making a transport system essential. (**M**)
3. Blood plasma is made of *water* with many *substances dissolved* in it (eg glucose, amino acids, mineral ions, blood proteins, O_2, CO_2). It may also have blood cells (red, white) and platelets. Its composition constantly changes, as substances move into and out of body cells. (***A***)
4. Oxygen-carrying pigments (haemocyanin, haemoglobin) are common as they can transport large amounts of O_2 around the body. (***A***) This is because the pigments can carry more O_2 than the equivalent volume of water (eg haemoglobin up to 60 times more). (**M**)
5. **a.** RBCs are non-living (no nucleus), *round, thin biconcave cells* containing *haemoglobin* (Hb) – this allows them to *transport O_2*. (***A***) The biconcave shape of the cells *increases the surface area*, allowing increased quantities of O_2 to enter/leave (per unit time); their packaging with Hb allows for *large quantities of O_2 to be transported*; their size and shape allow them to travel through the tiny capillaries to supply the cells with O_2. (**M**)

   b. Phagocytes are living (have nucleus and cytoplasm) cells which are able to *move and change their shape*, allowing them to *engulf pathogens*. (***A***) Phagocytes need energy to move, find and engulf pathogens. The presence of the nucleus (to control activities), mitochondria (to supply energy), the nature of the membrane, are all essential in allowing the phagocyte to move, leave the blood capillaries and engulf (by flowing around) pathogens. (**M**)
6. Arteries take blood away from the heart, veins return the blood to the heart. Therefore:
   - Arteries have a thick layer of elastic and connective tissue to allow for expansion of vessel to withstand high blood pressure; veins have a thinner layer, as blood is not under high pressure.
   - Arteries have a thick layer of elastic tissue and smooth muscle to allow arterial contractions to even out blood flow from heart pumping; veins have a thinner layer, as blood flow is much more even / no surging.

- Veins have valves to prevent the backflow of blood, as the flow is sluggish in veins compared with that in arteries.
- Arteries have a smaller lumen to keep the blood under pressure; veins have a large lumen, to deal with larger volumes of blood under low pressure.

(***A*** – function of both arteries and veins given, with at least three descriptive differences; ***M*** – differences explained; ***E*** – full explanations, with clear comparisons and contrast)

7. The heart is four chambered; two, thin-walled atria to collect blood from the body (right) and lungs (left); two muscular-walled ventricles to pump blood to lungs (right) and body (left). Veins bring blood to the atria; arteries take blood away from the ventricles. (***A***) Atria have thin, non-muscular walls, so that they can expand to fill with blood brought in from the body by the vena cava and the lungs by the pulmonary veins. The ventricles have thick, muscular walls because they are the pumping chambers. The right ventricle pumps blood to the lungs, via the pulmonary vein; the left ventricle pumps blood to the rest of the body, via the aorta. The right ventricle is less muscular than the left, as the blood is under less pressure so that it does not damage the delicate tissue of the lungs – which are only a short distance away. The two sides of the heart are separated by a septum, ensuring that oxygenated and deoxygenated blood do not mix. (***M***)

8. Fish have a single-loop system with blood being pumped by a two-chambered heart. Blood passes through the gills to collect O_2 and offload CO_2, then continues around the body. This is sufficient to meet the needs of fish living in an aquatic environment. Birds and mammals have a completely separate double system (ie two separate loops – one heart-lungs, other heart-rest of body). No mixing of oxygenated and deoxygenated blood occurs with a four-chambered heart. This keeps oxygenated and deoxygenated blood separate. One loop circulates to the lungs to get O_2 and offload CO_2, the other circulates around the body supplying the cells' needs. A completely separate double system is needed to meet the high energy demands of these animal groups. (***A***) In fish, blood loses a lot of pressure as it flows through the gills. This leaves only low pressure to continue the circulation of the blood around the body and back to the heart. Muscular movements from swimming are needed to help the circulation. Buoyancy of water assists, as fish do not have to pump blood upwards against gravity (unlike terrestrial animals). The double circulation of birds and mammals with two circuits involving the heart assists in maintaining pressure for efficient circulation around the body. This allows for an active terrestrial lifestyle; the low-pressure system of fish could not cope with double circulation. (***M***)

9. Three clearly different taxonomic (eg earthworms, insects, mammals) or functional groups (eg open blood system, single circulation closed system, double circulation closed system) need to be selected. For *each* group:

- Need to *describe* the ways in which the animals *transport the substances* around their bodies, from where substances enter the system to where they leave the system. Includes the *structures* concerned (eg heart, vessels (arteries, veins, capillaries), haemocoel, plasma, oxygen-carrying pigments, blood cells) as well as *names* the substances carried (eg O_2, CO_2, urea, mineral ions, nutrient, glucose, amino acids, hormones, blood proteins).
- Need to *explain* the ways in which the animals transport the substances around their bodies, from where substances enter the system to where they leave the system. Includes how the structures *function* in transporting substances (eg how blood circulates – pumping of the heart, operation of valves, muscular movements), need for oxygen-carrying pigments, composition of plasma and changes to it, oxygenation and de-oxygenation of blood (where and how it occurs), changes in blood pressure.

- Need to *discuss* the ways in which the animals transport the substances around their bodies, from where substances enter the system to where they leave the system. This *compares* the groups of animals and gives *reasons for similarities and differences* – eg whether the animals are aquatic or terrestrial, whether they obtain oxygen from water or air and different gas exchange sites and structures, comparative activity levels of the animals, comparative size of the animals.

(***A*** – descriptions of the structures for the three groups; ***M*** – explanations of how they function; ***E*** – discussions with comparisons/contrasts and reasons for differences)

10. Three clearly different taxonomic (eg earthworms, insects, mammals) or functional groups (eg open blood system, single circulation closed system, double circulation closed system) need to be selected. For *each* group, the *structures* associated with the transport of substances and the circulation of the blood need to be *described fully and carefully*, *using correct terms*, eg:

- The pumping device ('heart') present, and its structure.
- The parts (eg arteries, veins, capillaries, valves, haemocoel, blood cells, plasma) of the circulatory system, and their structure.
- Composition of plasma; what substances are transported.
- Blood cells.

For *each* group, *how* these structures perform their particular function needs to be *explained*, eg:

- How the heart pumps blood; how backflow is stopped.
- How blood pressure affects circulation.
- The role of pigments in transporting oxygen.
- How substances enter and leave the blood (and where).
- The role of blood cells, platelets.
- The role of the blood vessels.

For *each* group, reasons for *differences* between the groups needs to be *explained*, eg:

- Why open systems meet the needs of insects, but not the needs of vertebrates.
- Why a single circulation system meets the needs of fish, but mammals need a double system.
- Why some groups have haemocyanin as their oxygen-carrying pigment, while others have haemoglobin.
- Differences in the role of the blood in carrying O_2 and CO_2 (eg insects compared with mammals). These differences need to be related to differences in habitat (eg aquatic or terrestrial); gas exchange requirements (eg aquatic or air, site of gas exchange, O_2 transport, activity levels of the animals (high or low), size of the animals (large or small), ecological niches (specific adaptations to their habitat and lifestyle).

(***A*** – descriptions of the structures for the three groups; ***M*** – explanations of how they function; ***E*** – discussions with comparisons/contrasts and reasons for differences)

Unit 11.3 Activity 4E: Blood diseases and the lymphatic system (Page 175)

1. 1F; 2D; 3G; 4H; 5B; 6C; 7A; 8E.

2. **a.** F **b.** F **c.** F **d.** F **e.** T **f.** F **g.** F
h. F **i.** F **j.** F

3. **a.** Coronary arteries supply oxygen and nutrients to, and remove carbon dioxide and waste from, the heart muscle. (***A***)

b.
- High-fat diet.
- Stress.
- Obesity.
- High salt intake.
- Lack of regular exercise.
- High alcohol intake.

(***A*** – any two)

c. Answers given for high-fat diet and stress.

A high-fat diet will have high levels of cholesterol and lead to the build-up of plaque deposits on the walls of the arteries. This will reduce the diameter of the artery (lumen), therefore increasing blood pressure, which increases the strain on the heart. The deposits also cause narrowing and, eventually, complete blockage of the vessels.

High stress levels lead to an increase in blood pressure, which in turn places an extra strain on the heart.

(***A*** – describes what one factor does; ***M*** – explains what one factor may lead to)

4. **a.** The figure 110 represents the systolic blood pressure (pressure when blood is pumped out from the ventricles), the figure 70 represents the diastolic blood pressure (when heart is resting). (***A***)

A blood pressure of $\frac{110}{70}$ is relatively low, indicating that the person is not at risk of disease from high blood pressure. (***M***)

b. Hypertension is the word used to describe high blood pressure. (***A***)

c. Hypertension can result from narrowing of the blood vessels (particularly those within the heart) or enlargement of the heart. (***A***)

5. The following points are required in a paragraph. You must make links between the ability to transport oxygen, the need for oxygen in cellular respiration, the supply of energy to the muscles and the consequences of lack of oxygen in the muscles.

- Red blood cells are required to transport oxygen around the body, in this case to the muscles required for running, and haemoglobin is required within the structure of the red blood cells in order for them to carry the oxygen. (***A***)
- If there is not enough iron in the diet, there will not be enough haemoglobin and this will affect the ability to transport oxygen to the cells which in turn will affect the ability to provide energy. (***M***)
- Oxygen is required in order to obtain useable energy for the body via cellular respiration. Oxygen + glucose → energy + carbon dioxide + water.
- The runner who is anaemic lacks iron and therefore lacks haemoglobin and his or her blood cannot transport sufficient supplies of oxygen to the muscle cells. This means that the muscle cells will not get enough energy because there is not enough oxygen for respiration in the cells. As a result of these factors, the runner will lack energy and feel exhausted. (***E*** – link to respiration)

6. C

7. **a.** Deoxygenated blood is blood that is returning to the heart from the body and contains a higher level of CO_2 and a lower level of O_2 than oxygenated blood that left the heart. (***A***)

b. When a blood vessel is cut, a substance called fibrinogen in the plasma forms sticky threads of fibrin. (***A***) These stick to the wound and trap the platelets and blood cells in the area around the wound, forming a clot and stopping the bleeding. (***M***)

8. D

9. E

10. D

11. A

12. Crossword puzzle

Across

3. AMPHIBIANS—Have a three chambered heart
5. LYMPHATIC—A system that helps fight infection
8. MAMMALS—Have a four chambered heart
9. CLOSED—Blood flow is restricted to vessels
13. ERTHYROCYTES—Carry oxygen in blood
14. HEART—The “pump” of the circulatory system
18. ARTERY—Big vessel that carries blood away from the heart
19. WHITE—Blood cells that fight disease
20. CAPILLARY—Tiny vessels involved in material exchange
21. VENULE—Small vein

Down

1. FISH—Have a two chambered heart
2. PLASMA—Liquid part of blood
4. MYOCARDIUM—Heart muscle
6. PULMONARY—Blood flow through gills or lungs
7. CARDIAC CYCLE—Single beat of the heart (two words)
10. OPEN—Blood circulates through body cavities
11. SYSTEMIC—Blood flow through the body organs
12. VEIN—Big vessel that carries blood to the heart
15. VENTRICLE—Delivers blood to the body or lungs
16. ARTERIOLE—Small artery
17. ATRIUM—Delivers blood to the ventricles

Unit 11.3 Activity 5A: Excretion in animals (Page 187)

1. Excretion is the removal of *metabolic wastes* and excess chemicals from the body, egestion is the removal of *food that cannot be digested* (forms the faeces). (**A**)

2. Excretion removes waste products from the body's metabolic reactions as well as chemicals in excess of the body's needs (eg H_2O, salts). (**A**) This allows *homeostasis* – ie maintaining a stable (within a narrow range) internal environment so that life processes/metabolic reactions/cell reactions can occur. (**M**)

3. Excretion acts in osmoregulation by *removing excess water* from the body. (**A**) High water levels will create osmotic pressure and water will enter cells from the surrounding fluid. To maintain a balance, excess water is filtered from the blood by the excretory organs (eg kidneys) and passed to the outside environment. (**M**)

4. The kidneys help maintain homeostasis by removing *excess chemicals and metabolic wastes* from the body/blood. (**A**) The kidneys *filter* excess chemicals from the blood and these then pass out of the body. Needed chemicals are *reabsorbed* from the kidney tubules back into the blood and stay in the body. These two processes (filtration and reabsorption) keep the levels of chemicals (eg glucose, salts, nutrients, ions) in the body/blood at stable levels, ie homeostasis. (**M**)

5. Excretory systems remove waste products and excess chemicals from the blood, which is why the two systems need to be in close contact. (***A***) The excretory organs need to be in close association with the blood because they have to *filter* the chemicals from the blood. Needed chemicals are then *reabsorbed* into the blood. Neither of these processes could work unless the blood was in close contact with the excretory organs. (***M***)
6. Small, relatively inactive animals with a simple body construction (eg cnidaria) are able to rely on *simple diffusion* from cells to the environment for the removal of their metabolic wastes. (***A***) As animals have become larger, more active and more complex, simple diffusion with the environment is no longer effective to meet excretory needs. Therefore, an excretory system is needed to filter wastes from the blood and then remove the wastes to the outside. (***M***)

Unit 11.3 Activity 5B: Excretion in humans (Page 188)

1. 1B; 2G; 3J; 4I; 5D; 6E; 7F; 8A; 9H; 10C.
2. **a.** Brings blood high in urea and salts to the kidney to be filtered. (***A***)
   b.
   - Filters waste products from blood.
   - Helps control blood water levels by removing excess water and salts from the blood. (***A*** – one function needed)
3. Blood is entering the glomerulus from the renal artery at high pressure and small substances (everything other than blood cells and proteins) are being forced into the Bowman's capsule. (***A*** – idea of substances moving into Bowman's capsule; ***M*** – due to high pressure and larger substances remain in the blood, identification of one substance)
4. **a.** Drinking more water will mean less water is reabsorbed from the collecting ducts in the kidney and more water in the urine. (***A***) This will mean that there is less chance of solids sticking to the walls of the kidney, a higher chance of solids being dissolved and therefore less chance of a kidney stone forming. (***M***)
   b.
   - Ureter may become blocked by the stone and therefore prevent urine from passing out of the kidney.
   - The stone may be passed out with the urine.
   - Pain in the lower back.
   (***A*** – one outcome needed)
   c. Urea is the substance produced in the liver as a result of the metabolism of proteins, whereas urine is the mixture of urea, water and salts produced in the kidney. (***A***)
5. **Similarities**
   - Blood in the glomerulus and the capillary network protein molecules have the same concentration of proteins and blood cells, because protein molecules and blood cells are too large to be forced from the blood into the Bowman's capsule.
   - The concentration of glucose is the same, because, although glucose is a small molecule that is forced out of the blood in the glomerulus, all glucose should be reabsorbed back into the blood in the tubules.

   Differences
   - Blood in the glomerulus is high in urea, whereas blood in the capillary network is lower in urea. This is because urea is an unwanted waste that is filtered out of the blood at the glomerulus and not reabsorbed.
   - Blood in the glomerulus is high in salts, and blood in the capillary network is lower in salts. This is because all salts are filtered out of the blood at the glomerulus, and only those salts required by the body are reabsorbed into the blood in the tubule. All excess salts are excreted in the urine.

- Blood in the glomerulus has more water than does blood in the capillary network. Water is filtered out of the blood at the glomerulus and water is reabsorbed by the tubules, to obtain the concentration required by the blood. The level of water reabsorbed will vary depending on the level of salts in the blood and the volume of water that has been lost as sweat. Excess water is excreted in the urine.
- Blood in the glomerulus will have more oxygen and less carbon dioxide than blood in the capillary network. This is because cells in the kidney will require oxygen for cellular respiration and will produce carbon dioxide as a waste product of respiration.

(***A*** – describes one similarity and one difference / two similarities / two differences; ***M*** – gives a reason for each; ***E*** – identifies and explains at least two similarities and two differences)

6. **a.** Glucose, urea, amino acids, water, mineral salts. (***A*** – three substances needed)

b. High pressure in the glomerulus forces substances from the blood into Bowman's capsule which leads into the tubule. (***A***) This pressure is a combination of blood coming from the renal artery which carries blood under high pressure from the heart, and being forced from the larger diameter artery into the smaller diameter tubes of the glomerulus. (***M***)

c. **i.** Red blood cells, white blood cells, platelets, blood proteins. (***A*** – two substances)

ii. Too large to pass through the capillary walls in the glomerulus and into the Bowman's capsule. (***A***) Due to selectively permeable holes in the blood vessel walls. (***M***)

7. In hot weather, the body needs to sweat in order to reduce body temperature and maintain a constant internal temperature. This means that there is a lower concentration of water in the blood and more water is reabsorbed back into the blood. This means there is less water in the urine, making it more concentrated and darker in colour.

In cold weather, the body does not produce as much sweat. This means that there is a higher concentration of water in the blood and less water is reabsorbed back into the blood in the kidneys. This means that there is more water in the urine, making it a lighter colour. (***A*** – description links fluid levels in body and urine; ***M*** – link to urine colour)

8. Answer should contain the following points in paragraphs, not bullet points.

- The renal artery carries blood that is higher in water, urea and salts. The renal vein carries blood that is lower in urea and has less water and salts.
- Blood enters the kidney via the renal artery, moving into the glomerulus at high pressure, where all substances (apart from blood cells, platelets and blood proteins) are forced through into the Bowman's capsule and into the kidney tubule.
- In the tubule, useful substances are reabsorbed back into the blood. Substances that are not required are retained in the tubule and excreted in the urine.
- All the glucose and amino acids, and most of the salts, are reabsorbed by active transport. Water passes from the tubule to the blood via osmosis. Urea and the remaining water are excreted as urine.

(***A*** – describes function of kidney; ***M*** – explains function of kidney and refers to a difference between renal artery and renal vein; ***E*** – refers to named substances to show how difference arose)

9. **a.** Delivers urine produced in the kidneys to the bladder.

b. Tube via which urine leaves body after being stored in the bladder.

c. Filters blood, regulating salt and water balance in the body and excretes excess salts and urea. (***A***)

10. This could be answered well using bullet points.

- Aorta brings blood from heart into the kidney via the renal artery.
- Renal artery branches into small arteries which supply each nephron.
- The small artery splits into capillaries (known as the glomerulus as it enters the Bowman's capsule).
- Blood leaves the glomerulus and enters network of capillaries that surround the kidney tubule.
- These capillaries all deliver blood into the renal vein, which takes filtered blood back to heart via the vena cava. (***A***)

11. The glomerulus is a network of capillaries in the Bowman's capsule. Their role is to filter blood. If they are inflamed as in nephritis, they cannot carry out this role efficiently. Normally, the pressure inside the glomeruli is very high, forcing fluid from blood through walls of these capillaries into the Bowman's capsule. The filtrate that passes into the capsule contains water, minerals (salts), glucose, amino acids, vitamins and urea. Red and white blood cells, platelets and plasma proteins remain in the capillaries.(***A***) If the glomeruli are damaged due to inflammation or scarring, this initial filtration process does not work as it should, resulting in the potential for failure of the kidneys to effectively filter toxic wastes such as urea from blood and to control water and salt balance in the blood. Possible consequences of this are blood and/or protein in the urine. This damage to nephrons can lead to kidney failure over time. (***M***)

12. C

13. D

14. D

15. B

16. Crossword puzzle

Across

1. OSMOREGULATON—The control of body fluids
3. PITUITARY—Endocrine gland in brain that assists osmoregulation
7. NITROGENOUS—This type of waste is handled by kidney
9. ECTOTHERM—Body temperature determined by environment
10. URINARY—This system regulates body salts and fluids
11. KIDNEY—Organ that handles nitrogenous wastes
13. ADRENALGLANDS—Secretes aldosterone for osmoregulation (two words)
14. THERMOREGULATION—The control of body temperature

Down

2. REABSORPTION—The recovering of critical salts by the kidney
4. NEPHRON—Functional unit of the kidney
5. GLOMERULUS—Capillary knot in the kidney
6. COUNTERCURRENT—Blood flow that assists with thermoregulaton
8. UREA—A common type of nitrogenous waste in terrestrial animals
12. ENDOTHERM—Controls body temperature using anatomical features

Unit 11.4 Activity 1A: Respiration and gas exchange in plants (Page 195)

1. A

2. **I.** lots of air space in spongy mesophyll layer to allow CO2 movement to palisade layer where chloroplasts are located very close to cell walls

II. thin shape of leaf means short distance(<0.5mm) for CO2 to diffuse through

III. leaf has large <u>surface</u> area compared with its volume

3. (A) glucose + oxygen → carbon dioxide + water + ATP + <u>heat</u>

(B) $\underline{C_6H1_2O_6}$ + $6O_2$ $6CO_2$ + $\underline{6H_20}$ + 38ATP + <u>heat</u>

4. D

5. a: like D, the underside of leaf were not covered/painted, so there was no interference with loss of water during transpiration.

b: A lost water like D(about 5mls) and B and C did not

c: the undersides were painted so water-loss was minimized

d: stomata are found on underside of leaf

6. A

7. C

8. D

9. D

10. **a.** true

b. true

c. true

Unit 11.4 Activity 2A: Photosynthesis and respiration (Page 202)

1. **a.** Mitochondria are the cell organelles which are the site of (aerobic) respiration; chloroplasts are the cell organelles which are the site of photosynthesis. (***A***)

b. Grana are the membrane stacks in the chloroplasts where the light-dependent reactions of photosynthesis occur; stroma is the liquid matrix of the chloroplast where the light-independent reactions of photosynthesis occur. (***A***)

c. The light-dependent reactions of photosynthesis trap solar energy and convert it into chemical energy (ATP) and split water into H and O; the light-independent reactions join H and CO_2 into glucose using the ATP produced in the light-dependent reaction and release (waste) O_2. (***A***)

d. ADP (adenosine diphosphate) has one fewer phosphate than ATP (adenosine triphosphate), so ADP has less stored (chemical or bond) energy than ATP. (***A***)

e. Aerobic respiration needs O_2 for the (complete) breakdown of glucose; anaerobic respiration does not require O_2 for the (incomplete) breakdown of glucose. (***A***)

f. Glycolysis is the first stage in respiration and breaks down glucose into pyruvate (occurs in the cytoplasm); the Krebs cycle is the second stage of (aerobic) respiration and continues the breakdown of the pyruvate into CO_2 and H (occurs in the matrix of the mitochondria). (***A***)

g. The Krebs cycle is the second stage of (aerobic) respiration and continues the breakdown of the pyruvate from glycolysis into CO_2 and H (occurs in the matrix of the mitochondria); the electron transfer chain is the third/last stage of respiration and uses the electrons from H to produce ATP then O_2 bonds with the H atoms to from H_2O (occurs on the cristae of the mitochondria). (***A***)

- **h.** The Calvin cycle is the second stage of photosynthesis and bonds CO_2 and H to form glucose; the Krebs cycle is the second stage of (aerobic) respiration and breaks down pyruvate (from glucose) to form CO_2 and H. (***A***)

2. **a.** Chloroplasts are large oval organelles, inside which are the grana and stroma – where photosynthesis occurs. (***A***) The stacks of thylakoid membranes in a chloroplast contain chlorophyll to trap solar energy and provide a large SA for the light-dependent reactions. The grana provides the matrix for the light-independent reactions. (***M***)

b. Mitochondria are sausage-shaped organelles, with an inner membrane thrown into folds called cristae – where respiration occurs. (***A***) The matrix of the mitochondria provides the site for the Krebs cycle of aerobic respiration, while the cristae provide a large SA for the hydrogen transfer chain of aerobic respiration. (***M***)

3. Aerobic respiration has an electron transfer chain, where most of the ATP is made; anaerobic respiration does not have this stage, so produces much less energy. (***A***) The electron transfer chain needs O_2 to bond onto H atoms and form H_2O at the end of the chain. Without O_2, the electron transfer chain cannot proceed, stopping the production of ATP. Only the 2 ATP molecules made in glycolysis result. (***M***)

4. Respiration can be said to be the reverse of photosynthesis because respiration *breaks down glucose* to produce CO_2 and H_2O and releases energy (***A***); whereas photosynthesis uses (solar) energy to *produce glucose* from CO_2 and H_2O. (***M*** – both processes compared)

5. The rate of photosynthesis is faster when the chloroplasts are close to the cell membrane. (***A***) This is because the diffusion distance for CO_2 to move through the membrane to the chloroplasts is shorter than if the chloroplasts were centrally placed in the cell. (Similarly for the osmosis of H_2O into the chloroplasts.) (***M***)

6. Increasing the light intensity and $[CO_2]$ increase the rate of photosynthesis up to a maximum (graphs a and c), after which there is no change in the rate; increasing the temperature increases the rate up to an optimum of approx 40°C, after which the rate drops rapidly to approach zero (graph b). (***A*** – refers to temperature and one of the other two factors)

Increasing the temperature increases the rate of photosynthesis as the speed of the particles increases and so the number of collisions (and therefore reactions) increase. Increasing the temperature above the optimum breaks the weak hydrogen bonds that maintain the shape of the catalysing enzymes. The denatured enzymes can't catalyse the reaction and the rate of photosynthesis drops to zero. Increasing the light intensity increases the rate of photosynthesis, to a point at which the rate is limited by other factors (eg $[CO_2]$; temperature). (***M*** – at least one factor explained and the other two described).

(***E*** – at least two factors explained, together with a description from graph c that $[CO_2]$, light intensity, and temperature may *interact* to increase the rate of photosynthesis – eg increasing the $\%CO_2$ at any particular light intensity will increase the rate of photosynthesis; similarly, for an increase in temperature (up to the optimum))

Unit 11.4 Activity 3A: Gas exchange in animals (Page 211)

1. Gas exchange surfaces need to be *thin*, *moist*, and have *a large surface area.* (***A*** – all three)

2. **a.** Respiration is a *chemical* process that occurs in cells to *release energy*; gas exchange is a *physical* process (diffusion) in which the gases *O_2 and CO_2 are exchanged* across a semipermeable membrane/cell membrane. (***A***)

b. Breathing is a *physical* process, involving *muscular movements* to facilitate the movement of gases to and from a gas exchange surface; gas exchange is a *physical process* (*diffusion*), in which the gases O_2 and CO_2 are exchanged across a semipermeable membrane/cell membrane. (***A***)

3. Gas exchange surfaces are located between the environment (air, water) and the body (eg alveoli, gills, skin) *and* between the transport system and the cells. (***A***)

4. High concentration gradients are needed for O_2 and CO_2 to diffuse in the correct direction / O_2 in and CO_2 out. (***A***) The *higher the concentration gradient the more rapid the diffusion* of the two gases, and this will increase/maximise the rate of respiration/release of energy. (***M***)

5. The gas exchange system is closely associated with the circulatory system, as the circulatory system is the transport system for the body and transports both O_2 and CO_2 around the body to all the cells for/from respiration. (***A***) In large multicellular animals, the direct diffusion of O_2 and CO_2 into/out of the body and from cell to cell is too slow to meet the respiratory/energy demands of the animal. A transport/circulatory system is needed to ensure that the gases are rapidly taken to and from the cells and to maximise respiration/energy release. (***M***)

6. A gas exchange system is needed to obtain O_2 from the surrounding medium/environment and take it into the body rapidly while removing CO_2 from the body and returning it to the surrounding medium/environment. (***A***) A system is needed because most multicellular animals are large, with their cells forming complex tissues and systems that are removed from contact with the surrounding medium. Therefore, multicellular animals cannot use simple diffusion to obtain O_2 and remove CO_2; it is too slow and the skin of the animal is typically impervious to gases (exception: earthworms and some amphibians). A gas exchange system is needed, and usually is associated with, a circulatory system, to transport the gases throughout the body (exception is insects). (***M***)

7. *Comparisons* of the two systems could include:

- Both have closable openings to air (spiracles in insects, mouth/nose in mammals).
- Both systems have branching tubes that penetrate the body (trachea in insects, trachea and bronchi in mammals), and get smaller and smaller (tracheoles in insects, and bronchioles in mammals).
- Large air tubes have strengthening bands to keep them open (chitin in insects, cartilage in mammals).
- Gas exchange process is diffusion (across SPMs), dependent on the concentration gradient; O_2 diffuses inwards, CO_2 diffuses outwards. Gas exchange surfaces are kept moist.
- Breathing occurs (muscular movements) to facilitate movement of gases into and out of the body.

Contrasts of the two systems could include:

- Insects use a tracheal system; mammals use a lung system.
- Insect tracheal system transports gases directly to/from cells; mammals use the circulatory system to transport gases to/from cells.
- Oxygen-carrying pigment present (haemoglobin) in mammals.
- In insects, the finest air tubes pass between the cells for direct gas exchange; in mammals, the finest tubes end in air sacs (the alveoli), which exchange gases with the capillaries of the blood. Alveoli provide a large SA for gas exchange with the blood capillaries.

- Breathing in insects uses pumping muscular movements of the abdomen; in mammals, breathing uses muscular movements of the intercostal muscles of the ribcage, along with muscular movements of the diaphragm.
- Tracheal system of insects is a factor that acts to limit their size (insects all small); lung system with circulatory system for transport allows for development of large body size.

Correct terms should be used in all answers (eg trachea, not 'air tubes'; alveoli, not 'air sacs').

(***A*** – descriptions with little attempt at comparisons/contrasts; ***M*** – more extensive explanations with weak comparisons/contrasts; ***E*** – comparisons and contrasts extended, with detailed descriptions and explanations, including both structures and functions)

Answers could be enhanced by clear, labelled, annotated diagrams – annotations could be descriptions (***A***), or explanations (***M***).

8. *Comparisons* of the two systems could include:

- Both systems are closely connected with the circulatory system, which transports the gases.
- Both have a large SA (gill filaments, alveoli) for gas exchange with the blood capillaries.
- Numerous blood capillaries are in close contact with the gills/alveoli to increase gas exchange.
- Haemoglobin in RBCs transports O_2 in both systems.
- Breathing movements facilitate the movement of gases into and out of the body.
- Gas exchange process is diffusion (across SPMs), dependent on the concentration gradient; O_2 diffuses inwards, CO_2 diffuses outwards.

Contrasts of the two systems could include:

- Fish system is external (gills lie free in the water), mammal system is internal (lungs contained in chest cavity).
- Gas exchange occurs between gill filaments and capillaries in fish, between alveoli and capillaries in mammals. Both filaments and alveoli provide a large SA for gas exchange.
- Gills supported by bony structures, gills protected by bony operculum.
- Trachea and bronchi in mammals supported by bands of cartilage.
- Counter-current system exists between blood and gills in the fish – maximises concentration gradient for the gases to facilitate diffusion (needed as O_2 content of water lower than that of air).
- Breathing – fish gulp in water and force it over the gills by closing mouth, mammals draw air in and out using muscular contractions of the intercostal muscles of the ribcage together with the muscles of the diaphragm.

Correct terms should be used in all answers (eg trachea, not 'air tubes'; alveoli, not 'air sacs').

(***A*** – descriptions with little attempt at comparisons/contrasts; ***M*** – more extensive explanations with weak comparisons/contrasts; ***E*** – comparisons and contrasts extended, with detailed descriptions and explanations, including both structures and functions)

Answers could be enhanced by clear, labelled, annotated diagrams – annotations could be descriptions (***A***), or explanations (***M***).

Unit 11.4 Activity 3B: Gas exchange (Page 216)

1. - Air is not cleaned/filtered by the hairs inside the nose.
   - Air is not warmed.
   - Air is not moistened.
   - Bacteria and dust cannot be trapped by mucus. (***A*** – any two)
2. Keeps the trachea and bronchi open. (***A***)
3. Breathing is the action by which air is moved into and out of the lungs. (***A***) This allows oxygen (required by all cells) to enter the body through the lungs, and carbon dioxide (a poisonous waste product of cellular respiration) to leave the body. (***M***)
4. Respiration is the chemical reaction in which glucose and oxygen are used to produce energy. Carbon dioxide and water are waste products of this reaction. (***A***; ***M*** – the energy is required for cellular/body processes)
5. Answer should contain the following points in paragraphs, not bullet points.
   - Oxygen – more in inhaled than in exhaled air. This is because oxygen is required in the body for cellular respiration. Oxygen diffuses from the alveoli into the blood and is transported to cells in the body where it is used in cellular respiration. Blood leaving the tissues and air leaving the alveoli have lower oxygen levels because the oxygen has been used in the cells.
   - Carbon dioxide – more in exhaled than in inhaled air. Carbon dioxide is produced as a waste product of respiration in the cells, giving a higher level of carbon dioxide in the blood leaving the tissues and consequently a higher level of carbon dioxide in the exhaled air.
   - Nitrogen – no difference in amount in inhaled and in exhaled air. This is because nitrogen gas cannot be absorbed and is not used in this form by the body.
   - Water – more in exhaled air than in inhaled air. This is because water is produced as a waste product of respiration. Exhaled air is one way in which excess water is removed.

   (***A*** – describes difference for three of the four gases; ***M*** – explanation for why there is a difference for three of the four gases; ***E*** – answer should include a description of the differences between the inhaled and exhaled air for all four gases, explanation of why these differences occur, discussion showing understanding of how differences between oxygen and carbon dioxide levels are linked)

Unit 11.4 Activity 3C: Respiration, breathing and gas exchange (Page 220)

1. 1C; 2B; 3I; 4H; 5O; 6L; 7K; 8G; 9E; 10J; 11F; 12A; 13D; 14N; 15M.
2. Diaphragm – contracts and flattens, lowering the air pressure.
   Intercostal muscles – contract and move the ribcage up and out. (***A***)
3. **a.** The goblet cells produce and secrete mucus. (***A***)
   b. The cilia beat constantly (***A***); moving mucus that traps unwanted substances (such as dust and bacteria) and moving these substances out of the lungs and airways. (***M***)
   c. Tar in cigarette smoke paralyses the cilia. This reduces the ability of cilia to function in order to clear mucus containing bacteria and dust. This will increase the risk of infections in the airways and increase coughing in an effort to remove the mucus.
   Smoke causes inflammation of the airways, which in turn causes an increase in mucus secretions. This increased mucus reduces the ability of the airways to function effectively, causing increased coughing and increasing the potential for infection.

(***A*** – one effect described; ***M*** – one effect with a reason for the effect explained)

d. Exchange of gases (oxygen and carbon dioxide). (***A***)

4. Answer should contain the following points in paragraphs, not bullet points.

- Damage is broken capillaries, destruction of the walls of alveoli, reduction in surface area available for gas exchange, loss of elasticity of alveoli.
- Broken capillaries and/or damage to walls of the alveoli lead to a reduction in gas exchange between the alveoli and the blood, leading to reduction in available oxygen for respiration in cells within the body, which results in reduced energy, increased tiredness and increased strain on the lungs as increased rate and depth of breathing are used to attempt to increase oxygen supply to cells.
- Over-inflation of the alveoli leads to a reduction in the folds in the alveoli wall and reduction in the surface area available for gas exchange. This causes a reduction in available oxygen for respiration in cells within the body, which results in reduced energy, increased tiredness and increased strain on the lungs as increased rate and depth of breathing are used to attempt to increase oxygen supply to cells.

5. **a.** **i.** Provides large surface area for gas exchange between air and blood.

ii. Carries oxygen around the body.

iii. Dissolves gases in order for them to diffuse across the cell membrane. (***A***)

b. *Any of two of the following:*

- Large surface area.
- Thin membrane – gases diffuse more efficiently across a thin membrane as there is less distance to travel between the air and the blood.
- Moist surface – this is required for gases to dissolve for diffusion. (***A***)

6. The function of a bronchiole is to deliver and return air to and from the alveoli. In the diagram, the bronchiole of the asthmatic has constricted to reduce the size of the tube through which the air can flow and the volume of mucus produced has increased, further reducing the size of the tube through which the air can flow. (***A***) This means that the volume of air flowing from the nose/mouth into the alveoli has been reduced, decreasing the ability of the respiratory system to efficiently deliver adequate oxygen supplies to the body via the blood. (***M***)

7. a

8. c

9. b

Unit 11.5 Activity 1A: Hormones and plant responses (Page 231)

1. A

2. D

3. **a.** Ethylene (or ethene).

b. Ethylene/ethene stimulates enzymes that break down cell walls, break down chlorophyll and convert starch to sugars.

c. The tomatoes are less likely to be damaged in transport.

Unit 11.5 Activity 2A: Sensitivity and co-ordination in animals (Page 242)

1. Neurons are nerve cells. Each has a *cell body* with a nucleus and cytoplasm (contains the organelles). Extending from the cell body is a long *axon* to transmit the nerve impulse; the axon may have a sheath of myelin. Short, branching *dendrites* come from the cell body; these collect the stimulus from sensory receptors or the impulse from the axons of neighbouring neurons. (***A*** – all three parts described)

2. Information is transmitted *electrically* along neurons in the form of the *nerve impulse*, and then *chemically* in the form of *neurotransmitters* across the synapses between neurons. (***A***) The nerve impulse is a *flow of charge*. The inside of the axon is negatively charged with respect to the outside; Na^+ ions are actively excluded. The impulse changes the *permeability of the membrane*, allowing Na^+ ions to flood in, reversing the standing charge of the axon (ie now positive inside). The Na^+ ions are then rapidly pumped out to return the charge to negative inside. This process happens very rapidly (about a millisecond), and sweeps along the axon (ie there is a flow of charge). When the impulse reaches the terminal dendrites, it causes the release of chemicals called neurotransmitters, which rapidly diffuse across the synapse. The arrival of neurotransmitters at the dendrites of a neighbouring neuron causes *changes in permeability of the membrane* and Na^+ ions flood in, starting an impulse. An impulse is initially triggered by a stimulus – this needs to be above *a threshold level* to cause an impulse. The larger the stimulus, the more impulses are triggered. The speed of the impulse does not change. (***M***)
3. **a.** Sensory organs (eg eyes, ears) *detect environmental stimuli*, such as sound, light, chemicals, heat, touch, texture. The stimulus received *triggers a nerve impulse* in associated sensory neurons. (***A***)

   b. Effector organs (eg muscles, glands) receive the nerve impulse from the motor neurons and *make a response*. The response may be a *movement* (from the muscles) or the release of *chemicals* (eg hormones from endocrine glands, tears from tear glands, sweat from sweat glands).
4. Cephalisation refers to the centralisation of sensory receptors/organs and nerve co-ordination/brain in a 'head' region of the body. (***A***) This centralisation allows for *rapid transmission of information* from the sensory receptors to the brain, and so a *rapid response*. In most animals, the head is at the anterior ('front') end of the body, which is typically the first part of the body into a new environment; therefore, the centralisation of sensory receptors at the anterior end is important for *rapid collection of information*, enhancing detection of things (eg predators, prey, harmful chemicals) and *assisting the animal's survival*. (***M***)
5. Compared with a network system, a (simple) CNS allows for *rapid*, *directional* transmission of the nerve impulse with the *co-ordination* of movements. (***A***) This is because the (simple) CNS has a central nerve cord (ventrally placed), which allows for the movement of the impulse from the receptors along the cord to the effectors (ie one way); whereas, in a network of nerves, the impulse can move in all directions, reducing speed of transmission as well as co-ordination. In the network system, axons are shorter than in the CNS and with more synapses. As movement across synapses is slower than along axons, the more synapses, the slower the transmission of information. Hence, the slower transmission in the network system. There is no centralisation (eg ganglia, brain) of nerves in the network system – this means a lack of co-ordination in this system compared with the CNS. (***M***)
6. A complex CNS has *complex* sense organs, effector organs, brain, feedback control. A PNS and ANS may be present. All these allow for more *effective collection*, *processing* and *transmission of information*, as well as *co-ordination of activities* compared with the simple CNS. (***A***) Sophisticated sense organs (eg mammal eye, ear) allow for detection of a large range of stimuli, enhancing awareness of the environment (eg greater chance of detecting predators, prey). A PNS allows for efficient collection of information from the receptors and its transmission to the CNS and brain. ANS allows for control of the internal organs separate from the motor system; brain subconsciously controls ANS, so the individual can 'ignore' the workings of the internal systems (eg digestion) and concentrate on the voluntary system (ie the control of skeletal muscles or effectors). The highly developed

brain allows for increased co-ordination, processing of information, learning and thought in animals such as humans. Endocrine glands allow for fine feedback control of responses and promote homeostasis. Cephalisation is advanced, with a high degree of centralisation of sense organs in the head, together with a brain. This gives more effective reception of stimuli, more rapid transmission to the (close) brain, processing and response (eg 'flight or fight'). (***M***)

7. The endocrine system is one of the effectors in humans. The endocrine glands release h*ormones* when a stimulus/message is received. These *control the internal environment by regulating processes* (eg storage of glycogen); *regulating levels of chemicals* (eg glucose, water); *promoting growth*; *controlling sexual development*, *cycles and gamete production; promoting emergency response* ('flight or fight'). All these functions are part of maintaining *homeostasis*. (***A***) The master gland is the pituitary – found under, and controlled by, the hypothalamus of the brain. Receptors detect levels of chemicals (eg glucose, water) in the blood and feed this information to the endocrine glands. The glands respond by either producing or stopping production of hormone(s), regulating the level of a chemical in the blood by starting or stopping appropriate chemical processes (eg if level of glucose in the blood is high, the pancreas (an endocrine gland) is stimulated to produce the hormone insulin – this targets the liver to convert the excess glucose to glycogen for storage until needed). (***M*** – explanation includes appropriate example) This control is known as *feedback control*, and is typical of how the endocrine system works in promoting homeostasis. The level of a chemical in the blood feeds back information to the brain, which responds by instructing the endocrine gland to start or stop production of the hormone. For example, high levels of glucose stimulate the production of insulin by the pancreas – as the glucose is removed from the blood (as it is converted to glycogen), its level in the blood falls; this information is fed back to the receptors in the brain, which then instruct the pancreas to stop production of insulin. The level of glucose constantly fluctuates in the blood as glucose is added (eg diet) or removed (eg respiration). Insulin production is therefore stimulated or inhibited, and the level of glucose maintained within a narrow range in the blood at all times. (***E*** – ***M*** extended to include feedback control, with example)

8. Three clearly different taxonomic (eg cnidaria, earthworms, mammals) or functional groups (eg network, simple CNS, complex CNS) need to be selected.

For *each* group, the *structures* associated with the *nervous system* need to be *described fully and carefully using correct terms*, eg:

- Descriptions of structure of sensory neurons, motor neurons, synapse, nerve networks, CNS, brain; their location and function.
- Description of receptors and effectors; their location and function.

For *each* group, *how* these structures perform their particular function needs to be *explained*, eg:

- Collection of information/stimuli.
- Transmission of the nerve impulse.
- Role of myelin sheath.
- Transmission across synapses.
- Role of brain.
- Presence and purpose of cephalisation.
- Co-ordination of stimulus – response.
- Role of hormones and feedback control.

For *each* group, reasons for *differences* between the groups needs to be *explained* in relation to differences in:

- Habitat (eg aquatic or terrestrial, what stimuli need detecting).
- Activity levels of the animals (sessile or mobile, slow or fast moving, herbivore or carnivore).
- Size and complexity of the animals (large or small, simple or complex body structure, presence/absence of vertebral column, cephalisation).
- Ecological niches (eg underground detritus-feeding earthworm versus freshwater sessile *Hydra* versus active predatory mammal).

(**A** – descriptions of the structures for the three groups; **M** – explanations of how they function; **E** – discussions with comparisons/contrasts and reasons for differences)

Unit 11.5 Activity 3A: Regulation of body water content – osmoregulation (Page 245)

1. 1C; 2F; 3B; 4A; 5D; 6E.

2. E, B, A, F, C, D. (**A** – all correct)

3. When a person becomes dehydrated, the body responds using the following mechanisms to return to a normal water concentration in the blood:

- The reduced water concentration in blood is detected in the hypothalamus which causes the person to feel thirsty. This should stimulate the person to increase his or her fluid intake which will help increase water in the body.
- The hypothalamus also sends nerve impulses to the pituitary gland to increase secretion of the hormone ADH. This hormone increases the volume of water that is reabsorbed by the kidneys, reducing the volume of urine produced. These two factors work together to re-establish normal water concentrations in blood.

(**A** – description of two responses; **M** – explanation relates to how each response restores fluid balance; **E** – includes reference to hypothalamus and pituitary glands)

Unit 11.5 Activity 3B: Temperature regulation (Page 248)

1. **a.** Dehydration. (**A**)

b. The blood vessels close to the face and skin dilate in order to increase the blood flow in vessels close to the skin, producing the flushed look. (**A**) As the blood travels through these vessels close to the skin, heat is lost to the air, helping to control body temperature. (**M**)

c. Vigorous exercise will produce dehydration as a result of increased sweating. This is detected by osmo-receptors in the hypothalamus as a low level of water in the blood; the osmo-receptors send a signal to the pituitary gland to increase secretion of ADH. This increase in ADH increases the rate of reabsorption of water by the kidney tubules. After a period of increased reabsorption, the normal water balance will be established; this change will be detected by the hypothalamus, which in turn will reduce ADH secretion and therefore reduce reabsorption in the kidney tubules.(**A** – both ideas of production of ADH and increased reabsorption by kidney; **M** – production of ADH linked to pituitary and low water levels in blood)

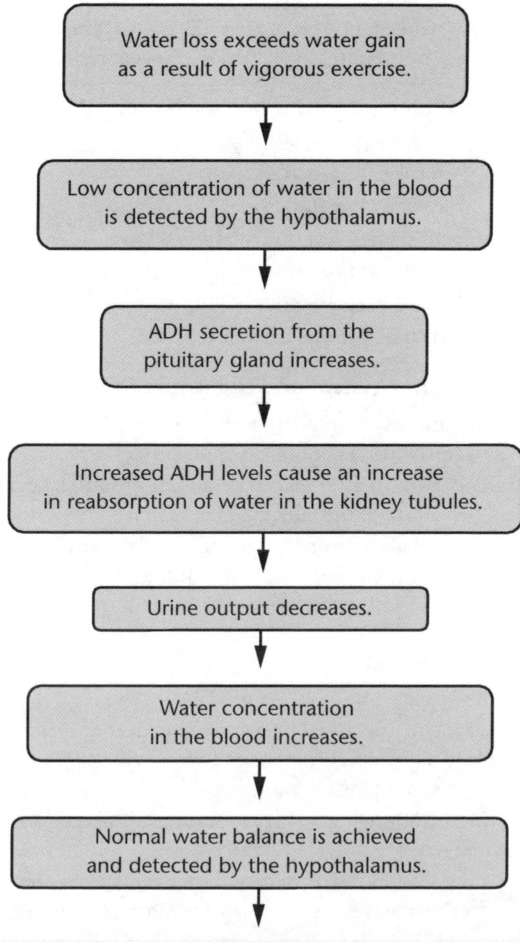

(***E*** – all ideas covered plus flow diagram)

2. **a.** The ability of the body to maintain a stable internal environment despite changes in the external environment. (***A***)

b. The hands and feet have a lower temperature than 37°C because they are a long distance from the core of the body, have a large number of blood vessels close to the surface of the skin, and expose a large skin surface area. (***A***) As blood passes through the vessels close to the skin, over this large surface area heat is lost, because the temperature inside the body is higher than the temperature outside the body and heat moves from hotter to colder places, therefore reducing the temperature of the hands and feet. (***M***)

c. When the body temperature is low, the blood vessels close to the surface of the skin constrict (get narrower), therefore reducing the volume of blood that passes close to the skin surface and therefore reducing heat loss – this is seen in pale-coloured skin at the extremities when the body is cold. Shivering produces heat from the muscles, in an attempt to increase the body's temperature. The metabolic rate increases in order to produce more heat, in an attempt to increase the body's temperature. (***A*** – any two points described; ***M*** – any two responses explained in terms of restoring temperature to normal)

Unit 11.5 Activity 3C: Maintaining blood sugar levels (Page 250)

1. Glucose drink increases blood-sugar levels as glucose is a small molecule absorbed directly into the bloodstream from the stomach. (***A***)

2. *The following points need to be made in the answer:*

- After the drink, blood-glucose levels increase.
- This increase is detected and insulin secretion by the pancreas increases. (***A***)
- Glucose is converted to glycogen under the influence of insulin, and stored in the liver.
- Level of glucose in the blood drops back down. (***M***)

Unit 11.5 Activity 3D: Endocrine system and challenges to homeostasis (Page 256)

1. a. Anti-diuretic hormone (ADH), growth hormone (GH), thyroid-stimulating hormone (TSH), luteinising hormone (LH), follicle-stimulating hormone (FSH). (***A*** – two correct)

b. The pituitary gland is the 'master gland' because it produces hormones that control the functioning of other hormone-producing glands in the endocrine system. (***A***)

Example: The pituitary gland secretes TSH (thyroid stimulating hormone) in response to detecting the level of thyroxine in the blood; the level of TSH is detected by the thyroid gland, which in turn produces thyroxine in response to the TSH level. (***M*** – example must be given)

c. i. Adrenal gland.

ii. On top of the kidney. (***A*** – both correct)

d. i. Increased heart rate, increased breathing rate, deeper breathing, constriction of blood vessels to skin and gut, increased blood glucose as a result of glycogen conversion, increased tension in muscles.

ii. Heart rate and breathing rate increase and breathing deepens in order to increase the rate of delivery of oxygen and glucose to the muscles.

Glucagon is released in order to increase the rate of conversion of stored glycogen to glucose, to provide increased energy for muscles.

Constriction of blood vessels to the skin and gut reduces blood flow to these areas in order to increase blood flow to the muscles.

Tension increases in skeletal muscles in order to prepare for rapid response.

(***A*** – one correct in part **d.i.**; ***M*** – ***A*** plus one correct in part **d.ii.**)

2. a. Relaxation, slowed reaction rate, dilation of blood vessels, loss of body heat, increased urine production, dehydration, nausea/dizziness, increased action of the liver in detoxification. (***A*** – any one)

b. Answers for both alcohol and tobacco are given – only one is required.

At least two of the points should be made, and the points should be made in paragraphs. For each point made, you must explain why the harmful effect occurs and discuss the effect of this on normal body function.

Alcohol

It is important to differentiate between low- and high-level use of alcohol in terms of effect on the body. The introductory sentence following makes this point clear.

Long-term, low-level use of alcohol (eg three glasses of wine or beer per week) will not have adverse effects on the body's ability to maintain normal function. However, long-term moderate- to high-level use of alcohol will have adverse effects on the body. This answer will address those issues.

- The liver has a central role in the removal of toxins from the bloodstream. Alcohol is one of these toxins. Drinking large quantities of alcohol will create a high workload for the liver and if this is continued for a long period, the liver will become damaged (cirrhosis, cancer, hepatitis) and will not be able to function properly.
- The brain and nervous system will be enhanced in the short term by low levels of alcohol, leading to relaxation and enhancement of mood (happiness). However, alcohol is a depressant drug and with increased consumption the ability of the brain to respond is slowed and aggression is increased. Over time, this can lead to loss of brain cells, memory loss, reduced ability to learn and reduced ability to make effective decisions. Alcohol can be an addictive drug, which in turn leads to increased consumption, increased tissue damage and increased reduction in normal body functioning.
- High levels of alcohol consumption can contribute to high blood pressure, heart disease and strokes. This potential damage of the heart tissue and circulatory system in turn lead to reduced quality of life, as the circulatory system is unable to effectively deliver oxygen and nutrients to the body.
- Alcohol has a very high energy (calorific) value, and consumption of alcohol may lead to weight gain and, potentially, obesity over time. High levels of alcohol consumption also lead to potential damage of the cells lining the gut, increasing the potential for stomach ulcers and impaired ability to digest food, and, again, reduced quality of life.

Tobacco

- Tar in tobacco cigarettes irritates the lining of the airways, paralysing the cilia and damaging the alveoli. This leads to increased mucus production, non-removal of mucus, reduction of gas exchange, shortness of breath, increased chance of airway infection, bronchitis and eventually emphysema. These factors all reduce normal body function. Asthmatic problems are made worse by smoking, due to the irritation of the airways combining with the narrowed airways resulting from asthma.
- The reduction in airway and gas exchange function leads to increased heart rate, and blood vessels constrict, leading to increased blood pressure. Carbon monoxide in the smoke binds more effectively to haemoglobin than does oxygen, leading to a reduction in the level of oxygen carried by the red blood cells. This results in a reduction in the ability of the circulatory system to deliver oxygen throughout the body. This, in turn, leads to an increase in demand for oxygen and more stress on the heart. These factors increase the potential for coronary heart disease and reduce the ability of the body to carry out normal function.
- Nicotine is a stimulant drug found in tobacco, which initially increases alertness and suppresses appetite. It is an addictive drug, making the brain respond by craving for more. This addictive nature leads to multiplication of the negative impacts of tobacco on the circulatory and digestive systems.
- Tar is a carcinogen, and as such increases the potential for multiple mutations in cells, which leads to the development of cancers. Cancers reduce the ability of the body to function normally and reduce life expectancy.

- Tobacco smoking causes irritation of the lining of the gut. This may lead to the formation of peptic ulcers and a reduction in the ability to digest food and therefore deliver energy to the body. This in turn will reduce normal body function and quality of life.

(**A** – description of long-term effects of either tobacco or alcohol, two needed; **M** – explanation of how at least one body function affected; **E** – discussion of impact on normal body function)

3. **a.** **i.** ADH – controls water balance.

ii. GH – controls growth rate.

iii. Thyroxine – controls metabolic rate.

iv. Insulin – controls blood glucose levels.

v. Adrenalin – prepares the body for fight/flight or emergency situations. (**A** – any three correct)

b. **i.** The thyroxine level in the blood is detected by the pituitary. (**A**) If the level of thyroxine is low, the pituitary gland increases the level of TSH released, which in turn causes the thyroid gland to release more thyroxine, thus increasing the thyroxine levels in the blood. Could also answer using the reverse scenario, starting with high thyroxine levels. (**M**)

ii. Lack of iodine means that the thyroid gland cannot make enough thyroxine. (**A**) Lack of thyroxine causes pituitary to produce more TSH, TSH stimulates thyroid gland, no feedback to stop TSH production. (**M**) This means that the feedback system cannot maintain the balance of thyroxine and therefore the appropriate level of thyroxine and therefore the correct metabolic rate. This will lead to low metabolic rate and potentially an increase in the size of the thyroid gland (goitre). Lack of iodine → Low thyroxine levels → Increased TSH production → As there is low iodine, the thryoxine will not be available and TSH levels will continue to increase, with no resulting increase in metabolic rate → Enlarged thyroid gland. (**E**)

4. **a.** Sudden increase in blood sugar/glucose levels resulting in increased production of insulin by the pancreas / excess blood sugar/glucose is converted into glycogen for storage (in the liver). (**A** – description of body's response)

b. Increased energy/alertness/irritability/nervousness/concentration/production of urine/production of stomach acid/stimulation of the central nervous system (CNS)/heart/breathing rate. (**A** – any two)

c. A sportsperson in training should drink water rather than energy drinks. Exercise results in sweating, sweat is composed mostly of water. Drinking water will directly replace water that has been lost, reducing the risk of dehydration and possible death. Energy drinks contain high levels of sugar and caffeine. These substances may have some short-term benefits, such as boosting energy levels needed for muscle action (both substances) and releasing stored glycogen for energy (caffeine). However, they may do more harm than good. High-sugar energy drinks may cause water to leave the bloodstream, resulting in dehydration. Excess sugar may not be used up by the sportsperson, resulting in weight gain over a period of time. Caffeine is a diuretic, resulting in more urine being produced and more water leaving the body. There is also the risk of caffeine addiction over a period of time. Limiting the energy drink intake to 500 mL for safety reasons may not be sufficient to replace lost water. (**A** – description of an effect of both water and energy drink with support for one; **M** – explains how water and energy drink affect the body; **E** – comparison of both water and energy drink, with reasons as to why water is preferable to energy drinks)

5. **a.** The two significant changes are:
   - Blood vessels dilate (become wider).
   - Greater volumes of sweat are produced. (***A***)

   These factors work together to increase rate at which blood is cooled. Dilation of blood vessels allows more blood to pass close to surface of the body where it is cooled. Increased sweat ensures that the skin kept moist. More heat will be lost from skin if it is moist than if it is dry.

   b. Reduction in body temperature detected by hypothalamus in the brain. As a result, the following factors work together to enable a cold body to return to normal temperature:
   - The blood vessels close to surface of skin and in the extremities (hands and feet) constrict (become narrower). This reduces volume of blood that travels close to surface of the body where it will lose heat to colder air outside the body via radiation. The constriction of these vessels makes the skin look very pale. The fingernails and toenails may take on a blue tinge.
   - The body decreases the production of sweat, ensuring that the surface of the skin remains dry and therefore reducing loss of heat via evaporation of the sweat. If the surface of the skin is wet, the rate of heat loss is increased.
   - Shivering is a result of muscles contracting to produce heat.
   - Thyroxine output increased, which in turn stimulates the metabolism, increasing metabolic rate and producing heat.

   (***A*** – description of two responses; ***M*** – explanation relates to how each response returns body temperature to normal; ***E*** – includes reference to hypothalamus)

Unit 11.6 Activity 1A: Reproduction in animals (Page 267)

1. **a.** Advantages: *Many offspring can be produced rapidly* to take advantage of good environmental conditions. (***A***) This increases the chances of individuals surviving when environmental conditions are favourable. Populations grow rapidly and spread/ distribute throughout the environment, often out-competing other species (eg for food, living space, water) – this assists the survival of the species as a whole. (***M***)

   Disadvantages: All offspring are *genetically identical* to each other and the parent. (***A***) This decreases the chances of the offspring surviving during environmental change, as there is no variation present; therefore, when the environment changes, it will affect all individuals – if individuals cannot survive environmental change, then the species may become extinct. Without the potential to adapt, the population may be wiped out / species may become extinct (eg, if a new (mutant) pathogen enters the population and kills one, then it will kill all, as none of the population will have genetic immunity to the pathogen). (***M/E***)

   b. Advantages: Offspring produced are all *genetically different* from one another and their parents. (***A***) This increases the chances of some members of the species surviving during environmental change, as when the environment changes some individuals may have the genetic potential to adapt and survive while others die out (eg if a new (mutant) pathogen enters the population, there may be individuals who are immune to it and will survive to breed (and continue the species) while the majority is wiped out). (***M***)

   Disadvantages: Offspring are (typically) *not reproduced rapidly or in large numbers*; individuals need to find a mate. (***A***) This decreases the chances of the species surviving, as the slower rate of reproduction means that individuals can't take advantage of good environmental conditions to increase their number of offspring and the population distribution. Loss of offspring has an increased negative impact on the population. Many of the endangered large mammals – such as the chimpanzee, gorilla – produce very few offspring during the course of their lifetime. (**M**)

2. In incomplete metamorphosis, the eggs hatch into nymphs which are small, sexually immature versions of the adult. These grow into the adult through a series of stages of growth called instars, between which each nymph sheds its exoskeleton. Adults and offspring occupy the same ecological niche (eg crickets, weta). In complete metamorphosis, the eggs hatch into a larval stage, which occupies a different ecological niche from the adult; its purpose is to eat and grow rapidly. A pupal stage follows, during which the larva metamorphoses (changes) into the adult. The adult (eg monarch butterfly, blowfly) disperses and reproduces. (**A** – two comparisons described)
3. Hermaphrodite animals have both male (produce sperm) and female (produce ova/eggs) sex organs. Individuals reproduce sexually with another individual by exchanging sperm. The advantage of this is that any two adults of the species can reproduce; there is no need to find a mate of the opposite sex. (**A**) This increases the chances of mating in species that live individually/are solitary, are small, move slowly, and have simple sensory receptors/organs (eg earthworms). (**M**)
4. Internal fertilisation allows animals to be *independent of water* for successful fertilisation. (**A**) Sperm no longer have to swim from the male to the female/ovum outside the female's body – this increases the chances of successful fertilisation; with internal fertilisation, animals can become completely terrestrial, as they are not dependent on water/moist environment for fertilisation. (**M**)
5. Fish and amphibians produce large numbers of eggs to *increase the chances of successful fertilisation and the chances of successful survival of some of the offspring* to become adults. (**A**) This is because, unlike mammals, these two groups have external (not internal) fertilisation, and there is no parental care of the offspring (or very minimal, eg some fish). Internal fertilisation increases the chances of successful fertilisation (eg sperm not at risk of predation), and parental care increases the chances of the offspring surviving to adulthood (eg parents provide food, protect from predators). (**M**)
6. Courtship occurs when a male advertises in some way (eg displaying, calling) for a female to mate with. This ensures that he reproduces with a member of the same species. (**A**) Courtship acts to bring members of the same species together in one area, increasing the chances of mating and therefore successful reproduction. It also provides a choice of mates, so a female can select the fittest male to reproduce with (which increases the overall fitness of the species). As courtship behaviour is species-specific, only females of that species respond to the male's behaviour. This ensures that members of one species remain reproductively isolated from other species (ie there is no mixing of the gene pool). (**M**) Courtship behaviour varies from one animal group to another and from species to species. Often, a male will establish a territory before advertising/calling for a female – common in birds (eg male the saddleback/tieke). In other groups, males may indulge in 'trials of strength', with the strongest male getting to mate with many females – common in herd mammals such as deer. In this manner, the strongest, healthiest males get to mate (often with many females) and the gene pool of the species is enhanced. (**E** – description of courtship with examples, and full, accurate reasons given for its occurrence, ie why it is significant)
7. Metamorphosis means 'change'. It is found in animal groups where there are stages in the life cycle that are completely different in form and function from the adult, such that the

adult and young occupy different niches. This occurs in insects, amphibians, and various invertebrate groups where aquatic members often have a free-swimming larval stage (eg molluscs such as shellfish, barnacles). (***A***) The different stages in the life cycle allow for *reduction in competition between adult and offspring*, as they are exploiting different habitats and foods using different adaptations (eg butterfly larvae feed by eating the leaves of plants using biting/chewing mouthparts; the adults may feed on nectar of flowers using sucking mouthparts (some adults do not feed)). The different stages also increase the chances of dispersal of the species, reducing competition between adults and offspring (eg the larvae of barnacles are swept in ocean currents until they find a suitable hard substrate to cement to before metamorphosing into the adult). (***M***) Complete metamorphosis is widespread in insects, with some groups (eg butterflies and moths) having larvae that feed on plants (growth stage) then pupate into the winged adult (reproductive and dispersal stage; may not feed); other groups (eg caddis flies, mayflies) have aquatic (freshwater) carnivorous larval stages that pupate into winged adults; others (eg mosquitoes) have filter-feeding aquatic larvae. In insects, the larval stage is the feeding stage; the adult reproduces and disperses the population. In other invertebrate groups, the adult is typically sessile (ie is fixed to a hard substrate – eg mussels, oysters, barnacles), and filter feeds from the surrounding water. Gametes are released into the water for fertilisation. The free-swimming larvae develop from the egg and disperse the species, finding a new substrate and then metamorphosing into the adult. Amphibians also typically have an aquatic larval stage ('tadpole') which breathes using gills; the adult lives on land and breathes, using lungs and skin. The larva is often herbivorous (eg frog tadpoles) while the adult is carnivorous. (***E*** – full explanation of the reasons for metamorphosis, together with good examples of its occurrence from two or three groups)

8. Three clearly different taxonomic (eg earthworms, insects, mammals) or functional groups (eg asexual reproduction, simple sexual reproduction, complex sexual reproduction) need to be selected. For *each* group, the *structure* of the *reproductive systems* need to be *described fully and carefully*, *using correct terms* (eg ovaries, testes, spermatheca, seminal vesicles, oviduct, etc) as well as the *function* of specific parts (eg sperm production, mating, courtship, egg laying).

- Description of *courtship* (if present); other aspects of reproductive behaviour.
- Description of *life cycle*; its stages, presence of metamorphosis.
- Whether reproduction is *sexual* and/or *asexual*, *hermaphrodite* or *heterosexual*.

For *each* group, *how* these structures work together needs to be *explained*, in, eg:

- Producing gametes.
- Mating.
- Fertilisation.
- Parental care.

For *each* group, reasons for *differences* between the groups needs to be *explained*, in relation to:

- Successful mating.
- Producing genetic variation.
- Survival of offspring.
- Reduction of competition.
- Enhancing fitness.

- Habitat (eg aquatic or terrestrial), internal/external fertilisation.
- Activity levels of the animals (eg sessile aquatic have external fertilisation and produce free-swimming larvae).
- Size and complexity of the animals (all large animals reproduce sexually; increasing complexity includes brain and behaviour, eg type and extent of parental care).
- Ecological niches (including changes during life cycle, eg insect and amphibian metamorphosis).

(**A** – descriptions of the structures for the three groups; **M** – explanations of how they function; **E** – discussions with comparisons/contrasts and reasons for differences)

Unit 11.6 Activity 2A: The reproductive systems (Page 278)

1. 1L; 2E; 3C; 4H; 5P; 6B; 7K; 8O; 9S; 10F; 11N; 12J; 13R; 14Q; 15M; 16A; 17D; 18T; 19G; 20I.

2. Oestrogen causes the uterus lining to thicken; progesterone causes the uterus lining to thicken further; FSH stimulates the development of follicle(s) in the ovary; ovulation occurs when LH levels peak. (**A** – description of function of any two)

FSH stimulates the development of follicles in the ovary and the secretion of oestrogen by the follicles / increase in oestrogen inhibits further production of FSH / increase in oestrogen stimulates production of LH causing ovulation when LH peaks / when corpus luteum secretes progesterone and oestrogen, these inhibit production of LH.
(**M** – explanation of how two hormones work together)

Menstrual cycle begins with menstruation, all hormonal levels are low. FSH levels rise as it is released by pituitary, causing development of follicles in ovary. Follicle cells produce oestrogen, which results in uterus lining starting to thicken. Increased oestrogen levels inhibit FSH production but stimulate LH production. When LH levels peak, ovulation occurs. Corpus luteum secretes progesterone, which causes uterus lining to thicken further, and inhibits LH production. As a result, corpus luteum dies and stops releasing progesterone. In addition, if implantation does not occur, progesterone and oestrogen levels fall, uterus lining breaks down, menstruation occurs again. (**E** – full discussion of how all four hormones work together to control menstrual cycle)

3. **a.** **i.** Production of sperm.

ii. Storage of sperm.

iii. Contributes secretions to seminal fluid that sperm swim in.

iv. Contributes secretions to seminal fluid that sperm swim in.

v. Carries sperm from testis/epididymis to penis. (**A** – any three)

b. *Any two from following list:*

- Growth of pubic, body and facial hair.
- Increased size of penis and testicles and production of sperm.
- Increased production of testosterone.
- Increase body size, including width of shoulders and chest and development of more muscular body shape. (**A**)

c. Physical changes seen during puberty are due to the secretion of hormones which will be active throughout the life of the adult male. The pituitary gland in the base of the brain secretes FSH and LH. (**A** – link to hormones) In the male, FSH stimulates production of sperm in the testes while LH stimulates production of testosterone from testes as they begin to grow and develop. The presence of these hormones in the body is responsible for changes seen at puberty. (**M** – explanation of effects of hormones)

Unit 11.6 Activity 2B: Mother and foetus (Page 286)

1. 1E; 2G; 3A; 4B; 5C; 6F; 7D.

2. **a.** The name of the process is ovulation. It is when the follicle ruptures and an egg is released. (***A***)

b. **i.** Progesterone.

ii. Maintains the lining of the uterus (endometrium). (***A*** – both correct)

c. In the Fallopian tubes. (***A***)

d. The sperm penetrates the egg, and the nuclei of the sperm and the egg fuse. (***A***)

e. **i.** To maintain a lower temperature than the core body temperature to make sperm. (***A***) If the temperature is maintained below 37°C, then the sperm count will be higher and there will be fewer sperm with abnormalities. (***M***)

ii. Place where sperm are stored / mature. (***A***)

iii. Many sperm are required as very few survive the journey from ejaculation through the uterus and into the Fallopian tubes. Factors that prevent sperm from reaching the Fallopian tubes include:

- There will be a proportion of defective sperm.
- Many sperm do not have enough energy to survive the required swim journey, or become trapped on the uterus lining.
- Acidic conditions in the vagina destroy many sperm.
- Normally, there is only one egg released from one ovary – a number of sperm will swim to the wrong Fallopian tube.

(***A*** – description of one way sperm numbers reduced / idea of few sperm reaching egg; ***M*** – reason why sperm numbers reduced and why many are needed)

3. **a.** **i.** Amnion: Chamber in which the foetus grows and is protected.

ii. Mucus plug: Increased protection of embryo/foetus from micro-organisms that may enter via the vagina. (***A***)

b. The foetus gets oxygen from the exchange of gases across the placenta from the mother to the foetus. (***A***) Oxygen diffuses from the capillaries of the mother to the placenta. From the placenta, oxygen diffuses into the capillaries of the foetus. (***M***)

c. Answer should contain the following points in paragraphs, not bullet points.

- Growth of the foetus from 20 to 40 weeks is achieved due to the ability of the placenta to supply nutrients to, and remove wastes from, the foetus; the uterus provides an ever-increasing safe space for this growth. (***A***)
- The placenta increases in size during this period to maintain the higher levels of nutrients required. This increase in size provides a greater surface area for the exchange of nutrients and wastes and a matching greater capillary network to allow the increased blood flow required to meet the increased nutrient demands.
- The uterus increases in size to accommodate the growing foetus and the increased amniotic fluid that is required to protect the foetus. (***M*** – reason for changes)
- Maintenance of the endometrium/placenta is controlled by hormones.
- Progesterone levels are high during pregnancy, enabling the endometrium (uterus lining) to remain thickened.
- Oestrogen is higher than normal during the pregnancy. This high level is associated with maintaining the uterus lining and increasing the blood flow to the uterus.

(***E*** – ***M*** plus linked to function of one hormone during pregnancy)

4. a. In the ovary. (***A***)
   b. Sperm and nutrient fluids. (***A***)
   c. A sperm penetrates the membrane of the egg. (***A***) The nuclei of the sperm and the egg fuse to create a zygote with a diploid ($2n = 46$) nucleus, containing 23 pairs of chromosomes. (***M***)
   d. Two embryos arrive in the uterus because the chances of successful implantation of the embryo in the endometrial lining of the uterus and development of a foetus are low (***A***); and this increases the chance that at least one will successfully implant/develop. (***M***)
5. a. If a pregnancy occurs, progesterone levels remain high in the second half of the menstrual cycle instead of dropping off. (***A***)
   b. Answer should contain the following points in paragraphs, not bullet points.
   If a pregnancy occurs:
   - During the first six weeks, the mother will notice that her period does not occur, her breasts swell and become tender, increased tiredness, possibly nausea. (***A***)
   - The corpus luteum continues to produce progesterone, which means that in the second half of the menstrual cycle, progesterone levels remain high. This enables the endometrium (lining of the uterus) to stay intact, allowing the embryo to implant when it enters the uterus. Once the embryo is implanted into the lining of the uterus, the placenta will start to develop and supply the growing embryo with nutrients and remove wastes.
   - Levels of oestrogen are also higher than normal, which, combined with the high levels of progesterone, prevent the maturing of an egg within the follicle and menstruation.
   - High levels of oestrogen stimulate the development of milk-secreting tissue in the breasts in preparation for breastfeeding. The high levels of progesterone prevent milk production at this stage.

   (***A*** – description of two effects; ***M*** – reason given for each effect; ***E*** – each explanation linked either to enabling embryo to implant or cessation of ovulation during pregnancy or future milk production)
6. a. Cushions/protects foetus as mother moves / reduces friction as foetus moves. (***A*** – one idea)
   b. Transports oxygen *and* nutrients/glucose/minerals/vitamins/amino acids (*not* food) from mother to foetus. (***A*** – both ideas)
   c. Placenta is where oxygen/nutrients from mother's blood pass through membrane into the blood vessel of foetus in the umbilical cord. (***A*** – description of function of placenta and umbilical cord)
   Foetus uses up nutrients as it grows and develops, meaning concentration of nutrients is low in artery in umbilical cord. Nutrients are rich in mother's blood. Because of concentration gradient, substances diffuse from mother's blood across placental membrane into foetus' blood. Placental membrane ensures the blood of mother and foetus never mix and maintains concentration gradient. (***M*** – one reason given for nutrient movement between foetus and mother; ***E*** – two explanations linked for nutrient movement)

Unit 11.6 Activity 2C: Human reproduction (Page 293)

1. a. Amniotic sac ruptures / amniotic fluid passes out of mother (***A*** – how waters break); this occurs because foetus pushing down on cervix causes amnion to break / contraction of uterus causes amnion to break. (***M*** – why waters break)
   b. Mucus plug expelled / uterus contracts / cervix dilates / oxytocin levels increase cause placenta to break away from uterus / increase in pain as uterus contracts / as vagina

stretches to let baby's head out / placenta / afterbirth expelled / progesterone levels decrease. (***A*** – any two descriptions)

2. **a.** Vagina → cervix → uterus → Fallopian tube. (***A***)

b. At ovulation, two eggs must have matured and been released. (***A***) As the twins are of different sexes, at fertilisation, separate sperm must have fertilised each of the eggs. (***M***)

3. **a.** Progesterone is produced initially in the ovaries and later in the pregnancy by the placenta. (***A***) The level of progesterone starts to increase at approximately day 14 (ovulation) and continues to increase to a peak in a normal non-pregnant cycle at approximately day 22. If an egg is fertilised, the progesterone levels remain high throughout the period of gestation. Progesterone stimulates the growth of the lining of the uterus (endometrium) and maintains this lining during gestation. (***M***) During gestation, progesterone has a role in stimulating the secretory tissues of the mammary glands; inhibiting milk secretion; stopping the muscles of the uterus from contracting; inhibiting FSH secretion (responsible for stimulating a follicle to develop and an egg to mature). (***E***)

b. Carbon dioxide and urea. (***A***)

c. The embryo is connected to the mother via the umbilical cord. Blood from the embryo flows to and from the placenta via the umbilical artery and vein where it comes very close to, but does not mix with, vessels carrying the mother's blood. (***A***) Carbon dioxide, oxygen, glucose and urea are exchanged between embryo and mother via diffusion from high to low concentration. Other larger substances are exchanged via active transport. In the diagram, the oxygenated blood entering the placenta will a have higher concentration of oxygen than the blood in the umbilical artery. Oxygen will diffuse from blood space in placenta to umbilical artery. Carbon dioxide will diffuse in opposite direction. (***M***)

d. The birth process has three stages:

- Stage one – contractions of muscles of the uterus wall gradually dilate the cervix; the amniotic sac may burst, releasing amniotic fluid.
- Stage two – contractions of the muscles of uterus wall increase. The baby is pushed head first through the birth canal. Once the baby has been completely delivered and is breathing, the umbilical cord is clamped.
- Stage three – the uterus continues to contract, causing the placenta to separate from the wall of the uterus and pushing it through the birth canal.

The hormone oxytocin is produced by the pituitary gland and stimulates the wall of uterus to start contractions. (***A*** – description of the three stages of birth; ***M*** – reason given for each stage; ***E*** – role of oxytocin included)

Unit 11.6 Activity 3A: Flowers, pollination and fertilisation (Page 312)

1. 1P; 2L; 3E; 4C; 5A; 6Y; 7W; 8Q; 9J; 10D; 11F; 12K; 13Z; 14G; 15M; 16R; 17V; 18U; 19B; 20H; 21O; 22X; 23I; 24N; 25T; 26S.

2. **a.**

Letter	Name of flower part	Function
A	Anther.	Production of pollen.
B	Sepal.	Protection of the flower at the bud stage.
C	Ovary.	Production of ovules, site of fertilisation of the ovule, site of development of the embryo.
D	Ovule.	Contains the female gamete (sex cell), which, after fertilisation, will develop into the seed.

(***A*** – any three correct)

b. Flower type 1 is insect pollinated, whereas Flower type 2 is wind pollinated. (**A**) The stigma of an insect-pollinated flower needs to be held protected within the petals. The insect will be attracted to the colour and/or scent of the flower, depositing the large sticky pollen grains that it is carrying onto the stigma. In contrast, the stigma of the wind-pollinated flower needs to be exposed to the wind rather than protected, so that as the light pollen from wind-pollinated flowers passes in the wind, it will be caught in the feathery structure of the stigma of the wind-pollinated flower. (**M**)

c. The pollen grain lands on the stigma; the sugar from the stigma stimulates it to germinate, producing two male gametes. A pollen tube grows down into the stigma and style, and then into the ovary, attracted by chemicals produced by the egg. (**A**) The male gamete (sperm) travels down the pollen tube, and once the tube reaches the female gamete (egg) in the ovary, the tip of the tube is broken down and the sperm can enter and fertilise the egg to make a zygote. (**M**)

3. When a pollen grain lands on the stigma, it germinates, forming male gametes or sperm. At the same time, a pollen tube starts to grow from the pollen grain. The tube grows down into the stigma and makes its way via the style to base of the ovule. Sperm formed travel down the pollen tube and when the tube reaches the ovule, one sperm fertilises one egg, forming a zygote. (**A**) Fertilisation is the process by which the nuclei of the male and female gametes (pollen and egg) are united in the ovule to form the zygote. The pollen tube provides pathway for sperm to reach egg for fertilisation. (**M**)

4. Reproduction via seeds is sexual reproduction whereas reproduction via tubers is asexual. Sexual reproduction involves the formation of gametes via meiosis and therefore produces variation within the population. Variation is important to ensure that the population is capable of coping with changing environmental conditions. It creates stability in the population. Asexual reproduction by contrast is a result of mitosis and therefore produces no variation whatsoever in the population. The parent plant will be identical to the offspring. The advantage of this is that the grower knows exactly what the offspring plants will look like and how they will behave. The disadvantage is that an entire population like this would lack stability as it would not be well equipped (via variation) to cope with potential changes in environmental conditions.

   Sexual reproduction via seeds will ensure that the offspring plants are distributed well away from the parent plant, reducing competition for resources and increasing the spread of the population. This is achieved via the dispersal of the seeds after pollination and fertilisation. Asexual reproduction via the tubers will not spread the offspring plants as far from the parent plant as sexual reproduction will, as the tubers are all attached via stems to the parent plant. This will increase the competition for resources around the parent plant and will potentially limit the growth of the plants. (**A** – description of two advantages *or* disadvantages of *either* seeds *or* tubers to plant; **M** – explanation of one advantage *or* disadvantage of either seeds or tubers, **E** – advantages *or* disadvantages related to *both* seeds *and* tubers outlined)

Unit 11.6 Activity 3B: Germination and seed dispersal (Page 318)

1. 1N; 2F; 3J; 4H; 5A; 6L; 7C; 8I; 9D; 10G; 11M; 12O; 13K; 14B; 15E.

2. A fruit contains/disperses seeds / develops from a flower. (**A**) Both pumpkins and lemons are fruit as both have developed from the enlarged ovary of a flower that matures into a fruit which disperses/contains seeds. (**M**)

3. a. Storage of food. (**A**)

   b. Water and warmth. (**A**)

c. Tough testa/cotyledons or endosperm/dehydrated/micropyle. (***A*** – one feature named or described) Testa is a tough impermeable coating around seed that protects seed/ prevents germination until enough water present / cotyledons provide energy source for dormant seed until first leaves can photosynthesise / dehydrated so metabolic activity very low, just enough to sustain life but maintain dormancy / micropyle allows oxygen to enter seed so some respiration can still occur. (***M*** – explanation for one feature)

4. **a.** **i.** The ovary of the flower swells and develops into the pea pod.

ii. The pea seeds develop from the fertilised ovules that are within the ovary. (***A*** – **i.** or **ii.** correct)

b. Flowering plants use fruit to enable seeds (which will develop into the next generation) to be dispersed. (***A***) Fruit such as the pea pod is eaten, and seeds are dispersed in the droppings of animals. This means that the seeds are dispersed away from the parent plant, reducing competition for space, light and nutrients with the parent plant. (***M***)

5. **a.** **i.** The radicle (young root) grows towards gravity. This helps to establish anchorage of the roots in the soil. The plumule (young shoot) grows away from gravity. This helps to bring the shoot up to the surface as quickly as possible.

ii. Once the plumule (young shoot) reaches the surface, it will grow towards the light, unfolding and lifting the leaves/cotyledons upwards. Once in the presence of light, the young leaves/cotyledons turn green and photosynthesis starts. (***A*** – both **i.** and **ii.** needed)

b. The carbohydrate store provides energy required for germination/growth (***A***) as the seedling cannot produce energy by photosynthesis until the first leaves have emerged from the soil. (***M***)

c. Both gravity and light are important for successful early seedling growth.

No matter how the seed is positioned in the ground, the radicle (young root) will grow downwards, towards gravity, and the plumule (young shoot) will grow upwards, against gravity. By growing towards gravity, the root is able to anchor firmly in the soil and locate water and nutrients. By growing against gravity, the shoot reaches the surface – this is essential as the seed only has a limited carbohydrate store to provide energy for growth for the seedling in the initial period when it is underground.

The shoot emerges with the bent hypocotyl coming out first to protect the cotyledons (young leaves). The hypocotyl responds to light by unfolding, lifting the cotyledons up to the light. This allows leaves to absorb light, as a result of which they turn green, enabling them to start photosynthesising. This provides energy for growth. The young shoot will now grow towards light, which will help it avoid competition from surrounding seedlings. (***A*** – no 'Achievement' opportunity in this question; ***M*** – gives a reason for importance of *either* light *or* gravity; ***E*** – outlines two things related to the importance of *both* light *and* gravity for seedling growth)

6. **a.** Animal – eating the berries and dropping the seeds after they have travelled through the digestive system.

b. Animal – hooks on seeds would allow seeds to stick to skin, feathers, coats of animals. Seeds would be deposited when brushed off the animal coat/skin.

c. Wind – structures on seeds allow them to travel easily in the wind.

(***A*** – need to explain difference between animal dispersal in Groups A and B; not adequate to simply put 'Animal' for each)

7. **a.** The seed consists of a hard protective coat called the testa within which is the embryo (D). The cotyledons (A) provide stored energy for growth. The embryo is made up of the radicle (C) which will form the first root when the seed germinates, and the plumule (B) which will form first shoots and the cotyledons. (**A**)

Question says to include names of parts indicated by letters A to D in your description – thus need to do more than just name the parts.

b. *Any one of the following would be an adequate answer:*

A Cotyledon – provides food supply for seed once it germinates and before young plant able to provide food via photosynthesis.

B Plumule – first shoot; will push hook-shaped stem tissue upwards out of the ground after the radicle has established the first root – as it pushes out of the ground it lifts cotyledons (seed leaves) with it.

C Radicle – first root; is first part of seed to emerge from the testa, growing down into the soil to provide anchorage and water for the young seedling. (**A**)

8. Three environmental factors required for seed germination are moisture, oxygen and warmth (or more correctly, a suitable temperature – exact temperature depends on species). (***A*** – identifies or describes all three factors)

When a seed germinates, it is breaking a period of dormancy where there has been no metabolic activity. In the last phase of development of the seed it dehydrates and the embryo stops growing. When the seed germinates, it takes in water so food store contained in the cotyledon can be used to provide source of energy for embryo to grow again – cells are again becoming active and for this to happen, cellular respiration will need to be activated in cells of the seed. This is why oxygen is required.

Oxygen + food → energy + carbon dioxide + water

Warmth is required to enable the process of respiration to take place. If the temperature is too low, chemical reactions involved in cellular respiration will be far too slow and the embryo will not get the energy required to grow.

Soil and sunlight are not required for germination because the food source for germinating the seed is contained within cotyledon of the seed. Soil and sunlight only required once seedling has established its first leaves above the ground. At this point, sunlight is required for photosynthesis and soil is required to provide minerals needed by the plant as it grows. (**M** – explanation for *each* factor needed for germination; **E** – *both* parts of the question must be correctly addressed, three factors needed for germination *and* soil and sunlight needed after germination)

9. a. i.

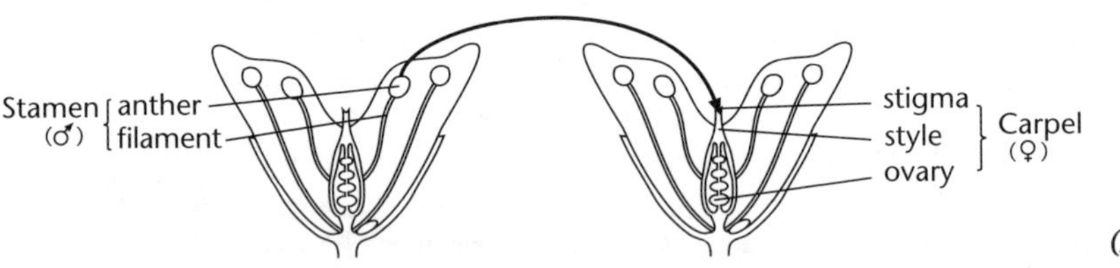

(A)

ii. Fertilisation is process by which male and female gametes (sex cells) fuse to form a zygote (fertilised egg). In sexual reproduction, joining of gametes doubles number of chromosomes in the cell, creating a 2*n* zygote from two different parents. (**A**)

iii. Wind-pollinated flowers need to produce large quantities of smooth, light pollen, whereas insect-pollinated flowers produce smaller quantities of rough, sticky pollen. (***A*** – describes difference between two types of pollen)

Pollen produced by wind-pollinated flowers is light and smooth so it can travel long distances in the wind. Large volumes of this pollen are produced because the process of pollination by this method is random (since much of the pollen does not reach its destination). Pollen produced by insect-pollinated flowers is rough and sticky because it needs to attach easily to insects that are used to transfer pollen from flower to flower. This method of pollination is less wasteful, therefore there is not the same need to produce large volumes of pollen – a larger proportion of the pollen reaches the intended destination. (***M*** – also explanation)

b. *An example answer follows.*

Fruit 1: Apple – a fleshy fruit.

Fruit 2: Dandelion – dry fruit.

An apple is a fleshy fruit dispersed via animals. Structure and features that encourage animal dispersal away from the parent plant are:

- Bright-coloured fruit to attract the animal.
- Succulent, sweet flesh easily eaten and digested.
- An indigestible seed that travels through the intestine of the animal and is dropped from the animal 'at the end of the journey through the alimentary canal' at some distance from the plant.

A dandelion is a dry fruit dispersed via the wind. Structure and features that encourage wind dispersal are:

- Feathery hairs that increase the surface area of the fruit and encourage it to travel long distances in the wind.
- Lightness of the fruit – enhances the ability to travel long distances from the parent plant in the wind.

It is to the advantage of the seed to be dispersed at a distance from the parent plant so that once germinated, the seedling is not competing with the parent plant for resources. Both fruits provide mechanisms that are appropriate to the means of dispersal used by the plant. For the apple, the key feature is the attractiveness of the fruit to the animal that will disperse it. For the dandelion, the key feature is the lightness and ability of the fruit to travel in the wind. These are very different features and demonstrate the way in which each plant is adapted to the specific requirements of its life cycle. (***A*** – descriptions of structures *or* features for *both* types of fruit; ***M*** – explanations related to structure *or* features for *one* type of fruit; ***E*** – structures *or* features for *both* types of fruits compared and contrasted)

Unit 11.6 Activity 3C: Plant reproduction (Page 327)

All answers should be full and lengthy whether descriptions ('Achievement' – ***A*** meaning 'satisfactory achievement') or explanations ('Merit' – ***M*** meaning 'high achievement') or discussions ('Excellence' – ***E*** meaning 'very high achievement'). You should provide evidence from the plant groups you have studied. The answers given below are for aspects of plant reproduction systems and are not indicative of the answers that may be required in the examination. Seek your teacher's guidance.

1. a. Asexual reproduction involves a single parent producing identical offspring through mitotic cell divisions, while sexual reproduction involves two parents producing varied offspring through fertilisation of gametes produced from meiotic cell divisions. (***A***)
   b. The gametophyte is the haploid generation in plants that produce gametes from mitosis, while the sporophyte is the diploid generation in plants that produce spores from meiosis. (***A***)
   c. Meiosis (in plants) is the cell division that halves the chromosome number to produce the *haploid* spores, while mitosis is the cell division that maintains the chromosome number, and produces the gametes and the general body cells in growth. (***A***)
   d. Meiosis is cell division which halves the chromosome number from diploid to haploid, while fertilisation is the fusion of the gametes which restores the chromosome number from haploid to diploid. (***A***)
   e. Pollination is the transfer of the pollen grains from the male sex organs to the female sex organs (conifers and angiosperms), while fertilisation is the fusion of the male and female gametes. (***A***)
   f. Spores are the light haploid grains produced by the sporophyte generation of mosses and ferns; on germination, they grow into the gametophyte. Seeds are the diploid structures produced from the fertilisation of the gametes in the gametophyte generation of conifers and angiosperms; on germination, they grow into the sporophyte generation. (***A***)
   g. Haploid (n) refers to the state when a cell/organism has one set of chromosomes, while diploid (2n) refers to the state when a cell/organism has two sets of chromosomes. (***A***)
   h. The seed develops from the fertilised ovule and contains the embryo, while the fruit develops from the ovary and encloses the seed, aiding in its dispersal. (***A***)
2. Mosses and ferns are restricted to damp habitats because they need water for fertilisation/sperm to swim to the egg. (***A***) Fertilisation is external in these two groups of plants; therefore, they are dependent on water from the environment for the sperm to swim in to reach the egg to be able to fertilise it. (***M***)
3. The trend for the sporophyte to become dominant is because the sporophyte is diploid while the gametophyte is haploid, so the sporophyte has twice the genetic information of the gametophyte. (***A***) This is advantageous, because the great increase in genetic variation in the offspring provides greater chance of combating environmental change (and, incidentally, more raw material for evolution/natural selection to act on). This increases the chances of the survival of an individual's genes when the environment changes (and, incidentally, increases the chances of the survival of the species). (***M***)
4. The gametophyte in mosses is a conspicuous, leafy plant; the sporophyte is parasitic on it. In ferns, the gametophyte is much smaller, becoming a heart-shaped prothallus about 1 cm^2; however, it still leads an independent existence from the sporophyte, which grows through it on germination. In angiosperms, the gametophyte does not lead an independent existence and has been reduced to a few cells (male gametes and tube nucleus in the pollen tube, female ovule with ovum and endosperm nuclei) that are contained within the flower of the sporophyte. (***A*** – minimum of one change described between three groups)
5. Cross-pollination is achieved by the transfer of pollen by wind or animals from the anthers of flowers of one plant to the stigma of flowers of another plant. (***A*** – relates to cross-pollination) It is needed to produce offspring that are genetically different from the

parents. (***A*** – how cross-pollination is achieved) This produces variation in the phenotypes of the offspring, which increases the chances of the survival of an individual's genes (and, incidentally, more raw material for evolution/natural selection to act on to increase the chances of the survival of the species). (**M**)

Supplementary Unit Activity 1A: Micro-organisms (Page 340)

1. a.

Type	Bacillus	Vibrio	Spirillum	Coccus
Shape	Rod shape.	Bent rod shape.	Spiral shape.	Spherical.

(***A*** – any two)

b. *Streptococcus* – causes throat infections.
Staphylococcus – causes pimples and skin infections.
Mycobacterium tuberculosis – causes TB (tuberculosis).
Salmonella – causes gastroenteritis or food poisoning. (***A*** – any one)

c. Saprophytic bacteria feed off dead or decaying matter, whereas parasitic bacteria feed off a living host. (***A***) Saprophytic bacteria cause no harm, while parasitic bacteria harm the host. (***M***)

d. i.

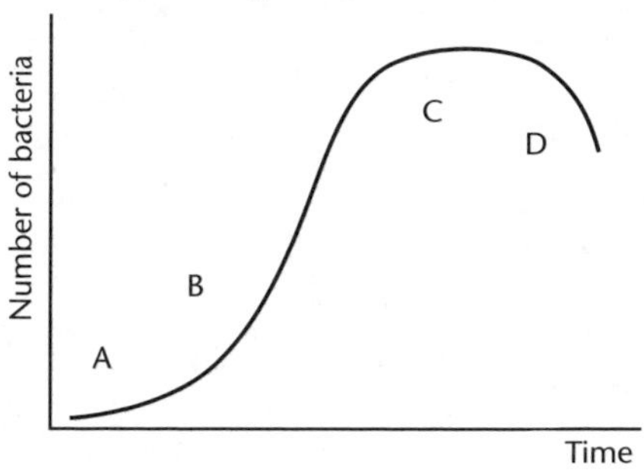

(***A***)

ii. Bacteria reproduce by binary fission. This means that they grow and then divide into two. (***A***) If there is sufficient food, water, space and appropriate warmth, many species of bacteria will reproduce every 20 minutes.

In the first part of the curve (A), there are no restraints on reproduction (plenty of food, no toxic wastes), so each bacterium reproduces at its maximum rate. The population increases by the same percentage each hour, and growth is exponential. In phase (B), each bacterium is reproducing more slowly because food begins to be in short supply and/or wastes begin to accumulate. Despite this, the curve continues to get steeper because the decreasing rate of reproduction of individual bacteria is more than offset by the increasing number of 'parents' (a 50% increase in a very large population may add more numbers than a 100% increase in a small population). In (C), the decreasing rate of reproduction of individual bacteria more than offsets the effect of the increasing number of 'parents' (the figures on page 178 illustrate this). In (D), the numbers of living bacteria actually decrease. (**E** – link to limiting factors)

Supplementary Unit Activity 2A: Antibodies, antigens, lymphocytes and lysozomes (Page 347)

1. 1B; 2C; 3N; 4F; 5A; 6J; 7O; 8L; 9K; 10M; 11H; 12I; 13E; 14D; 15G.
2. A person will never become immune to the common cold, because the virus that causes the common cold mutates very quickly / there are many types of cold virus. (***A***) When a person is exposed to a common cold virus, their body makes antibodies to that virus, and, if exposed to that particular strain of the common cold again, would have immunity. But, because the cold virus mutates, a person would not be immune to a new strain of the virus and therefore it would take time for the body to learn to make new antibodies to that specific strain of the virus. (***M*** – reason linked to time taken to produce antibodies)
3. **a.** Active immunity is achieved when the body is capable of producing antibodies against a pathogen. (***A***) The polio vaccination contains inactive or weakened forms of the polio pathogen. These will stimulate the body to produce its own antibodies – giving active immunity. (***M***)

   b. The child's body has produced antibodies against measles. (***A***) This is because, having suffered from the disease, the child's body has produced antibodies specific to this pathogen, and retains the knowledge (via memory cells) of how to do this very rapidly if infected in the future. (***M***)

   c. During pregnancy, the foetus receives antibodies from the mother across the placenta. These provide the foetus with passive immunity for diseases that the mother has been exposed to. After birth, if the baby is breastfed, this passive immunity will continue as antibodies from the mother are found in breast milk. During the first months of life, the baby relies on the immunity from the mother. As it is exposed to pathogens, the baby starts to make its own antibodies, reducing its reliance on the antibodies provided by the breast milk. (***A*** – description of baby receiving antibodies either across placenta or in breast milk; ***M*** – ***A*** plus linked to idea that as a result, baby has a high level of immunity; ***E*** – ***M*** plus idea that baby starts to make own antibodies)
4. Examples of the chemical defences include:
   - Tear glands in the eyes – contain an enzyme called lysozyme which kills bacteria.
   - Acid in the stomach destroys pathogens that enter via the mouth.
   - Nose contains mucus which collects pathogens as they enter the airways.
   - Cilia in the airways beat, moving mucus up the airways until it is either coughed out or swallowed. If swallowed, the pathogens within the mucus are destroyed by the acid in the stomach.
   - Genital openings – in the female, the acidic environment in the vagina prevents reproduction of pathogens; in the male, pathogens that enter the urethra will be destroyed by phagocytes before reaching the bladder.

   The body openings shown in the diagram all provide potential entry points for pathogens. However, at each of these points there are chemical defence mechanisms which destroy pathogens at the point of entry, *reducing the potential for the pathogens to enter the bloodstream via these openings, reproduce, and cause disease.* (***A*** – descriptions of three features linked to idea of pathogens being destroyed; ***M*** – reason given as to how each feature results in pathogens being destroyed; ***E*** – plus *italicised* part of sentence, idea that destruction of pathogens means they can't reproduce and cause disease)

Supplementary Unit Activity 2B: Defence against disease (Page 351)

1. a. Around the wound – heat/increased temperature, swellings, redness/itching, pain. (***A*** – any two)

b. Inflammation is triggered by damaged cells which release a chemical called histamine / plasma leaks from capillaries into the damaged area / phagocytes move into damaged area/engulf pathogens / immunoglobulins produced. (***A*** – any two responses described) Histamine dilates (increases the diameter of) the blood vessels at the injury site, therefore increasing the blood flow to this site. Histamine also attracts white blood cells (lymphocytes and phagocytes) to the site. Lymphocytes will produce antibodies which will bind to pathogens at the site, making them inactive and preventing them from reproducing and potentially causing infection. Phagocytes remove pathogens and non-microbial debris via phagocytosis. Plasma also leaks from capillaries at the injury site, causing swelling. (***M*** – gives a reason as to how one inflammatory response helps prevent spread of pathogens; ***E*** – links together three explanations for one inflammatory response, answer given is for release of histamine)

2. a. More than one injection is required in order to stimulate the body to build up enough antibodies/memory cells. (***A***)

b. The hepatitis B vaccination contains a weakened or inactive form of the virus/pathogen. (***A***) This stimulates the body to produce lymphocytes which will produce antibodies that are specific to the virus and memory cells that will retain the knowledge of how to produce these if the body is exposed again at a later date. If a person who has had the vaccination is exposed to the virus, the memory cells will act promptly to produce antibodies which will bind to the viruses, making it posible for the viruses to be engulfed and destroyed by phagocytes. This means that the antibodies are able to prevent a virus from reproducing to a level high enough to cause symptoms of the disease. (***M*** – one explanation for vaccination; ***E*** – three explanations linked together as a discussion)

3. a. i. A pathogen is a disease-causing micro-organism. (***A***)

ii. Fever, inflammation or swelling (eg redness and pain in the throat), runny nose, pain, vomiting. (***A*** – any two)

b. i. When pathogens enter the blood, lymphocytes (white blood cells) recognise the pathogens as being foreign and produce specific antibodies that will bind to the pathogens. Phagocytes then recognise the pathogens with the antibodies bound to them, engulfing and destroying the pathogen-antibody complex. (***A*** – description of a response; ***M*** – links responses to pathogens as foreign)

ii. Lymph nodes are the site where lymphocytes and phagocytes are found in large numbers. When an infection enters the body, lymph nodes swell due to an increase in the number of lymphocytes and phagocytes. (***A***) The cut is in John's hand. The lymph nodes in the armpit are closest to the site of the cut – therefore, as the infection travels up the arm, it will reach the lymph nodes in the armpit first and therefore this is where the response will first be seen and pathogens will be destroyed. (***M***)

iii. Tetanus is a bacterial infection transmitted in puncture wounds (such as those caused by rusty nails). This means that John is at risk of getting the infection. Even if John has been vaccinated in the past, the antibodies provided by the vaccination will break down and therefore reduce over time. (***A***) The new vaccination will stimulate the production of more antibody. (***M***)

4. **a.** Form of immunity obtained via injection of ready-made antibodies into the body. (***A***)

   Type of immunity provides short-term protection against disease.

   b. The tetanus vaccination provides passive immunity. This means it contains ready-made antibodies. (***A***) This vaccination does not provide body with ability to make tetanus antibodies and the antibodies break down after time, therefore vaccination needs to be renewed after a period of time. (***M***)

   c. Active immunity is gained from being exposed to either the disease or a vaccination which contains a form of the disease that has been made harmless. When the body is exposed in this way, it responds by making antibodies. (***A***) A memory of how to make these antibodies is retained so that when exposed in the future, the antibodies are produced very quickly, destroying microbes before they are able to reproduce and cause symptoms of the disease within the body. (***M***)

   d. **i.** B lymphocytes develop into plasma cells that produce antibodies which bind to the surface of the virus (antigens), preventing them from entering host cell and reproducing. (***A***)

   ii. When antibodies are produced by the lymphocytes, memory cells are produced that remain active in the body and 'on watch' for the presence of the virus in the future. (***A***) If the virus enters the body in the future, these memory cells activate, enabling the body to produce large amounts of antibody very quickly and destroying virus before it is able to reproduce and cause symptoms of the disease within the body. (***M***) Because of these memory cells, once the body has been exposed to an infection such as chicken pox, it is very unusual for individual to get disease a second time. (***E***)

5. **a.** Saliva contains enzymes / cilia lining throat / sticky mucus. (***A*** – description of two mechanisms)

   The enzymes in saliva can destroy some pathogens / cilia beat to move trapped pathogens out of the throat / sticky mucus traps pathogens which can then be removed by cilia/coughing. (***M*** – explanation of how each mechanism works)

   b. **i.** *Any of following answers would be acceptable:*

   - Storing food covered and in a refrigerator.
   - Fully cooking food – ie the food must be cooked at temperature high enough to kill all microbes and all parts of the food must be cooked.
   - Not reheating food after it has cooled.
   - Washing hands after using the bathroom and prior to handling food.
   - Ensuring all cooking utensils and surfaces kept clean. (***A*** – one method described)

   ii. *Any two of following answers acceptable:*

   - Vomiting.
   - Diarrhoea.
   - Nausea.
   - Fever.
   - Stomach cramps. (***A***)

   Note that the answer was given in introduction to question.

   iii. The body could not stop *Salmonella* from causing food poisoning because the bacteria were contained within the food that was eaten. (***A***) Once inside the body, *Salmonella* bacteria grow and reproduce, producing toxins that cause the symptoms of gastroenteritis. (***M***) If the symptoms are seen in the person who ate the infected food, this person did not have any memory cells that were capable of recognising the bacteria as a foreign antigen that it had previously encountered (and would have produced large quantities of antibodies very quickly which would have locked onto the outside of the bacterial cells, stimulating the phagocytes (a type of white blood cell) to engulf and destroy the bacteria). Antibodies will also neutralise toxins (poisons) produced by bacteria which cause the symptoms in a patient. On initial exposure to the bacteria, the individual will take longer to make antibodies for this specific strain of bacteria; this will increase effect of the infection as the bacteria will have grown and multiplied further before antibodies are able to control the infection. (***E*** – link to antibody response)

abiotic (229): non-living environmental factor, such as light intensity, temperature.

absorption (98, 119): taking in the proteins of digestion.

acrosome (270): tiny compartment at the tip of a sperm, containing enzymes that enable the sperm to enter the egg.

active site (37): the place on an enzyme where a substrate fits, as described in the lock-and-key method.

active transport (10, 25, 103): movement of a substance across a membrane from a lower to a higher concentration, using energy in the process.

adenosine diphosphate (ADP) (198): when ATP loses a phosphate group to produce energy it also forms into ADP.

adenosine triphosphate (ATP) (198): the main energy-carrying chemical in all organisms.

ADH (antidiuretic hormone) (187): controls water loss in the kidneys.

adipose tissue (247): fat storage tissue, present especially under the skin.

adrenal gland (249): ductless gland above the kidney, secreting adrenalin and other hormones.

adrenalin (156, 157, 249): hormone secreted by the inner part (medulla) of the adrenal gland.

aerobic (331): in the presence of oxygen.

aerobic respiration (19, 197, 200, 331): the use of oxygen to release energy from food.

afferent arteriole (185): tiny blood vessel supplying one of the million or so microscopic filters in the kidney.

agar (337): jelly derived from seaweed and used to grow bacteria and fungi.

air sacs (205): muscular chambers in insects that assist in air movement in and out of an insect's body.

alimentary canal (97, 115): tube connecting mouth and anus, in which digestion and absorption of food occur.

allergen (219): normally harmless chemical that provokes a strong immune response, such as inflammation and/or mucus secretion.

alveoli (singular alveolus) (209, 213): minute sacs that form the ends of the air tubes in the lungs.

amino acids (99): chemical building blocks from which proteins are made.

ammonia (180): toxic excretory product NH_3, formed from the breakdown of amino acids.

amnion (281): membrane containing protective fluid that surrounds the foetus in the womb.

amniotic fluid (281): protective fluid surrounding the developing foetus.

anabolism (37): chemical reactions that synthesise large molecules from small ones (eg protein synthesis).

anal pore (62): opening in the cell membrane of some unicellular organisms through which excretory products are expelled.

anal sphincter (119): circular muscle that contracts to hold faeces in the rectum.

anaerobic (198, 331): without oxygen.

anaerobic respiration (197): respiration (release of energy) in the absence of oxygen.

androecium (306): male part of a flower, consisting of the stamens.

angina (172): pain caused by impaired blood flow to the heart muscle.

angioplasty (172): a surgical procedure in which a balloon is inserted into a blocked artery and inflated at the site of the blockage.

angiosperms (125, 126, 323): flowering plants.

animals (44): multicellular heterotrophs with a great diversity of life forms, life styles, feeding types, habitats.

annuals (305): plants that complete their life cycle in one year or less.

antennae (114): 'feelers' – sensory detectors on insects' heads.

anterior pituitary gland (274): front part of the pituitary gland; produces growth hormone and hormones controlling the ovaries, testes and certain other endocrine glands.

anther (306, 326): part of a stamen that produces the pollen grains.

antibody (151, 152, 346): protein that helps to protect the body against invading organisms.

anticoagulant (112): substance that stops blood from clotting.

anti-diuretic hormone (ADH) (243): hormone produced by the posterior pituitary gland promotes retention of water by the kidney.

antigen (152, 346): a substance, usually foreign to the body, that stimulates the production of an antibody.

antitoxin (346): an antibody that neutralises a poisonous antigen.

anus (119): the opening through which indigestible material is passed.

aorta (144): the largest artery; takes blood directly from the heart to the body.

aortic valve (156): valve at start of the dorsal aorta.

apical meristem (225): growing tip of a stem or root, consisting of rapidly dividing cells.

appendix (115): pocket of gut tissue situated between the small and large intestines.

arachnida (53): class of arthropods that includes spiders, mites, ticks.

archaea (44): the domain of bacteria that are the evolutionary oldest; includes all the extremophiles.

archebacteria (44): oldest group of bacteria, all inhabit environmental extremes; belong in the archaea domain.

artery (149): vessel that carries blood away from the heart.

arteriole (159, 160): tiny artery.

artificial insemination (296): the process of making a woman or an animal pregnant by an artificial method.

asexual (259): not involving sex; mitotic cell division produces offspring identical to the parent.

assimilation (98): absorption of nutrients by cells and their conversion into the constituents of the cells.

atherosclerotic plaque (171): deposit of cholesterol and other fatty materials in the wall of an artery.

ATP (198): most common energy-carrying molecule.

atrioventricular valve (156): valve between a receiving chamber (atrium) and a pumping chamber (ventricle) of the heart.

atrium (146, 147, 154): chamber of the heart that receives blood from the veins.

atrophy (70): degeneration or wasting away of a muscle as a result of lack of use.

autoclave (335): kind of pressure cooker used to sterilise equipment used in microbiology.

autosome (35): chromosome that is not concerned with the determination of sex.

autotrophic (69, 253): 'self-feeding' – an autotroph is an organism that can make its own organic substances from inorganic materials.

auxin (228): plant growth substance involved in phototropism and other activities.

axial skeleton (53): skull, backbone and ribcage

axil (224): the angle between a stem and the base of a leaf.

axillary bud (224): bud at the base of a leaf (ie in the axil).

bacilli (330): plural of bacillus – a rod-shaped bacterium.

bacteria (44): one-celled prokaryotes; may be autotrophs or heterotrophs.

basal metabolic rate (BMR) (247): rate of chemical processes needed to provide energy for minimum activities fundamental for life.

bicuspid valve (156): valve ensuring one-way flow from left atrium to left ventricle.

biennial (305): plant with a two-year life cycle, flowering occurring in the second year.

bile (102, 118): alkaline digestive juice, produced by the liver, emulsifies fats.

bile pigment (151): yellow-green waste product of haemoglobin breakdown, excreted in bile.

binary fission (66): process by which a unicellular organism divides into two.

biotic (229): relates to other living organisms, such as predators, competitors.

bladder (182, 184): muscular bag in which urine is temporarily stored.

blastocyst (279): early stage of development of humans and other mammals, consisting of a hollow ball of cells.

blood (149): the material within which substances are transported.

bone (54): very hard tissue that is the main component of the human skeleton.

bone marrow (344): tissue within cavities inside bone, site of blood cell production.

botulism (331): frequently deadly form of food poisoning, caused by the bacterium Clostridium botulinum.

Bowman's capsule (185): the blind ending of one of the microscopic kidney tubules, site of filtration.

breathe (203): the use of muscles to move water or air over gas exchange surfaces.

breathing (201, 205): physical process involving muscular movements to take O_2 and CO_2 to and from gas exchange surfaces.

bronchi (209, 213): plural of bronchus, one of the larger air tubes in the lungs, with walls supported by cartilage.

bronchial tree (212): branching system of air tubes within the lungs.

bronchiole (209, 213): microscopic 'twigs' of the bronchial tree, lacking cartilage in the walls.

bryophyta (57): mosses and liverworts.

caeca (117): plural of caecum.

caecum (115): associated with digestion of plant material in herbivores.

callus (304): mass of plant cells produced in response to injury or in response to treatment by plant hormones.

Calvin cycle (94): see light-independent reaction.

calyx (306): collective name for the sepals, the outermost layer of parts of a flower.

cambium (127, 226): layer of actively dividing cells, responsible for growth in thickness in stems and roots of many seed-bearing plants.

canine (100): tooth immediately behind the incisors.

canine teeth (113): 'dog teeth'. Long pointed teeth used by carnivores to kill their prey.

capillaries (145, 160): smallest blood vessels.

capillary (103, 149): microscopic blood vessel, in which materials enter and leave the blood.

capillary action (132): the movement of water up a narrow tube such as a capillary tube.

capsid (337): layer of protein surrounding the nucleic acid of a virus.

cardiac cycle (156): rhythmical muscular contractions of first the auricles then the ventricles of the heart.

cardiac infarction (172): death of heart muscle cells resulting from blockage of coronary artery.

cardiac muscle (157): muscle making up the bulk of the heart.

cardiac veins (157): veins draining the muscle of the heart.

carnassial teeth (113): large flesh-cutting molars that can crack bones.

carpel (306, 326): female equivalent of a stamen of a flower, usually several carpels being joined together.

cartilage (209): slightly elastic tissue found at the ends of bones as well as in other parts of the body, eg the nose.

catabolism (37): chemical reactions that break down large molecules into smaller ones (eg digestion of proteins into amino acids).

catalysts (37): substances that greatly speed up the rate of chemical reactions.

cell (43): basic units of all living matter; made up of cell/plasma membrane surrounding cytoplasm, usually with a nucleus.

cell membrane (13): bilipid layer that surrounds every cell; controls movement of materials in/out of a cell.

cell organelles (43): structures inside cells.

cell theory (5): theory that cells form the basic units of living matter.

cell wall (6, 13, 20): outermost layer of a plant cell, consisting mainly of cellulose.

cellular respiration (201): breakdown of glucose to release chemical energy (ATP) and heat energy: occurs in the cytoplasm and mitochondria of cells.

cellulose (6, 20): complex carbohydrate forming the major constituent of plant cell walls.

cement (101): bone-like material that binds tooth into the socket.

centriole (21): involved in cell division, in animals cells producing spindle fibres.

centromere (15): part of a chromosome that attaches to the spindle during cell division.

cephalisation (236): development of a recognisable head and brain in animals.

cephalopoda (53): class of molluscs that includes the octopus and squid

cervix (274, 289): neck of the womb.

chemosynthesis (44): using energy from chemical reactions to produce complex organic compounds from simple (eg CO_2, H_2O) compounds.

chewing (98): chopping of food into smaller pieces.

chitin (205, 335): nitrogen-containing polysaccharide that makes up the tough skin of arthropods. Also found in the cell walls of fungi.

chitin (335): complex carbohydrate, similar to cellulose except contains nitrogen; forms major part of the cell walls of fungi and the exoskeleton of insects and other arthropods.

chlorophyll (7, 20, 83): green pigment that absorbs light in photosynthesis.

chloroplasts (7, 20, 62, 70, 83): organelles in which photosynthesis occurs.

chromatin (14): diffuse material that chromosomes appear as when a cell is not dividing.

chromosomes (6, 14): thread-like structures within a cell nucleus made of DNA (and protein) that contain genes.

chyme (116): creamy fluid in stomach; results from the action of hydrochloric acid and stomach movements.

cilia (21, 209, 215, 274, 342): hair-like extensions of the surfaces of many animal cells, used to beat against the surrounding liquid to create a current (plural of cilium).

circulatory systems (143, 203): the heart, blood vessels, blood, lymphatic vessels, and lymph, which together serve to transport materials throughout the body.

clitoris (274): small female reproductive structure containing erectile tissue.

clotting (152): when platelets gather at an injury to stop blood flow.

cocci (330): plural of coccus, a spherical bacterium.

co-enzymes (39): complex organic molecules that help certain enzymes to function correctly.

cohesion (132): the result of the attractive forces that exist between water molecules.

coleoptile (227): tubular sheath protecting the young foliage leaves of a grass seedling.

colon (118): large intestine; absorbs minerals and water from undigested food.

collecting ducts (184): tubes that deliver urine into the pelvis, the cavity of the kidney that is drained by the ureter.

colostrum (292, 348): milk produced during the first few days after birth, especially rich in antibodies and other proteins.

commensal (331): organism that obtains benefit by living in close relationship with another of a different species, which is neither benefited nor harmed by the relationship.

common ancestry (43): organisms have evolved by natural selection from a common ancestor.

companion cells (134): cells in phloem tissue containing nuclei; probable function is to control sieve cells.

compensation point (80): light intensity at which photosynthesis is just balanced by respiration.

complete metamorphosis (264): method of reproduction in insects where the fertilised eggs hatch into larvae/caterpillars which then pupate ('the metamorphosis') into the adults.

concentration gradient (25): change in concentration between one area and another.

cones (323): the reproductive organs of gymnosperms.

contractile vacuoles (29): specialised vacuole in unicellular organisms; has an osmoregulatory function, as it removes excess water entering in osmosis.

cork (227): dead layer of cells protecting stems and roots of many seed-bearing plants.

cork cambium (227): layer of actively dividing cells responsible for the production of cork.

corolla (306): collective name for the petals of a flower.

coronary arteries (157): vessels that deliver blood to the heart muscle.

corpus luteum (273): 'yellow body', formed in ovary after release of an egg, responsible for secretion of hormones progesterone and oestrogen.

corpuscles (149): the cells in the blood.

cortex (184): outer layer of a young stem or root, or of a kidney, or of the cerebrum in the brain.

costal cartilage (55): cartilage that attached the ribs to the sternum, allowing the expansion of the chest when a deep breath is taken.

cotyledon (314, 316): a leaf in the embryo of a seed – there are two in dicotyledons, one in monocotyledons.

Cowper's glands (271): small glands opening into the male urethra, make a small contribution to the seminal fluid.

cross-pollination (307): transfer of pollen from an anther of one plant to the stigma of another, genetically different, plant.

cristae (19, 199, 200): infoldings of the membranes of mitochondria.

crop (117): chamber in digestive system for storing food; present in insects, birds.

crustacea (54): class of arthropods that includes the crabs, shrimps, crayfish.

cyanobacteria (44): group of bacteria in the domain eubacteria that are capable of photosynthesis.

cytoplasm (6, 15, 199): where most of the chemical reactions of a cell occur; consists of water, protein, cell organelles and other cell structures.

cytoplasmic streaming (64): movement of the fluid component of the cytoplasm.

cytosis (30): the movement of large amounts of substances into/out of cells by the folding of membranes.

cytosol (15): part of the cytoplasm of a cell that is liquid – mainly water containing many dissolved substances.

cuticle (76): waxy layer secreted by the epidermis of a leaf or young stem.

deamination (179, 182): removal of ammonia from amino acids prior to the latter being used in respiration.

deciduous (teeth) (100): juvenile teeth that are shed are said to be deciduous.

denature (38, 94, 100): to break down the chemical structure of a molecule so that it no longer functions (eg high temperatures will denature DNA and many proteins by breaking chemical bonds).

dentine (101): part of a tooth surrounding the pulp; susceptible to decay.

dermis (246): deeper layer of the skin, beneath the epidermis, and containing blood vessels and nerve fibres.

diaphragm (211): sheet of muscular tissue that separates the gut and lung cavities.

diastolic pressure (173): pressure of the blood when the ventricles are relaxing.

dicotyledon (60, 125, 224): one of the two groups of flowering plants, with two seed leaves (cotyledons) in the embryo.

differentiation (225): process by which cells specialise to take on different functions.

diffusion (165, 203): movement of a substance from a region of higher concentration to a region of lower concentration by random movement of its particles.

digestion (98, 114): breaking down of complex food substances into simpler ones by enzymes.

diploid *or* 2n (261, 322): having a double set of chromosomes.

dizygotic (twins) (293): twins formed from different fertilised eggs.

DNA (6, 14, 43): deoxyribonucleic acid; the genetic material which makes up chromosomes and codes for proteins.

domains (43): the three evolutionary lineages that all organisms are placed into (archea, eubacteria, eukarya) in modern classification.

donor (296): a person who gives part of her or his body to someone else, to be used by doctors for medical treatment.

duodenum (102): first part of the small intestine, into which the bile and pancreatic ducts open.

early onset (diabetes) (250): diabetes caused by destruction of insulin-secreting cells, and hence lack of insulin.

ecological characteristics (253): organisms within a community, together with the physical attributes of their habitat.

efferent arteriole (185): tiny arteriole taking blood away from Bowman's capsule in the kidney.

effector organs (233): the muscles and glands that produce a response to a stimulus.

egestion (98): getting rid of undigested food via the anus.

egg (305, 307): a female gamete.

ejaculation (299): process by which seminal fluid is squirted out through the penis.

electron transfer chain (199): the final biochemical pathway of aerobic respiration, producing ATP molecules. Also known as hydrogen transfer chain.

embryo (260, 280, 314, 316): very early stage in development of a multicellular organism.

emulsion (102): a very fine suspension of fat or oil in water.

enamel (101): hardest substance in the body, covering the crown of a tooth.

endocrine gland (233, 238, 251, 270): a gland that releases its secretion directly into the blood, rather than into a duct.

endocrine system (249, 251): the system by which parts of the body communicate with each other via the blood, and consisting of the various endocrine glands.

endocytosis (30): the taking of substances into the cell by the infolding of the cell membrane.

endodermis (126): single layer of cells forming a sheath around the vascular region or stele.

endometrium (274): lining of the womb, to which an embryo mammal attaches itself.

endoplasmic reticulum (6, 15): complex system of membrane-bound cavities in the cytoplasm of a cell, concerned with the production of proteins.

endosperm (307): storage tissue found in many seeds, containing protein and either starch or fat.

endosperm nucleus (307): nucleus that divides many times to give rise to the nuclei of the endosperm.

endospore (335): stage in the life of certain kinds of bacteria that is highly resistant to drying and to heat.

enzyme (18, 37, 100, 114): protein catalyst that speeds up the chemical processes of life.

epidermis (76, 246): outer layer of the skin of a mammal; the outermost layer of cells of a leaf, young stem, or very young root.

epididymis (270): part of the testis of a mammal in which sperm mature prior to their release.

epiglottis (101, 209, 212): flap of tissue behind the base of the rear of the tongue, which helps to prevent food going down into the air passages during swallowing.

epithelia (342): plural of epithelium – tissue that forms a covering or lining.

EPO (167): erythropoetin; a hormone secreted by the kidney which stimulates red blood cell production.

erectile tissue (271): tissue in the penis containing spongy blood spaces that can become gorged with blood, causing the penis to become stiff during sexual excitement.

erector muscle (247): muscle attached to base of hair, whose contraction raises the hair.

erythropoetin (167): see EPO.

eubacteria (44): the domain of bacteria that are most recently evolved; includes the bacteria and cyanobacteria (but not the extremophiles).

eukarya (44): the domain that contains all the large groups of organisms that have eukaryotic cells (protista, fungi, plants, animals).

eukaryote (8, 11, 43, 330): organism whose cells have a true nucleus.

eukaryotes (43): organisms made of eukaryotic cells, ie their cells all have a true nucleus.

excretion (179): getting rid of waste products that have been made in chemical processes in cells.

exhaling (205): breathing out.

exocytosis (30): removal of substances from the cell in the reverse manner to endocytosis.

exponential growth (332): growth in which numbers increase by a constant percentage each given time interval.

extra-cellular (331, 335): outside the cell.

eyespot (22, 62, 65): light-sensitive pigment in some unicellular organisms and invertebrates.

facilitated diffusion (26): movement of small molecules through channel proteins.

faeces (104, 119): undigested food, together with large numbers of bacteria; got rid of via the anus.

fatigue (157): tiredness that means a muscle can no longer contract.

fatty acid (98): one of the chemical constituents of fat.

fermentation (198, 331, 353): process by which energy is obtained from carbohydrate by converting it into other organic products, such as lactic acid or ethanol, there being no net oxidation.

fertilisation (260, 305): joining of egg and sperm to form a zygote.

fibrin (152, 166): protein that precipitates during blood clotting.

fibrinogen (152, 166): soluble plasma protein that precipitates as fibrin when blood clots.

filaments (207, 326): stalk of a stamen.

fimbriae (274): tentacle-like structures surrounding the opening of the oviduct in female mammals.

flaccid (29): limp.

flagella (21, 330): plural of flagellum; a structure essentially similar to a cilium, but longer and present in smaller numbers.

flower (305): shoot of a flowering plant, modified for sexual reproduction.

foetus (282, 283): pre-birth stage of a mammal that has developed recognisable organ systems.

follicle (272): cluster of cells surrounding a developing egg in the ovary of a mammal.

follicle stimulating hormone (FSH) (274): hormone secreted by the anterior pituitary, stimulating the growth and development of eggs in the ovary.

food vacuole (62): a sac in a unicellular organism that contains ingested food particles.

fraternal (twins) (293): twins that develop from different fertilised eggs, and thus are genetically different.

fronds (83): leaves of ferns.

fruit (314, 315): ripened ovary of a flower.

fungi (43): heterotrophs that may be single or multicellular; cell walls are made of chitin, feeding is extra-cellular and fungi are parasites or saprophytes.

gall bladder (102, 118): organ that stores bile.

gamete (5, 260, 261, 269, 305): cell that can only develop further by joining with another gamete to form a fertilised egg.

gametophyte (322): haploid plant that produces gametes by mitosis.

ganglia (236): concentrations of nerve cells/neurons.

gas exchange (74, 201): diffusion of oxygen and carbon dioxide in opposite directions across the surface of an organism or cell.

gastric juice (342): digestive juice secreted by the stomach wall.

gastropoda (53): class of molluscs that includes the slugs and snails.

gene (6): a hereditary unit – a length of DNA carrying the information for making a particular protein.

genus (47): a classification group made up of a number of similar species.

gestation (282): period between fertilisation and birth in a mammal.

gills (206, 207): gas exchange surfaces of aquatic organisms.

gizzard (117): part of gut/stomach modified as a strong-walled grinding chamber.

glomerular filtrate (185): part of the plasma that is filtered by the glomeruli in the kidney.

glomerulus (185): one of about a million clusters of capillaries in each kidney, from which liquid is filtered under pressure of the blood.

glucagon (249): hormone that acts to raise blood glucose concentration by stimulating the release of glucose into the blood.

glucose (93, 98): simple sugar that is used by cells for energy, and is the building unit of starch and cellulose.

glycerol (98): one of the chemical constituents of fat.

glycogen (248): complex carbohydrate made up of repeating glucose units.

glycolysis (199): first stage in respiration; occurs inside the cell cytoplasm.

Golgi body (6, 17): organelle concerned with the modification of proteins following their manufacture.

grana (20, 93): flattened discs of thylakoid membranes inside chloroplasts that hold chlorophyll.

growth ring (226): cylinder of secondary xylem produced in a tree or shrub over a period of a year.

guard cells (85, 138): cells that surround a stoma in the epidermis of a plant.

gut (97): tube in which food undergoes digestion and absorption.

gymnosperms (323): group of plants that reproduce using cones.

gynoecium (306): female part of a flower, consisting of one or more carpels.

haemocoel (144): blood cavity.

haemoglobin (143, 144, 145, 151): red blood pigment that carries oxygen.

haemophilia (166): hereditary condition in which the blood cannot clot.

haemorrhoids (162): varicose veins around or near the anus; also known as piles.

haploid *or* n (261, 322): having only one set of chromosomes.

haustoria (91, 92): modified roots of plant parasites that grow into the host plant using digestive enzymes and join to the host's xylem and phloem.

heart (143, 149): muscular pump that propels blood round the body.

hemiparasite (91): *see* partial parasite.

hepatic portal vein (103, 119): blood vessel that carries food from the gut to the liver.

hermaphrodite (260): individuals that have both male and female reproductive organs (eg earthworms, snails).

heterotroph (111): organism that cannot make its own organic matter from inorganic raw materials (all animals and fungi and most bacteria).

histamine (350): substance released in tissues during inflammation.

homeostasis (180, 243): maintenance of near-constant conditions in the body.

homeostatic mechanisms (180): self-regulating systems that maintain a constant internal environment.

hormone (180, 233, 251): chemical carried in the blood and which carries information.

human chorionic gonadotrophin (284): hormone that results in the prevention of the breakdown of the lining of the uterus during early pregnancy.

hybrids (47): the offspring produced from the breeding of two closely related species; typically infertile (eg mule from horse-donkey breeding).

hydrophytes (141): plants that were originally terrestrial plants that have now adapted to living in an aquatic environment.

hyperglycaemia (249): abnormally high level of blood glucose.

hypertension (173): high blood pressure, usually considered to be greater than 140/90.

hyperthermia (246): abnormally high body temperature.

hypertonic (27): solution that has a higher concentration of solute than another solution.

hyphae (335): plural of hypha, a fungal filament.

hypotension (173): low blood pressure.

hypothalamus (244, 274): part of the brain controlling body temperature, body water content, and certain other vital functions.

hypotonic (27): solution that has a lower concentration of solute than another solution.

ileum (118): last section of the small intestine; joins onto the large intestine.

implantation (280): process by which the fertilised egg (or rather, the blastocyst that develops from it) attaches to the lining of the uterus.

incisor (100): chisel-shaped tooth at the front of the mouth.

incisor teeth (113, 114): chisel-shaped teeth at the front of the jaw used to sever portions of food.

incomplete metamorphosis (264): method of reproduction in insects where the fertilised eggs hatch into a juvenile adult (not a larvae); the juvenile grows and becomes sexually mature through a series of moults (eg crickets).

infertile (47): state in which an organism is unable to reproduce or have offspring.

infertility (296): not being able to have babies.

ingestion (98, 111): taking in of food by an animal.

inhaling (205): physical process of taking air into the body.

inorganic nutrients (minerals) (83): ionic substances.

insects (53): class of arthropods that includes grasshoppers, flies, butterflies, moths.

insulin (249): hormone that promotes the uptake of glucose by cells, thus lowering blood glucose level.

integuments (307): protective layers round an ovule.

interferon (342): protein produced by white corpuscles that inhibits the reproduction of viruses.

internode (224): region of a stem between two nodes.

interstitial fluids (144): liquid that is found between the cells of the body; this liquid makes up most of the liquid in our bodies.

invertebrates (54): animals that lack a vertebral column/backbone.

in-vitro fertilisation (296): the fertilisation of an egg outside the body, by artificial methods.

ischaemia (172): reduction of blood supply, and thus oxygen supply, to a tissue as a result of partial blockage of artery supplying it.

islets of Langerhans (249): microscopic patches of cells in the pancreas responsible for secreting the hormones insulin and glucagon.

kidneys (182): paired organs which excrete urea and help to maintain a constant internal environment.

kingdom (43): one of the (old) five large classification groupings of organisms (monera, fungi, protists, plants, animals) – organisms are now grouped into three large domains.

Kreb cycle (199): middle stage of aerobic respiration; occurs inside the mitochondrial matrix space.

labial palps (113): small organ in bivalve that sorts and directs food trapped from the gills.

labium (114): lower lip (modified) of insects.

lactation (292): secretion of milk from mother's breasts.

lactic acid (198, 201, 211): mildly toxic product of fermentation (anaerobic respiration) in muscle and in some bacteria.

lamellae (207): projections on gill filaments that increase the surface area for gas exchange.

larynx (212): 'voice box' at the opening to the windpipe, containing the vocal cords.

late onset (diabetes) (250): diabetes caused by loss of sensitivity of cells to insulin.

leaves (83, 84): the photosynthetic organs of higher plants.

leucocyte (345): white blood corpuscle.

leukaemia (152): disease in which white blood cells fail to mature properly.

light-dependent reaction (93): the biochemical pathway in photosynthesis in which solar energy interacts with the electrons of chlorophyll molecules and is converted into chemical energy (ATP); occurs on the thylakoid membranes of the grana of chloroplasts (at the same time, water molecules are split into oxygen and hydrogen).

light-independent reaction (94): the biochemical pathway in photosynthesis in which carbon dioxide and hydrogen are bonded to produce glucose; occurs on the stroma of chloroplasts; also known as the Calvin cycle.

lignin (76, 131): complex material used to strengthen cell walls in plants, making them resistant to compression.

lipase (102): fat-digesting enzyme.

liver (118, 182): large organ situated at the top of the gut cavity; performs many essential functions.

loop of Henle (185): hairpin loop along nephron, involved in reabsorption of water.

lungs (206, 209): gas exchange surfaces of air-breathing vertebrates.

luteinising hormone (LH) (274): hormone that promotes ovulation and development of the corpus luteum in females, and secretion of testosterone in males.

lymph (168): colourless liquid contained in the vessels of the lymphatic system.

lymph nodes (168, 344, 345): small swellings in the lymphatic system packed with lymphocytes and other white cells.

lymph system (119, 169): series of vessels throughout the body that drain tissue fluid back to the heart. Lymph also contains leucocytes for defence against disease.

lymphatic capillaries (168): microscopic, blind-ending, terminal branches of the lymphatic system.

lymphatic system (168): system that drains surplus tissue fluid and plays an important part in defence against microbes.

lymphocytes (151): white corpuscles with small amount of cytoplasm; one kind can develop into cells that make antibodies.

lysozyme (342): enzyme present in various body fluids; breaks down bacterial cell walls.

Malpighian tubules (182): excretory organs of insects.

mandible (112, 114): lower jaw.

marsupials (55): the group of mammals where the embryo is born in a very immature state and completes development in the mother's pouch (eg possums, koala).

masticate (113): to chew.

maxillae (112): part of jaw/mouthparts of an insect.

medulla (184): inner region of the kidney.

meiosis (305): process in which a diploid nucleus divides twice to produce four haploid, genetically different, nuclei; in animals, occurs in the ovaries and testes and results in the formation of gametes.

menopause (276): period marking the end of fertile life of a woman.

menstruation (274): breakdown of the lining of the uterus, occurring at monthly intervals.

mesophytes (139): plants found growing under average conditions of water supply.

metabolic wastes (180): waste products from metabolic processes.

metabolism (37, 245): collective name for all the chemical processes going on in cells.

microbe (329): microscopic organism.

micrometre (μm) (329): millionth of a metre, or thousandth of a millimetre.

micronutrients (84): trace elements.

micro-organism (329): microscopic organism.

micropyle (307, 316, 323): small hole in the integuments surrounding an ovule in flowering plants.

microtubules (21): provide an 'internal' skeleton, help in the transport of materials in cells, and form the spindle fibres for mitosis and meiosis.

microvilli (103): minute finger-like extensions of the surface of cells lining the small intestine and kidney tubules (plural of microvillus).

mitochondria (6, 19, 200): organelles in which aerobic respiration occurs, releasing energy.

mitral valve (156): valve between the left auricle and left ventricle.

molar (100): adult tooth with more than one root and which does not replace an earlier tooth (in contrast to premolars, which have one root and are second teeth).

molars (113, 114): large crushing teeth found at the back and sides of the jaw.

monera (43): classification term (old; one of the five kingdoms) that includes all bacteria.

monocotyledon (224): one of the two groups of flowering plants, with one seed leaf (cotyledon) in the embryo.

monocotyledons (60, 125): group of angiosperms that has one cotyledon in the seeds; includes the grasses.

monotremes (55): the group of mammals that reproduce by laying eggs (eg platypus).

monozygotic (twins) (293): genetically identical twins formed by division of a single fertilised egg.

morula (279): ball of cells formed by repeated divisions of the fertilised egg of a mammal.

mucus (342): slimy substance secreted by the lining of the breathing passages and alimentary canal.

multicellular (5): many-celled.

mutualists (331): organisms that live in a relationship in which both obtain benefit.

mycelium (335): body of a fungus, consisting of many threads or hyphae.

myometrium (274): muscular layer in the wall of the uterus.

myriapoda (54): the class of arthropods that includes the centipedes and millipedes.

nanometre (nm) (327): billionth of a metre, or millionth of a millimetre.

nasal cavity (209): area inside the nose.

nectar (306): sugar solution secreted by flowers, serving to attract insects for pollination.

nectary (306): patch of glandular tissue in a flower that secretes nectar.

negative feedback (218): process in homeostasis, in which the greater a change in an internal condition, the stronger is the corrective response.

nephridia (180): tubular organs present in several invertebrate groups, usually used to transport excretory products form the body cavity to the outside.

nephridiopore (180): opening of the nephridium, either to the outside of the body or to the gut.

nephron (184): one of a million or so units of the kidney, each consisting of a filter and a long tube from which useful substances are selectively reabsorbed.

nephrostome (180): opening of the nephridium into the coelom (internal body cavity) of an organism.

nerve impulse (234): an electrical charge that travels along the axon of neurons, transmitting information.

nervous system (233): body system that receives, transmits, and responds to environmental stimuli.

neurons (233): nerve cells.

niche (43): the role or way of life of an organism in a community.

nuclear envelope (6): double-walled membrane separating nucleus from cytoplasm.

nucleic acid (337): nitrogen and phosphorus-containing polymers of two kinds – DNA and RNA.

nucleolus (14): dark staining region inside the nucleus that controls the production of RNA for ribosomes.

nucleus (6, 14): control centre of a cell; contains chromosomes and the nucleolus.

nutrients (98): substances taken in from the environment and used as raw materials for growth.

oesophagus (116): tube that takes food from the mouth to the stomach.

oestrogen (272): female hormone controlling the development of female secondary sexual characteristics and the monthly development of the uterine lining in the menstrual cycle.

oogenesis (273): egg production in an animal.

operculum (207): a bony cover that protects the gills in fish.

oral groove (22, 63): part of the cell membrane of *Paramecium* specialised for ingestion.

organ (73): number of tissues which cooperate to carry out a function that none of the individual tissues can do by itself.

organelle (6, 11): region of a cell specialised to carry out a particular function, eg nucleus, chloroplast.

organism (43): an individual living thing.

osmoregulation (28, 243): the maintenance of correct body fluid levels in organisms.

osmosis (26): movement of water across a semi-permeable membrane from an area of high water concentration to an area of lower water concentration.

ova (260): plural of ovum.

ovaries (272): reproductive organs of the female that produce eggs and also female sex hormones.

ovary (260, 307, 326): egg-producing organ of the female.

oviparous (260): form of reproduction where the female releases ova which develop externally.

ovulation (273): process of release of an egg from the ovary in a mammal or other back-boned animal.

ovule (307): structure in seed-bearing plants that contains a female gamete.

ovum (260, 269): the female sex cell/gamete; commonly known as the egg.

oxygen debt (211): the lack of oxygen that occurs in cells after vigorous activity; means rapid breathing occurs after exercise stops.

oxy-haemoglobin (154): haemoglobin carrying oxygen.

oxytocin (289, 292): hormone that stimulates contraction of uterus during birth, and contraction of milk ducts during feeding of the infant.

pacemaker (156): piece of nervous tissue that sends out an impulse that causes the auricles and ventricles of the heart to contract.

pancreas (118): gland which secretes digestive enzymes into the duodenum and also produces insulin.

palisade layer (75): major photosynthetic layer of a leaf, consisting of elongated cells hanging from the upper epidermis, rich in chloroplasts.

pancreas (102): organ that produces digestive enzymes and also hormones concerned with regulation of blood sugar concentration.

parasite (91, 331, 341): organism that feeds off another (the host), harming it but without killing it.

parthenogenesis (260, 311): development of an egg without fertilisation.

partial parasite (91): parasitic plant that obtains its food through photosynthesis but obtains their water and minerals from a host plant (eg NZ mistletoes).

partially permeable membrane (10): membrane allowing the passage of small molecules such as water, but not larger molecules.

passive transport (25): transport of substances across a membrane without the need for energy (eg diffusion, osmosis).

pathogen (151, 341): organism that causes disease.

pelvis (184): urine-filled cavity in the kidney, drained by the ureter.

penis (271): male organ that introduces semen into the vagina of the female.

pepsin (102): protein-digesting enzyme present in gastric juice.

perennial (305): plant that survives from year to year.

pericardium (154): membrane round the heart containing lubricating fluid.

pericarp (314): wall of a fruit.

peristalsis (101, 116, 184): muscular waves that propel material along a tube such as the gut, ureter or oviduct.

petal (306, 326): flower part modified to make the flower conspicuous to pollinating animals.

petiole (224): leaf stalk.

Petri dish (337): dish used to grow micro-organisms on agar.

phagocyte (151, 152): white blood cell that can eat bacteria.

phagocytosis (30, 31, 62): the process by which the cell membrane surrounds large particles to take them into a cell.

pharynx (101, 212): region of the gut through which both food and air pass.

phloem (76): tissue in plants that transports sugar and amino acids.

photosynthesis (20, 44, 45, 83, 93): process whereby plants manufacture food from raw materials.

pinocytosis (30, 31): the process by which the cell membrane surrounds fluid to take it into the cell.

pituitary (240, 243, 247): endocrine gland beneath the brain, producing a number of hormones, some of which control other ductless glands.

placenta (274, 280, 282): organ joining the foetus to the wall of the uterus during pregnancy.

placentals (56, 267): group of mammals in which the young develop in the mother's uterus nourished by a placenta during gestation.

plants (44): multicellular organisms that photosynthesise (are autotrophs); cells are enclosed in cell walls of cellulose.

plasma (147): liquid part of the blood.

plasma cell (151, 345): an antibody-secreting cell, developing from a B lymphocyte.

plasma membrane (6, 330): membrane forming the outer boundary of the cytoplasm.

plasmid (334): a 'mini-chromosome' in bacteria.

plasmolysed (29): the state of plant cells when the cytoplasm has pulled away from the cell wall as a result of loss of water from the cell.

platelets (150): involved in blood clotting.

pleural cavity (212): cavity surrounding each lung, and containing lubricating fluid.

pleural membrane (212): membrane covering each lung and lining the wall of the chest.

plumule (314, 316): embryonic shoot of a seed.

polar bodies (273): minute, functionless meiotic products of egg production in the ovary of an animal.

pollen grains (305, 323): tiny cells within which male gametes are produced in seed-bearing plants.

pollen sac (306): pollen-producing parts of a stamen in a flowering plant.

pollen tube (227): tube that grows out from a pollen grain as it germinates, carrying the male gametes in its tip.

pollination (305): transfer of pollen from an anther to a stigma on a flower of the same species.

polypeptide (102): chain of amino acids joined together by peptide bonds.

premolar (100): tooth with two cusps and one root, used in crushing food.

primary cell wall (20): a newly formed cell wall, usually made of a substance called cellulose (some plant cells retain the primary cell wall at maturity).

primary growth (225): in plants, growth in length of a stem or root.

proboscis (112): the extension of the mouthparts of an insect to form a piercing tube.

producers (96): plants: organisms that manufacture their own food.

progesterone (272): female sex hormone, concerned with preparing for, and maintenance of, pregnancy.

prokaryote (8, 11, 330): organism with cells in which the DNA is not separated from the rest of the cell by a nuclear envelope.

prokaryote cells (43): cells that do not have a true nucleus (ie contained in a membrane) or membrane-bound organelles.

prolactin (292): pituitary hormone that influences milk secretion.

prostate (271): one of the glands in the male reproductive apparatus whose secretion contributes to the seminal fluid.

proteins (37): large organic molecules that have either a structural (eg collagen) or a regulatory (eg enzyme) function in organisms.

protista (43, 57, 61): unicellular organisms with a true nucleus (prokaryotes).

protoctista (57, 61): *see* protista.

puberty (276): changes in the body that occur as the body becomes sexually mature.

pulmonary artery (155, 213): vessel that takes blood from the heart to the lungs.

pulmonary valve (156): valve in pulmonary artery.

pulmonary vein (155, 213): vessel taking blood from the lungs to the heart.

pulp cavity (101): innermost part of a tooth, containing blood vessels and nerves.

pus (298, 351): whitish liquid developing in infected tissue, consisting of white blood corpuscles.

pyramids (184): conical bulges into the pelvis of the kidney, on the surface of which open the collecting ducts.

pyrogens (342): chemicals responsible for raised body temperature during fever.

radicle (314, 316): root of an embryonic plant.

radula (113): rasping tongue found in herbivorous molluscs.

receptacle (306): tip of a flower stalk, to which the flower parts are attached.

rectum (119): last section of gut; stores faeces.

red blood cells (150): carry oxygen in the blood; also called erythrocytes.

reflex arc (251): a nerve pathway that controls a reflex action, initially bypassing the brain.

renal arteries (183): vessels that supply blood to the kidneys.

renal tubule (185): part of a nephron in which liquid filtered from the blood is modified to produce urine.

renal vein (183): vessel taking blood away from the kidney.

residual volume (217): that part of the lung volume that cannot be deflated of air.

respiration (179, 197, 198, 331): process in the mitochondria in which pyruvic acid is oxidised to carbon dioxide and water, yielding useful energy.

respiratory centre (211): part of the brain that affects the rate of breathing by monitoring blood CO_2 levels.

reticulum (118): second 'stomach' of a ruminant in which food is mixed and fermented.

Rhesus system (167): blood group system that can be important in pregnancy.

ribosomes (6, 14, 15): tiny organelles involved in protein synthesis.

root hairs (225): threadlike extensions of epidermal cells of a root, greatly increasing the surface area.

ribosome (14, 15): cell organelle that attaches to mRNA to provide a site for tRNA and amino acids in protein synthesis.

root pressure (132): osmotic pressure that develops in roots: helps movement of water up a plant's stem.

rough ER (15): endoplasmic reticulum with ribosomes attached.

rumen (118): first part of ruminant digestive system; the first of four 'stomachs'.

sac-like gut (115): primitive 'blind' gut with a common entrance and exit.

saliva (116): fluid produced in the mouth that contains enzymes; lubricates the passage of food.

salivary gland (101): gland in the mouth that secretes saliva.

saprophyte (331): term applied to micro-organisms that live on dead matter.

saturated fat (172): fat that contains as much hydrogen as it can (ie no double covalent bonds).

scrotum (270): muscular bag containing the testes.

sebum (342): greasy substance secreted by glands opening into hair follicles.

secondary cell wall (20): develops inside the primary cell wall, due to a process called lignification, which makes the cell walls strong and rigid.

secondary data (2): known information which has been obtained by someone other than the writer.

secondary growth (225): growth in thickness of a stem or root, due to activity of the cambium.

secondary phloem (226): phloem produced by the cambium during secondary growth.

secondary xylem (226): xylem produced by the cambium during secondary growth.

seed (307, 314, 323, 325): embryo plant surrounded by a protective coat of parental tissue.

self-pollination (307): transfer of pollen from an anther to a stigma of a flower on the same plant, or another plant that is genetically identical.

semilunar valve (156): valve at base of aorta or pulmonary artery; similar valves are present at intervals in most veins.

seminal fluid (271): liquid secreted by glands in the male reproductive system, carrying sperm into the female during ejaculation.

seminal vesicles (271): glands that secrete a component of seminal fluid.

seminiferous tubules (270): tubes in the testis, the walls of which produce sperm.

semi-permeable (14): allowing only certain substances to pass through; all cell membranes are semipermeable.

sensory palp (114): jointed section of feeding organ of insects that detects food.

sepals (306, 327): protective 'leaves' that protect a flower when a bud.

sexual reproduction (322): fusion of haploid gametes from two individuals to form a diploid zygote.

sieve tube cells (134): specialised conducting tissue of phloem in angiosperms.

sinus (146): space or chamber filled with blood.

slime capsule (330): protective layer round some bacteria.

smooth ER (16): endoplasmic reticulum with no ribosomes attached.

species (47, 107): a classification group made up of a individuals that freely interbreed to produce fertile offspring.

sperm (260, 269, 305, 310): male sex cell or gamete.

spermatogenesis (270): sperm production.

sphincter (102, 184): ring of muscle that can prevent movement of liquid down a duct.

spindle (21): contractile fibres that separate chromatids in cell division.

spiracles (205): holes in the thorax of insects through which air is taken in.

spirilla (330): plural of spirillum, a spiral-shaped bacterium.

spleen (344): organ below the stomach, which stores red corpuscles and is also part of the immune system.

spongy layer (75): lower layer of mesophyll of a leaf, containing very large air spaces.

sporangia (336): plural of sporangium, a structure within which spores are produced.

spore (322, 323, 336): resistant stage of a bacterium; also a single-celled stage in the life cycle of plants.

sporophylls (323): the reproductive leaves (look like woody scales) of cones in gymnosperms.

sporophyte (322): diploid plant which produces spores by meiosis.

stamen (326): male parts of a flower; made up of an anther and a filament.

starch (87): polysaccharide which stores the products of photosynthesis.

stent (172): a wire mesh tube that keeps the artery open, improving blood flow.

stigma (plants) **(306, 326):** tip of the carpel to which pollen becomes attached.

stigma (unicellular organisms) **(65):** alternative word used to describe the eyespot.

stimuli (227): plural of stimulus, a change in the environment to which an organism can respond.

stoma (20, 85): small hole in the epidermis usually on the underside of a leaf.

stomata (76): minute holes in the epidermis of a leaf or young stem, through which gas exchange can occur (singular: stoma).

stroma (94): liquid matrix inside chloroplasts containing starch grains and enzymes.

style (306, 326): region of the female part of a flower that is just below the stigma.

stylet (112): sucking mouthparts of an aphid.

subspecies (47): group of organisms very closely related, but not distinct enough to be called a 'new' species.

substrate (37): substance that an enzyme acts upon.

sugar (98): soluble, sweet-tasting carbohydrate.

synapse (233): gap that separates the dendrites of one neuron and the axon of an adjacent neuron.

systolic pressure (173): peak blood pressure, when the left ventricle is contracting.

technology (3): any knowledge of human-made things or methods which can have a useful purpose to people.

terminal bud (224): bud at the tip of a stem.

testa (314, 316): seed coat.

testes (260, 269): male reproductive organs that produce sperm and testosterone.

testosterone (270): male sex hormone.

thrombosis (166, 175): blood clot formed without injury to blood vessels.

thylakoids (20): membranes that make up the grana of chloroplasts, contain chlorophyll.

thymus (334, 346): organ in the chest in which T-lymphocytes mature; a vital part of the immune system that progressively decreases in size after birth.

thyroid (247): endocrine gland in the neck, secreting the hormone thyroxine.

thyroxine (247): iodine-containing hormone that stimulates metabolism.

tidal volume (217): volume of air exchanged each breath during breathing.

tissue (8, 73): group of cells organised to carry out a function the individual cells cannot do.

tissue fluid (168, 243): colourless liquid that seeps out of capillaries and bathes the cells.

toxoid (348): bacterial toxin made harmless but still able to stimulate the immune system to produce antibody.

trace elements (84): essential nutrients, only needed in very small amounts.

trachea (205, 209, 212): in insects – large vessels that carry air throughout the body; in mammals – the windpipe, the tube that connects the mouth and nose to the lungs.

tracheal system (205): the system of air tubes (trachea) in insects.

tracheoles (205): small vessels that carry air directly to individual cells in insects.

tracheophyta (57): the vascular plants (ferns, gymnosperms, angiosperms).

translocation (134): movement of glucose throughout a plant.

transpiration (76, 137): evaporation of water from the leaves and stems of a plant.

transpirational pull (132): the drawing of water up the xylem vessels from the roots as a result of evaporation of water from the leaves in transpiration.

tricuspid valve (156): valve that prevents blood flowing from right ventricle to the right atrium.

trophoblast (280): outer layer of cells of the blastocyst stage of development of a mammal; develops branched outgrowths that absorb nutrients.

tropism (227): growth response of a plant organ to a stimulus; direction of response depends on direction of stimulus.

TSH (247): thyroid stimulating hormone.

turgid (29): swollen by water pressure.

turgor pressure (29): pressure within a cell that results from osmosis.

umbilical arteries (281): vessels taking blood from foetus to the placenta.

umbilical cord (281): cord containing the umbilical arteries and umbilical vein.

umbilical vein (281): vessel taking blood from placenta to the foetus.

unicellular (5): consisting of a single cell.

urea (180, 182): excretory product of mammals; formula is $CO(NH_2)_2$.

ureter (184): tube carrying urine from kidney to bladder.

urethra (184, 271): tube carrying urine from bladder to the exterior.

uric acid (180, 182): semi-toxic, largely insoluble excretory product of insects, birds, and reptiles.

urine (183, 185): excretory liquid produced by the kidney.

uterus (267): organ in female mammals where the embryo develops; also know as the womb.

vaccination (348): introduction of a weakened, harmless, pathogen into a person to induce immunity.

vacuole (7, 18): cavity within a plant cell, containing watery sap.

vagina (274): muscular tube into which penis is inserted in intercourse, and through which baby is later born.

vacuoles (18): organelle in cells which is a membrane-bound sac, commonly used for storage.

valves (145, 154): flaps of tissue which prevent the backflow of blood in veins and heart.

varicose (162): permanently dilated.

vas deferens (271): tube carrying sperm from testis.

vascular bundle (76): strand of xylem and phloem in a leaf or young stem.

vascular system (143, 235): transport systems.

vein (149): vessel carrying blood towards the heart.

vena cava (155, 169): largest vein in the body; connects directly to the right auricle.

ventricle (146, 147, 154): pumping chamber of the heart.

vertebrate (54): organism with a backbone.

vibrio (330): bent, rod-shaped bacterium.

villi (103, 119): minute, finger-like extensions (of the lining of the small intestine or the trophoblast; re trophoblast – the growths are on the baby's side of the placenta through which food and oxygen are obtained from the mother's blood).

virion (338): virus particle.

vital capacity (217): largest volume of air that can be breathed in or out in one breath.

viviparous (260): form of reproduction in which the young are retained and develop in the female's body (ie internally).

vulva (274): fold of skin at opening to the vagina.

waxy cuticle (85): waterproof, waxy outer skin of epidermal cells.

white blood cells (150): involved in disease defence; also called leucocytes.

xerophytes (139): plants able to endure arid/dry conditions

xylem (85): vascular tissue that transports minerals and water throughout a plant.

zygote (260, 261, 269, 305): fertilised egg.